THE MELON

Amy Goldman

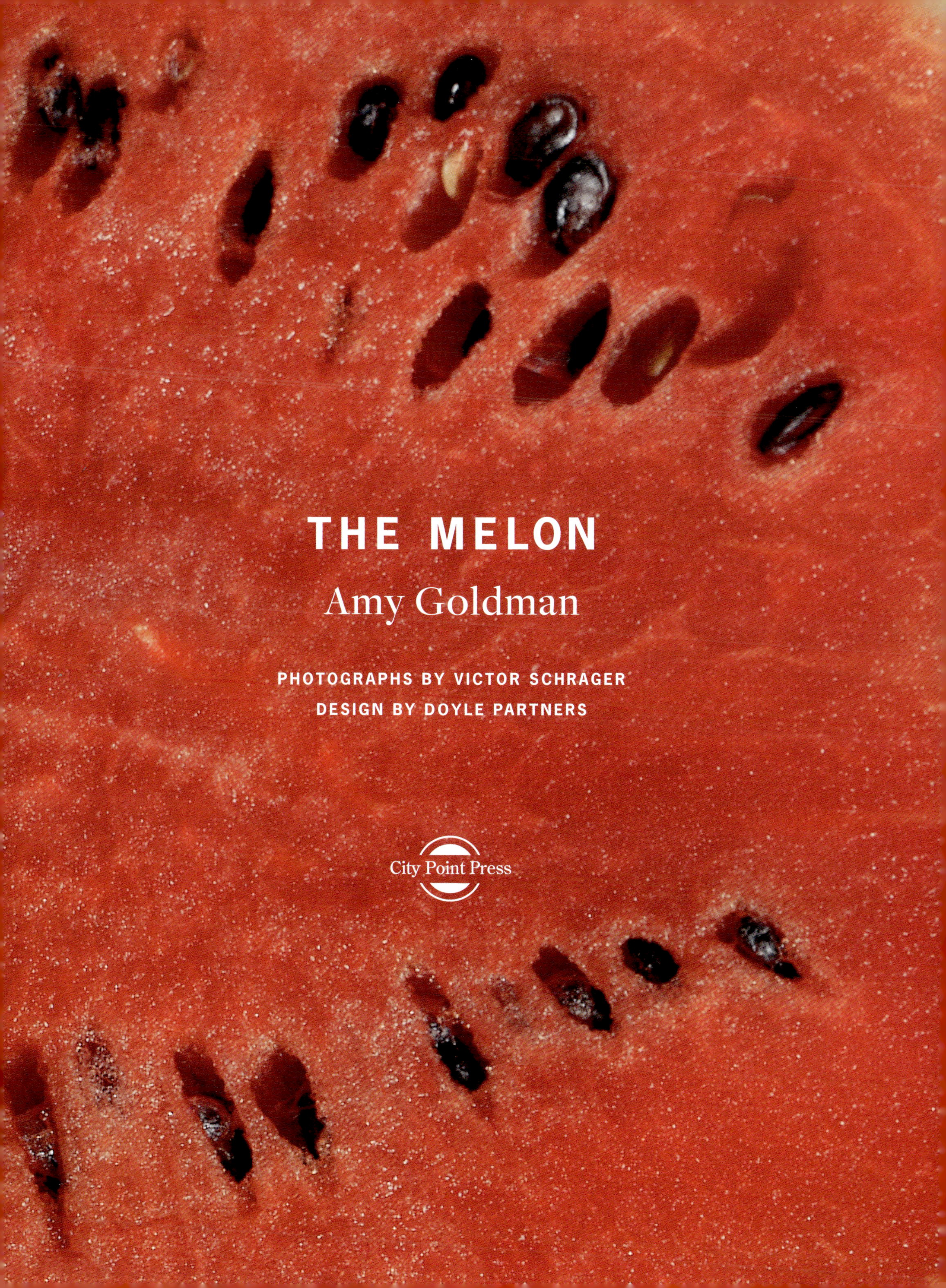

THE MELON

Amy Goldman

PHOTOGRAPHS BY VICTOR SCHRAGER

DESIGN BY DOYLE PARTNERS

City Point Press

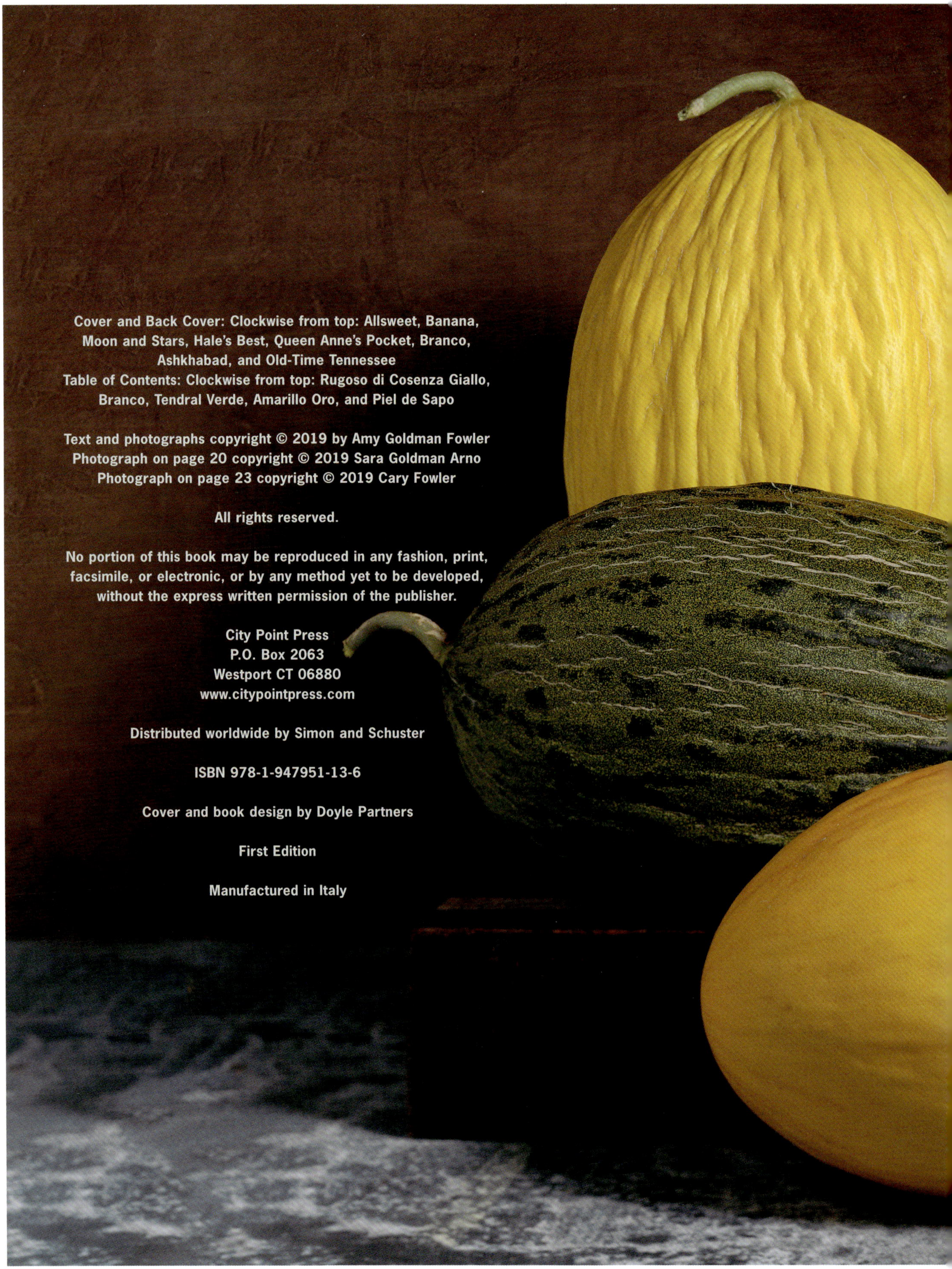

Cover and Back Cover: Clockwise from top: Allsweet, Banana, Moon and Stars, Hale's Best, Queen Anne's Pocket, Branco, Ashkhabad, and Old-Time Tennessee
Table of Contents: Clockwise from top: Rugoso di Cosenza Giallo, Branco, Tendral Verde, Amarillo Oro, and Piel de Sapo

City Point Press
P.O. Box 2063
Westport CT 06880
www.citypointpress.com

Distributed worldwide by Simon and Schuster

ISBN 978-1-947951-13-6

Cover and book design by Doyle Partners

First Edition

Manufactured in Italy

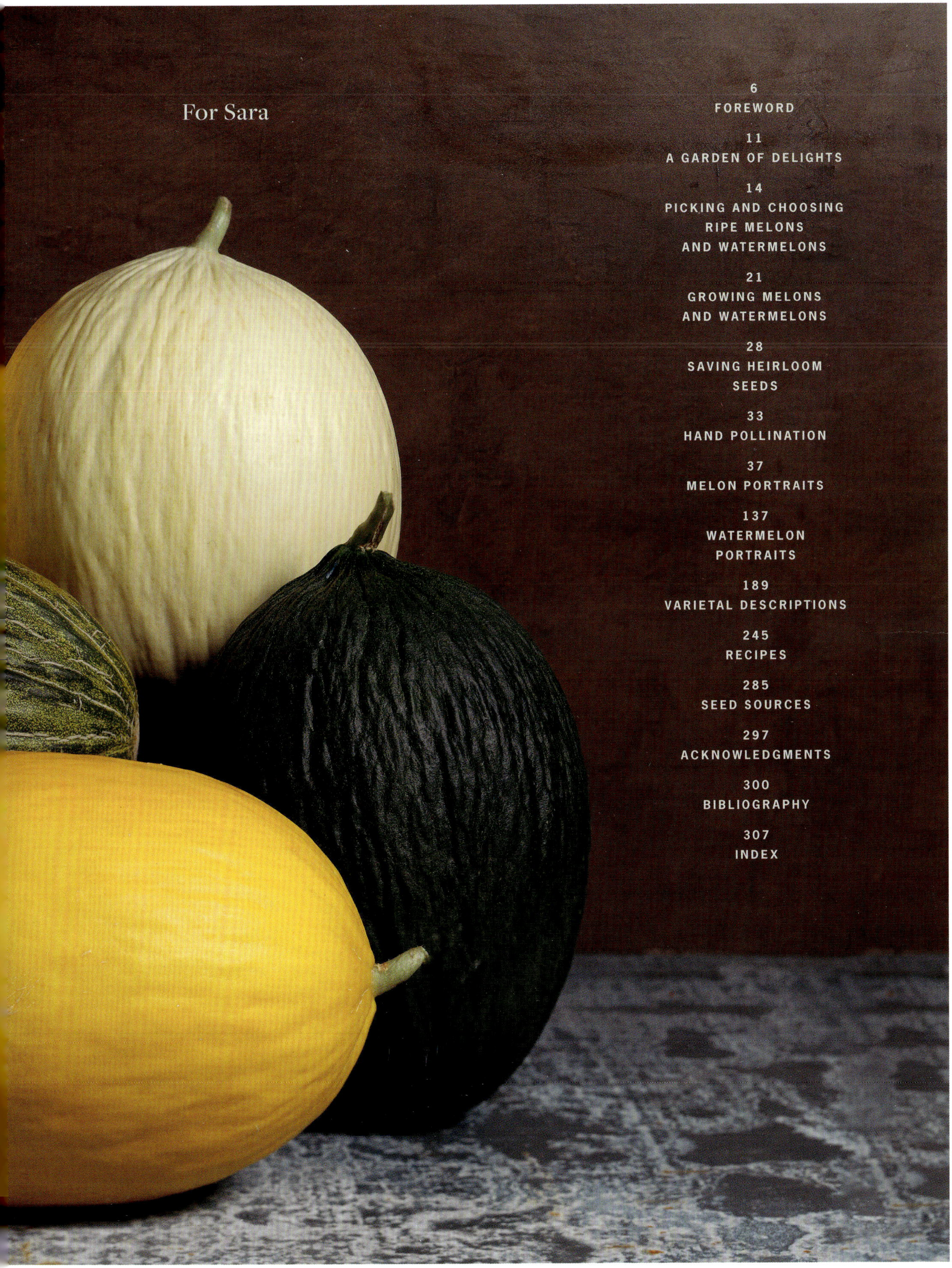

For Sara

Foreword

elons are a lifelong love and calling. They stir passions and memories in me. When I taste the cream of the crop, I can picture my parents swooning over similarly luscious fruits at meals during my childhood. Lillian and Sol were melon mavens: Mom had a thing for watermelon while Dad went for green-fleshed casaba. At our home on the North Shore of Long Island, I planted my first vegetable patch when I was eighteen. Paul, our Italian gardener, gave me lessons on how to sow melon seeds in the greenhouse, when to usher plants out into cold frames, and which techniques to use in the open field. I have so many happy memories of melon-filled summers. It was enormously gratifying to put homegrown food on the table, and I came to realize that I have a knack. I took a few years off from gardening during my graduate school days in Oklahoma, but ever since I've had melons in my life and garden. Today I work three vegetable plots on our place in the Hudson Valley of New York. And I keep my family—Cary, Sara, Martin, and Thomas—well supplied with heirloom foods. In late August, when we've had our fill of melon, we deliver surplus fruit in carefully arranged gift baskets to nearby loved ones and a local food pantry.

Much like practicing psychology (which I did for seven years), and raising children, gardening is about nurturing growth. In 1998, having won more than enough blue ribbons in vegetable growing competitions, and being at loose ends, I naturally decided to try my hand at garden writing. With fingers crossed, I sent one of my famous melon-filled baskets to an acquaintance at *Garden Design*. She must have liked the contents since the magazine sent a renowned photographer up to my garden a short time later, and I received a contract to write a 250-word piece on melons and their culture. "Globe Theater" was published the following year with

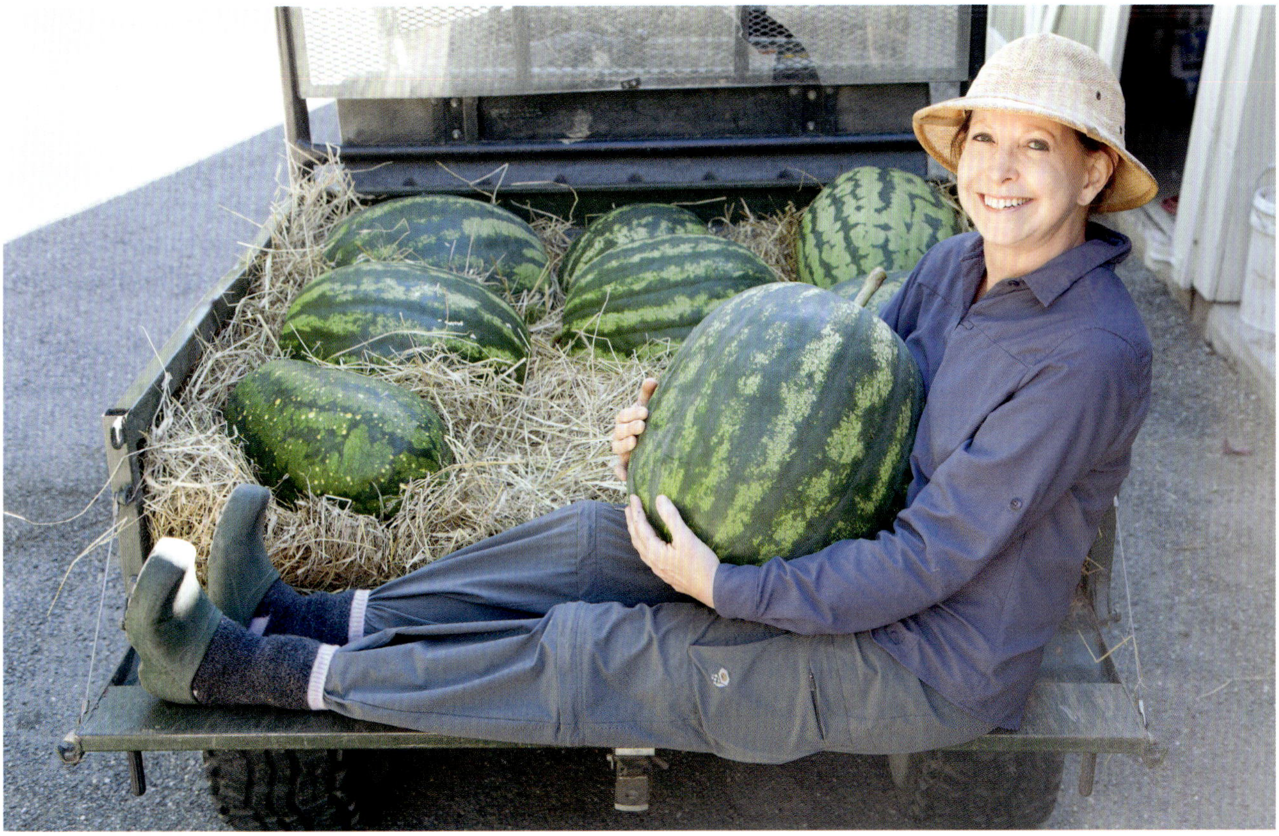

Mick Hales's images of my babies on the vine: Queen Anne's Pocket Melon, Prescott Fond Blanc cantaloupe, and Moon and Stars watermelon. (You haven't heard the last of them.) Friends planted a seed in my head about writing a book on my favorite topic. So I gave it a shot and wrote a proposal. Emma Sweeney, my literary agent, found a happy home for the book at Artisan: They published *Melons for the Passionate Grower* in 2002.

In Rhinebeck there are melons—whole motorized mule-carts full—and one happy gardener.

Why write another book about melon? The short answer is that I have grown and learned a lot more. Ten years after the first book was published, I decided to do another. From 2012 through 2018, I took great pleasure in cultivating the fruits, having them photographed, and doing the detective work on varietal descriptions for *The Melon*. I delved deep into the historical record including my own antiquarian book and catalogue collection (amassed for just such a purpose), made ample use of

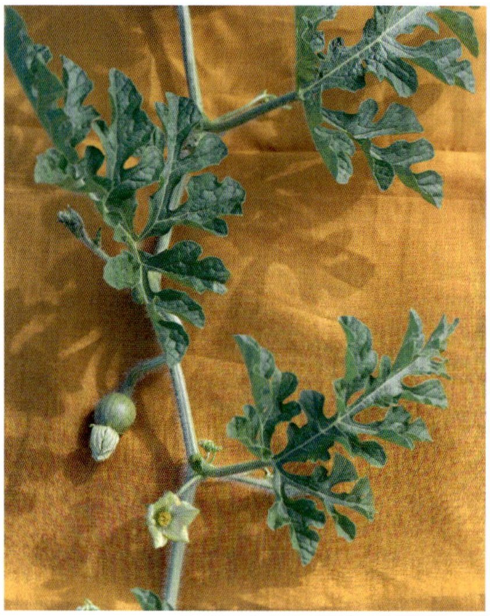
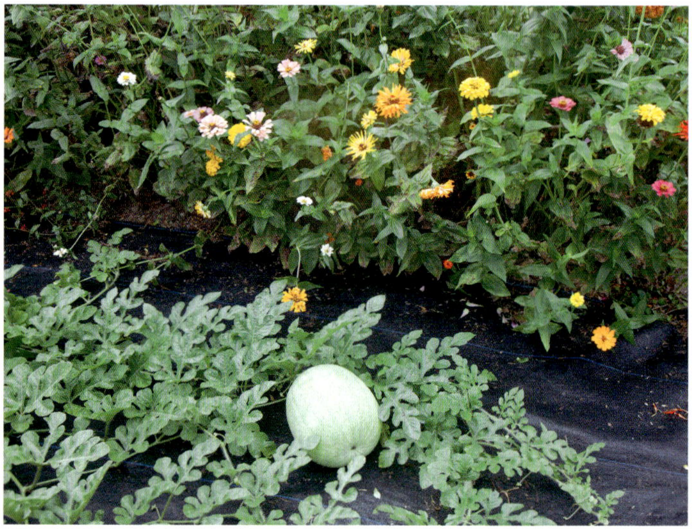

online resources unavailable at the turn of the century, and called upon talented archivists in a number of horticultural libraries. The writings of nineteenth-century authorities such as Sageret, Loisel, Naudin, Downing, Brinckle, Berckmans (all are referenced in the book) inspired me. Several plant scientists of my acquaintance lent their expertise, and I took advantage of a bonanza of recent research findings that informed my thinking on crop history and best cultural practices.

This book is more well rounded. It includes additional horticultural groups of melon, and I give watermelon—which is less genetically diverse and lacks horticultural groups—more of a fair shake. Much like the word "cantaloupe," which is used colloquially and erroneously by Americans to describe muskmelon, the word "melon" is commonly used in the United States to refer to both melon and watermelon. These vining crops belong to two different species within the Cucurbitaceae or gourd family of plants. I'm neither a botanist nor a stickler about these things—case in point, the title of this book. Recent phylogenetic studies show that the cucurbits most likely originated in Asia in the Late Cretaceous period (over 70 million years ago). Various lineages found their way time and again to different continents by transoceanic long-distance dispersal. Picture gourds afloat! Watermelon's more recent place of origin is in Africa; the wild progenitor of melon has been found growing in India, and its closest relative lives in Australia. Melon and watermelon are now among the world's most important vegetable crops. *The Melon* has gone global too: The 125 varieties illustrated and described here comprise 85 melons and 40 watermelons from all over the map. Half are gems retained from the first book and half are new. Their stories are as diverse as the melons themselves.

And what lookers they are! Victor Schrager made sure of that. He's the man who created the extraordinary images for this book. Our first photo shoot—on July 29, 2000—turned out to be the beginning of a collaboration spanning nineteen years. We produced four books together during that time and have two more in the works. On that long-ago Saturday in July, we were excited to begin, but the pouring rain thwarted our plan to do garden shots of melons growing on the vine. We holed up in the hot and steamy greenhouse and hatched another idea. Victor set up his wooden Deardorff view camera while I dashed off to the garden, clipped some melon and watermelon vines with their flowers, tendrils, and incipient fruit, and rushed them back inside lest they wilt or get beaten down by the rain. Two of the resulting images

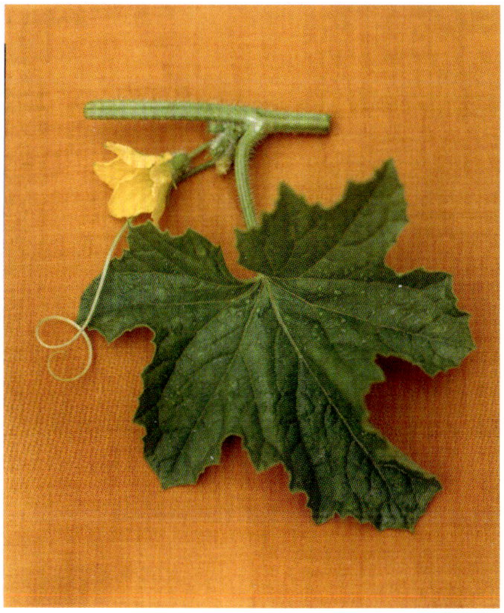
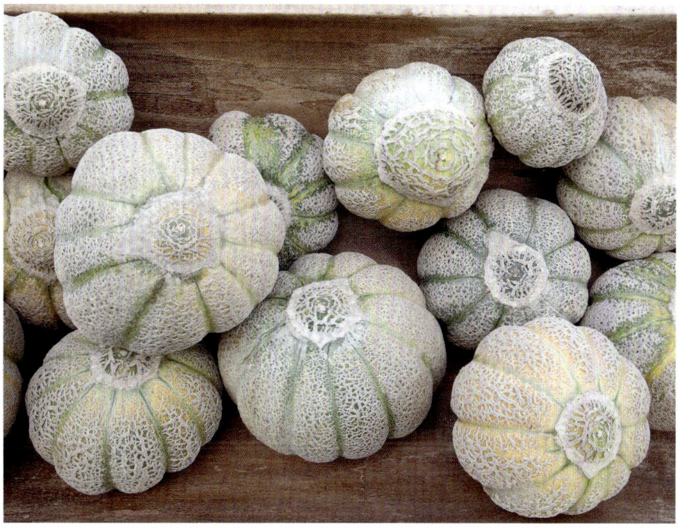

were used by our book designer, Stephen Doyle of Doyle Partners, for the endpapers in the first melon book, and two others appear here, as a beautiful reminder of that auspicious day.

It would take us nine years—nine growing seasons' worth of photo shoots—before I felt it was a wrap. Not only do we have a more representative sample of the diversity that exists in cultivated melon and watermelon, but we have images for every chapter and lots of beauty shots for your viewing pleasure. And I was able to replace some sore thumbs from previous years. A case in point is the muskmelon known as Jenny Lind, originally photographed in 2000. She looked sad and wan to me. I obsessed, we reshot, and reshot again until Jenny looked her best in 2015 (we went digital on melon in 2012). Time also allowed us to play with potential cover shots. The last one, done on the Friday of Labor Day weekend 2018, was in response to Stephen Doyle's idea: Create a group shot for the front cover with melons cut open to display their wares, and for the back cover, a shot of those same melons uncut. Not easy to achieve, considering that I'd already pulled up most of the melon plants. The ripped-up plant material had been composted and compacted with our new tractor the day before. I scrambled to salvage the Old-Time Tennessee, Banana, and Queen Anne's Pocket melons that had been left behind in one of the gardens and retrieved Hale's Best, Pollock Rocky Ford, some Mango melons, one Branco, four Ashkhabads, and a number of Allsweet and Moon and Stars watermelons from the walk-in cooler.

Twelve days later, on September 19, we photographed our last solo portrait: Grover Delaney watermelon. I cut six Grovers from the vine in the garden nearest the house, cradled the specimen melon in my lap, and slowly drove in a motorized mule a half-mile down the driveway to the barn studio. It was bittersweet to see Victor in the usual corner of the barn setting the stage for the still life atop a piece of plywood resting on two sturdy sawhorses. It could've been a scene from nineteen years earlier. Our wooden and marble props and backgrounds designed for melon photography are showing their age and fraying at the edges, but Victor remains amazingly fresh in his approach. The challenge for him has always been to provide elements of surprise while celebrating the obvious. He explains it's like a wonderful pas de deux, a balance between aesthetics and practicality. And the opposite of tricks and illusions, because you can see his props and how he did it. It still boggles my mind.

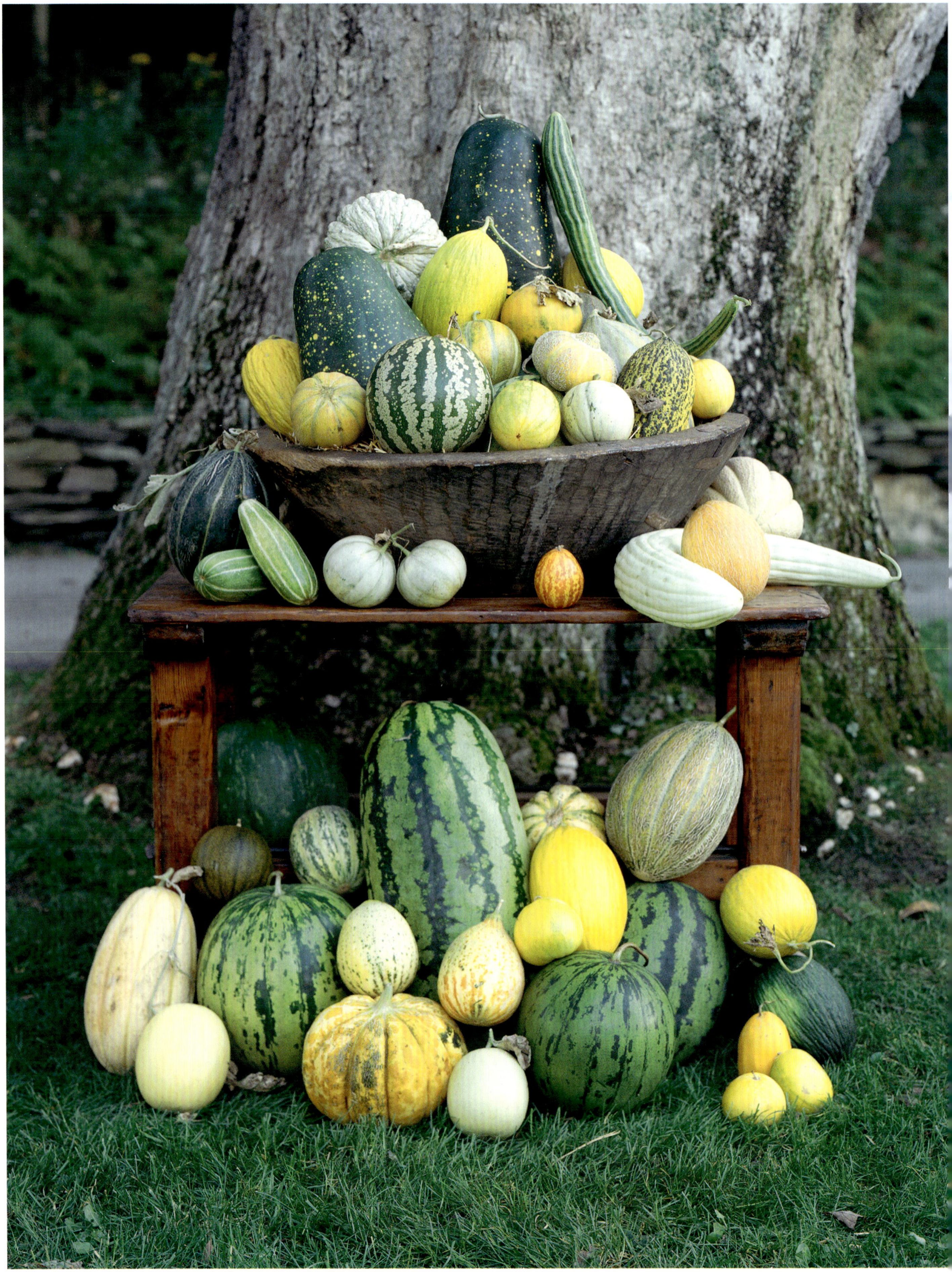

CHAPTER ONE

A Garden of Delights

The taste of melons at their peak, oozing honey, is incomparable, as is the air, sweet-smelling with muskmelon on an August night. And no other homegrown "fruit-vegetable" seems to kindle as much love and laughter as a good old-fashioned watermelon. Picture yourself with friends and family at the farmhouse kitchen table or in the backyard, picnicking in the shade of a catalpa tree. Everyone is sated from Sunday dinner. There's a pause in activity, and a somnolent lull. But soon enough it's time for dessert: an heirloom watermelon, harvested fully ripe and allowed to cool overnight in a tub of water. When thumped, it makes a dull *punk* sound. When cleft with a knife, it snaps and rumbles, cracking open to reveal an expanse of crisp and juicy red flesh with glistening black seeds. Everyone grabs a slice and dives in. Kids make mischief and spit seeds. This same scenario is impossible to imagine with a seedless watermelon.

The Santa Claus casaba was my first melon love. For weeks I had waited for the fruit to ripen, and one morning it was ready, lolling in the garden like some outlandish hot air balloon, its wrinkled yellow skin covered with dark green splotches. I dropped to my knees and cut it open to taste its heavenly flesh. Since then, I've formed passionate attachments to many other melons and watermelons. I'm devoted to Old-Time Tennessee, Zatta, Branco, Chamoe, Crimson Sweet, Wilson Sweet . . . and the list goes on.

In May or June, scores of heirloom melon and watermelon transplants crop up in my garden, blanketed by tents of spun polyester cloth atop black plastic mulch. They are protected from the elements as well as marauders by an application of diatomaceous earth (an organic pest control), a spritz of Bt (*Bacillus thuringiensis*), and TLC (tender loving care). What more can I do but pray for sun? After plants have put on growth and the first hatch of insect pests has passed, I remove the coverings and let the plants sprawl and flower, at ease, until the garden becomes a verdant sea of vines bearing fruit.

The seedless (sterile triploid) bowling balls that pass for watermelons and the less-than-honeyed honeydews in grocery stores across America don't begin to describe the diversity of these crops. Heirloom fruits and vegetables are treasures worth knowing, growing, and saving. While supermarkets devote whole aisles to modern diploid hybrids—many of which are wonderful—heirloom varieties are hard to find. Untold numbers have become extinct. I can't count how many times over the years someone has tasted one of my sweet dessert melons or watermelons and said, "This brings back memories of my childhood," or "I'm in ecstasy." Until they sampled heirlooms, they didn't know what they were missing. It's time to bring back some of the simple culinary pleasures of the past and introduce them to a new generation. And, in order to make further crop improvements and keep us well supplied, breeders need heirlooms for their quality components, the genetic bases of which are not as yet well defined or understood.

"IT'S TIME TO BRING BACK SOME OF THE SIMPLE CULINARY PLEASURES OF THE PAST AND INTRODUCE THEM TO A NEW GENERATION."

Fortunately much remains of our vanishing vegetable heritage. Many countries maintain genebanks with valuable germplasm for use by breeders. And the Svalbard Global Seed Vault, located above mainland Norway, provides the ultimate safety net for genebanks conserving different crop varieties. Nonprofit organizations like the Seed Savers Exchange (SSE), this country's premier nonprofit seedsaving organization, have a role to play as well. Its mission is to conserve and promote America's culturally diverse but endangered garden and food crop heritage for future generations by collecting, growing, and sharing heirloom seeds and plants. If we sow and grow those seeds, we are nourished; and if we harvest more seed, we ensure next year's bounty. But it's not the way we commonly feed ourselves today.

Most of these old-time varieties are not native to North America, but became part of our common heritage when immigrants brought the seeds here, hidden in their suitcases or sewn into their hems, aprons, or hatbands. These precious portable possessions spelled breakfast, lunch, or dinner and the comforts of home in an uncertain New World. And among "the finest varieties of fruit, native and foreign, cultivated in this country" by 1845 were melon and watermelon—the sole herbaceous annuals included in *The Fruits and Fruit Trees of America*, a treatise written by the illustrious American horticulturist Andrew Jackson Downing (1810–1852).

Seed Savers Exchange gave me the chance to play a small part in agricultural history when they sent me two cardboard boxes of melon and watermelon seed by overnight mail in the spring of 2000. Opening them, I realized I had in my hands a gift of vast magnitude, an irreplaceable wonder: the seeds of yesterday. SSE had only a few seeds of some varieties left, and they wouldn't have given them to me if they felt I couldn't handle them. Still, I was afraid something would go wrong and even had a nightmare about the theft of plants from my garden. I've since regained my composure and have harvested hundreds of fruits from those seeds—and hundreds from seeds sourced elsewhere. This book attempts to portray them in all their glory.

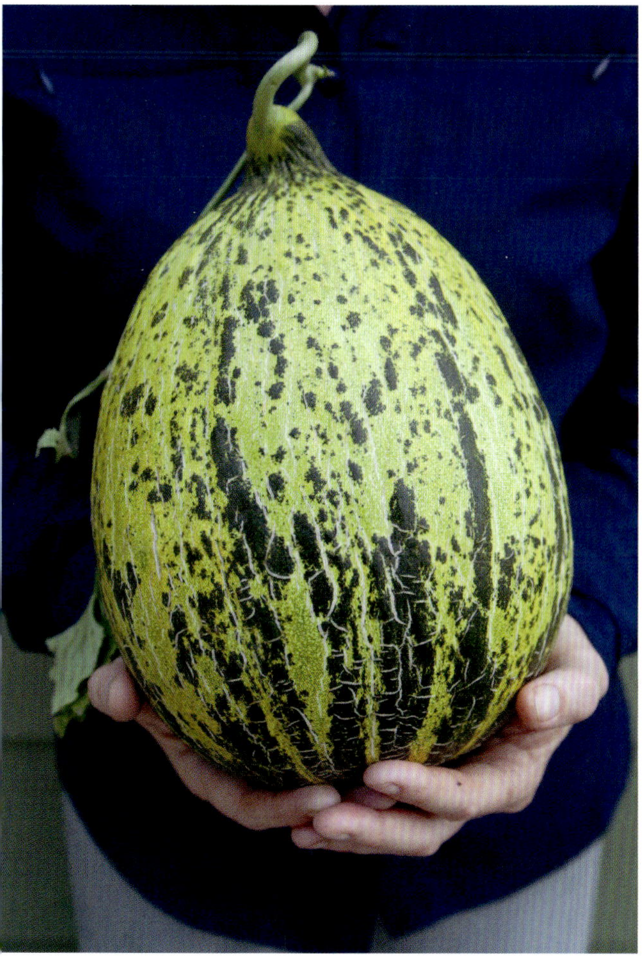

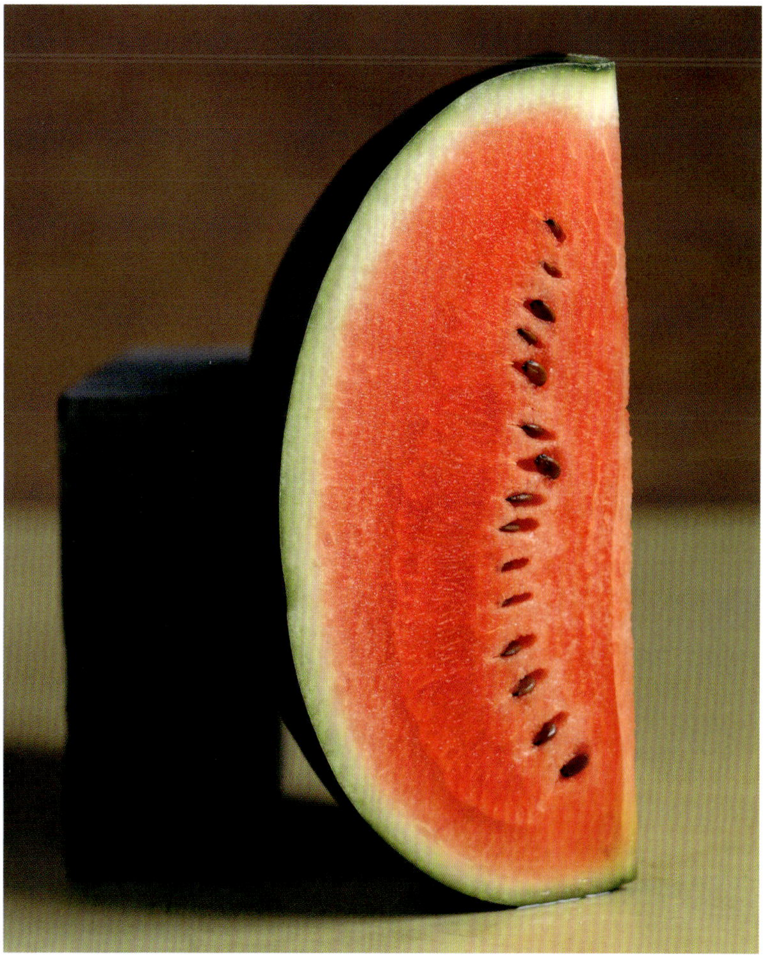

Left: Amy's first melon love: Santa Claus casaba. Right: Quintessential watermelon: Blacktail Mountain.

CHAPTER TWO

Picking and Choosing Ripe Melons and Watermelons

eirlooms taste good. The eating part comes easy—it's the picking and choosing that needs to be taught. If you want to know how to judge when a melon or watermelon is ready to be harvested from the vine, as well as how to select a fruit of top quality from the market, you've come to the right place. Thumping and sniffing alone don't do it; you need to take other factors into account.

I can't guarantee that you'll be right every time. There is no sure way—apart from nuclear magnetic resonance imaging, used on intact melon and watermelon in Japan—to select a winner before cutting it open and tasting it. There are, however, certain correlates of quality that, when taken together, can help you make your best guess.

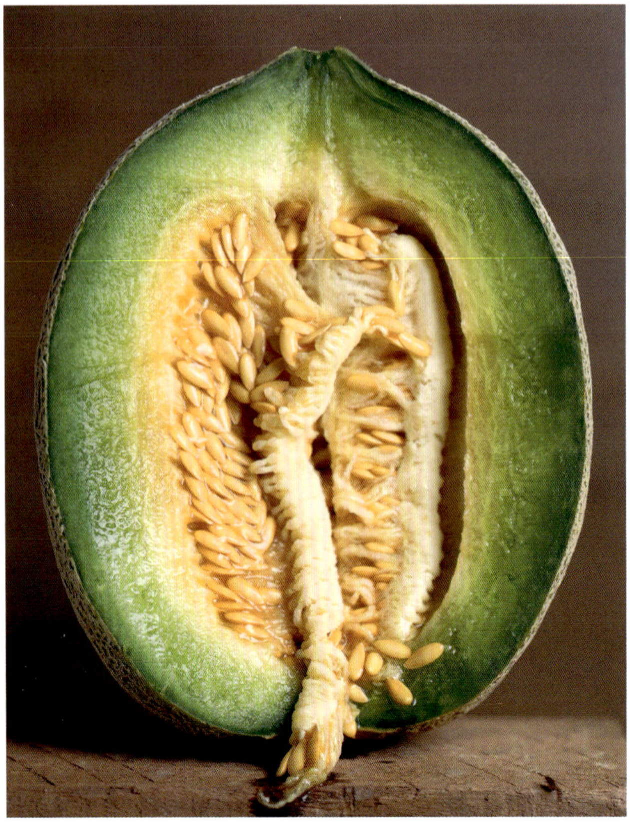

Fordhook Gem, fully ripe, is one of the best of melons.

According to Dr. James D. McCreight of the Agricultural Research Service (ARS) of the United States Department of Agriculture (USDA) in California, sweet dessert melon flavor is a complex interaction of sugar, pH, texture, and volatile compounds. Vine-ripened melon, harvested at maximum sugar content, has the best flavor and texture. Green melons, picked before their time, can become juicier, but they never catch up on flavor. Unlike their cousins pumpkins and squash, melons and watermelons don't have much in the way of carbohydrate reserves; in fact, most of their carbohydrates are sugars, not starch. That being the case, little or no additional conversion to sugar can occur. Melons and watermelons, unlike pumpkins and squash, do not get any sweeter once they've been harvested. What you pick is what you get.

The instrument most often used to assess sweetness is a spoon. However, the total soluble solids (TSS) content of melon and watermelon juice, measured in degrees Brix with a handheld refractometer, provides a more objective measure. Good-quality fruits have a minimum TSS of 10 percent.

Sweet melons start to ripen about thirty days after anthesis (flowering): Total sugars increase along with pH (which translates to low organic acid content). Sugars such as fructose, glucose, and sucrose make up most of the total soluble solids. The late Don Maynard, who taught me so much about watermelon, wrote that sucrose in dessert watermelon is first detectable at 20 days after pollination and continues to increase; fructose and glucose increase up to 24 days after pollination and then decrease. The relative concentration of these sugars (which is dependent upon cultivar and stage of maturity) affects perceived sweetness. Fructose tastes sweetest. I used both spoon and refractometer to assess the sugar content of every melon and watermelon in the book, including the non-sweet types, and most of the time there was agreement between my taste buds and the Brix reading. I'm not urging you to buy a refractometer, but it's an especially useful tool when you're doing taste tests and comparisons.

The other important index of internal quality is flesh texture (mouthfeel). Look for sweet melons that have neither the hardness associated with immaturity nor the softness associated with being overripe. Unlike sugar content, melon texture (including succulence) and aroma can improve after harvest. Well-nourished melons generally have conspicuous fibers in their flesh. If fibers are conspicuous and tender, the flesh has a melting quality; if they're tough, the texture is stringy. When fibers are large and abundant, without being tough, the melon seems coarse. Mealiness occurs when the flesh is dry and fibers are not evident. Fully ripe fruits do differ in texture among cultivars—and consumers differ in their preferences. Most people prefer their casaba quite firm, almost crunchy, and muskmelons a bit softer. The best watermelons have a fine-grained texture rather than coarse. They are firm and crisp (nonfibrous) with flavor that is sweet and mild—neither bitter nor caramelly (the latter sometimes associated with dark red flesh color).

"A DESSERT MELON CAN BE SOFT AND RIPE AND EVEN SMELL AMBROSIAL, YET STILL DISAPPOINT CONNOISSEURS—INCLUDING BENJAMIN FRANKLIN, WHO WROTE (UNDER THE PEN NAME POOR RICHARD) IN 1733: 'MEN AND MELONS ARE HARD TO KNOW.' SOME CULTIVARS HAVE BETTER FLAVOR. MELONS AND WATERMELONS ARE NOT ALL CREATED EQUAL."

Judging the worth of many sweet dessert melons is literally right in your hands and under your thumb. Using your sense of touch to detect nuances of softening in a melon is what the distinguished French botanist Loisel (1774–1849) called *tâter un Melon* (to feel out or explore a melon) in *Traité Complet de la Culture des Melons* (circa 1853). By gentle manipulation, we can all tell fairly easily when a peach or pear is ripe. Dessert melon, with its thicker skin, poses more of a challenge. Go straight to the melon's blossom end: the base of the fruit, opposite the stem. Immature melons do not yield to pressure. A melon is ready to pick or eat when the skin is fairly easy to depress. If your finger finds little resistance or elasticity or even breaks the skin, the fruit is kaput. This method does not work with most winter melons, though some casabas will yield a bit to thumb pressure.

Judging watermelon may make you wish you had extrasensory perception. According to the "thump test," an unripe watermelon when rapped with the knuckles supposedly has a high, musical, or metallic sound. Mark Twain said it says *pink* or *pank*. A ripe one, on the other hand, makes a dull *punk* sound. To some it sounds hollow. Unfortunately, they all sound pretty much the same to me. The "press test" doesn't work either—watermelon rind and flesh are too thick to respond to mild pressure, and if you push hard with your palms, what you hear is the rumble of tender rind and damaged flesh giving way. The "sniff test" is a nonstarter since watermelon rind has no external aroma. For the supermarket or farmers-market shopper, the most reliable indicator of ripeness is a yellow rather than white ground spot on the underside. If the darker green stripes contain too much yellow, the fruit is overripe—or has been out in the sun too long.

OH SO SWEET

Sugar content and flavor are nearly synonymous in sweet dessert melons and watermelons, and both suffer unless the fruit is allowed to mature on the vine. Take the sad case of Honeydew (which matures naturally about seven weeks after flowering): With one of the highest sugar contents of any melon if well grown, it is routinely harvested mature but unripe for commercial production, then zapped with ethylene to artificially induce ripening during shipping.

Happily, the gardener has several reliable clues to go by. Is this a watermelon of substance, weighty, generally rounded (with well-filled ends), and of normal proportions? Tendril withering is also a dependable sign: When the tendril closest to the fruit stem turns brown and dry, the fruit is likely to be fully developed. Don't rely on just one sign, though. Consult your calendar for average days to maturity to see if you're in the ballpark (plants usually begin to flower 40 to 60 days after sowing; add another 25 to 40 days from flowering to fruit maturity). Small-fruited varieties ripen earlier. Examine the ground spot on the belly of the beast—yellow is encouraging; white is not. The rind should be dull rather than shiny. Dark-skinned fruit will feel a little rough to the touch. Columbus took a chance . . . why not you? Harvest the biggest one first.

If you really need to keep track of growth, use a scale in the garden to measure daily weight gain. This is especially helpful when growing exhibition watermelons. Cut the fruit from the vine when it stops putting on weight—but not in the early morning, when fruits are bloated and likely to crack; wait until later in the day. Consumers, who arrive on the scene after harvest, must rely most on the "eyeball test" for intact watermelon, sizing up form, weight, volume, and condition of the rind. They can only hope that the picking was good—and that no internal tissue damage was caused by compression or impact injury during shipping and handling. Watermelon is a big bruiser and more fragile than you think.

A dessert melon can be soft and ripe and even smell ambrosial, yet still disappoint connoisseurs—including Benjamin Franklin, who wrote (under the pen name Poor Richard) in 1733, "Men and Melons are hard to know." Some cultivars have better flavor. Melons and watermelons are not all created equal. Consumers are at a decided disadvantage; at the mercy of suppliers, they have no one to ask, "Did the grower plant fine-flavored varieties? How well were they grown, and under what conditions? Was irrigation reduced prior to ripening to help increase sugar content? Were they harvested from dead or diseased vines? Were they picked at the proper degree of ripeness, or were they harvested too soon or too late? How were they handled after harvest? Were they refrigerated too long?" Home gardeners are lucky—they can choose melons and watermelons of fine quality to grow, and they have the luxury of eating vine-ripened fruit too perishable for long-distance shipping.

Whether you're a grower or a buyer, there are a number of signs that can help you select the good and reject the bad. You want a sweet dessert melon whose form is regular and normal. Those that are rounder, more symmetrical, and have a filled-out appearance generally taste better. A well-cultured melon has a small ground spot. That's the portion that lies on the soil and is usually discolored, flattened, and inferior in quality. Neglected melons with prominent ground spots weren't propped up or turned by their growers.

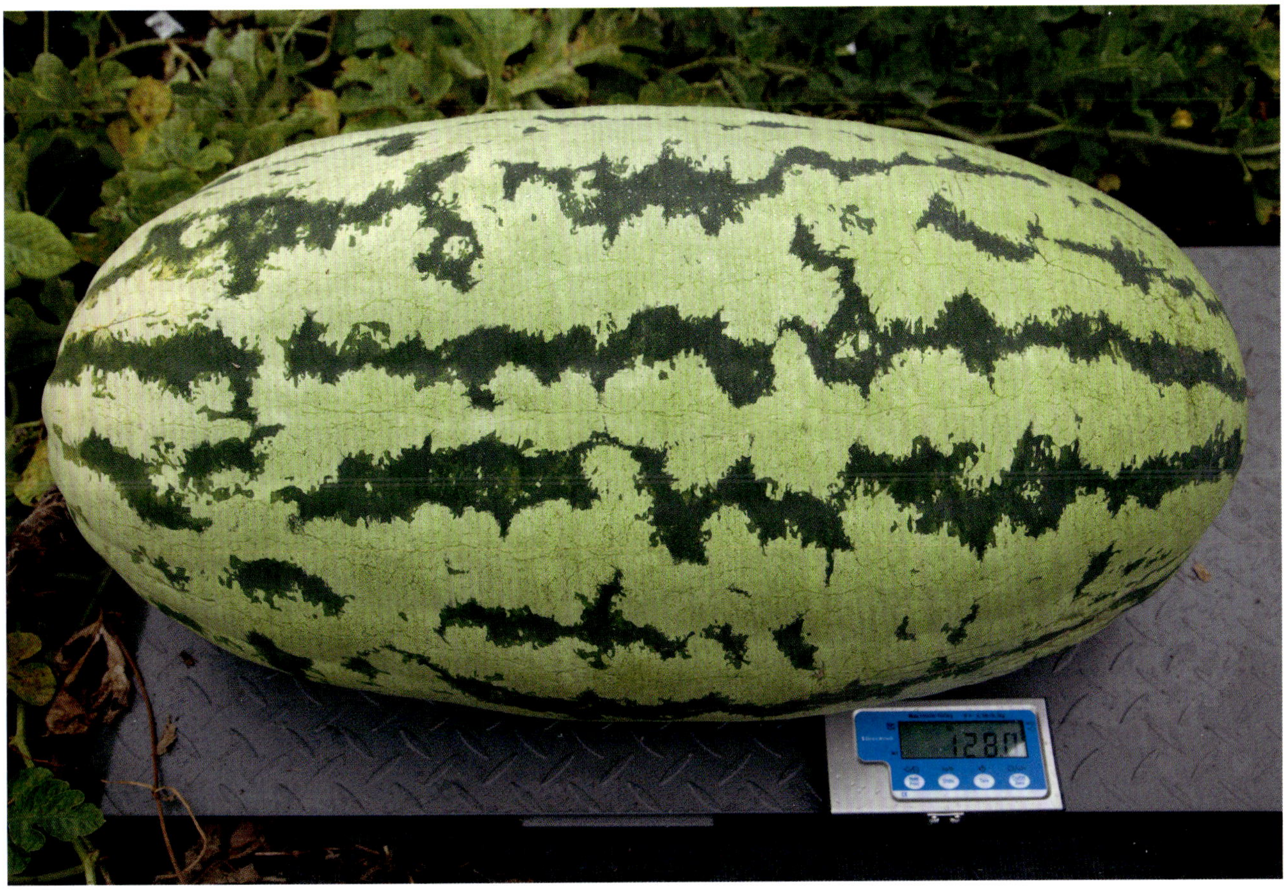

Carolina Cross on the vine clocked in on the scale at 128 pounds and was harvested at one pound heavier after it stopped gaining weight.

Size and weight are two other reliable indicators. Sweet melons that seem heavy for their size have more abundant juicy flesh. Bigger is often better, but it's generally a good idea to avoid giants, which might have been fed too much fertilizer and water. Don't consider bruised or damaged goods (with sunscald, hail spots, cuts, or abrasions) even if they're on sale. Internal tissue damage caused by impact injury is not easy to detect from the outside. Select those melons that appear fresh and not faded.

Some melons abscise or slip from the vine when ripe; others don't abscise until they're overripe. "Persistent" fruit varieties don't abscise at all and must be cut from the vine when ripe. When melons are at full slip, a crack completely encircles the stem. At half or forced slip the quality is not quite as good (there is a lower concentration of aroma compounds), but the shelf life is longer. To harvest at prime time, growers need to pay attention to certain changes in the melon's character. As maturity approaches, the primary color of the fruit skin warms, usually changing from greenish to yellowish. Experienced pickers rely more on color change than the appearance of the slip.

Many growers and consumers think that a good American cantaloupe has a crack around the stem, but that is not always true. Better for the consumer to consider the color first, and then examine the stem or its cuplike remains to see if it is fresh or faded. Muskmelons, which generally slip, have a netted skin. The more raised, hard netting or "corking" that is dense and uniformly distributed, the better-tasting the melon. Melons that don't slip, and watermelons, should be cut (leaving part of the stem attached to the fruit) rather than pulled from the vine when ripe, to avoid damaging plant and fruit and inviting stem-end rot.

The perfume or odor of some melons becomes apparent as the fruit ripens. Whether it's musky or floral, for example, depends on many volatile compounds, including alcohols, acids, and their esters; the amount and ratio of these substances vary from one type to another. Aroma volatiles are formed during ripening, and beyond harvest, from free amino acids in the fruit. A sweet fragrance is a good sign. However, some melons are harvested green and their flesh normally smells like cucumber.

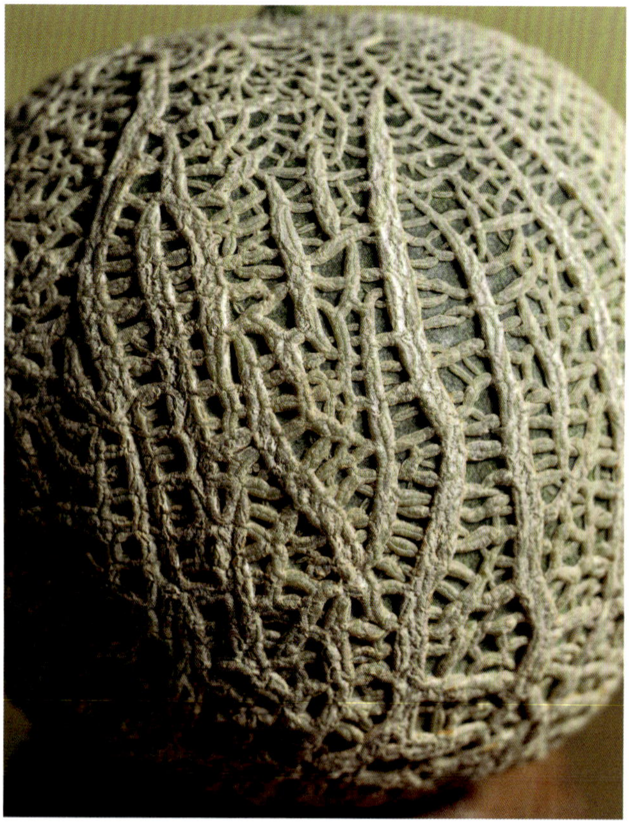

Above: Close-up of French heirloom Sucrin de Tours, with its distinctive overlapping layers of ropy netting. Opposite: The watermelon gods were good to us in 2015; witness "Watermelonfall" cascading down our front steps.

Climacteric melons including muskmelons and true cantaloupes exhibit an autocatalytic ethylene burst during ripening, which is responsible for the production of volatile compounds, fruit abscission, degreening of the skin, and flesh softening. These rapidly ripening melons have a shorter post-harvest or shelf life. Nonclimacteric melons experience no such intense burst of ethylene at fruit maturity and, consequently, their shelf life is longer. These melons generally lack aroma, netting, and abscission zone. But since sugar and carotenoid accumulation are ethylene independent, casabas, which mature late, can accumulate very high sugars. Their ripeness can be frustrating for the melon lover to decipher: Be alert to cracks in the thick skin and color changes (sometimes subtle) that signal maturity.

Melons and watermelons, like all living things, go through inevitable decline and senescence. After harvesting, muskmelons keep their sweetness and texture for three or four days at ambient temperatures; casabas can easily store a week or two and, in some cases, months. As a dessert melon grows old, it loses flavor and sugar and can develop stem-end cracks and water-soaking of the flesh. Overripeness can bring an acetone aroma as well as loss of color, sugars, and beta-carotene in orange-fleshed melons. In the market, avoid overripe melons, which are more susceptible to decay-causing fungi and bacteria; these melons often sound sloshy when shaken. Rattling is okay—it's just loose seeds. Watermelons exhibit a nonclimacteric respiratory pattern and deteriorate gradually over a period of weeks: Flesh softens and the color of red-fleshed varieties fades. You can slow down the ripening process, and keep decay at bay, by removing fruits from the hot sun and immediately cooling them down.

Melons and watermelons will last longer if chilled at the right temperature and relative humidity. Muskmelons and true cantaloupes can be kept for two to three weeks at 40 degrees Fahrenheit; watermelons store well for two or three weeks at 60 degrees Fahrenheit; and casaba, crenshaw, and honeydew varieties can keep three or four weeks at 45 to 50 degrees Fahrenheit. Relative humidity should be around 85 to 95 percent. It's prudent during storage to reduce exposure to other ripening climacteric fruits that produce ethylene, such as apples. If you keep melons and watermelons on ice too long, chilling injury can occur, including surface pitting, discoloration, watery tissue, and loss of flavor. Use them or lose them!

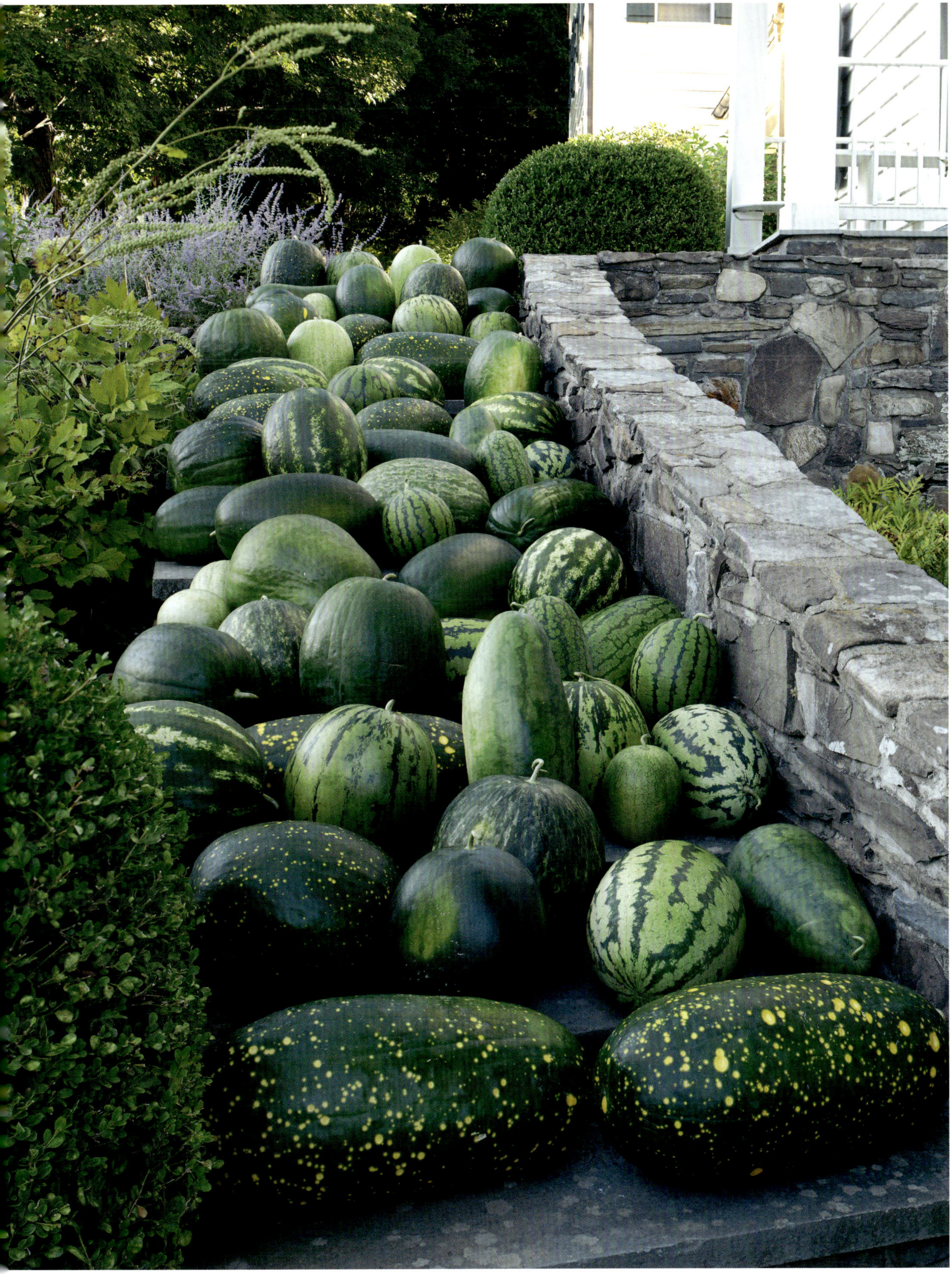

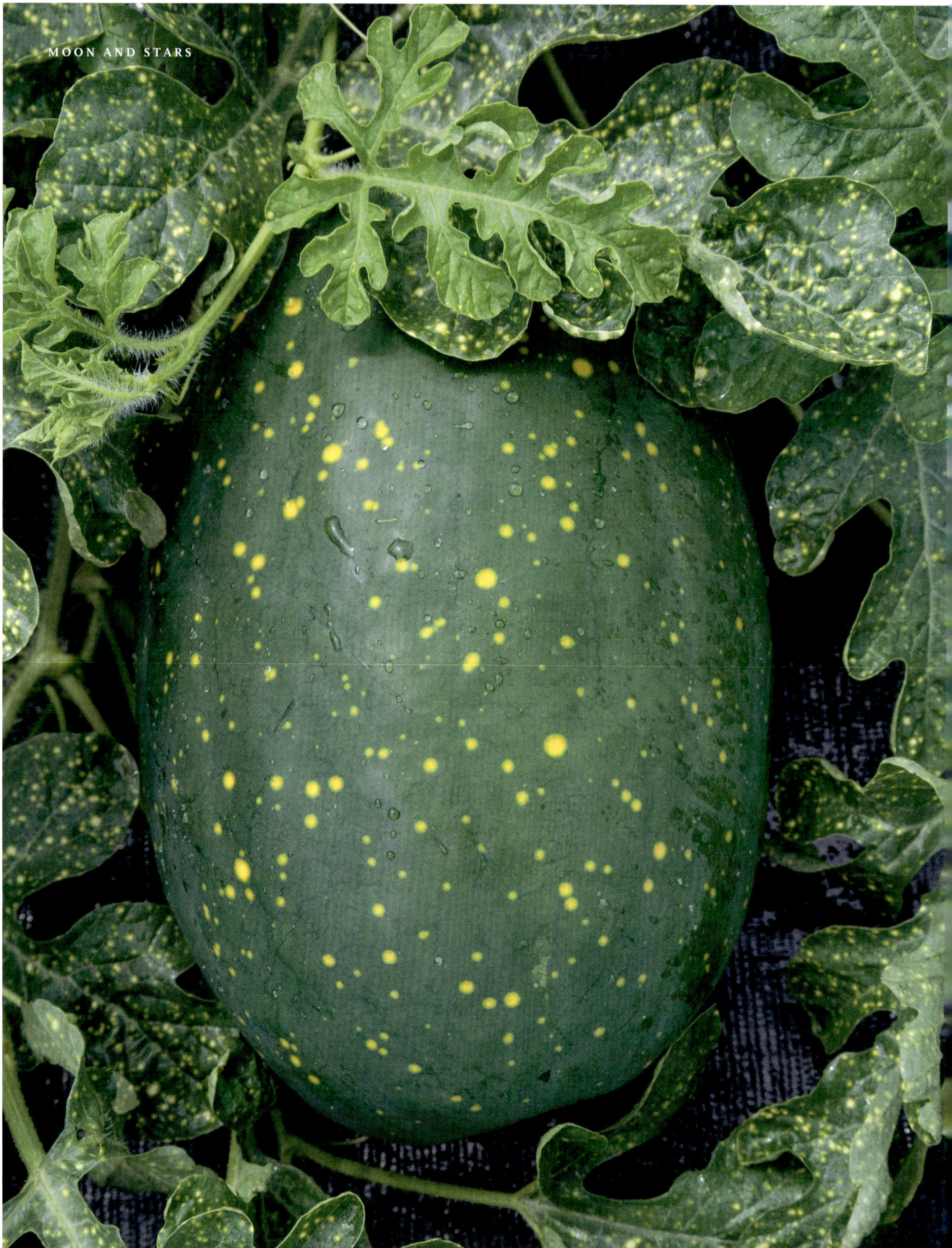
MOON AND STARS

CHAPTER THREE

Growing Melons and Watermelons

I heartily recommend that you grow heirloom melons and watermelons. The rewards are enormous. These types of plants may not be as easy to grow as tomatoes, but you don't need a staff of gardeners and a greenhouse, either. If you are blessed with a long, warm, dry summer, and a garden with lots of room to sprawl, plants produce flavorful fruit in just a few months. One of the main ingredients for success is what the astute seedsman and melon connoisseur Jacquin long ago called the vigilance of the gardener. Gardeners need to be more vigilant in some areas of the country than in others. Liberty Hyde Bailey, one of America's premier horticulturists, referred in 1909 to a certain "line of perfection" at the latitude of Georgia and Texas (Orcutt 1982). Every deviation north or south of this golden mean could spell trouble for would-be growers. Geography needn't determine destiny, though. I'm willing to test the limits, and you should be, too.

Through the wonders of modern technology, and a little know-how, you can improve your odds. Melons and watermelons are very tender and like it warm and bright: sunny days in the eighties, nights in the sixties. Dessert melons generally mature 75 to 110 days after seeding; most dessert watermelons mature 75 to 95 days after seeding (big-fruited cultivars generally ripen later). Neither one of these crops takes kindly to frost, but if you don't have a decent interval between spring and fall freezes, there are ways to stretch the season.

The first step is to make a comfortable bed for your vines to lie on. If you're lucky, your soil is light, deep, and well drained; sandy loam is best. If not, you can amend the soil in the fall by incorporating leaf mold and a generous helping of compost, or sowing a green manure or non-host cover crop, such as winter rye, then tilling it in come spring. A field that slopes gently southward has the distinct advantage of warming more quickly as well as shedding excess water. Melon and watermelon root systems are large but relatively shallow. These species abhor wet feet (waterlogging) and soil compaction, which severely limit growth of the roots.

A BOTANY LESSON

Melons and watermelons are members of the Cucurbitaceae, the gourd family. The scientific name of melons is *Cucumis melo* L. (the "L" abbreviation is for the Swedish botanist Linné, who invented this name to designate melon). Watermelons are quite distinct and belong to the genus *Citrullus* Schrader.

Like most other cucurbits, melons and watermelons are annuals that grow on long, trailing, tendril-bearing vines. The fruits are a specialized type of berry with a leathery skin. Melon plants are softly hairy and have ridged stems and simple tendrils; their leaves can be round, oval, or kidney-shaped, and angled or shallowly lobed. Watermelon stems are hairy, angular, and grooved with tendrils that are branched, and leaves deeply divided into pairs of lobes (pinnatifid).

Since melons and watermelons belong to separate species, they cannot cross with one another.

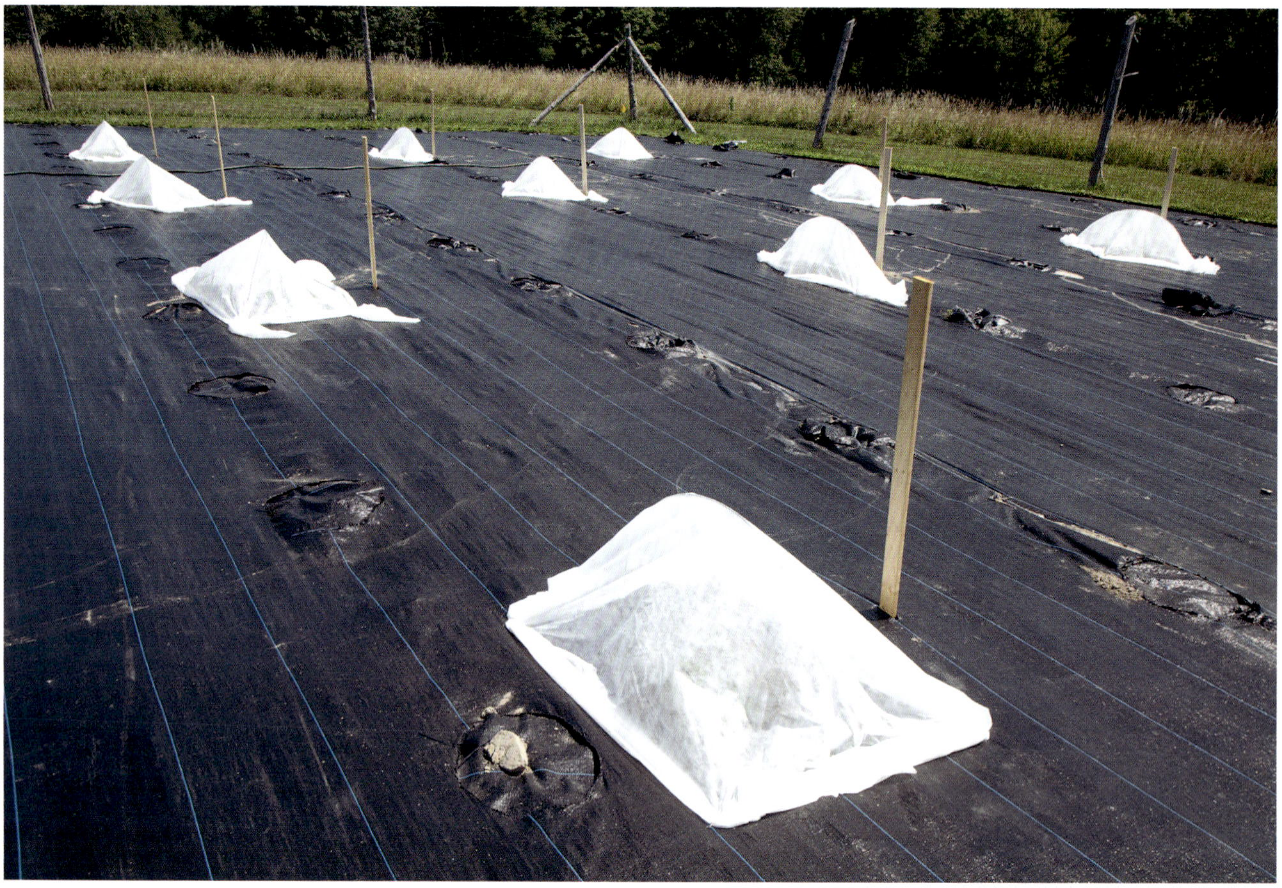

Flowers are naturally self- and cross-pollinated by bees. Depending on cultivar and environmental conditions, melon plants usually begin flowering 35 to 60 days after sowing; an additional 30 to 70 days is required from pollination to first ripe fruit. Watermelon plants begin flowering 40 to 60 days after sowing; count on an additional 35 to 40 days from pollination to first ripe watermelon fruit.

Above: Spunbonded polyester row covers protect vulnerable transplants when weather is unsettled. Black plastic mulch spurs growth. Opposite: Overview of melon patch at full tilt. Note the rampant vines of Citron watermelon in the foreground.

Soil pH affects nutrient availability and for that reason it's smart to take a soil test yearly to assess pH and nutrient levels. If done several months before the start of the growing season, you'll be able to make any adjustments before consigning your plants (or seeds) to the ground. The optimal pH range for growing melon and watermelon is 6.0 to 6.5. You can use sulfur to acidify the soil (lower the pH) or lime to "sweeten" it (raise the pH). We generally apply lime as needed in fall, and fertilizer at planting time in spring. Your local Extension office or soil testing lab can advise on the particulars of testing and remediation.

Practicing good field sanitation (removal of spent plant materials) and a three-year crop rotation can help reduce the incidence of certain pests and diseases. These methods are not terribly effective if the soil is infected with root-knot nematodes (parasitic unsegmented roundworms that form galls) or fusarium wilt (a fungal vascular wilt disease), which persist for many years in the soil and can cause significant losses in melon and watermelon crops. Grafted resistant rootstocks offer a solution not commonly employed by home gardeners; this technique has proven successful with heirlooms such as Petit Gris de Rennes.

One of the biggest mistakes beginners make is rushing the season. Transplant or direct-seed after the threat of frost has passed and the soil has warmed to at least 60 degrees Fahrenheit. The average safe planting-out date in my Zone 5b area is around Memorial Day weekend. Because I can count on only 120 frost-free days, direct-seeding in the ground—where germination is subject to the vagaries of cold soil and weather—is not an option. I get a head start by sowing seeds inside three to four weeks in advance of the safe planting-out date. Those who live in northern latitudes

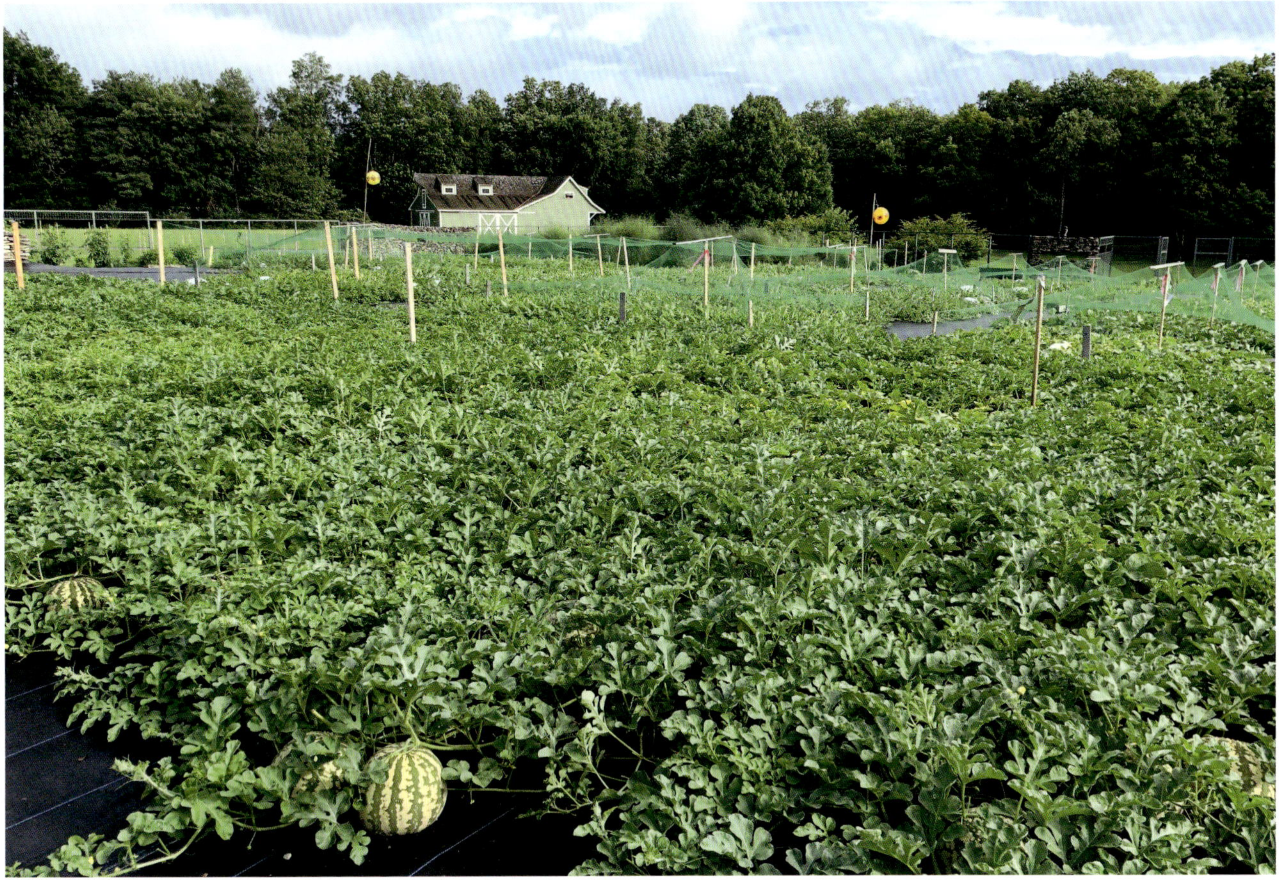

might want to concentrate on short-season varieties that mature in seventy-five days or less. Consult the USDA Plant Hardiness Zone Map (http://planthardiness.ars.usda.gov/) if you're unsure of your planting zone.

Start seeds in pots or trays filled with a sterile potting mix, labeled by variety, and water gently using a can with a rose. Keep sown containers covered in a dark, warm place until plants emerge. Check moisture levels and ventilate to prevent build-up of condensation and damping-off disease. To avoid leggy seedlings, check for emergence at regular intervals. Discard any elongated ones. Seedlings grown in three- or four-inch-diameter plastic containers transplant beautifully, with their root balls intact; seed starting trays with smaller cells are also popular. Melons germinate in about a week with soil temperatures between 75 and 85 degrees Fahrenheit and even faster at 90 degrees Fahrenheit. Watermelons generally take longer; they are tougher nuts to crack, so nick or file the seed coat gently (or even carefully clip off a tiny bit of the radicle or tapered end) and presoak the seed briefly in water before sowing.

Treatment with gibberellic acid (GA) is another option: soak for about 30 minutes in a mix of GA and water (to get 1000 ppm GA), then air dry the seeds before planting. Optimal soil temperature for watermelon germination is 95 degrees Fahrenheit. To gauge temperature just right, a soil thermometer is a must. In order to boost germination, simple and inexpensive plastic trays with covers (and vents if possible) serve well as propagators to house sown pots and/or smaller trays. Seedling heating mats and seed propagators with thermostats, which provide bottom heat, are excellent investments. For me they are indispensable. If seeds fail to germinate, sow, sow, again!

After germination, rig a light stand or ordinary grow lights with a timer to provide sixteen hours of light a day, and place seedlings two or three inches below the light source. Before seedlings begin to look overcrowded in the pot, tray, or cell, you can thin them (nip at the soil line with a scissors) to the three most vigorous per container or, if grown in smaller cells, gently prick out young plants that have their first set of true leaves and pot up individually into larger containers. Sprouts should be fed with a diluted fertilizer mix. When the time comes, plants can be "hardened off" in a protected place with natural light by curtailing water and raising (or lowering) the temperature before setting out.

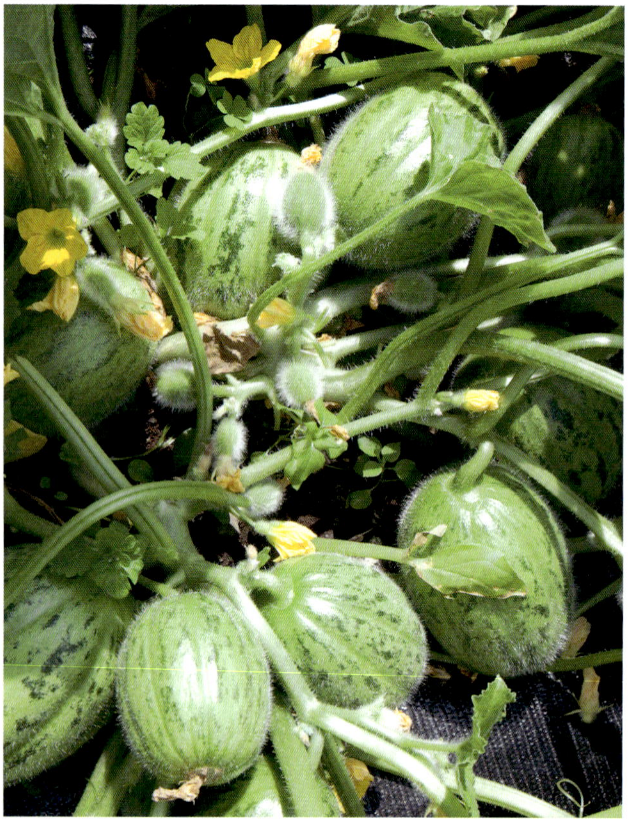

Tondo di Manduria melons on the vine. Carosellos such as these have a birdnest growth habit that produces a concentrated fruit set on short vines early in the season. They are best enjoyed at the immature stage.

I've been growing melons since I was a teenager, and black plastic mulch or ground cloth has always carpeted my melon and watermelon patch. True, it's not a thing of beauty or an emblem of sustainability, but it's the way to go if you want a garden that really produces when the climate is sub-optimal. The rationale: It raises soil temperature, conserves moisture, eliminates weeds—and weeding—and keeps fruits clean and free of rot. Melons and watermelons grow faster, flower sooner, and produce more fruit. Reflective mulches also promote growth and effectively deter aphids. Since north and west winds can wreak havoc with trailing vines, especially when they're on plastic mulch, use earth staples, stones, and bricks to anchor the mulch to the ground; the edges can be buried. (In the spring some growers plant a winter rye or another tall-growing grass, then till in all but windbreak strips between beds or rows.) Cut planting holes in the mulch two feet in diameter. A wooden or cardboard template comes in handy if you're planning on lots of varieties. Many growers plant melons and watermelons more densely than I do: two to three feet between plants in the row; five to eight feet between rows. Space plants or sow seeds more widely if you live in the South, where things grow bigger. Watermelons can also be given wider berth, since their vines are usually longer.

With the advantage of an acre and a half devoted to vegetables, I space "hills" (a discrete grouping of three plants in the same hole) at wide intervals to promote plant health and reduce interplant competition for water: ten feet between melon hills, and fifteen to twenty feet between watermelon hills. I grow dozens of large-fruited watermelon varieties, which can have vines as long as thirty feet. Some of the more compact melon cultivars with an early and prolific fruit set (see Adzhur and Makuwa groups, pages 113 and 127, respectively) are well suited to small home gardens; these plants can even be supported with a trellis or grown in containers.

Mix a shovelful of well-rotted manure, or a capful or two of pelleted time-release fertilizer, into the soil (it's best done just before planting), and plant melons and watermelons in hills, or singly in rows. If your soil tends to become compacted or waterlogged, it's a good idea to mound the soil before planting.

Since the advent of spunbonded polyester row covers, melons and watermelons

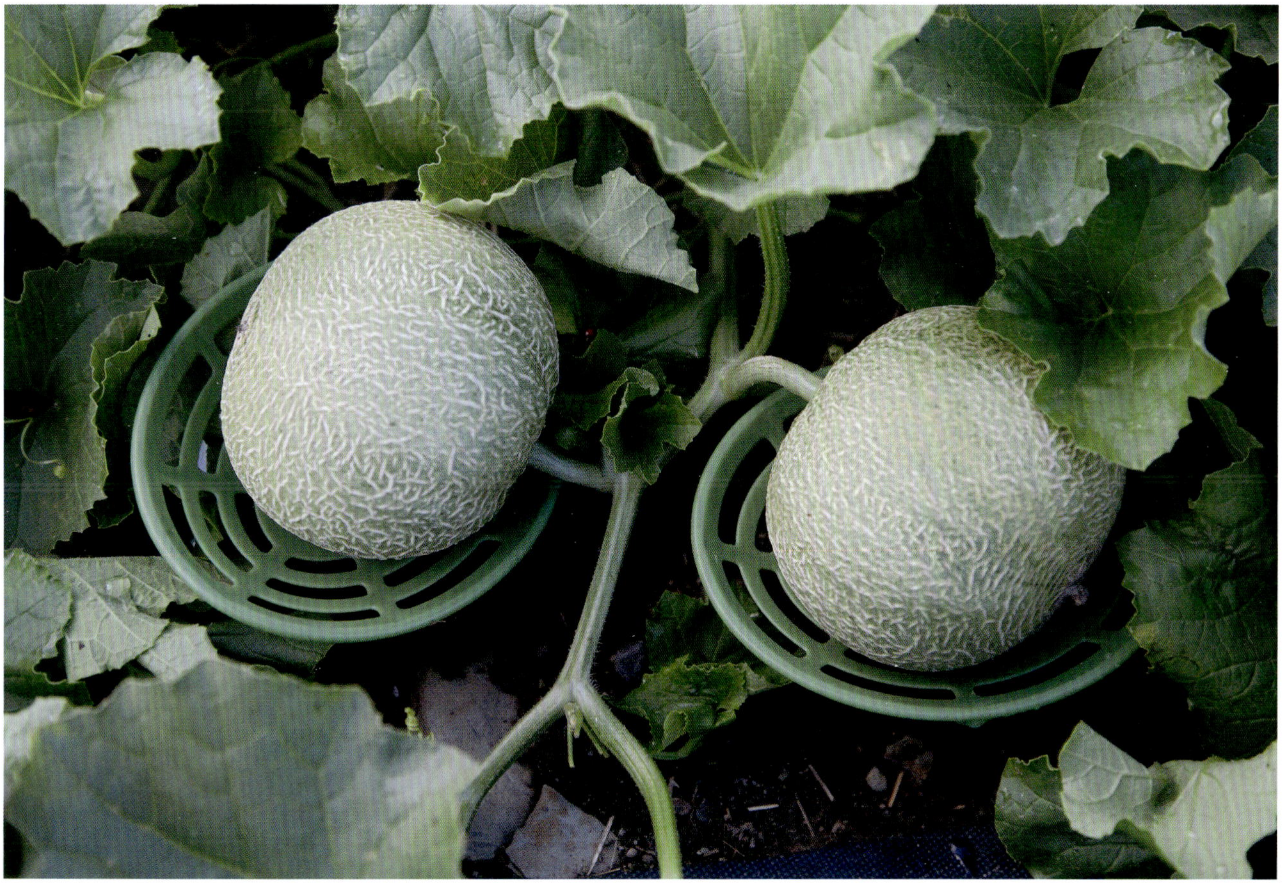

Hero of Lockinge melons on cradles, elevated above the field surface to keep them clean and free of rot.

have never been happier. After planting, I dust diatomaceous earth around the young plants to kill slugs and spray *Bacillus thuringiensis* (Bt) to eliminate caterpillars. These “tents” protect the vulnerable transplants from drying winds, cold, and insect predators like the dreaded aphids that transmit *Zucchini yellow mosaic virus* (ZYMV). Two or three weeks later, when plants can no longer be contained—and pollinators no longer excluded—I remove the covers, securing vines loosely with earth staples.

Melon and watermelon plants will not produce fruit without help from bees and other insect pollinators. As bees forage for pollen and nectar among flowers, they transfer the large sticky pollen grains from anther to stigma. Within hours, the pollen grains germinate and, by the next day, fertilization occurs. For better fruit set and seed production, you might consider keeping a hive of bees. Honey is an added bonus.

The best way to fight pests and diseases is to stop them before they get started. I'm very careful to minimize physical contact with the plants to prevent spread of pathogens, but I am resolute early in the season about removing single leaves (and entire plants if necessary) if they show signs of Alternaria leaf spot. It's best to sterilize clippers with a spritz of rubbing alcohol before moving between plants and to discard diseased materials in a plastic bag (rather than composting them). Foliar fungicides can help prevent and contain the damage. For more about pathogens and their management, consult the authoritative second edition of *Compendium of Cucurbit Diseases and Pests*.

No garden is free of pests, but you can fight back without poisoning them. Brush removal around the melon patch in early spring eliminates hiding spaces for rabbits and deters them from feeding on tender vines and fruits. As fruits size up, it's surprising how effective the modern-day version of a scarecrow can be: Large yellow balloons, with eyes painted on and silvery streamers trailing behind, can scare off crows. (Crows know a good thing when they see it: They can distinguish between dessert watermelons and insipid citron watermelons, and leave the latter alone.) When crows persist in dive-bombing ripening fruits in my garden, we erect netted tents over hills. Our farm manager, Michael Ryan, devised this clever solution: One-inch non-tangling nylon netting, sufficient to keep out pests yet allow bees through to pollinate, is supported by wooden stakes about eighteen inches high with screws on their sides to grab the netting and keep it taut.

"IF YOU REDUCE FRUIT LOAD ON THE VINE, YOUR DESSERT MELONS AND WATERMELONS WILL BE BIGGER AND SWEETER."

Opposite above: Amy harvesting a crown set of Crimson Sweet watermelons.
Opposite below: Large yellow balloons flapping on poles and netted tents over melon hills help ward off marauding crows.

As fruits approach maturity, loosely wrapping them in spun polyester cloth prevents sunscald and damage by marauding cucumber beetles and a new scourge in my garden: leaf-footed bugs, which suck juices from plants. Adults of both species can deface the rinds of watermelon, leaving them tan and scarred. The cucumber beetle, one of melon's mortal enemies, transmits bacterial wilt when it feeds—injury that's more than skin deep.

Some experts advise pruning the growing tips of melon vines to produce earlier and more abundant crops in the vegetable garden. Melon plants produce female or perfect flowers—which potentially become fruit—only at the first one or two nodes of branches. To encourage early branching and fruiting, main stems and initial branches are nipped, so that new branches with female/perfect flowers appear earlier. However, the technique isn't effective in the open field after fruit set has begun. Nipping or pinching was a commonplace American and European practice in the nineteenth century for growing melons under glass or in frames. It does not yield more melons and can impair quality.

Once fruits have formed, you can prune off the "cull" or defective melons and watermelons. If you reduce fruit load on the vine, your dessert melons and watermelons will be bigger and sweeter. Practice vigilance and routine maintenance as needed. Position fruits so they have enough slack and remain shapely without being strangled by vines. Cucurbits need a regular supply of water as they get established and fruits size up. But don't overwater. One to two inches per week is usually enough. Drip irrigation or hand irrigation of plants is preferable to overhead sprinkler irrigation, which can waste water and increase foliar disease. Water should be tapered off as fruits near maturity to enhance flavor and reduce splitting. Rainfall or waterlogging during late fruit ripening is apt to reduce the sugar content of dessert fruits. Low light intensity will do the same. If the weather cooperates and conditions are favorable, melon plants (although not watermelon) can produce a second wave of fruit. During the last two weeks in August, I harvest three times a day: morning, noon, and night.

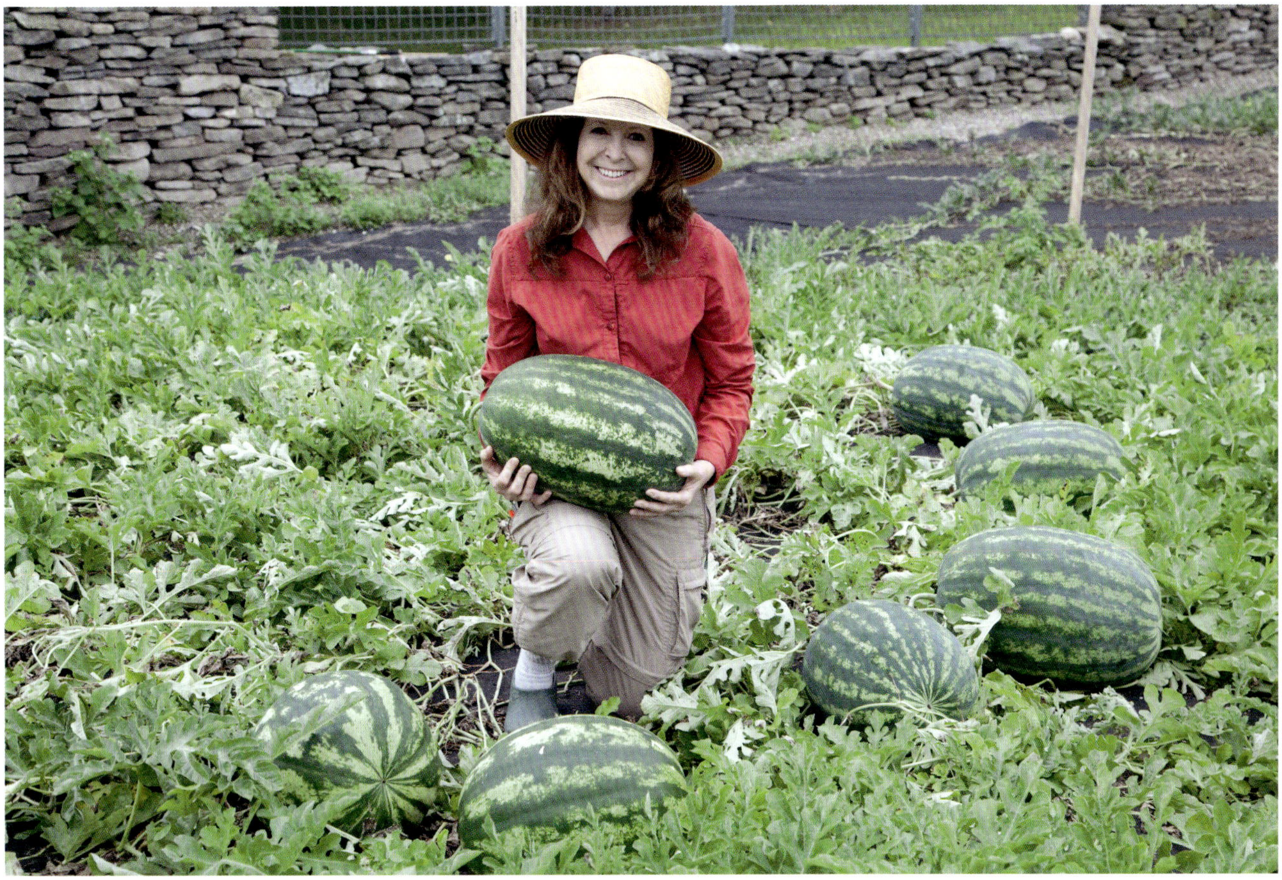

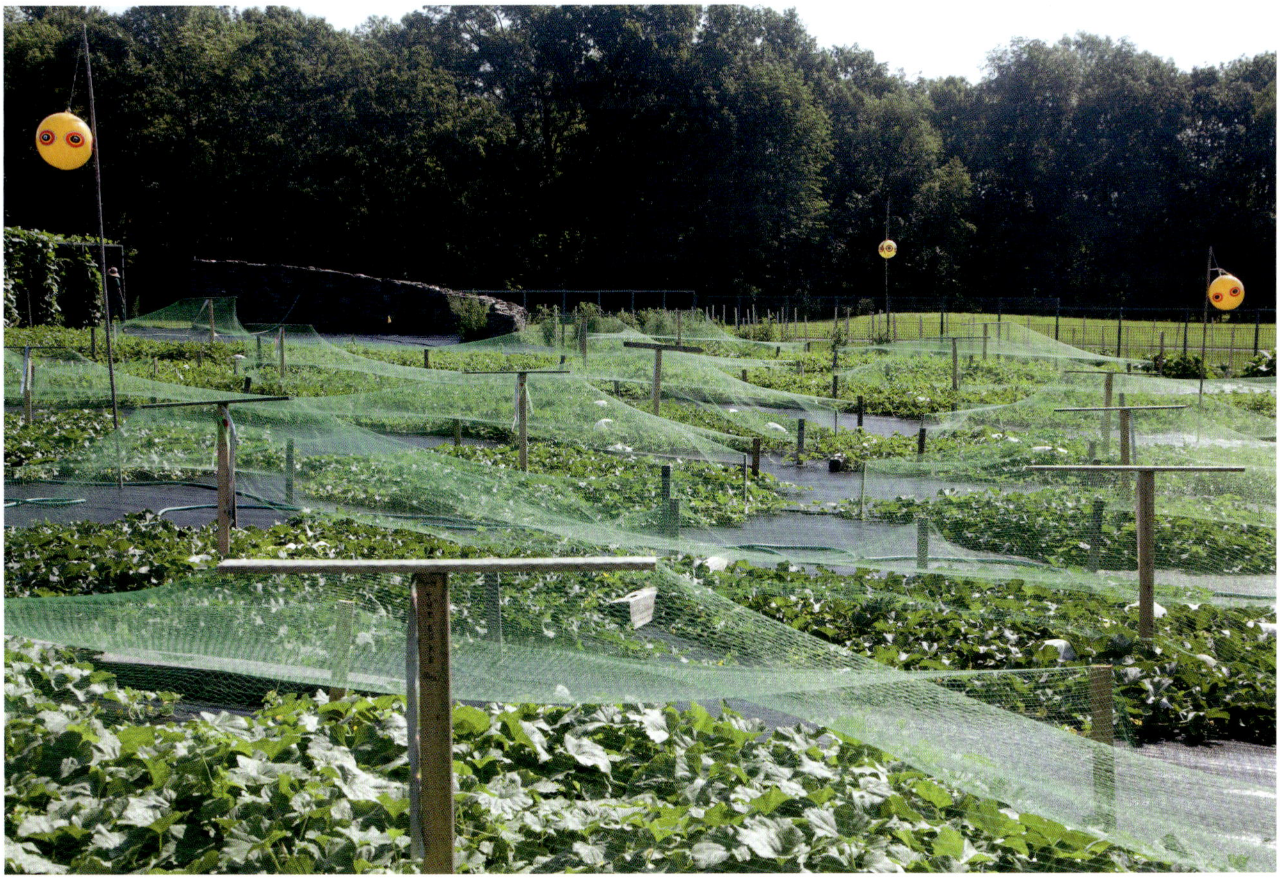

CHAPTER FOUR

Saving Heirloom Seeds

f you want to maintain a particular variety that you love and save pure harvested seed for replanting, you'll need to prevent bees from transferring pollen from other varieties of the same species. If heirlooms are not allowed to cross, they produce plants with fruits that look and taste like their parents. Known as open-pollinated or standard varieties, heirlooms differ from modern F1 hybrids, whose offspring don't breed true.

There are two basic ways for the home gardener to control pollen flow between varieties of the same species: isolation by distance, and hand pollination. Since melons and watermelons belong to different species and don't cross with each other, I recommend that you begin with isolation. Limit yourself to one variety of melon and one variety of watermelon that you value and know to be heirlooms rather than F1 hybrids. Plant them in the same garden but make sure that they are at least eight hundred feet to a half mile from other varieties with which they could cross-pollinate. Check with your neighbors to see if they're planning to grow these species. If not, you can let nature take its course.

If you can grow melon and watermelon for the fruit, you can grow it for seed, too. In most cases, when your fruit is ready, so are the seeds. It's easy to scoop them from the cavity of a muskmelon or honeydew. But because watermelon seeds are scattered and embedded in the flesh, they present a bit of a challenge. Either collect the seeds after the fruits have been eaten and seeds set aside or pick them out in advance and put them in a colander, reserving the flesh for fresh eating or juicing. Rinse seeds under running water, gently rubbing to remove stickiness and tissue; then shake off the water, blot extra moisture from the bottom of the colander with a paper towel, and turn the seeds out onto sturdy, absorbent paper plates (or coffee filters) that are labeled with the variety name. If your main concern is to maximize mature seed production, allow the intact harvested fruit to ripen past "market maturity" before processing seeds. (Don't wait too long though: watermelon seeds can begin sprouting in vivo.) Some growers prefer to soak seeds for a few hours—or even ferment them in water for a day or two—in order to remove attached pulp before rinsing. Dry the seeds at room temperature, occasionally turning them over, until they can be cracked or snapped rather than bent.

DRYING SEEDS

Air temperature for drying melon and watermelon seeds should not exceed 110 degrees Fahrenheit. In humid climates, use a dehydrator or the low heat of a lightbulb or pilot light, keeping it at about 90 degrees. Or, simply turn on a fan to move the air in the room. Never dry the seeds in the oven: They may shrink or crack and die.

Opposite: A mélange of melon and watermelon seeds. Unlike melon and other cultivated cucurbits, watermelon exhibits a wide variety of seed coat colors and patterns.

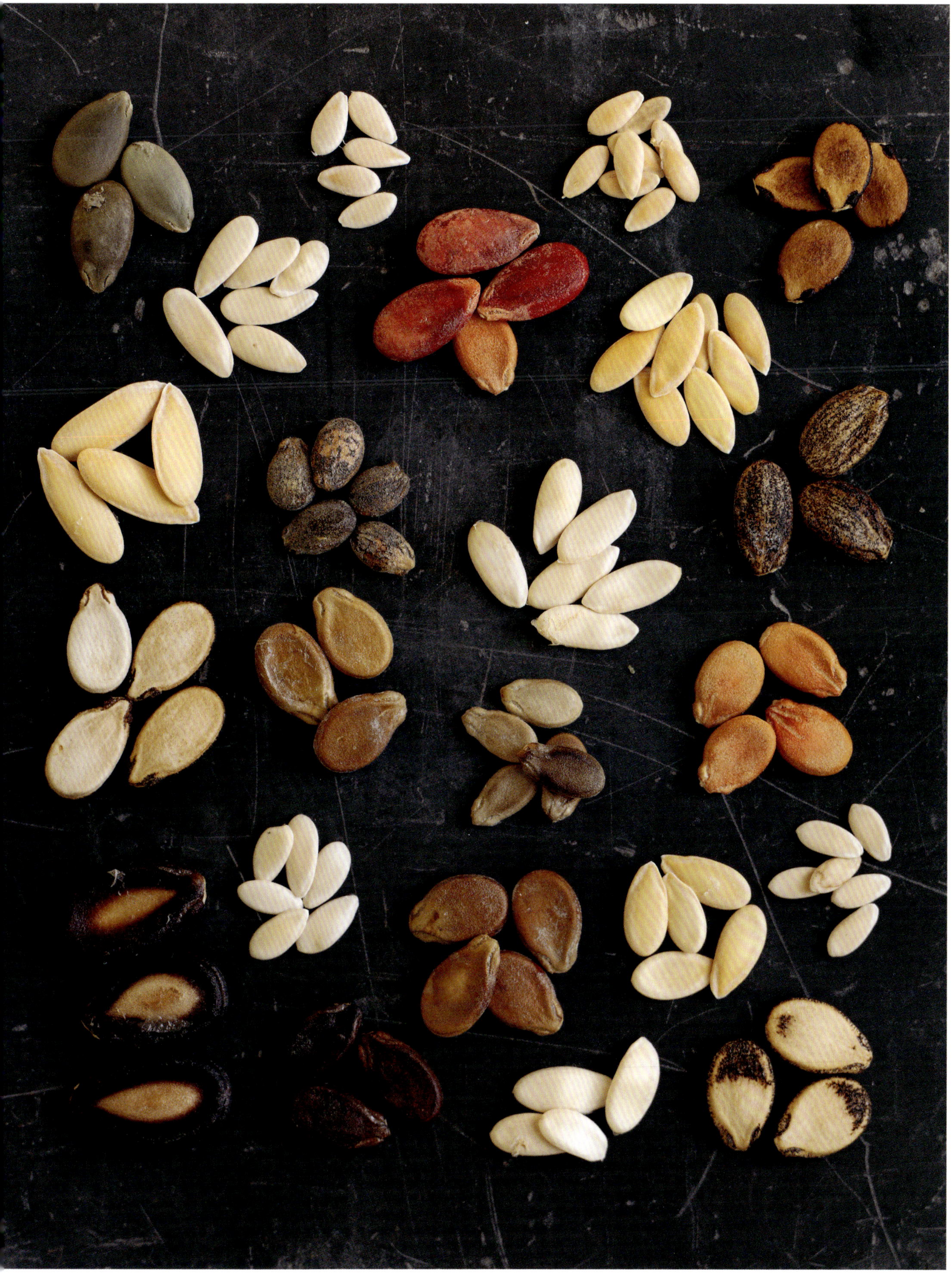

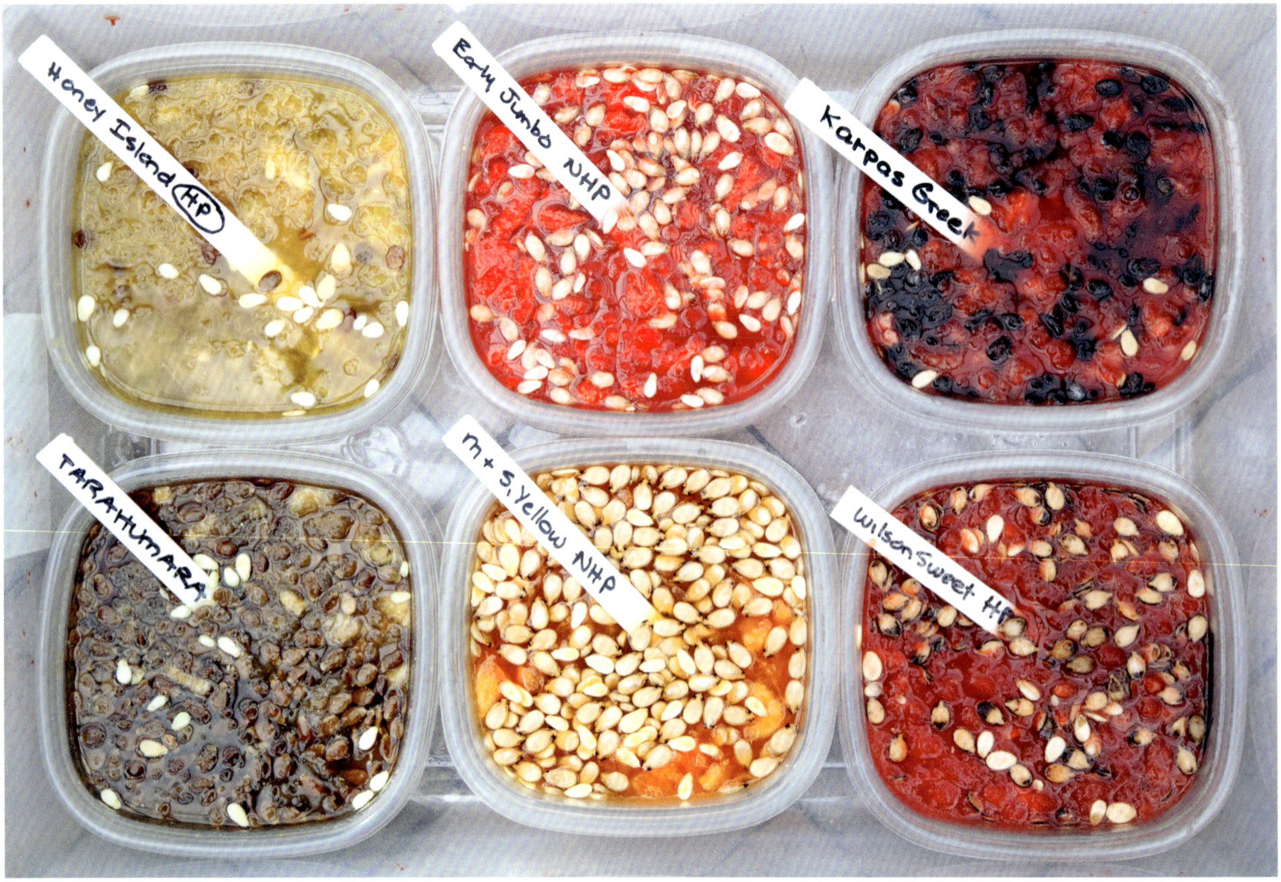

Briefly soaking or fermenting watermelon seeds makes it easier to remove attached tissue before rinsing and drying. Opposite: Dried watermelon seeds ready for packaging. Note the many seed coat colors and blunt hilum ends.

Since heat and humidity cause dried seeds to lose vigor, store them in packets in airtight containers, in a cool, dark, dry place. Seeds that are well cured (with a moisture content of 4 or 5 percent) and stored this way can remain viable for five years or more; if stored in the refrigerator (at about 40 degrees Fahrenheit and low relative humidity), they can remain viable twice as long. But don't let them languish on the shelf. If you share your seeds with other dedicated gardeners, you'll be doing your part to ensure that heirloom melons and watermelons don't vanish from our gardens and tables.

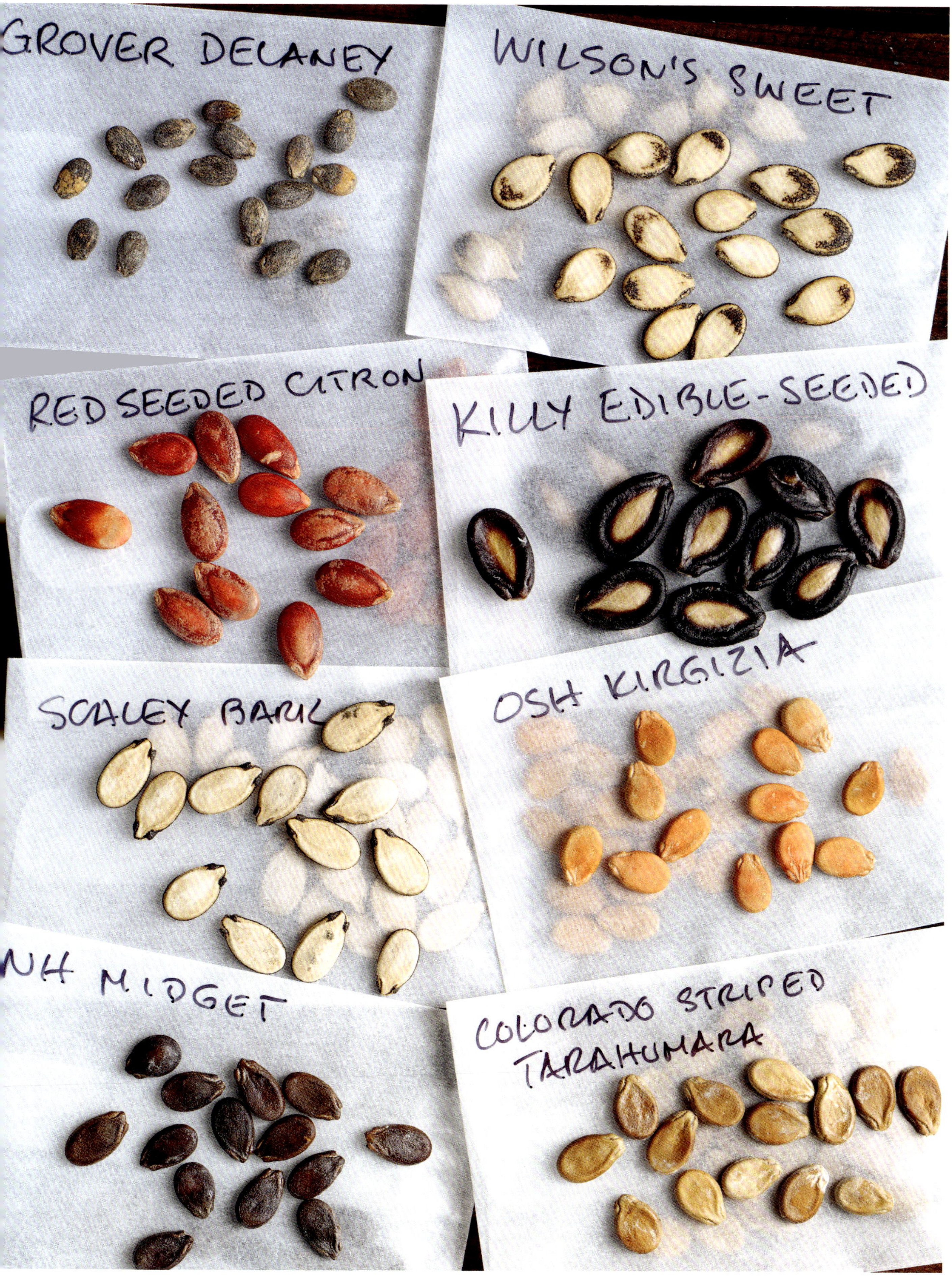
GROVER DELANEY
WILSON'S SWEET
RED SEEDED CITRON
KILLY EDIBLE-SEEDED
SCALEY BARK
OSH KIRGIZIA
NH MIDGET
COLORADO STRIPED
TARAHUMARA

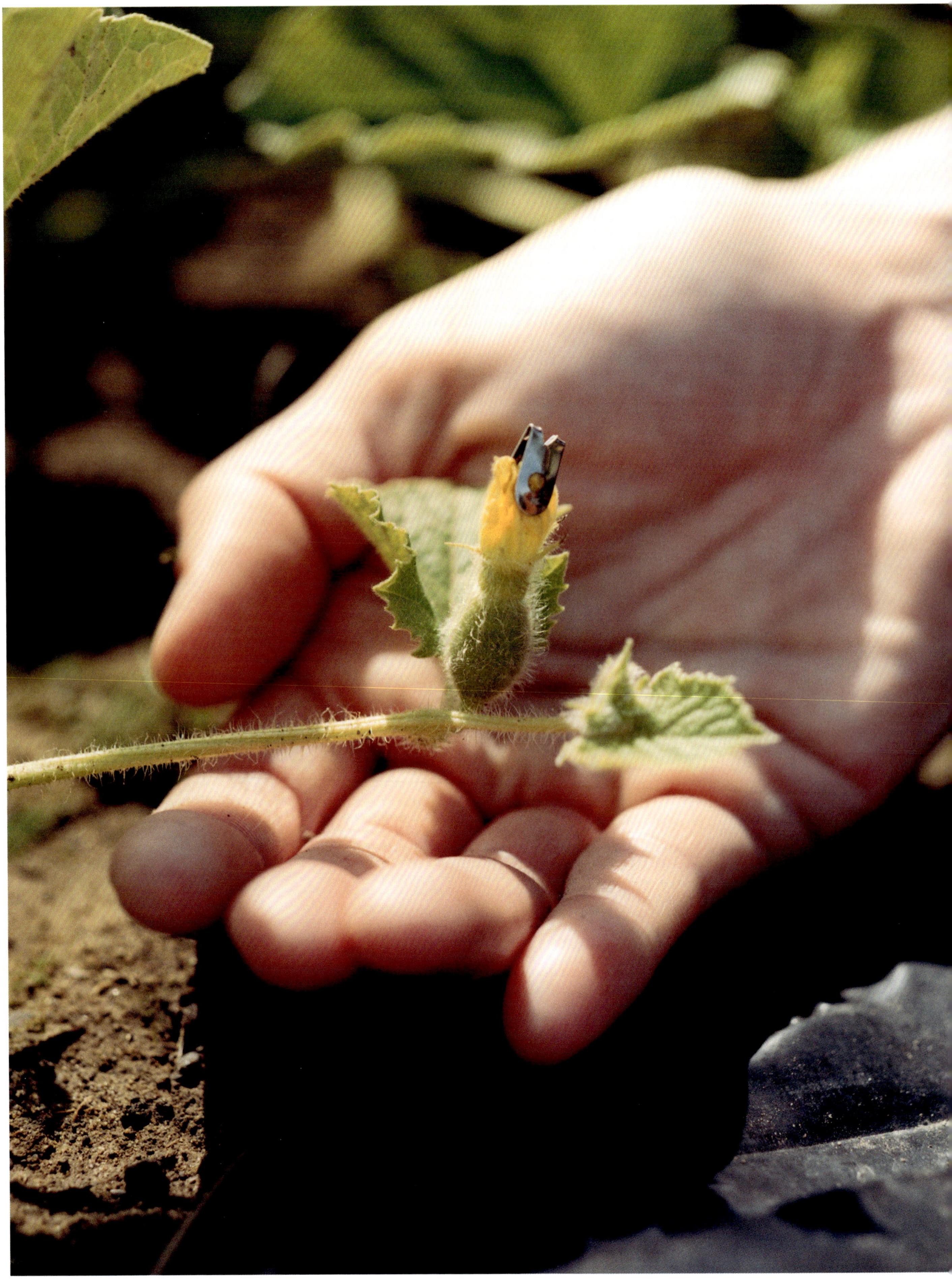

CHAPTER FIVE

Hand Pollination

Although melon (*Cucumis melo*) and watermelon (*Citrullus lanatus*) belong to different genera and species, and cannot cross-pollinate, cultivars within each species will cross. So, when you grow more than one cultivar of a species at the same time and want to maintain maximum seed purity, hand pollination is the way to go. You'll control pollination, excluding bees with foreign pollen, by transferring pollen by hand from the male flower to the female flower.

Melon and watermelon flowers are fragile and tiny and require a gentle touch. You'll need patience and high frustration tolerance, because you'll fail more often than you succeed. Bees themselves are successful with melons just 20 percent of the time. Only about 10 percent of hand pollinations of melons and 50 percent of watermelon pollinations "take" and set fruit.

Keen powers of observation are essential. You'll need to know all about the growth habits of flowers: when males are ready to release their pollen—indeed, what fresh pollen looks like (it's bright yellow, but you'll need good eyesight to see the grains)—and when females are receptive. Watermelons are mainly monoecious, with separate staminate (male) and pistillate (female) flowers. Most melon varieties are andromonoecious and have staminate and hermaphroditic (perfect) flowers. Hermaphrodites have male parts that need to be emasculated by removing the anthers before hand pollination, so they can function as females.

Female flowers are vastly outnumbered by males. You'll recognize the female by the immature ovary (which looks like a miniature fruit) at the base of the flower; males have a straight, simple stem. After plants start flowering in early summer (males appear first), producing new flowers that open early in the morning, there's a brief opportunity for the hand pollinator. You'll be more successful if you conclude by mid-morning, noon at the latest, especially in hot, dry weather. Don't pollinate on a rainy day, when pollen is unlikely to shed.

Every good pollination kit should contain forceps with fine, pointed tips, rubbing alcohol, ¾-inch masking tape, clips or paper fasteners, surveyor's or pollination flags in two colors, surveyor's tape, clear empty gelatin capsules, and notebook and pen for recording data and observations. You're now ready to proceed.

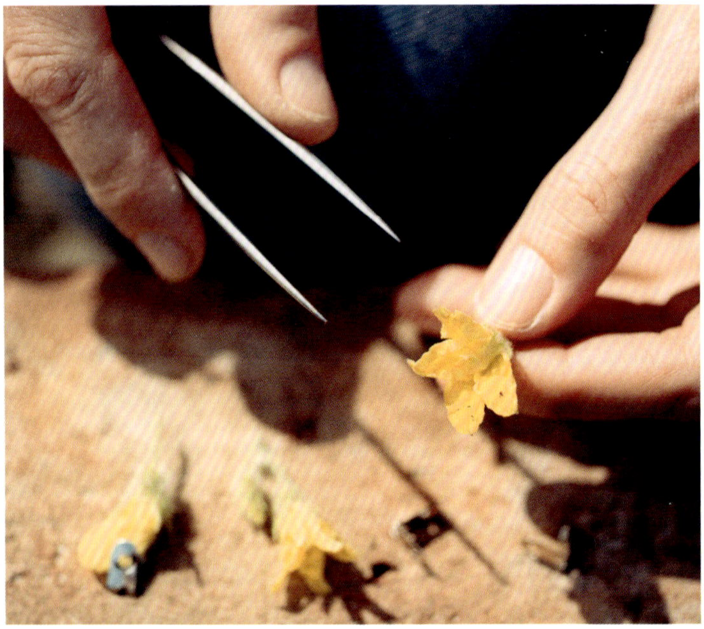
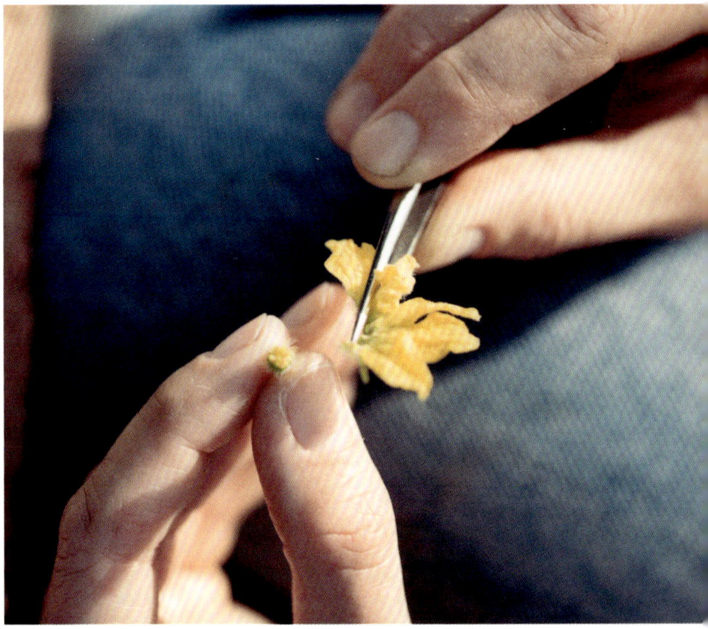
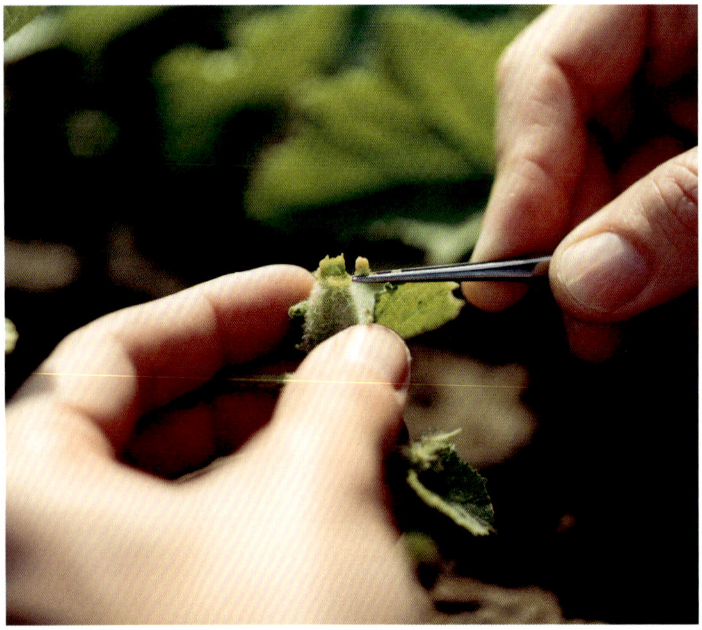
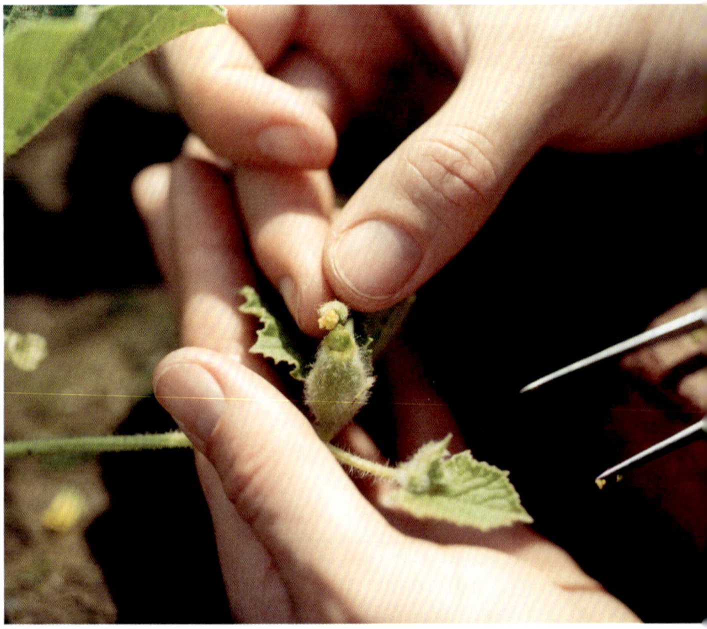
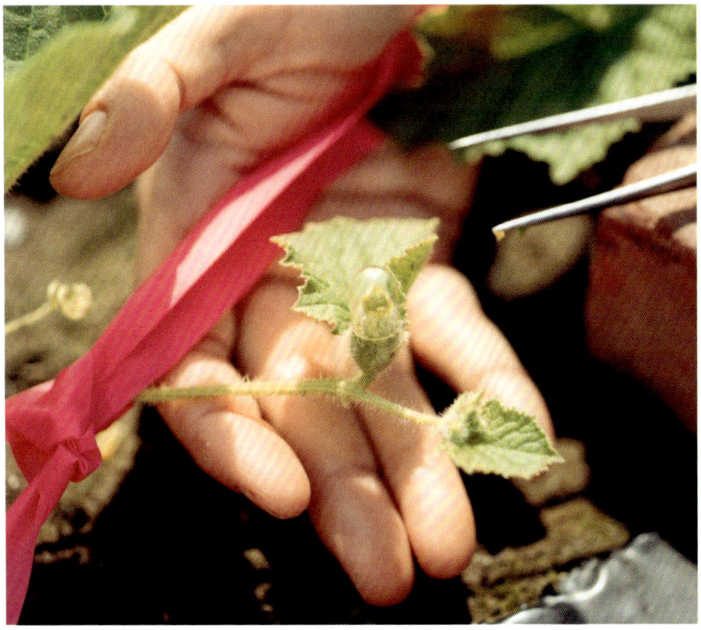
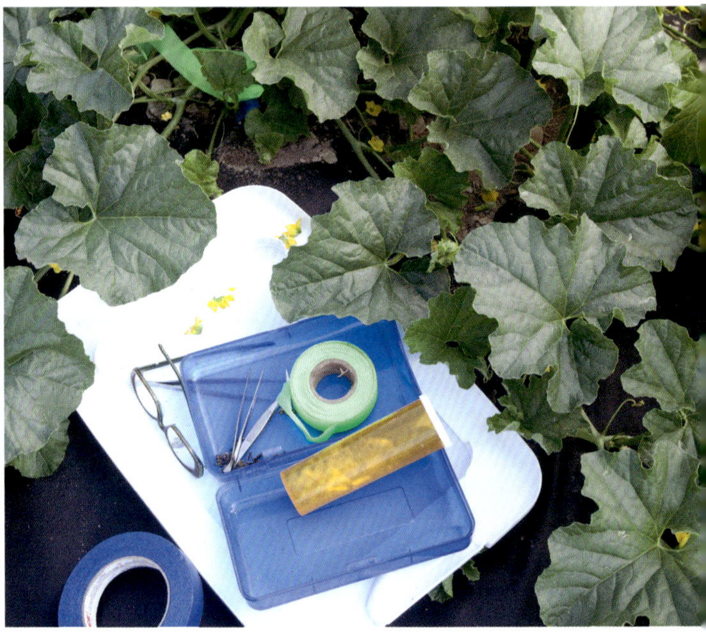

The following protocol for hand pollination of melon and watermelon is based in part on the work of the late Henry M. Munger, plant breeder extraordinaire, of Cornell University, who generously shared his trade secrets with me.

EARLY EVENING DAY ONE

1. Any fruits that have already set, and any open female or hermaphrodite flowers, must be removed, thereby "stripping" the plants and eliminating the competition.
2. Find both male and female flowers of the same variety that are likely candidates to open for the first (and only) time the next morning. Size and color changes (from green to yellowish) provide a clue; look for buds that appear fresh and not faded. Try to select two males for every female.
3. Tape the buds shut with a small piece of masking tape, leaving the ends separated to make it easier to remove in the morning. Clips can be used instead to keep the blossoms closed. The trick is to get all of the petals inside without inflicting damage to the pollen-bearing anthers or the stigma (that part on the female where the pollen lands).
4. To make blossoms easier to find the next morning, place brightly colored pollination flags beside them. Use different colors for male and female.

EARLY MORNING DAY TWO

1. Pluck the male flowers, leaving stems intact if possible to serve as handles, and make a beeline to the females.
2. Remove the tape or clip from two male blossoms. Using forceps, carefully slit them, one by one, down to the base and remove the petals and the entire calyx (ring of sepals). It's something like peeling an orange. This will expose the sides of the anthers where the pollen is borne. Set the males aside and cover with the petals for protection from bees.
3. Remove the tape or clip from the first female; when receptive, the petals will slowly expand. Use forceps to remove the petals. If you see any yellow anthers covering the green stigma, remove them cleanly with the forceps and discard by brushing against a leaf.
4. Watch out for bees and work quickly and carefully. Transfer the loose pollen from the anthers of the two males by gently applying them directly to the receptive stigma. Be sure to completely coat the stigma in pollen, but don't damage it with an overzealous application.
5. Protect the fertilized female with a fresh piece of tape or a gelatin cap, being careful not to constrict the stigma. Tie surveyor's tape near the adjoining leaf. I like to keep two pollination flags in place for easy identification.
6. If more blossoms are available, repeat the process. Don't do more than three or four pollinations until you see if there are results.
7. If you intend to pollinate more than one variety that morning, make sure to sterilize your hands and equipment with rubbing alcohol before proceeding with each.
8. Be encouraged if after a few days ovaries appear slightly enlarged and the stem attachments are still green. Fertilized fruits can abort spontaneously, however; don't count your successes until later.

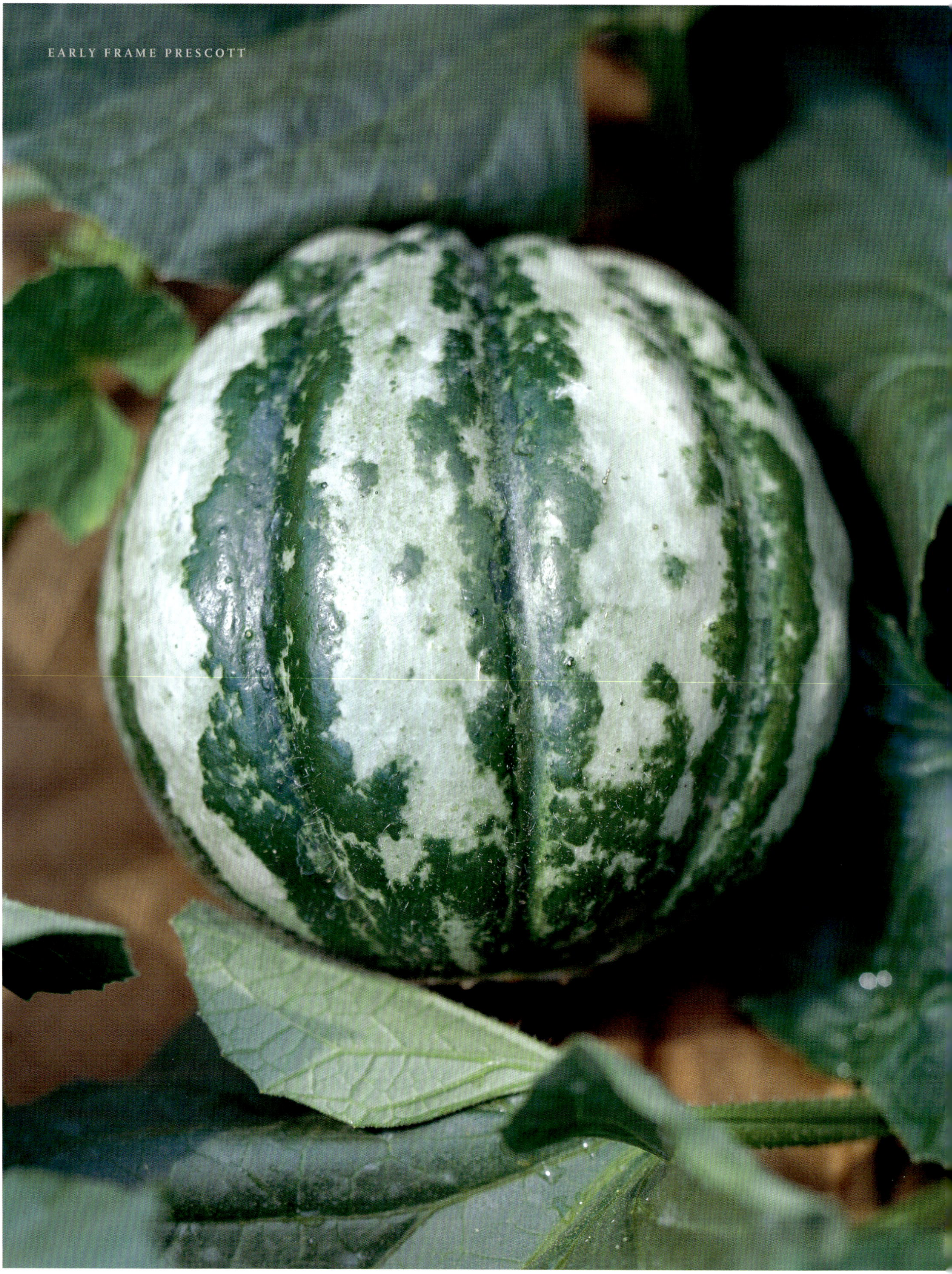
EARLY FRAME PRESCOTT

CHAPTER SIX

Melon Portraits

We are heir to a vast fortune: the gene pool of diverse melons developed over the centuries by our farming and gardening forebears. Melons are not one thing to all people. And to the uninitiated, it's almost incomprehensible that a snake melon and a muskmelon belong to the same species. Rarities from foreign lands can seem quite unmelonly to Americans. Some varieties have cornlike cobs or a cucumber-like crunch; others impersonate apples, sized just right for the lunchbox. Snap melons are cooked in stews or baked in breads. Oriental pickling melons are brined in salt. We've hardly begun to understand their potential in cookery.

Opposite: An immature Early Frame Prescott, or Prescott à Châssis, on the vine. Can be grown in cold frames or in the open field.

Just as melon fruits take on many shapes and forms, so too do the classification schemes designed by botanists over the last two hundred years to describe and account for melons' variability. As a horticulturist (and not a plant scientist), addressing a general audience, I find Burger et al.'s (2010) summary to be the most useful, as it relies on Pitrat's classification (Pitrat, Hanelt, and Hammer 2000) but speaks in the language of cultivated plant taxonomy (cultivar groups), rather than botanical varieties. Of those sixteen cultivar groups—with exotic-sounding, evocative names—twelve are featured and described in this book. Present and accounted for are Cantalupensis, Reticulatus, Inodorus, Ameri, Chito, Flexuosus, Adzhur (also known as Chate), Dudaim, Khandalak, Momordica, Makuwa, and Conomon. It's worth noting that the first nine groups belong to subspecies *melo* and the last three belong to subspecies *agrestis*. (The others are virtually unknown to home gardeners in the United States: Chinensis, Acidulus, Adana, and Tibish.)

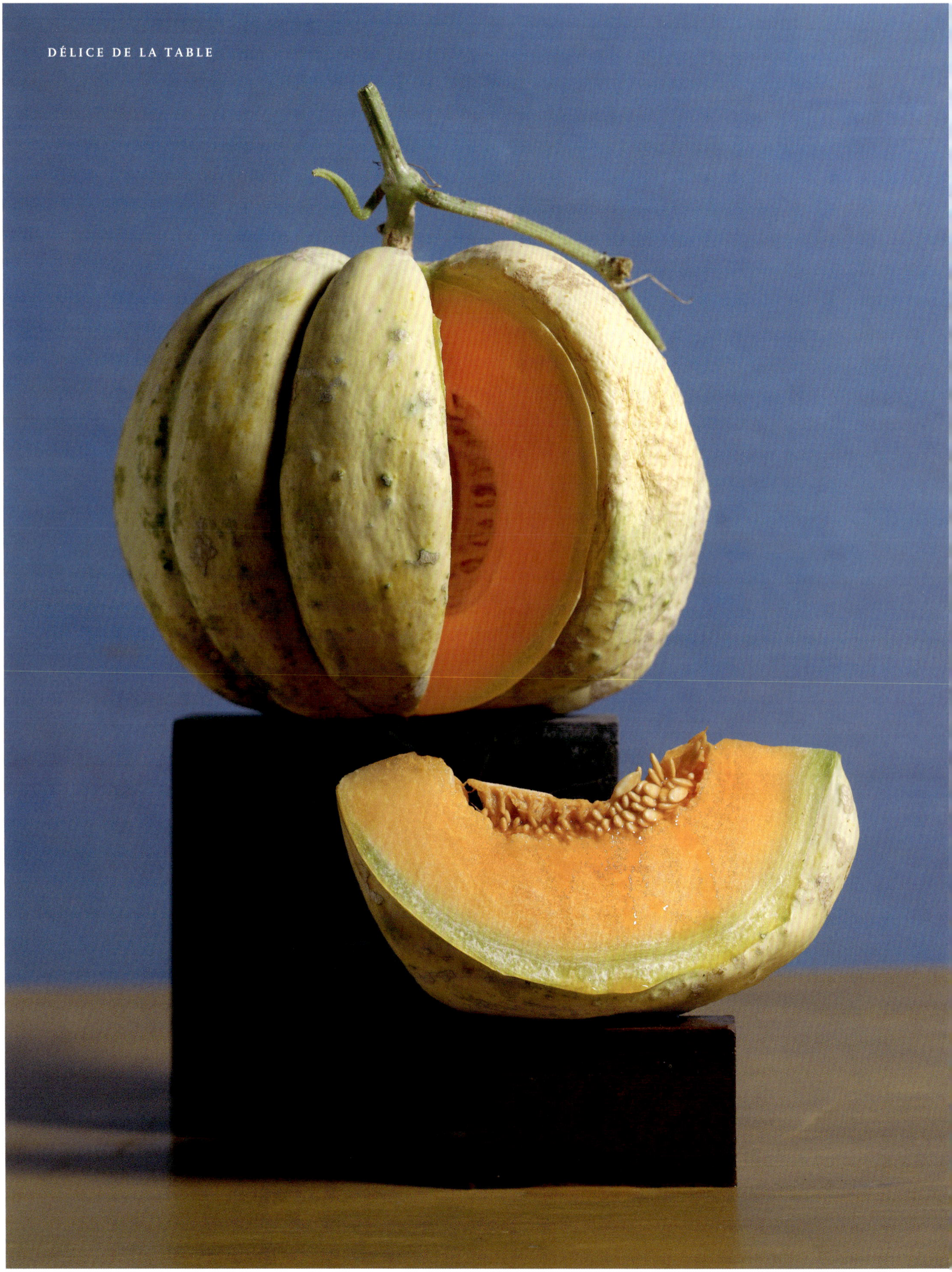

Cantalupensis Group

CANTALOUPE MELON Cantaloupe as Americans know it is a misnomer for muskmelon (Reticulatus group). The true cantaloupe, seldom seen in the United States, is much more common in Europe. In the fifteenth century, cantaloupes were introduced to Italy from Armenia (now eastern Turkey), apparently by missionaries, and named for the papal country house at Cantalupo (Singing Wolf). Images of them appear in many early Renaissance herbals. Melons resembling Charentais and Noir des Carmes appear in festoons in the loggia of Cupid and Psyche in the Villa Farnesina, in Rome, painted by Giovanni Martini da Udine, between 1515 and 1518. Jacquin, author of the indispensable and truly complete *Monographie Complète du Melon* (1832), was quite certain that the *Melons Cantaloups* or *Italiens*, after becoming acclimated, were brought from Florence to France by Charles VIII, in 1495.

Also known as rock melons, cantaloupes come in all sizes and shapes and often have prominent alternating longitudinal lobes and furrows. Sometimes the boundary between cantaloupe and muskmelon blurs, with the appearance of sparse or fine netting on the lobes, between the vein tracts.

Cantaloupe skin can be smooth and thin like the Charentais's, or thick, warty, and bumpy like Prescott Fond Blanc's. Many cantaloupes have green spots or galls that dim with time. Most are orange-fleshed, succulent, and aromatic. Skin color change signals maturity. Most cultivars do not slip from the vine when ripe; some slip when overripe. This horticultural group includes some of the melons most dear to my heart: Petit Gris de Rennes, Ha'Ogen, and Zatta. Cantaloupes are sweet dessert melons par excellence.

Opposite: Délice de la Table, also known as Delight of the Table, has the telltale signs of a true French cantaloupe: thick and warty rind and prominent longitudinal lobes and furrows.

Prescott Fond Blanc

Once a favorite of French market gardeners, the Prescott Fond Blanc is now a favorite of mine. This ethereal, deeply furrowed, puffy-looking melon reminds me of dough rising. I love to run my fingers over its deeply furrowed, warted, rock-hard shell and to smell true cantaloupe—a blend of a wonderful floral bouquet, wild muscat grapes, and a squeeze of verjuice. A rare commodity these days, the Prescott deserves to be brought under wider cultivation. Although it is less meaty than the more efficient modern melons, tasting one is an experience not to be missed.

The white-skinned Prescott is one of a number of similar varieties that range in color from whitish to blackest green. Like all the Prescotts, it has a recessed stem and a large round scar at the blossom end; it grows on vigorous vines with dark toothlike leaves. Prescott has the signs of a true rock melon: It is thick-skinned, warted, and bumpy, displays no netting, and does not slip from the vine until overripe.

Prescott's savory flesh is dense and sweet, and the fruit itself is strikingly attractive. Artists have been inspired by this melon—antique French ice cream molds in the shape of Prescott are collectors' items. Prescott à la mode: food for thought. I've been inspired to cast these rare forms in bronze.

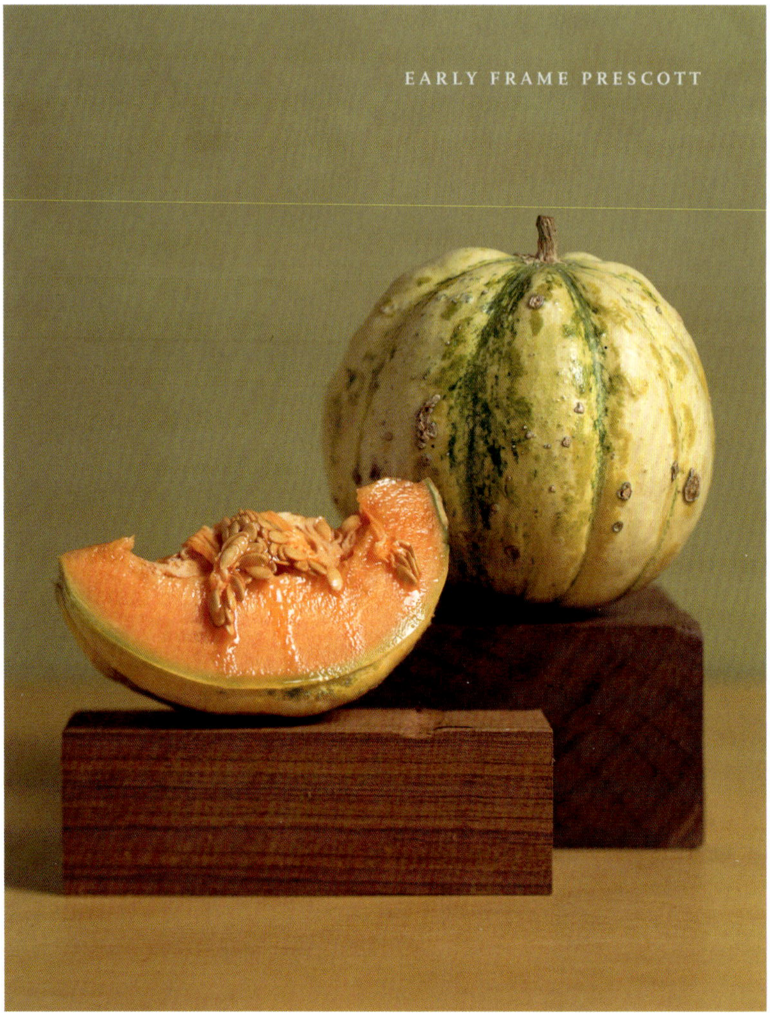

An Early Frame Prescott fit for the table. The fruit skin colors and patterns change at maturity from pale green with dark green galls and striping to light yellow and grayish green, respectively. An immature Early Frame Prescott appears on page 36.

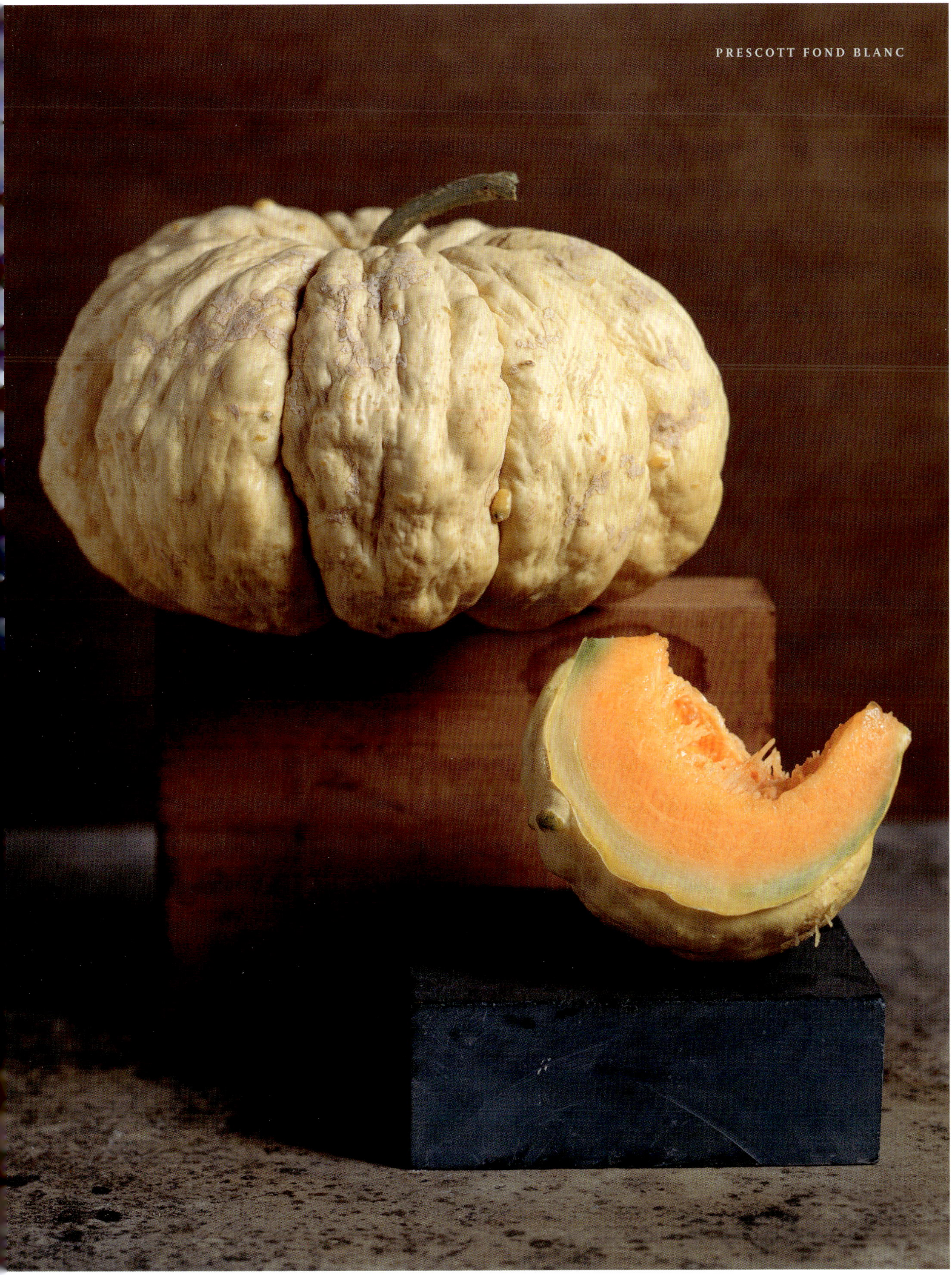
PRESCOTT FOND BLANC

Ha'ogen

This cherished melon is named for Kibbutz Ha'Ogen in Israel. Natan Fuchs (1915–1984), an immigrant to Israel from a Hungarian part of Slovakia, was one of the founders of Ha'Ogen, an agrarian community established in 1939 in the Hefer Plain of central Israel. His brother sent him several varieties of melon seed in an envelope from Hungary, probably in the 1940s. That melon has now been identified as the Muskotály by the Hungarian breeder Csaba Szamosi. Muskotály types were described and painted in Hungary as far back as 1664 (*Posoni kert* by J. Lippay). The Hungarian Muskotály itself resembles a melon from Uzbekistan by the name of Makhalliy. I can imagine a logical scenario in which the melon originates in Uzbekistan (Khorasan), makes its way to Asia Minor, then Hungary and beyond. That "beyond" may include America: Muskotály and Ha'Ogen bear a striking resemblance to Ananas d'Amérique à chair verte (see page 61).

With passion and determination, Fuchs took the Muskotály in the 1940s and "mass-selected" by growing lots of plants and selecting with a discriminating eye—for improved adaptation to local soils and climate, superior quality (highly aromatic, excellent flavor, soft melting flesh, small seed cavity), and greater uniformity.

By 1956, Ha'Ogen was commercialized and being shipped abroad. The melon provided an important source of income for its namesake kibbutz. It proved so well suited to full irrigation that it would usher in an era of intensive melon cultivation. And although it is no longer commonly grown in Israel, Ha'Ogen inspired breeders to make additional improvements. One of its derivatives, the powdery mildew–resistant Galia F1 hybrid, initiated a new and widely popular market type.

Petit Gris de Rennes

The Petit Gris de Rennes is so good it gives me the chills. As wonderful as Charentais is, the petit Rennais goes a baby step further, making it *la crème de la crème* of French melons. You will blink your eyes with disbelief when you sample its sweetness, which is more like brown sugar than white. It will melt on your tongue, and your mouth will water for more.

This melon is a *tableau vivant* of Pointillistic mustard and olive, and it inspires devotion in France. It may descend from the unnamed "*melons forts bons*" (very good melons) noted by François-Nicolas Baudot Dubuisson-Aubenay (1590–1652) in the garden of the Bishop of Rennes in 1636. Until the twentieth century, Petit Gris thrived in the mild climate and horse manure of the garrison town, Rennes, but is now cultured in hothouses and polyester tunnels. Despite the best efforts of a handful of farmers and chefs, and the Rescan family of Cesson-Sévigné, who grew the melon for more than seventy-five years, Petit Gris is menaced by market forces. For this group of preservationists, quality control is critical, so they've made a science of what once came naturally. Melon plants are grafted onto squash rootstock to confer fusarium wilt resistance, and then espaliered to keep them off the ground.

Grown this way, "the little gray melon"—so named for its appearance before

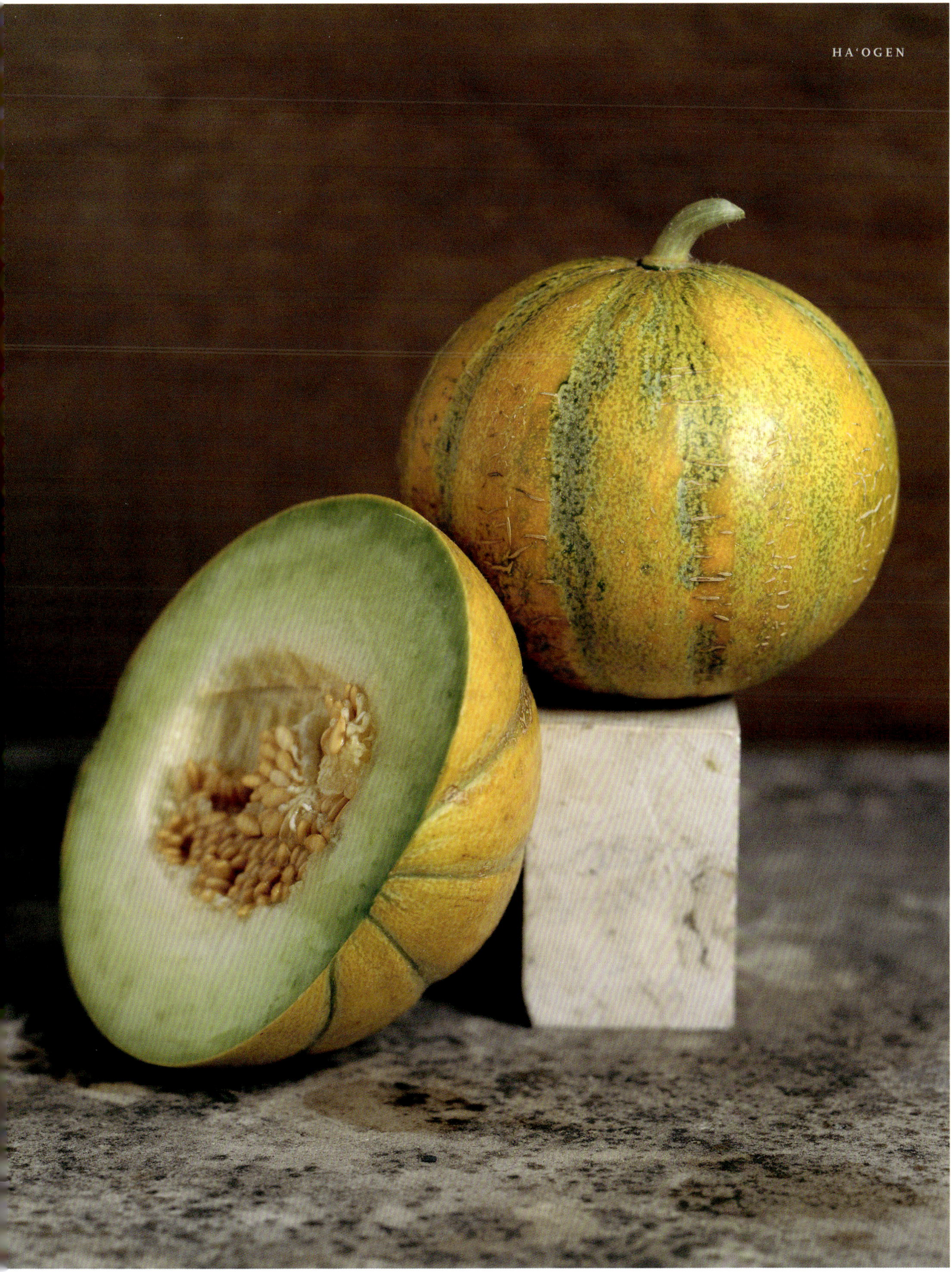
HA'OGEN

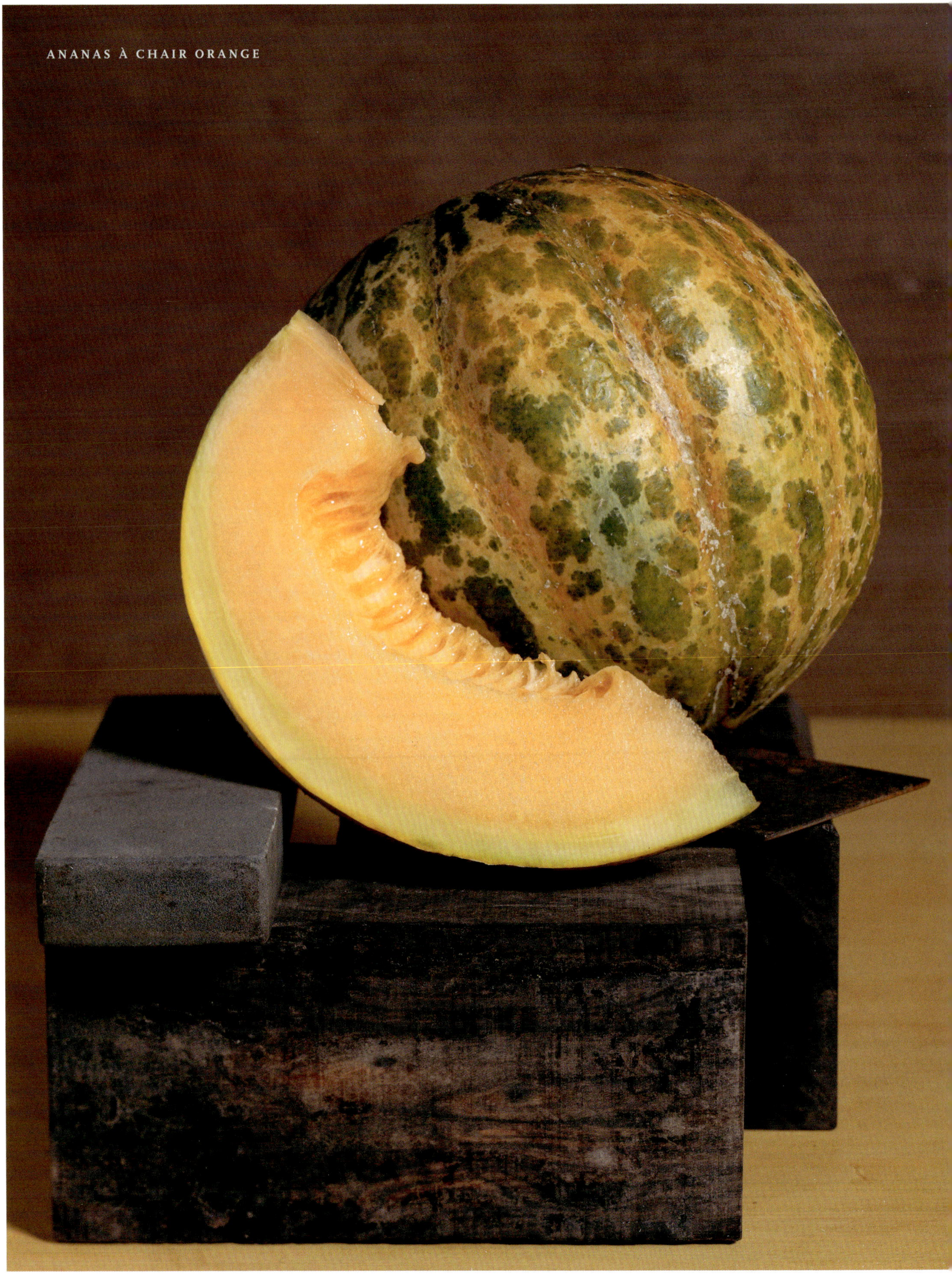

ANANAS À CHAIR ORANGE

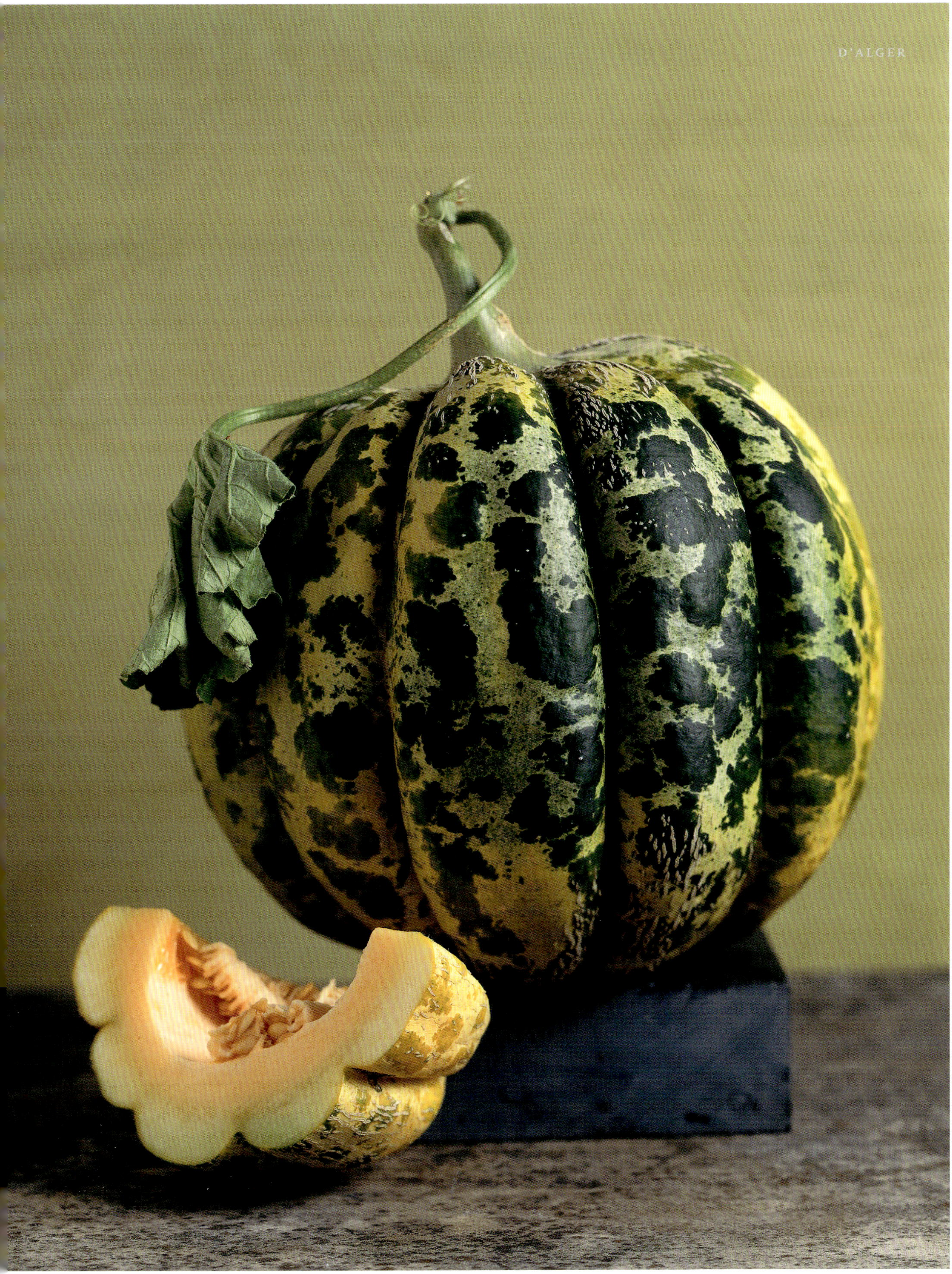

ripening—doesn't see the light of day or touch the earth, but it's hardly factory farming. Kid gloves guide the process every step of the way. Maybe it's not as easy to grow as other melons, but you can cultivate it in your own garden. Prop fruits up with a melon cradle—a newfangled plastic device that elevates the melon above the field surface—and protect them from pests with spun polyester cloth. It's worth the extra effort.

Zatta

I'm crazy in love with this melon, which clearly belongs in the Prescott family of cantaloupes. Plants throw off an occasional Early Frame Prescott type (see page 40). *Zatta* is Italian for cantaloupe (plural *zatte*). This variety may taste even better than Prescott Fond Blanc (see page 41)—but must we quibble? When overripe it can be dry and cottony, I admit.

I am fascinated by the unusual fruit color change as Zatta matures. Unlike Prescott Fond Blanc, whose skin goes from whitish green to light yellow when ripe, Zatta turns from dark green to orange yellow or yellowish orange. Jacquin (1832) was right: The more green-black the cantaloupe's skin in youth, the more intensely yellow orange it becomes at maturity.

Thomas Jefferson sowed and planted 18 hills of Zatte di Massa cantaloupe melons in 1774. Presumably he had a harvest, but melons didn't fare well for him at Monticello. Peter Hatch, former Director of Gardens and Grounds at Monticello, explains the underlying issues: "Whether because of the Virginia humidity, the heavy clay soil, or the skills of the gardener, fine and tasty melons are only occasionally harvested from the restored Monticello garden today." Nonetheless, Monticello does produce and sell seed of this historic variety.

Zatta di Massa is named for a town in central Italy where these melons grew and matured near the sea. The variety is also known as Zatta Arancina (orange-colored Zatta) and Melone Cotogno (quince-tree melon). It was described in 1828 as the least warted, and the most yellow orange of the various Zatte by Moretti and Chiolini.

People love to call this cantaloupe ugly. As early as 1972, Gleckler's Seedmen, of Metamora, Ohio, catalogued it as Mr. Ugly Muskmelon. In their words: "From Italy. One look at the fruit and all agree they are the most horrible looking, with their deep ribs, big warts and rough all over; look like the rough looking Japanese Squash. But the salmon-netted flesh is so sweet." Italians call it ugly but good.

Charentais

When Montaigne said, "I am not excessively fond of salads or of fruits, except melons," he might well have been talking about the Charentais. It's the melon the French adore above all others, and for good reason. The scent is divine, the flavor ambrosial. Who can resist this melon reverently nestled in fancy tissue paper on a bed of straw in the marketplace? Where Frenchmen disagree is on how best to serve it: Slightly chilled or not? Alone, savory with sliced ham, or saturated with port, muscat, or Ricard? One can almost hear the cacophony of cutlery and opinion.

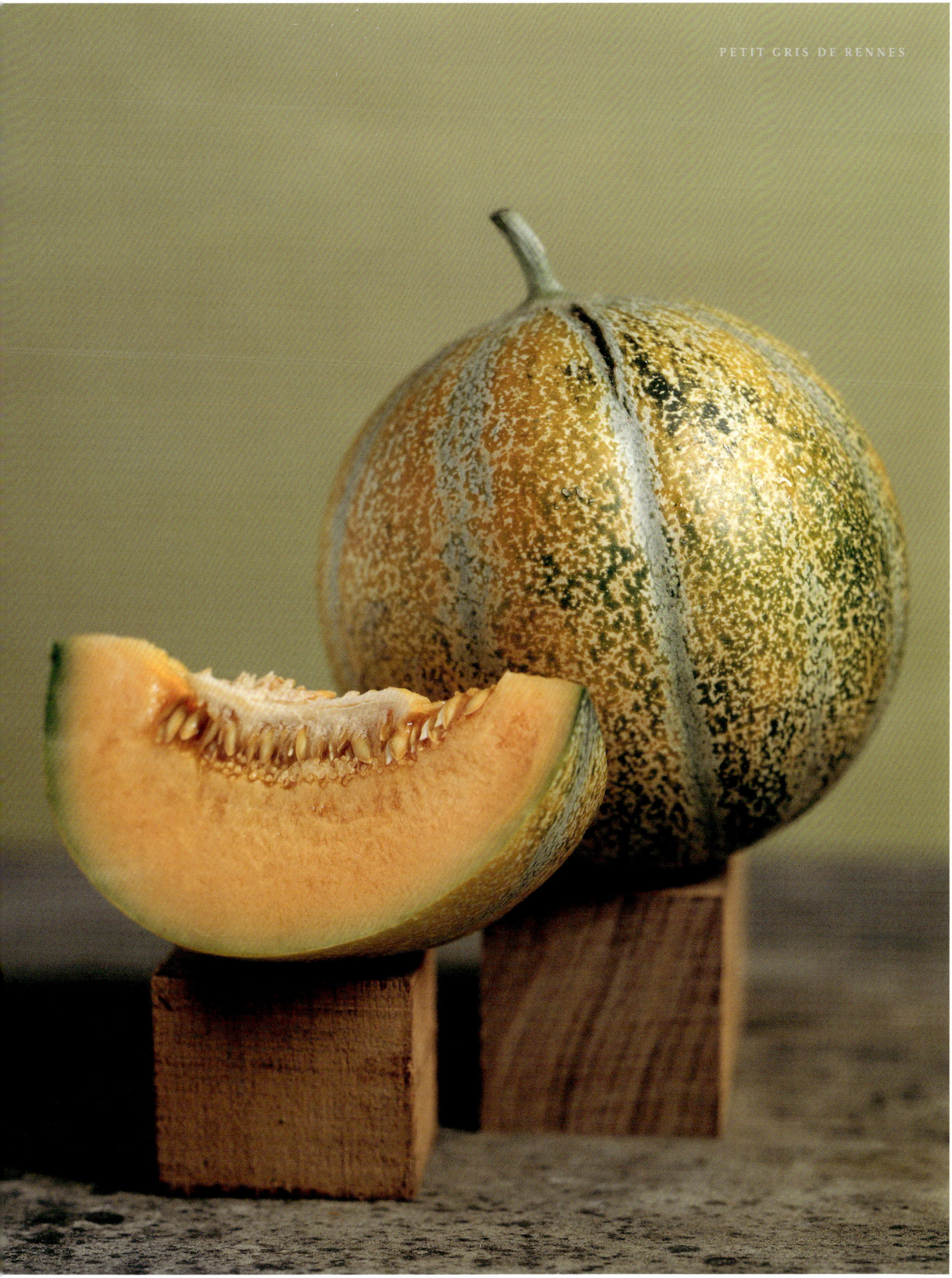
PETIT GRIS DE RENNES

Sweet cantaloupes arrived in Europe from Armenian Turkey in the fifteenth century and diffused westward. Four Charentais look-alikes were painted between 1515 and 1518 in festoons at the Villa Farnesina in Rome. Charentais was known and grown in the Poitou-Charentes region of western France around 1920. And by 1946 it had become the leading market type in France. Charentais is a refined cantaloupe, smooth-skinned and free of blemishes and warts. Round or sometimes flattened at the poles, with tan or grey lobes, its dark green furrows guide the knife where to cut. The melon is sometimes called a Cavaillon, which creates a bit of confusion. Cavaillon, a town (far from Charentes) in the South of France noted for the quality of its melons, has lent its name to many melons simply because they were grown there. It's like the *Good Housekeeping* Seal of Approval. But you don't have to go to France to enjoy the Charentais (although it is worth the voyage). It can be grown well in the United States.

Noir des Carmes

Named for the Carmelite monks who once tended it in France, Noir des Carmes has been known and grown in Europe for hundreds of years. But, like many other true cantaloupes, it has never been popular in the United States. Nineteenth-century gardeners routinely "forced" it under glass on hotbeds in January or February to meet the demand for out-of-season fruit. Cultivating melons in this manner is an effort that only the melon-obsessed can appreciate. Count me in.

There's no need for you to test the limits, however, especially if your climate is warm and sunny. Simply grow Noir des Carmes in season in the open air as you would any muskmelon. The reward will be a crop of more melons than you can eat. Dark as a monk's frock when unripe, Noir turns tantalizingly yellow as it nears full maturity. Harvest it a bit green, before cracks widen and invite dryness and decay.

Le petit cantaloupe noir des carmes was precious to the eminent French botanist Augustin Sageret (1763–1851) because of its remarkable flavor and earliness; he succeeded with it like no other cantaloupe. Sageret grew hundreds of varieties and experimented with hybridization, outlining his observations in two fascinating memoirs on the Cucurbitaceae (1826 and 1827). He had hoped to write a more comprehensive work on melon but failing eyesight apparently prevented him from doing so. Still, across the years, I heed his plea to preserve the integrity of the best of the past for the future. (Sageret is viewed as a predecessor to Gregor Mendel.)

Jacquin (1832) pronounced Noir des Carmes very good if eaten within a day of harvest. The house of Vilmorin-Andrieux (from 1856 on) touted the melon's thick orange flesh as sugary, perfumed, and excellent. I echo their sentiments. Slice Carmes along the furrows and serve it in wedges (with yellowish skin left on) with cheese and walnuts. Gilding this lily is totally sinful: try it for dessert with sugar sprinkled on top.

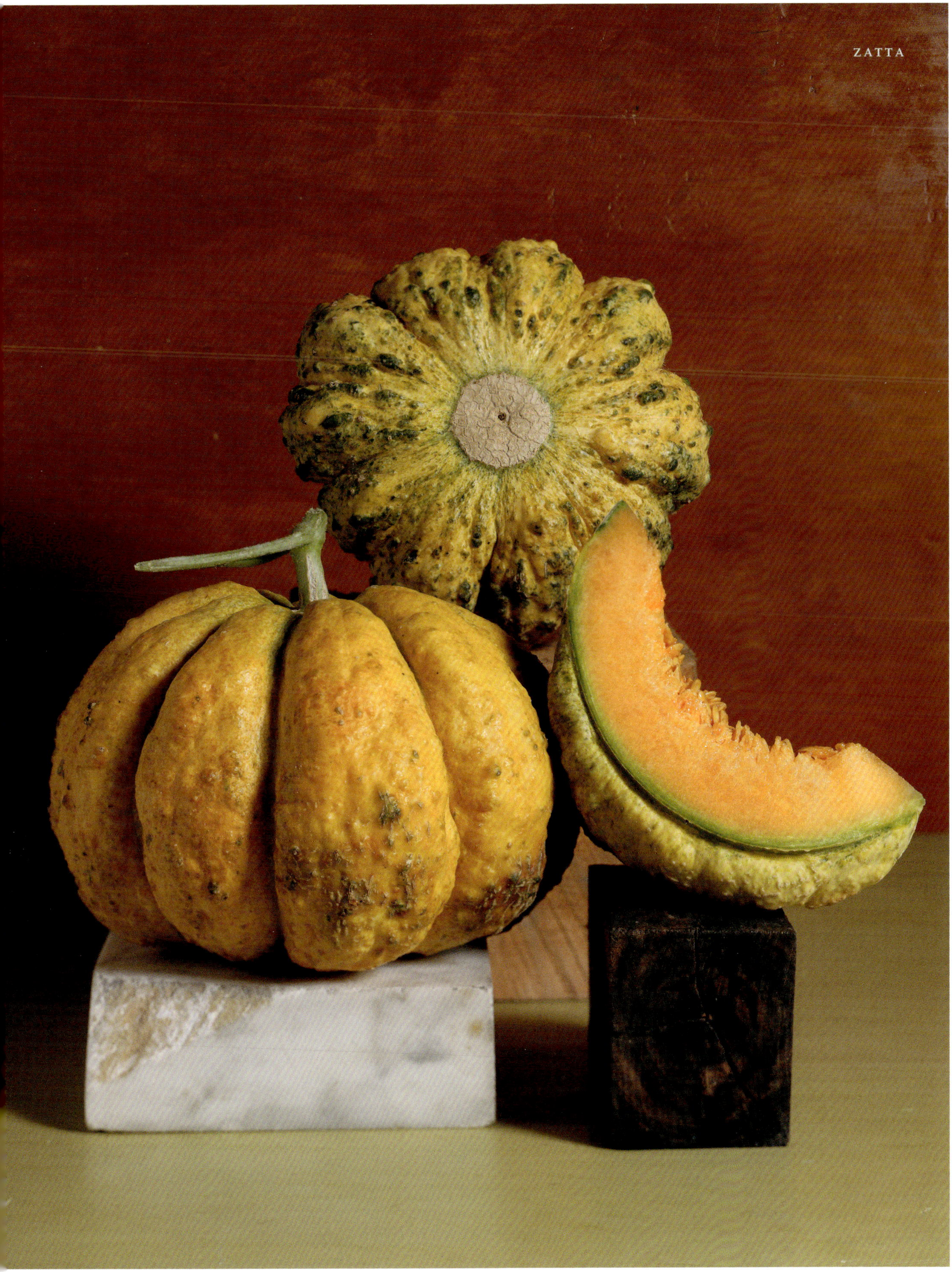

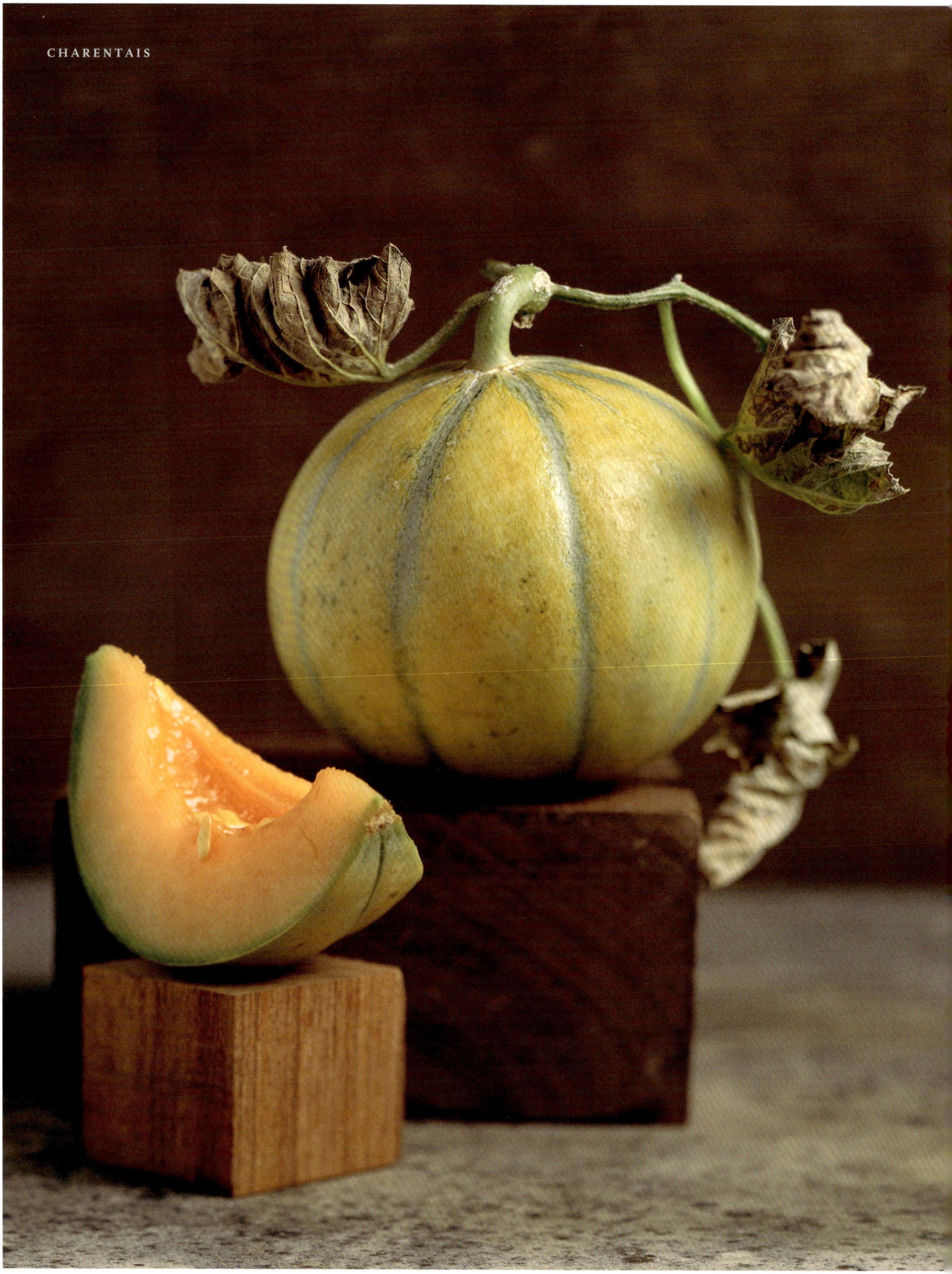

CHARENTAIS

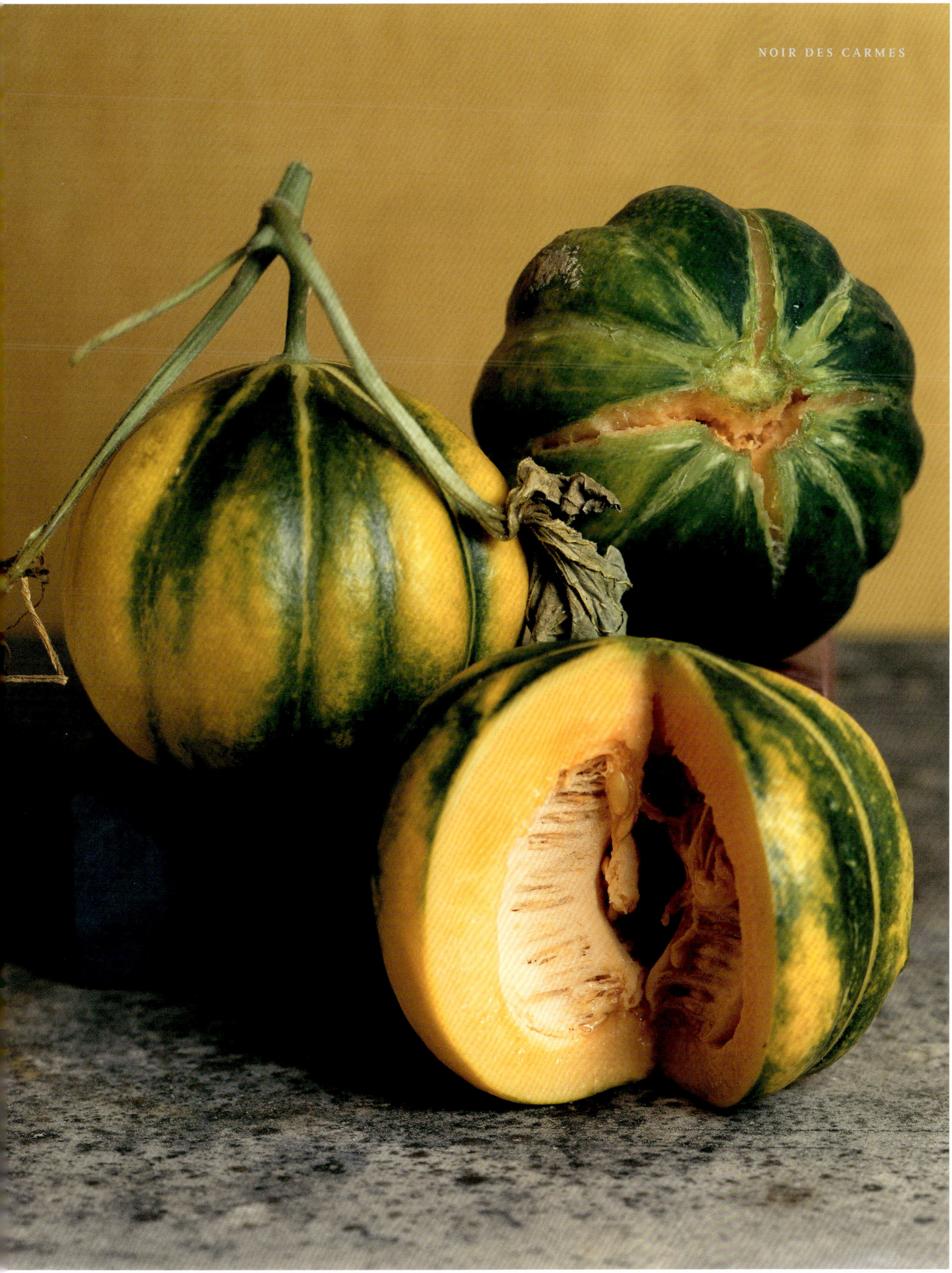

OLD-TIME TENNESSEE AND BLENHEIM ORANGE

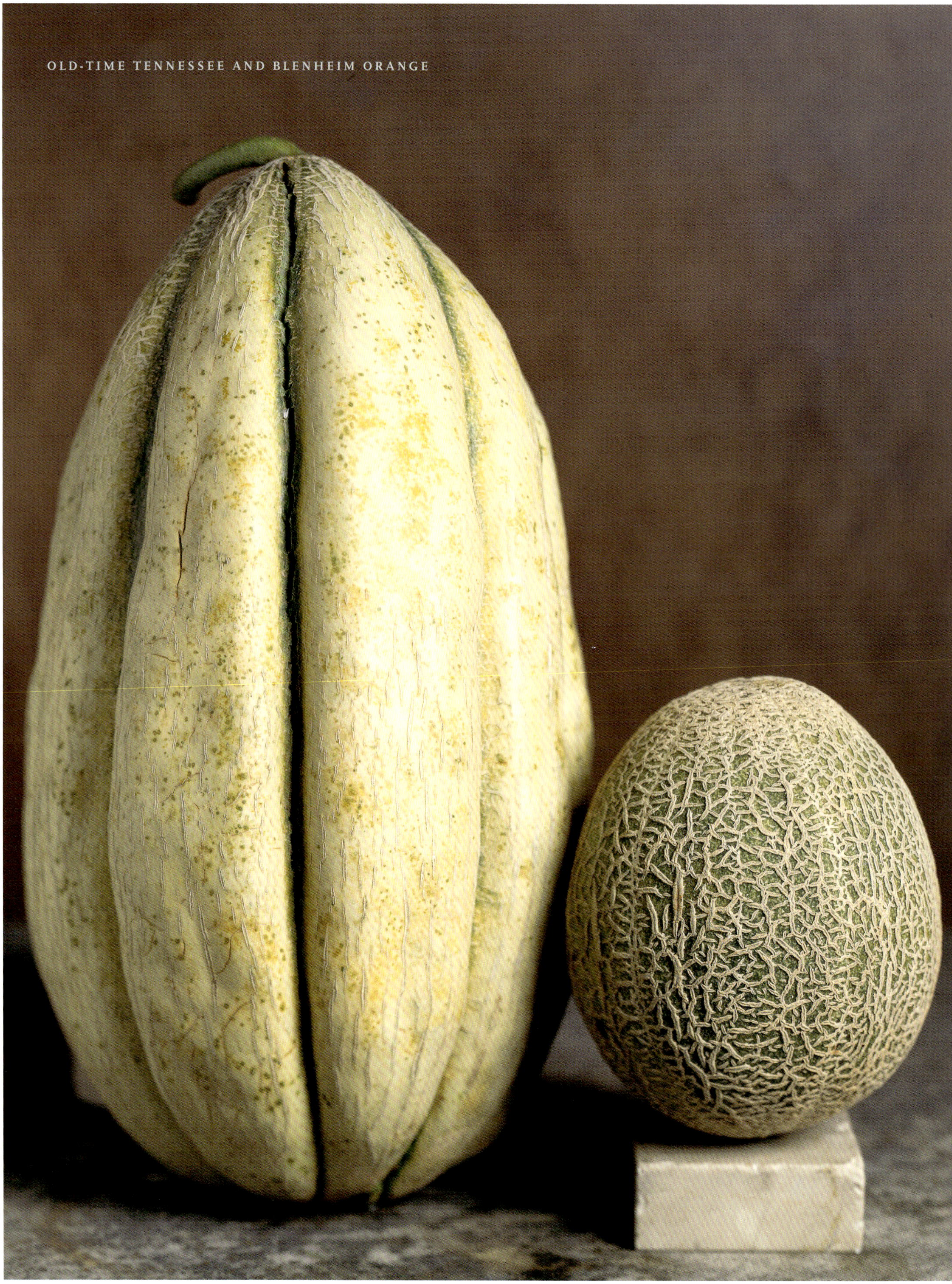

Reticulatus Group

MUSKMELON In the United States muskmelons are colloquially referred to as cantaloupes or American cantaloupes but they differ from the true types (see Cantalupensis group, page 39). The hallmarks of this group of melons are the reticulated or netlike tissue that covers the skin and the musky flavor of their ripe orange flesh. The French refer to reticulated melons as *melons brodés* or *melon écrits*, melons with embroidery or writing. We are less familiar with the exquisite green-fleshed muskmelons that once flourished in nineteenth- and early twentieth-century home gardens. Andrew Jackson Downing (1845) considered the muskmelon to be "the richest and most luscious of all herbaceous fruits." And, among muskmelons, the green-fleshed ones were at the top of the heap. American farmers or "practical breeders" created a diverse array of varieties with distinct personalities and delicious aromatic flesh, well adapted to home and market gardening or short-haul shipping. According to Whitaker and Jagger (1937), "natural crossing in mixed plantings furnished these growers with an abundance of material for selection." Professional breeders took the ball and ran with it by making mass selections and deliberate crosses to enhance melons' shipping and eating quality and disease resistance. By 1937 several hundred muskmelon cultivars were grown in the United States, although many were similar or synonymous. New forms gradually replaced the old. Today Americans buy and consume more muskmelons than any other type. The majority featured in this book are best suited for home growing or local markets.

You'll never think about a reticulated melon in quite the same way again after reading the stories of Ananas d'Amérique à chair verte, Jenny Lind, Fordhook Gem, or Hero of Lockinge. These varieties are living vestiges of grandmother's garden. Handed-down heirloom muskmelons are usually lobed and furrowed, and generally slip or fall from the vine when mature. Most are orange fleshed and musky flavored; the green ones are sweeter, yet more delicately flavored. Crate melons with uniform size and solid net such as Pollock Rocky Ford and Eden Gem, originally designed for shipping, fare well in the home garden, too. Persian melons are finely and abundantly netted, globe shaped, not lobed or furrowed. All are prized for their sweetness.

Opposite: Netted or reticulated melons come in a range of shapes and sizes. These orange-fleshed muskmelons, Old-Time Tennessee (left) and Blenheim Orange (right), were harvested a bit green in a year with unfavorable weather.

Schoon's Hard Shell

Schoon's Hard Shell is an excellent melon for fresh eating. I eagerly watch it mature on the vine as it changes from celadon green to warm ivory and acquires its telltale netting. Introduced as Schoon's Hard-Shell Bender in 1939 by F. H. Woodruff and Sons of Milford, Connecticut, it has a fine flavor and aroma. The hard shell, interwoven willy-nilly with coarsely laced ropelike netting, bolsters shelf life. Bender was the number one type of melon grown in New York State in the 1930s for local markets and short-haul shipping. There were many names for and strains of Bender—including Schoon's Hard Shell.

At the foot of the Helderberg Mountains near Albany, New York, not far from where I live now, in a place where Schoon is a common last name, Charles Bender began growing melons in 1884. Using three muskmelon varieties—Surprise (introduced 1876), Irondequoit (1889), and Tip Top (1892) (see page 62), all descended from Sill's Hybrid (1870)—he bred a melon with the best qualities of each. He named the resulting variety for himself in 1900. It was a melon earlier in maturity than Surprise, more heavily netted, and less susceptible to cracking than the other two. Its parents were noted for flavor but, for me, Schoon's surpasses them all.

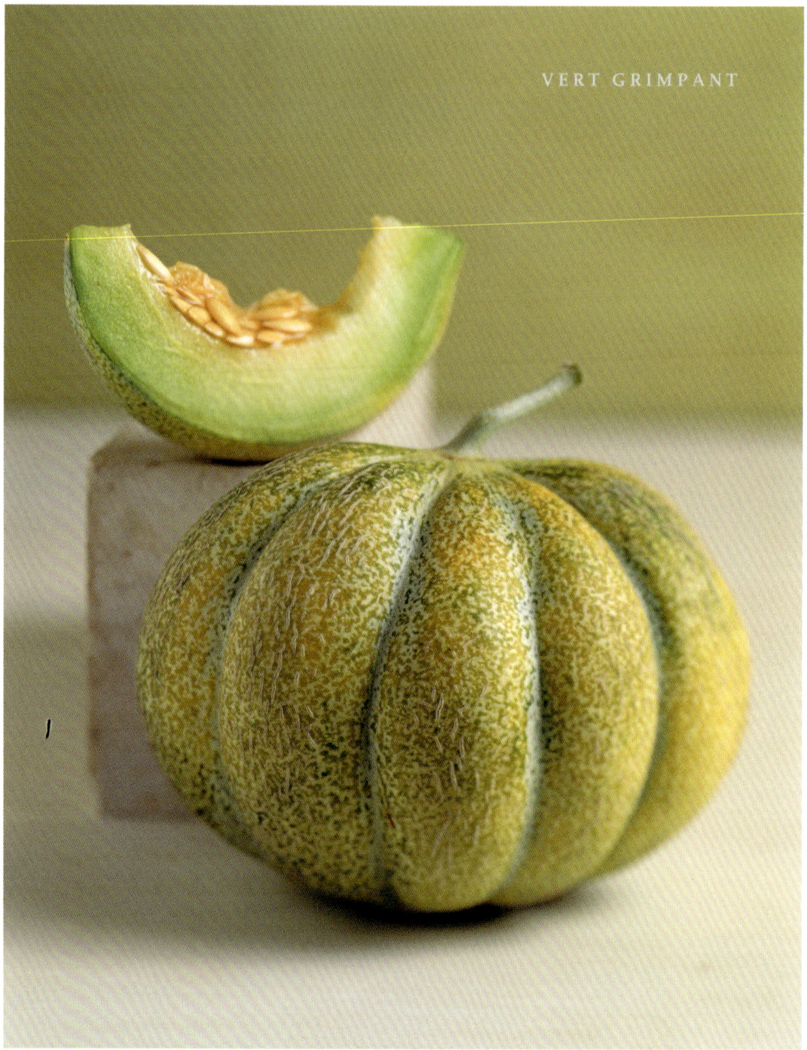

Vert Grimpant, or Green Climbing, like that other favorite pineapple melon, Ananas D'Amérique à chair verte (see page 58), can be grown on a trellis.

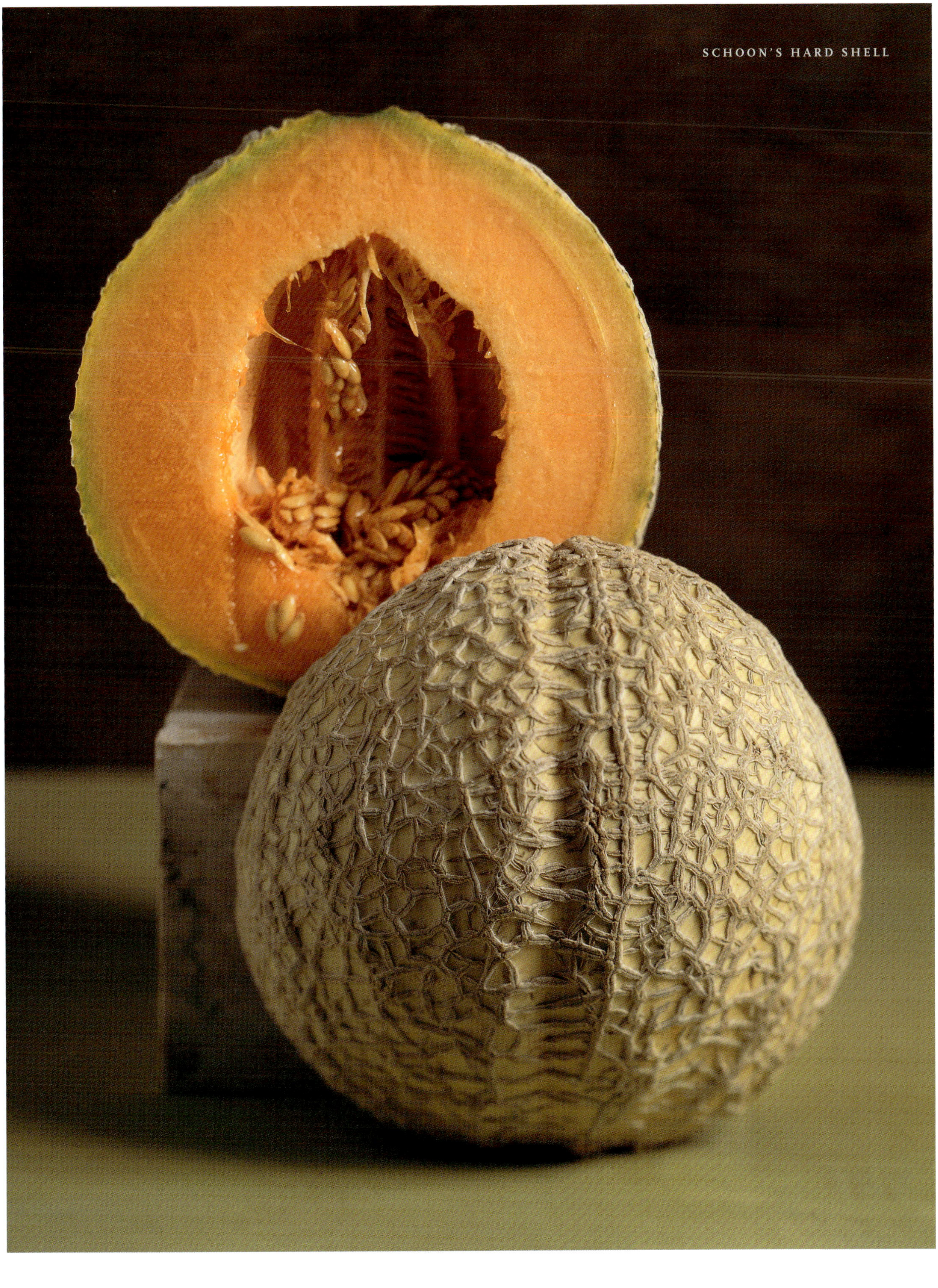

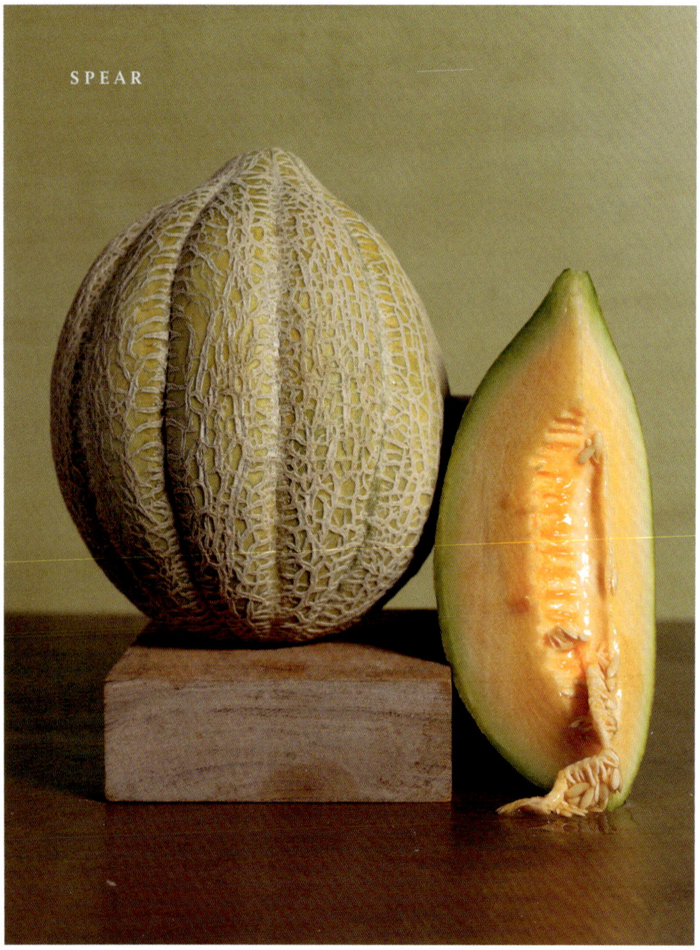
SPEAR

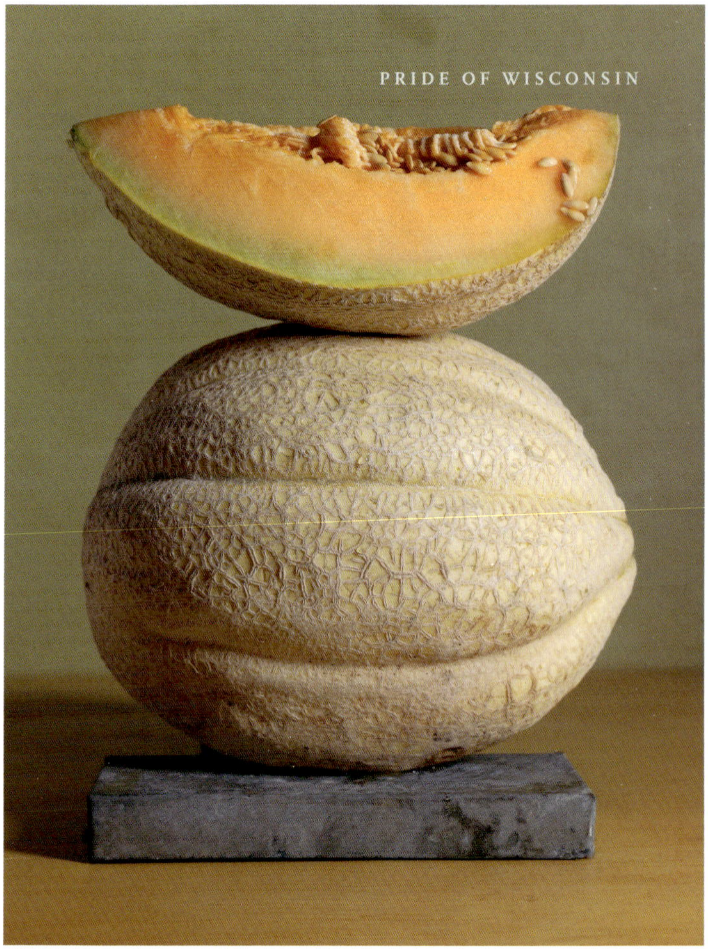
PRIDE OF WISCONSIN

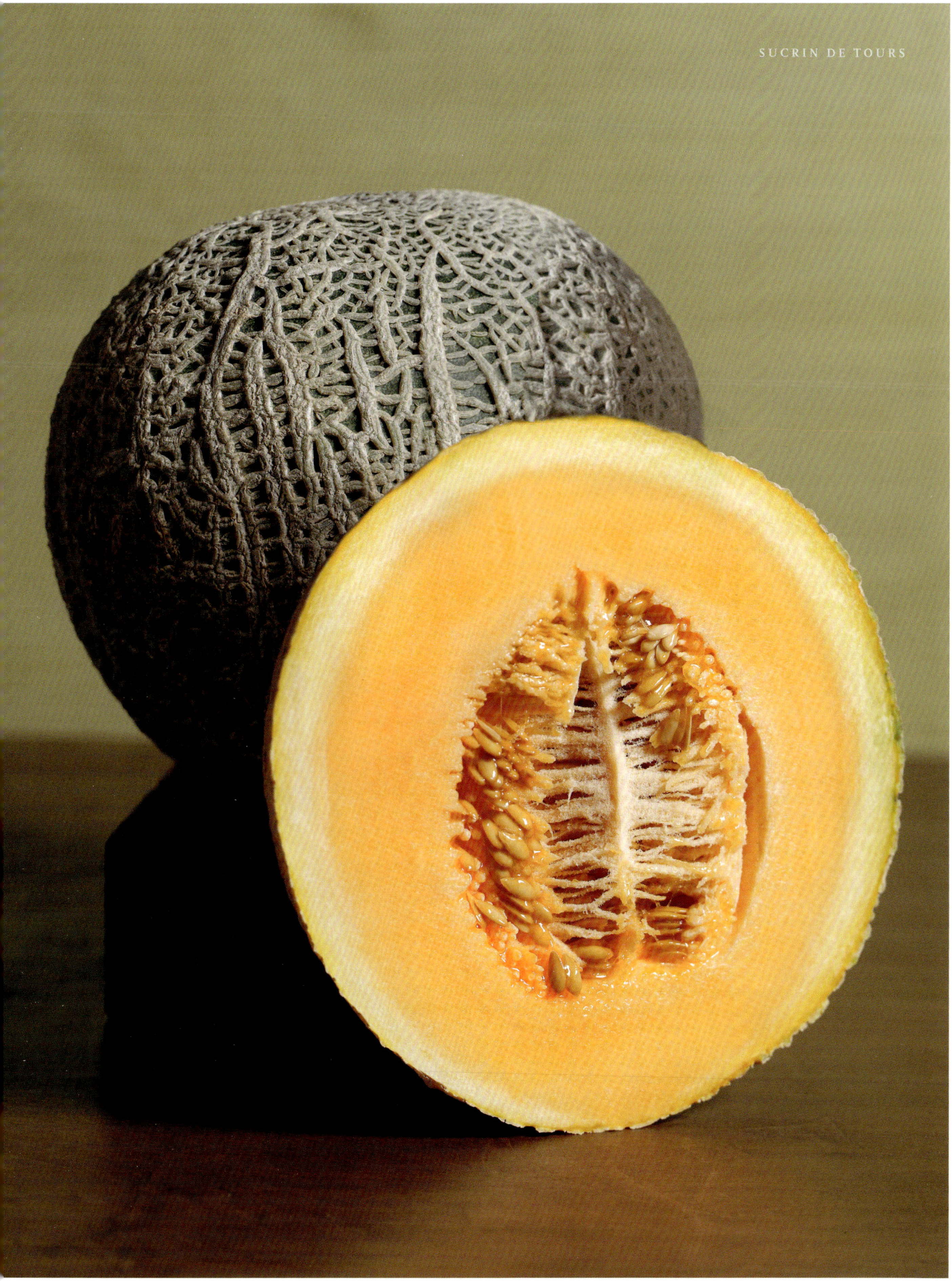

Hero of Lockinge

The Hero of Lockinge melon is no longer politically correct in its native land. Restrictive licensing laws in England make it a crime to sell vegetable seeds that are not on the approved National List. Belonging to the in-group means paying dearly: Lockinge meets the prerequisites as a distinct, uniform, and stable variety, but the exorbitant registration and annual fees are beyond the means of its vendors. The melon was dropped from the list in 1995 and, like so many other unsung heroes, has been replaced by modern varieties that bring in prettier pence.

This sorry state of affairs would furrow the brow of Scotsman Robert Loyd-Lindsay (1832–1901), perhaps better known as Lord Wantage of Lockinge, a founder and the first chairman of the British Red Cross. He is the hero of this story, and the man behind the melon. His greatest claim to fame was earning the Victoria Cross, bestowed upon him by Queen Victoria herself in 1857, for acts of bravery in the Crimean War.

The melon was introduced to the public in 1881 by Sutton's Seeds. The House of Sutton, founded in 1806, was one of the first mail-order seed companies, and it retains its royal warrant today "for seeds of the highest purity." Lord Wantage's head gardener, Mr. Atkins, developed the melon at the family seat, Lockinge Park, the biggest estate in Berkshire, by crossing Read's Hybrid, Colston Bassett, and Victory of Bath.

With such a noble heritage, the Hero of Lockinge should be winning blue ribbons—just as it did in Victoria's day. Its pint size and delicate lace are appealing; the translucent white flesh hints of fig. Thanks to Europe's leading organic gardening organization, Garden Organic (formerly known as the Henry Doubleday Research Association), the melon seeds can now be sourced legally in Britain, through membership. Seeds are commercially available in the United States.

Ananas d'Amérique à chair verte

This ravishing little finely netted melon, best known in English as Pineapple, has a long history of cultivation in the United States. A melon called pineapple was grown as early as 1794, in Virginia, by Thomas Jefferson. Pineapple melon was named for the sweet native American tropical fruit of an herbaceous perennial whose original name was "ananas," derived from the Tupi word "nanas" for excellent fruit. The perennial's Latin name became *Ananas comosus*, its French name ananas, and in English it became known as pineapple because its fruit resembles a pine cone.

The pineapple, with its spiked form, became a sensation after it was found on Guadeloupe by Christopher Columbus in 1493 and disseminated to Europe and North America. At the time, intensely sweet fruits were in short supply in the north, and refined cane sugar was expensive. The Spanish King Ferdinand II of Aragon was the first to taste pineapple brought back from the Americas by Columbus in 1496. Peter Martyr, tutor to the Spanish princes, wrote: "He has eaten another fruit brought

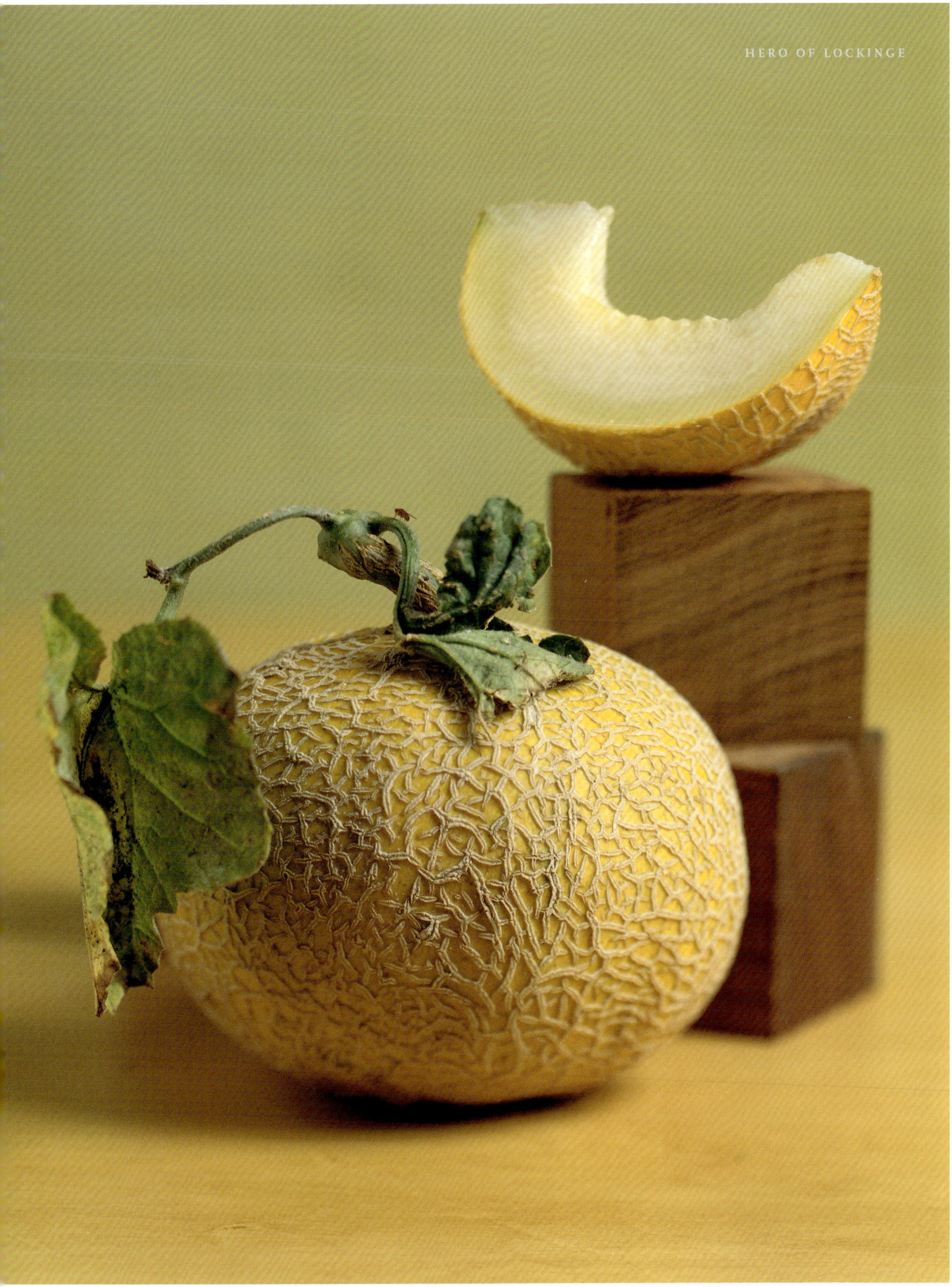

from those countries. It is like a pine-nut in form and color, covered with scales, and firmer than a melon. Its flavor excels all other fruits." The pineapple would become a symbol of status, hospitality, and friendship. How fitting that Pineapple melon bears such a name. Ananas is the kind of melon you'd want to present and share at the table with loved ones. So sweet, so melting, so excellent.

I love tracing the history of favorite melons. My collection of American seed catalogues allows me to observe the ebb and flow of popular varieties, to pinpoint introduction dates, and to peek into the mind of nineteenth-century seed and nursery growers. Grant Thorburn's *The Gentleman and Gardener's Kalendar,* published in New York in 1821, is the earliest catalogue or calendar I have that lists pineapple melon. Under "Melons" in the month of April, it reads, "The best kinds are the nutmeg, Romana, green flesh, pine-apple, large rock cantelope, &c." I'd say the man had good taste. My treasure is the *1834 Annual Catalogue of Esculent Vegetable and Other Seeds of the Choicest Imported and American Varieties,* from Wm. Prince & Sons, Proprietors, Flushing, Near New-York. This interleaved catalogue has annotations in William Robert Prince's own handwriting for changes to the next edition. He offered the following melon and watermelon varieties, including our star attraction: Green Citron, Pine-apple, Persian, Nutmeg, Bingham's green fleshed, Small oval Romana (*perfumed*), Large Yellow Canteleupe, Round Pomegranate or sweet scented, Long Pomegranate or sweet scented, Minorca or Netted Canteleupe, Skillman's fine netted, Star (*very late*), Large Musk, Egg (*beautiful*), Long Island water, Carolina water, Apple seeded water, Citron water, Snake, and Malta winter. In Prince's cursive writing, opposite: "Small oval Romana is either Long Pomegranate or the Egg Melon; Netted Romana is same as Skilmans; French Muscade is our Nutmeg, see Bongard [for synonymy]; Skilman's netted smaller than Pine apple but will do for it."

Ananas d'Amérique was widely grown and enjoyed in France during the nineteenth century. Sageret (1827) pronounced it sugary, melting, perfumed, and delicious—and perhaps the best of fifty varieties of melon he grew in 1826. He wondered if it was a true cantaloupe ("Est-ce un vrai cantaloup?"). This is a question I also ponder. Ananas d'Amérique resembles the true cantaloupe Ha'Ogen (see page 43). Jacquin (1832) illustrated it in living color—might that illustration have been the first? Naudin (1859) grouped the green-fleshed Ananas, and orange- and white-fleshed variants, with the highly esteemed and deliciously perfumed Sucrin melons. No other seed house did more than Vilmorin-Andrieux to promote Ananas d'Amérique (both green- and orange-fleshed versions): They catalogued them both for nearly a hundred years (from 1856 through at least 1946). According to Whitaker and Jagger (1937), Pineapple was the supposed progenitor of the famous Netted Gem melon, renamed Rocky Ford in 1897, which kickstarted the modern American cantaloupe industry.

Fordhook Gem

A rush of superlatives comes to mind when I envision Fordhook Gem. This is a melon that's easy to love. Looks, taste, and pedigree combine to make it the winningest. Green-fleshed muskmelons were once adored, but they've been supplanted by orange melons designed for the long haul. What a pity. As our

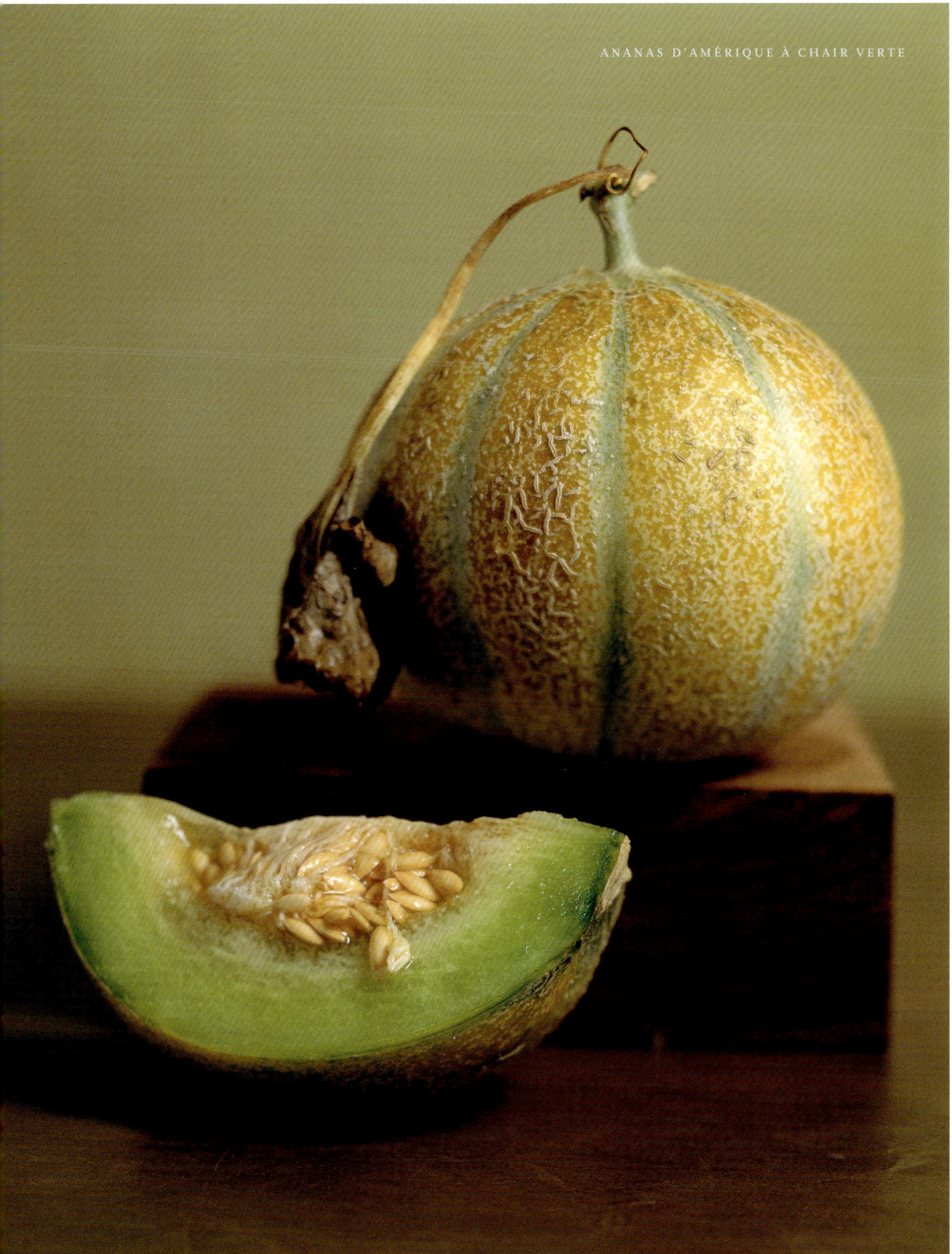

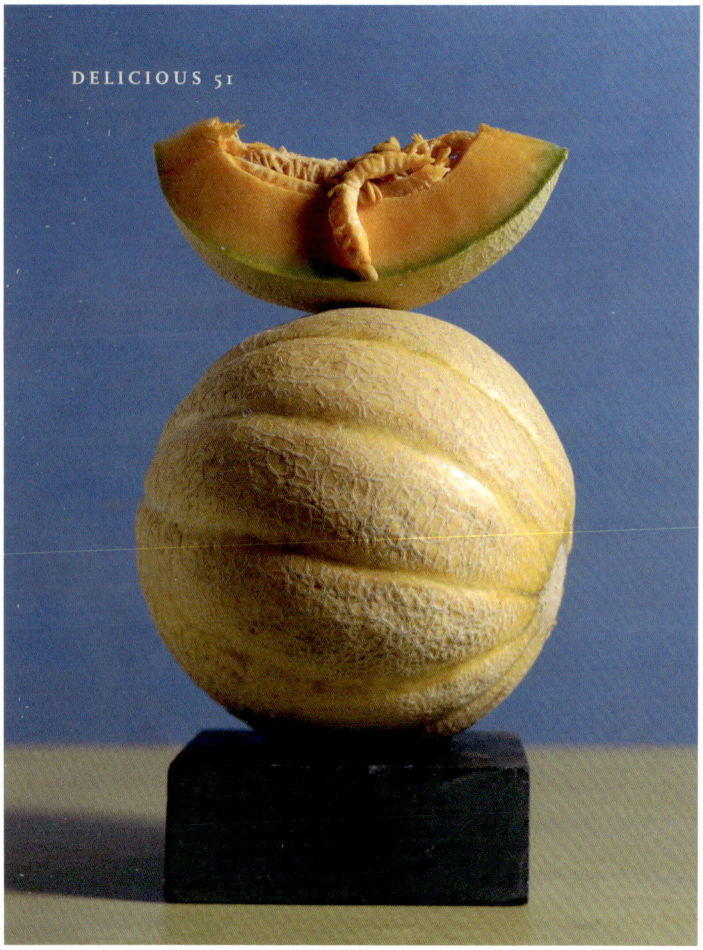
DELICIOUS 51

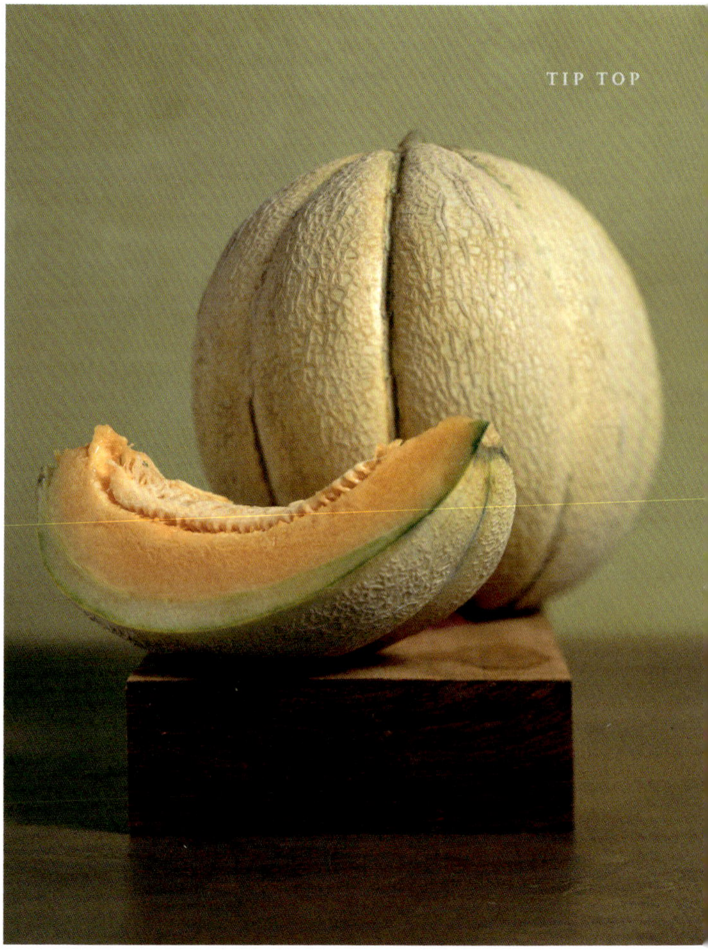
TIP TOP

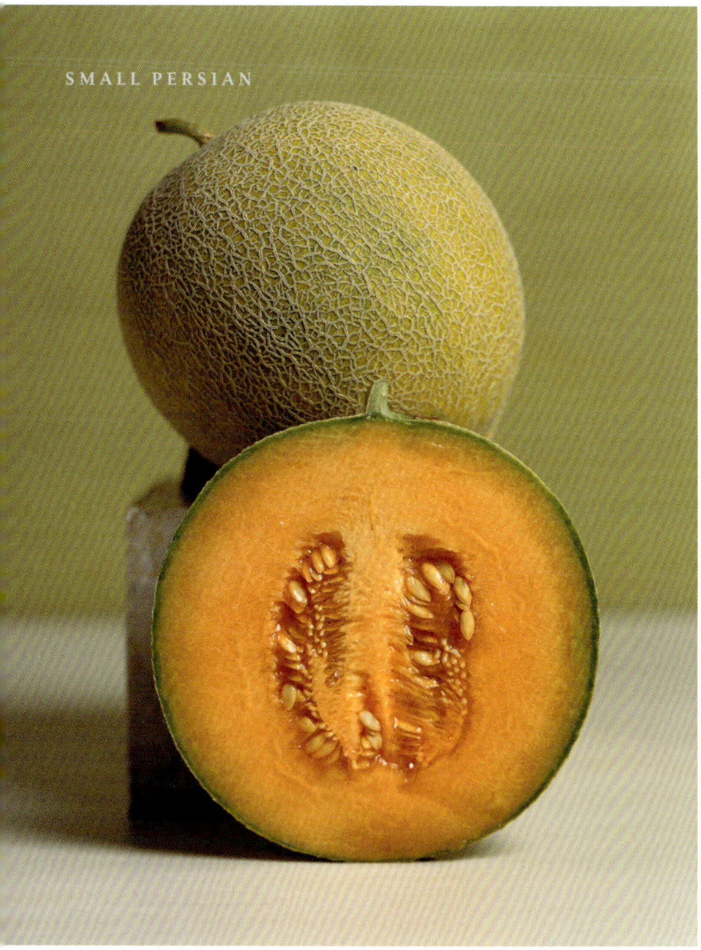
SMALL PERSIAN

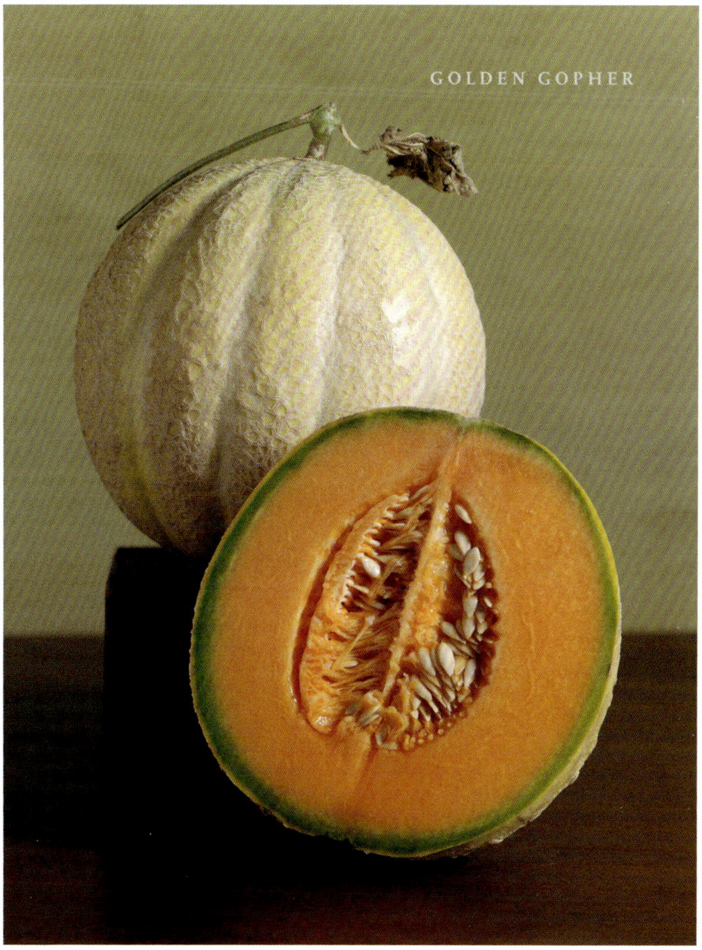
GOLDEN GOPHER

grandparents knew, the green has no musk and tastes sweeter. Fordhook's green flesh, encased in silvery filigree, varies between shades of sea foam and growing grass, with a salmon lining for emphasis. Your eyes conjure melting, sugary, peachlike flavors. Your nose discovers Riesling. This Gem is dessert material; lime juice is optional.

The Fordhook Gem belongs to a class of muskmelon known as Nutmeg. Its shape, netting, and aroma mimic the seed of that evergreen tree (*Myristica fragrans*). It may actually descend from a melon named Nutmeg or from the similar Pineapple (see Ananas d'Amérique à chair verte, page 61); both were well-known types grown in the United States prior to 1850. Sageret (1826) rated the nutmeg (*melon muscade*) as one of the best melons, the earliest and most productive. His seed stock had come from America via M. Vilmorin of Vilmorin-Andrieux.

Introduced by Burpee in 1967, and named for the family farm, Fordhook Gem is a cross between two noteworthy heritage varieties. Netted Gem, an 1881 Burpee introduction, is the celebrity parent. Considered the most important melon of its time, it was the progenitor of the Rocky Ford melons and the basis for the modern muskmelon industry. The other parent was Extra Early Knight, quite possibly a large-fruited selection from Netted Gem. Extra Early Knight was first offered by George Tait in 1908, but its popularity was limited to the Chesapeake Bay area.

Fordhook Gem combines the best of both parents, which were similar in appearance, flavor, and earliness. It owes its nutmeg shape to Extra Early Knight and its prodigious output and thicker skin to Netted Gem. Burpee dropped it from their catalogue in 1983 for reasons unknown to me, and since then there has been no other reliable commercial seed source. Would anyone like to adopt an orphan?

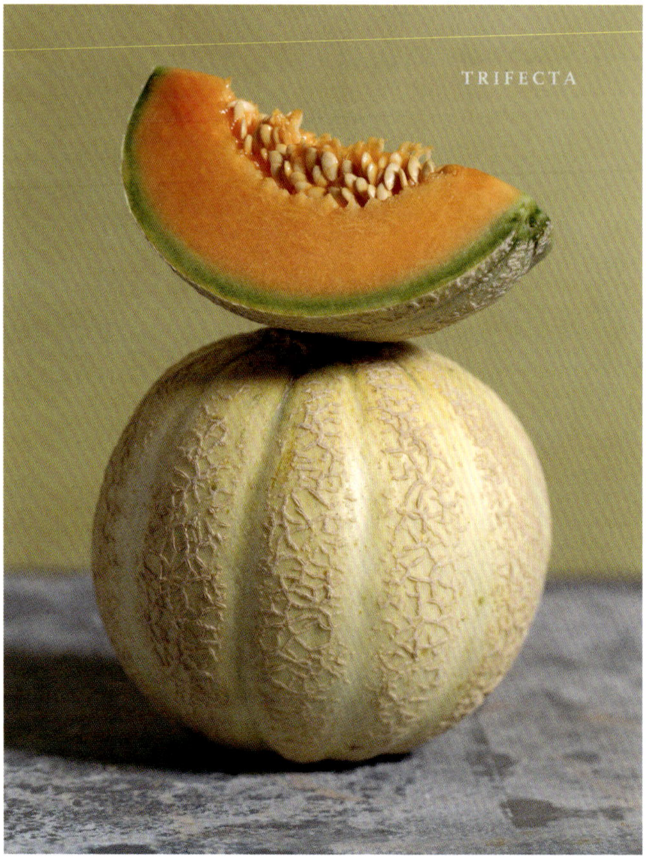
TRIFECTA

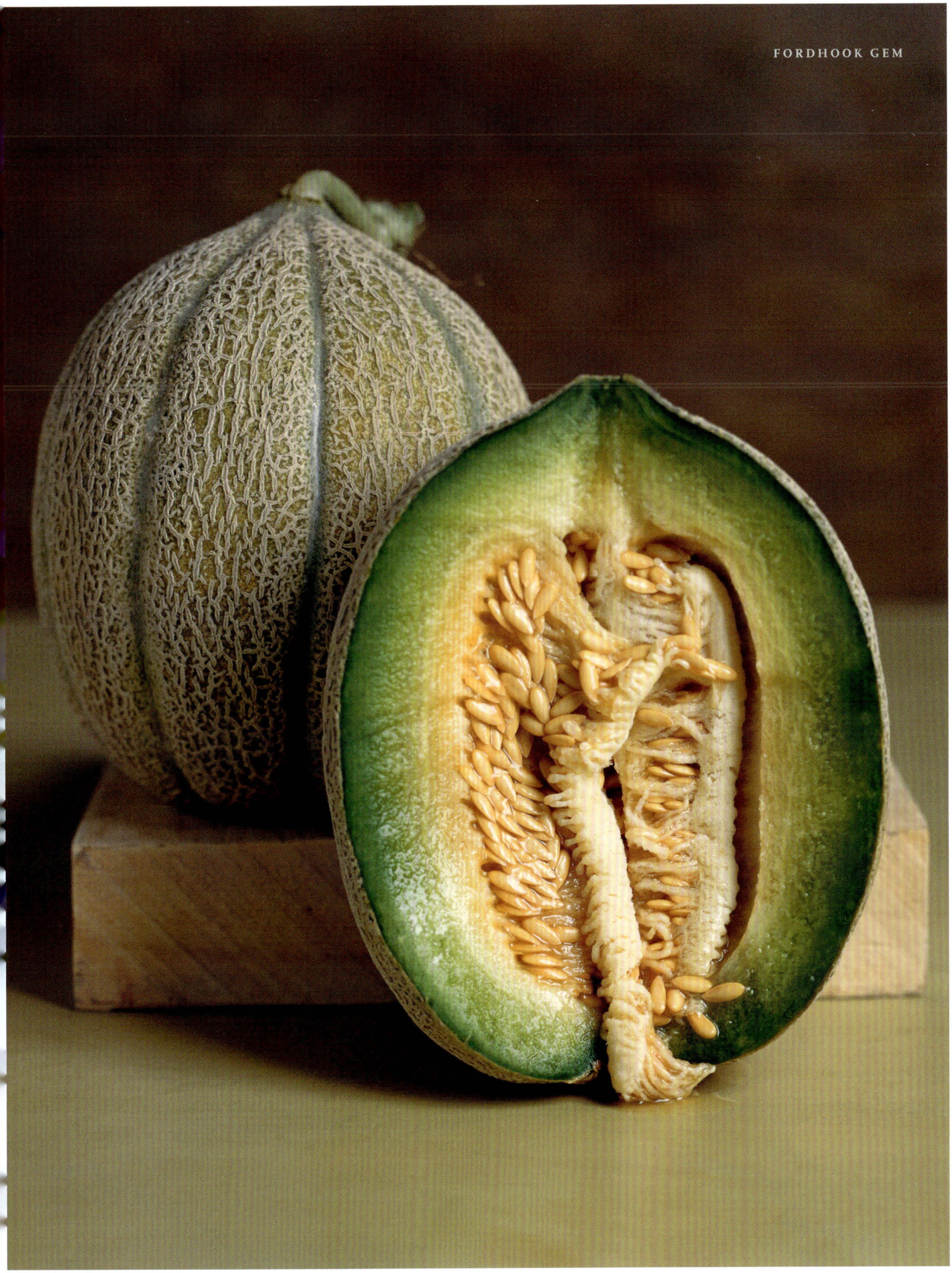
FORDHOOK GEM

Jenny Lind

This melon was named after "the Swedish Nightingale," Jenny Lind (1820–1887), a coloratura soprano promoted to stardom in the United States during the early 1850s by the prince of showmen, P. T. Barnum. Already a sensation in Europe, Miss Lind bewitched American audiences a hundred times over, from New York to Natchez. She was exalted for her voice, piety, and charitable giving; but when she broke her contract with Barnum and married her pianist, the spell broke too.

Although Jenny Lind's glory faded, memorabilia flourished. She was immortalized in song and poetry, and her bust, with bertha collar and brooch, adorned everything from shawls to glass whiskey bottles in the shape of a calabash. A strawberry, potato, and carnation also bore her name. According to Lind's contemporary, Dr. Robert P. Harris of Philadelphia, the Jenny Lind melon was named and introduced around 1846, and may have been selected from an old local variety called Center, or from an unnamed melon of Armenian origin.

Yet since Jenny Lind didn't sing in Philadelphia until the fall of 1850, and was little known in the United States prior to that date, it's unlikely a new melon would have been named, or an old one renamed, in her honor until after 1850. Indeed, Jenny Lind melon does not appear in the 1848–1850 Landreth seed catalogues of Philadelphia (nor does it appear as late as 1856). Landreth noted in 1847, "The Citron Melon, sometimes called Nutmeg, is almost the only one of its kind ever seen in our market, it is exquisitely flavoured, and perhaps, the best of its class." Ethan Platt, current owner of Landreth's, recently showed me the company's 1858 Rural Register and Almanac wherein Jenny is noted to be "of diminutive size but high flavor" and like the green-fleshed Nutmeg and Citron "of the old Musk or Canteleup variety." (Relevant pages from Platt's 1857 catalogue are missing and I haven't been able to locate another copy.)

One of the melon's claims to fame is the "outie" belly button at its blossom or stylar end. Burpee touted the button as the best part (of a new and improved strain known as Jersey Button)—"a delicious morsel"—in its 1902 farm annual. This bit of whimsy helped make the variety a best seller for years. These days, extremely protruding or broad stylar scars like Jenny's are rare. Considered undesirable, the trait has been bred out of modern melons.

Banana

The Banana seems to defy most people's definition of melons. Not even vaguely round, it conjures up images of loaves and fishes as well as bananas. The melon is yellow, elongated, and finely tapered at both ends, just like a banana, but those who say it tastes and smells like one too are stretching the truth a little. Its taste is heaven-scent, soft and sweet, with an external aroma that recalls true cantaloupe.

The Banana is a looker and has been pleasing audiences for more than a hundred years. In 1883, when seedsman James J. H. Gregory of Marblehead, Massachusetts,

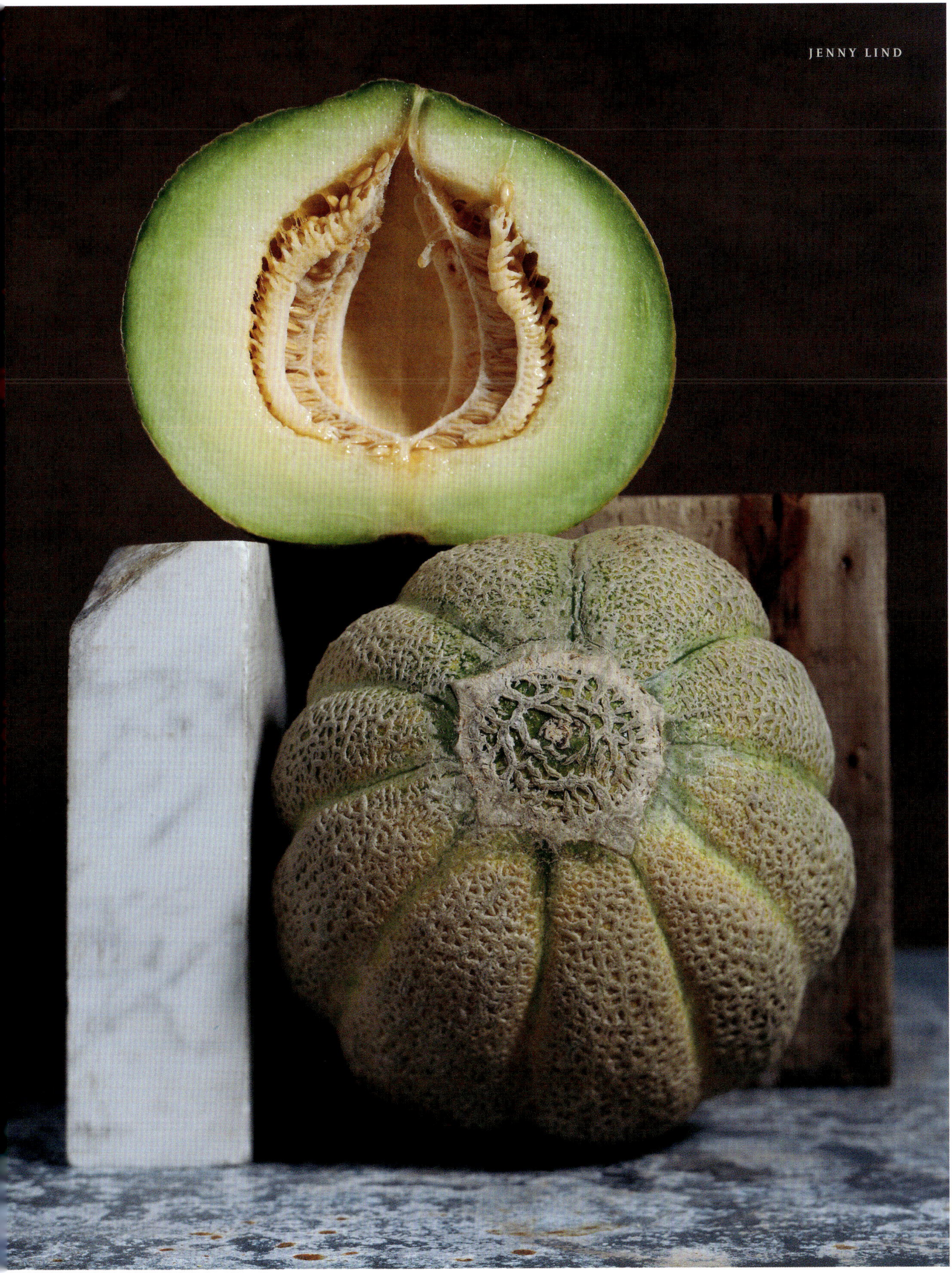

exhibited it at the Essex Agricultural Society's fall show and created a sensation, most Americans had never even laid eyes on a banana or plantain. Three years later, banana plants were shown at the Centennial Exposition in Philadelphia, and only a sentry could dissuade pilferers from taking home souvenirs. At a more recent Dutchess County Fair exhibit in Rhinebeck, New York, my Banana melon stopped fairgoers in their tracks to ask questions I was happy to answer.

The Banana does not readily fit into any one cultivar group of melon, sharing features common to several (although it does resemble some of the spindle-shaped melons grown in Central Asia). The authors of *The Cucurbits of New York* (1937) noted that the Banana is "rather unique" and intermediate between the muskmelons and the snake melons. Their physical descriptions still hold, although the Banana, split, that we see here is a bit more zaftig. Some specimens have longitudinal netting or deep furrows and prominent lobes; in others those traits are completely lacking. Whatever its cosmetic variations, it's easy to understand why the Banana has always been a crowd pleaser.

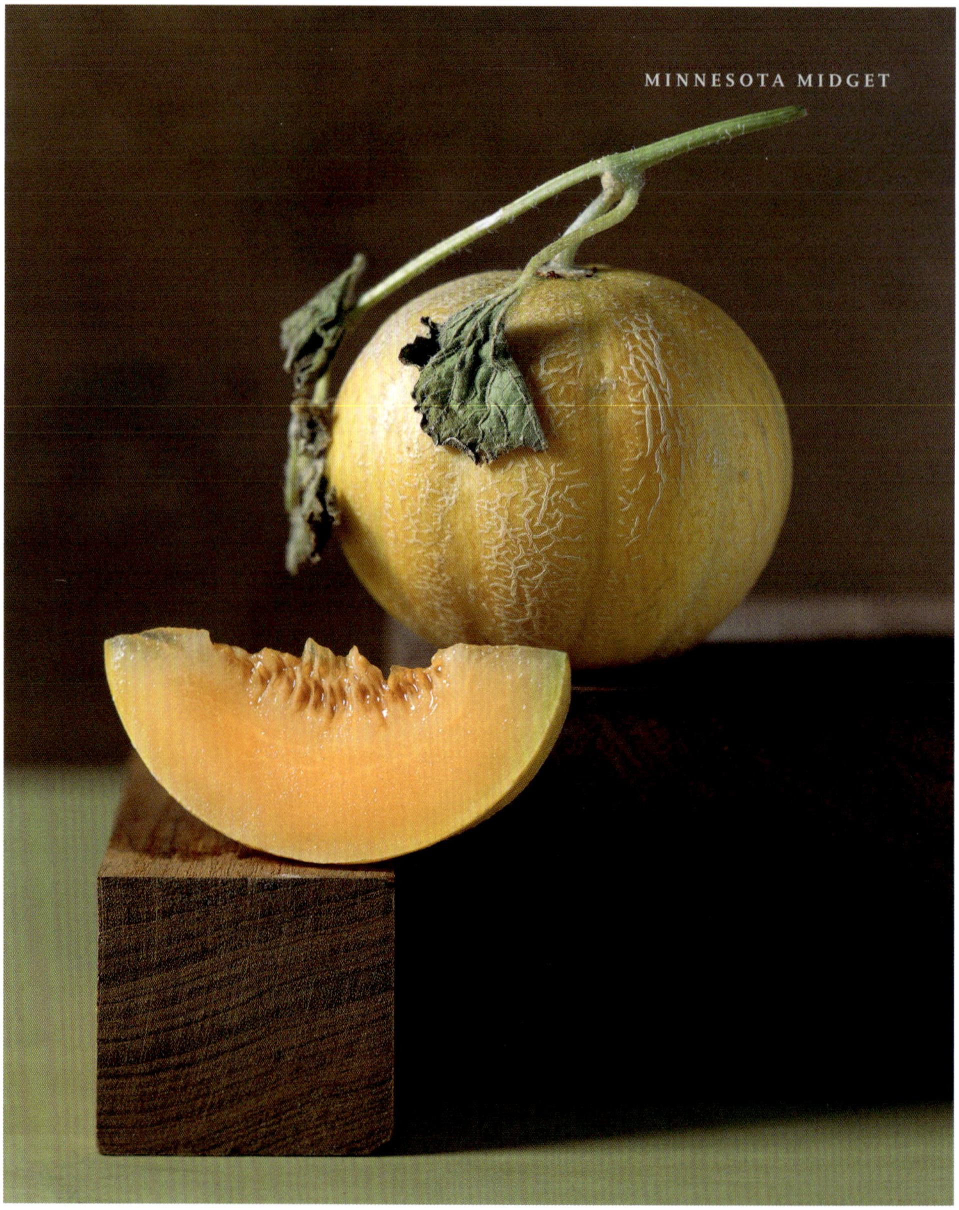

MINNESOTA MIDGET

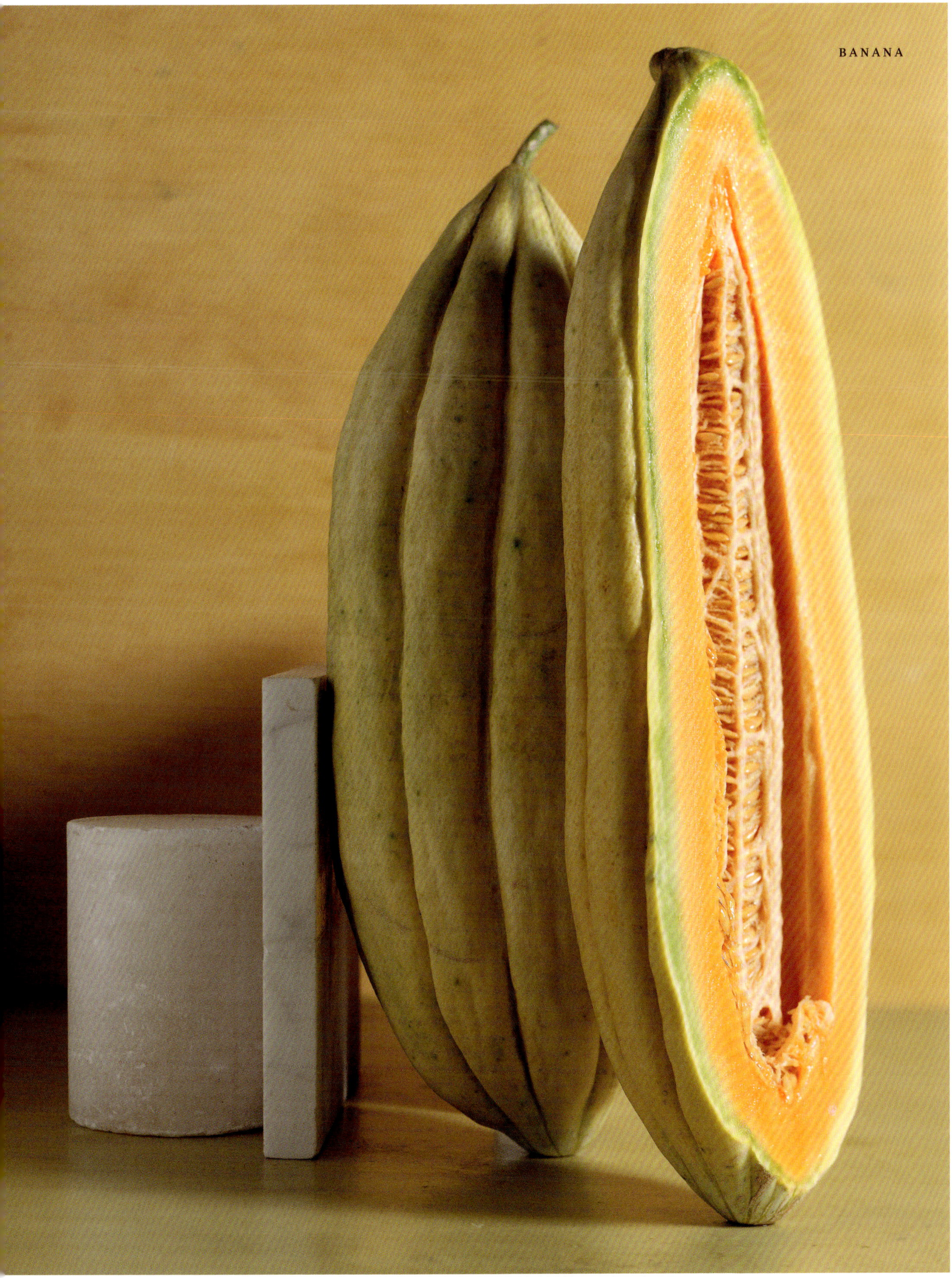

Cob

This obscure melon is full of surprises. Though the heart of it looks like cotton candy, the melon is anything but sweet. The flesh is devoid of water, very mealy, almost crystalline. Underneath the fluff is a fibrous cob with seeds clutched tightly together. Cob is hard to place and perhaps unique: it shares features with a number of melon horticultural groups.

Americans owe our acquaintance with Cob to individual seed savers. Pat Crill, the breeder of the Morgan melon, kept the Cob seed alive, from stock his great-great-great-grandfather John brought from Germany more than two hundred years ago. According to Lillian Remmick, another cache of Cob seed came to our shores in 1910 on the ship *Main*, in the apron pocket of her mother, Christina Schweikert Hein (1885–1982). An immigrant from Borodino, Ukraine, near Odessa on the Black Sea, Christina wanted a taste of the Old Country, where melon was served (like everything else) with bread. She settled near Terry, Montana, raised a family, and grew a garden. Grover Delaney, who lived in nearby Baker, got hold of some of her seeds and passed them to August F. Schuman in Bassett, Nebraska, in 1960. Active in the Seed Savers Exchange, August disbursed Cob melon—and the Grover Delaney watermelon (see page 176)—to countless members.

Emerald Gem

Washington Atlee Burpee (who preferred to be known as W.) introduced Emerald Gem to the public in 1886 with justifiable pride. He penned copy for his mail-order seed catalogues in longhand by the light of a coal-oil lamp. "Sweet and luscious beyond description," he wrote, and the melon became the most popular variety of its time.

Emerald Gem, a mutant or natural hybrid, was discovered by William G. Voorhees of Benzie County, Michigan. He sent the seed to Burpee, who had a reputation for paying top dollar. In those days, farmers and gardeners were more active than seedsmen in selecting and developing new varieties. The old-time practical breeders didn't attempt to control pollination, but like their forebears, they were careful observers and saved seed from genetic accidents.

W. knew to save his best for seed. Applying lessons learned in his youth as a renowned breeder of pedigreed fowl, he excoriated those seedsmen who cut and marketed melon seed from rejects while selling short the most excellent for fresh eating. This was an all too common practice at the time, and it was, of course, self-defeating: seeds from inferior stock beget inferior stock.

One of the best-kept secrets in melon world is that Emerald Gem does not remain emerald at maturity. A green-skinned Emerald Gem is immature, low sugar, dry, and not fit for the table. If you were to see it next to a fully ripe one—brownish green and netted—you'd wonder if the two were the same cultivar. The Emerald Gem that

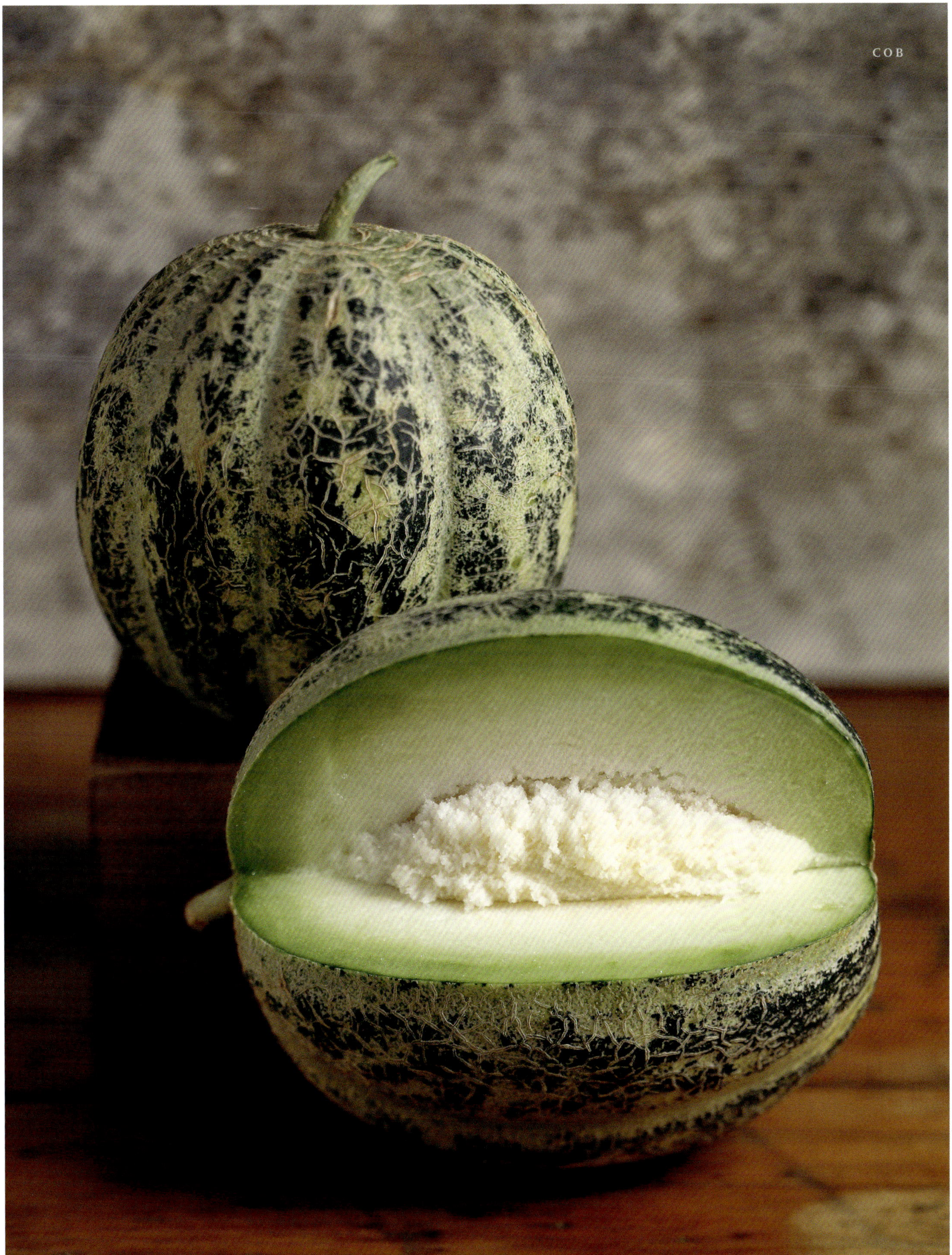

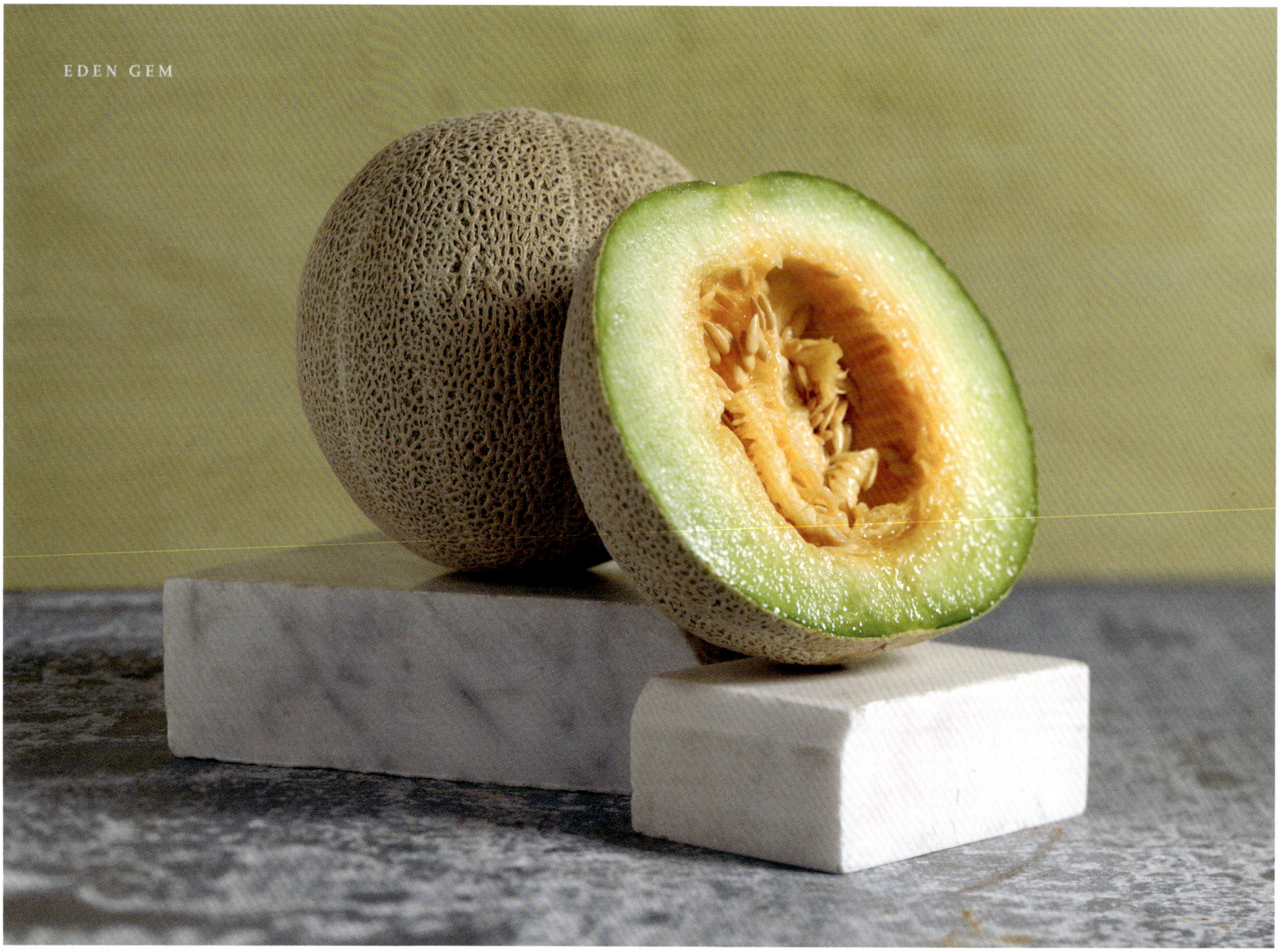

Burpee described in 1887 had a furrowed but smooth (non-netted) skin that was very deep emerald-green at maturity. A year later, Burpee published a color plate painted from nature of a still-green yet netted Emerald Gem with orange flesh that was ready to eat. Apparently the Emerald Gem has gone through a progressive change in type over the past 130 years. I harvest today's version at forced slip before it softens and becomes overripe.

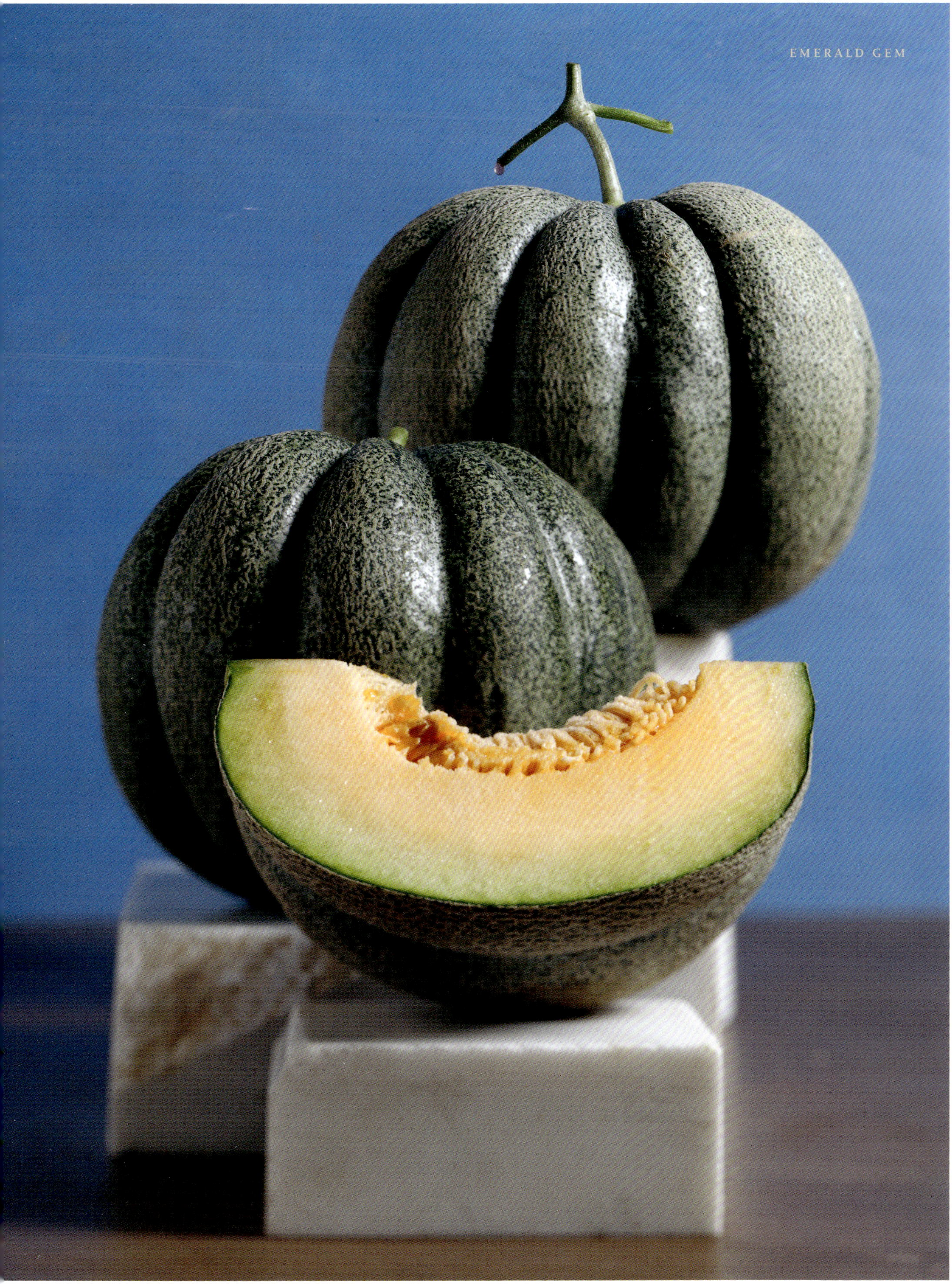

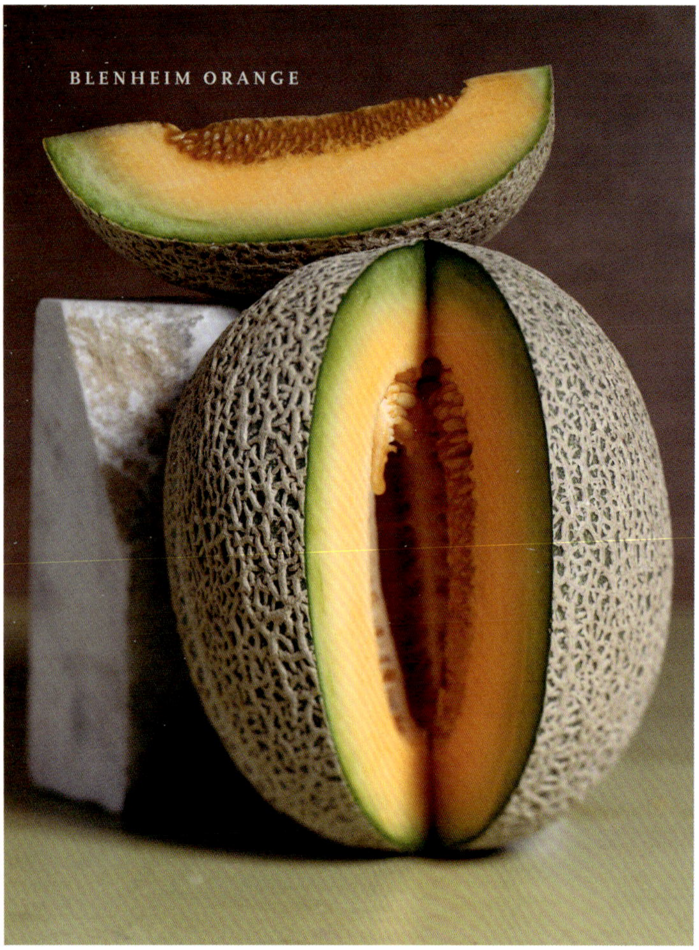
BLENHEIM ORANGE

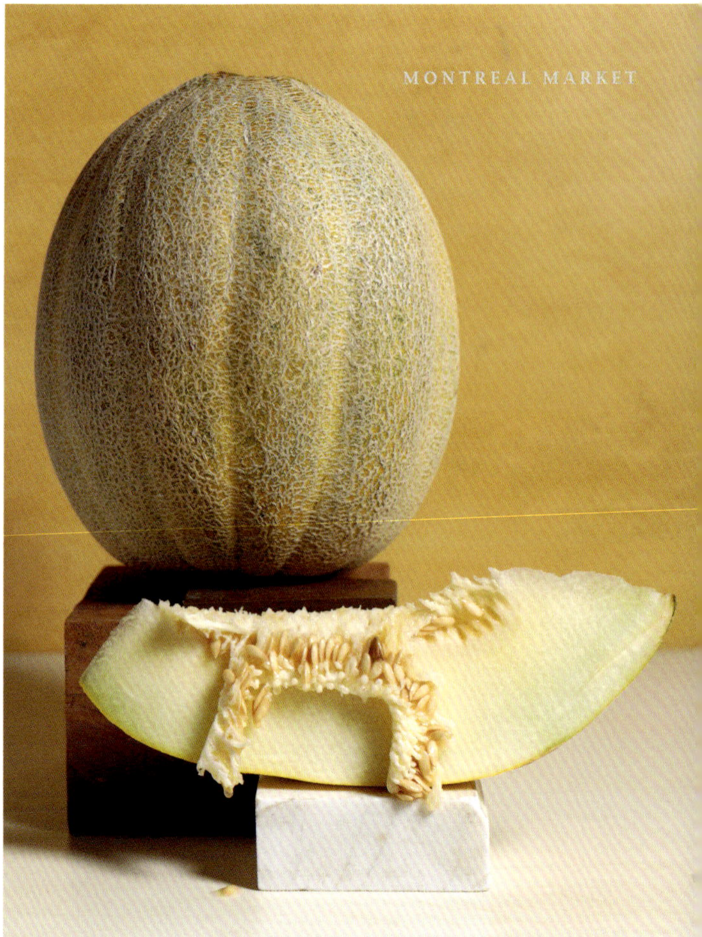
MONTREAL MARKET

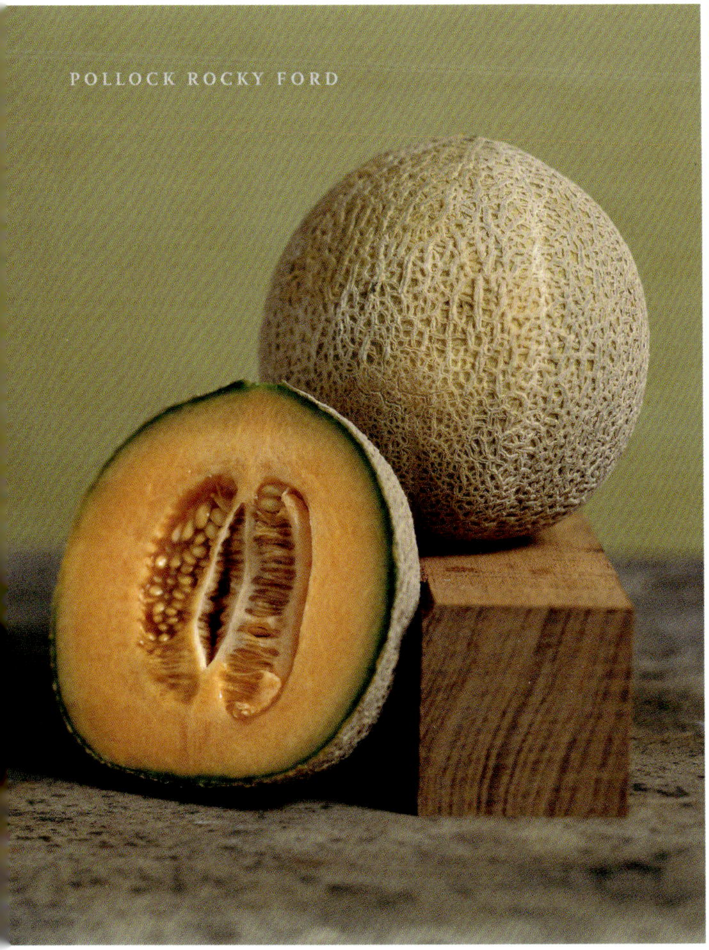
POLLOCK ROCKY FORD

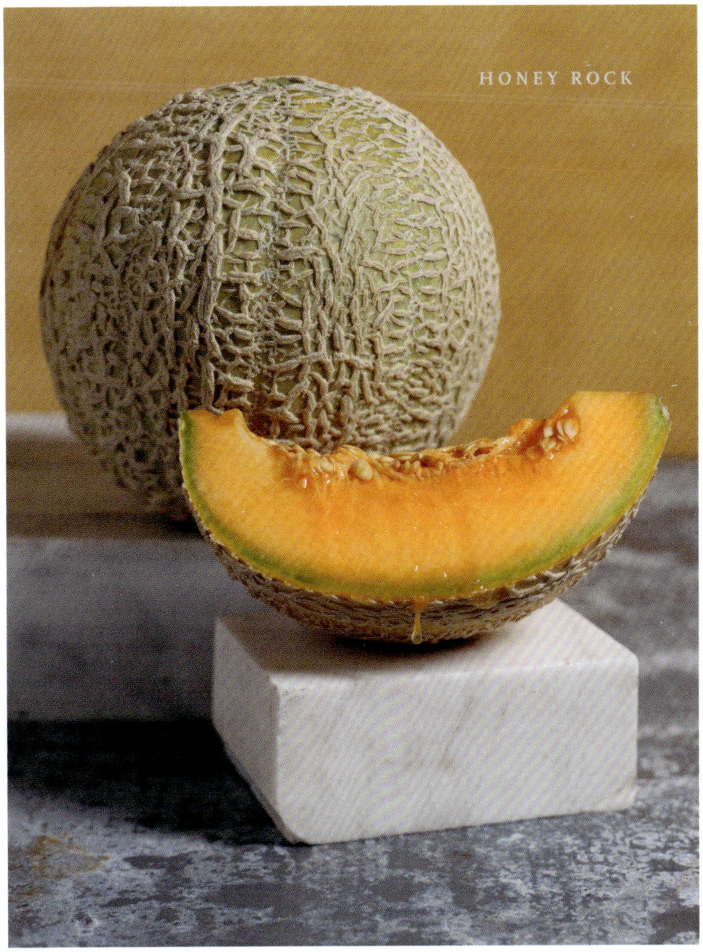
HONEY ROCK

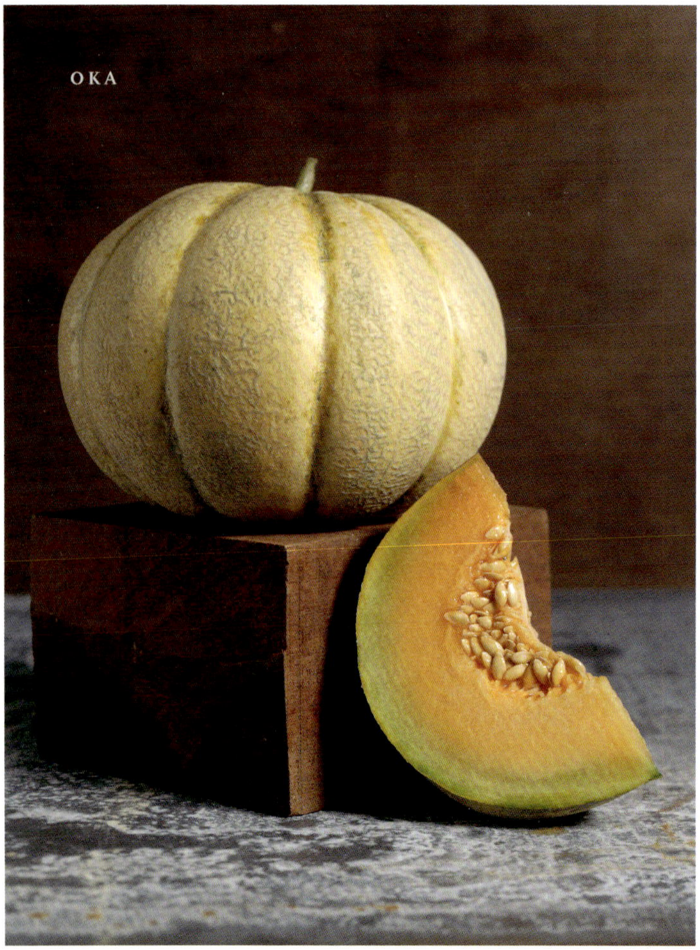
OKA

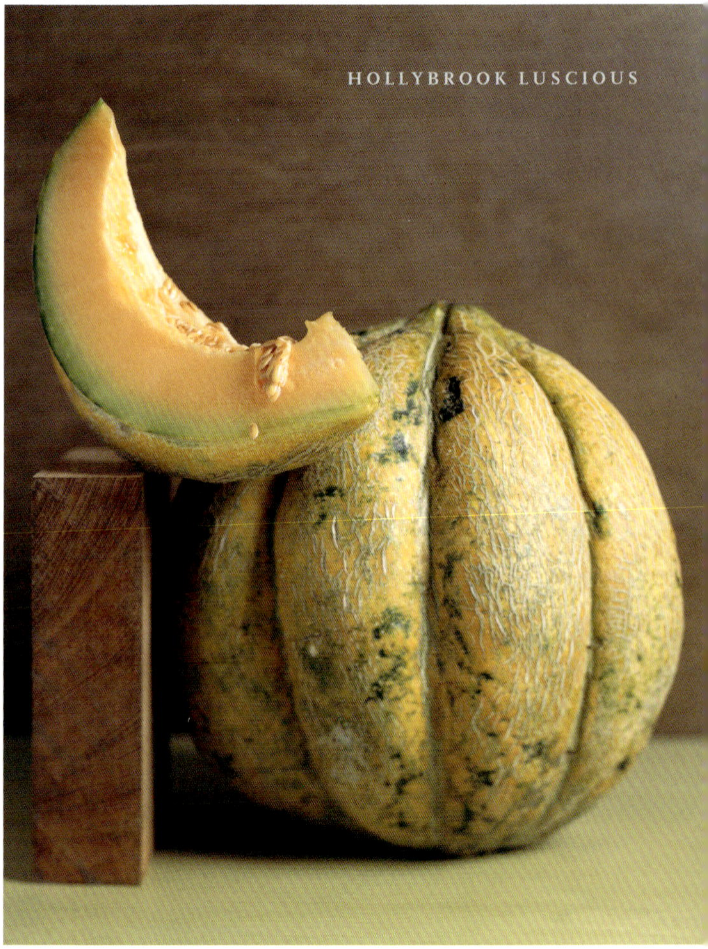
HOLLYBROOK LUSCIOUS

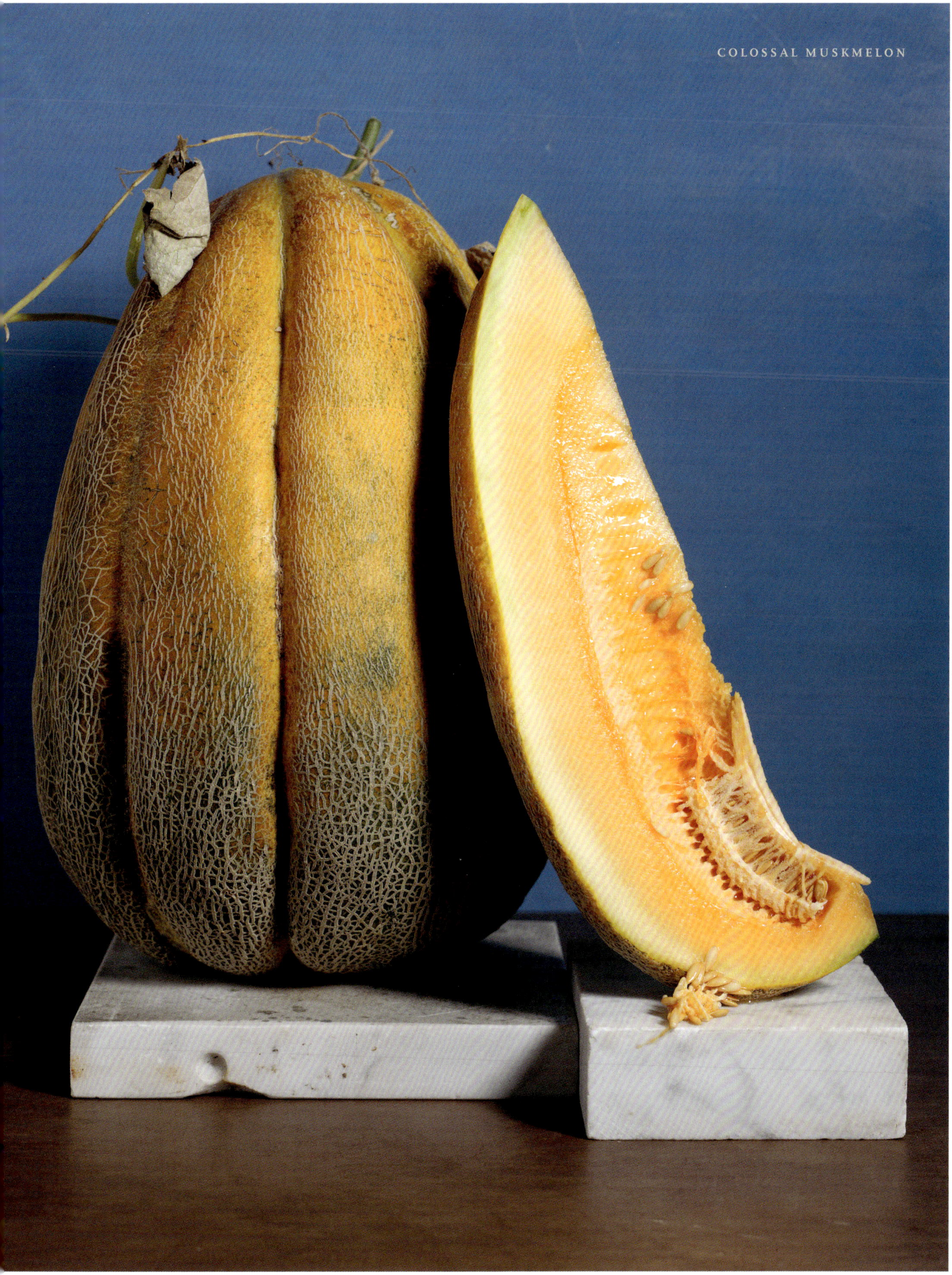

Anne Arundel

When the Anne Arundel melon was introduced commercially by Griffith and Turner of Baltimore, in 1890, it was already well known to the denizens of Anne Arundel County. Both the melon and the place where it grew were named in honor of Countess Anne Arundell of Wardour (circa 1610–1649). Anne became Lady Baltimore after marrying Cecil Calvert, the first Lord Proprietor of the Maryland Colony.

The precise origin of the Anne Arundel melon is lost to history. For all we know, it might have arrived in Anne Arundel County with the first European settlers in 1649; it could have been grown even earlier by the Algonquin tribes along the Patuxent River. Elsewhere in Maryland, muskmellons were grown by the natives, according to *A Relation of Maryland*, published in London in 1635.

Tobacco was the main crop in the Chesapeake colonies until the Civil War. When slavery ended, farmers diversified and Anne Arundel County became the nation's strawberry capital. Muskmelons were grown in private gardens but didn't appear much in the marketplace until after 1870. Starting with improved green-fleshed nutmeg or muskmelon varieties, such as Burpee's Netted Gem (1881), Landreth's Baltimore Market or Acme (1884), and Anne Arundel (1890), these melons were then more widely grown and shipped farther afield.

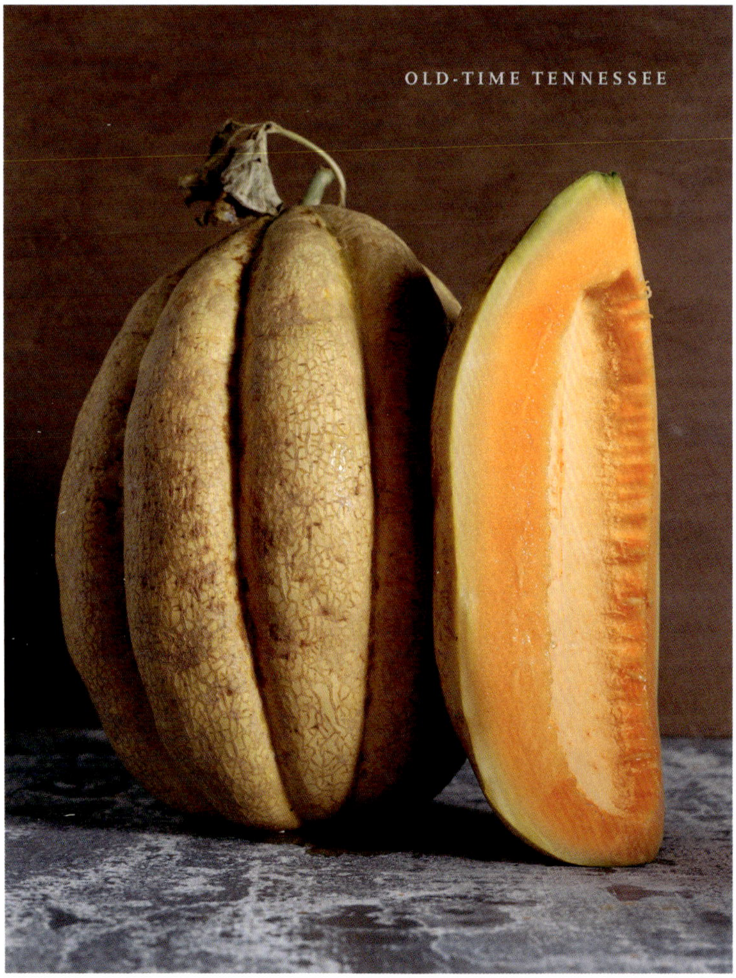

OLD-TIME TENNESSEE

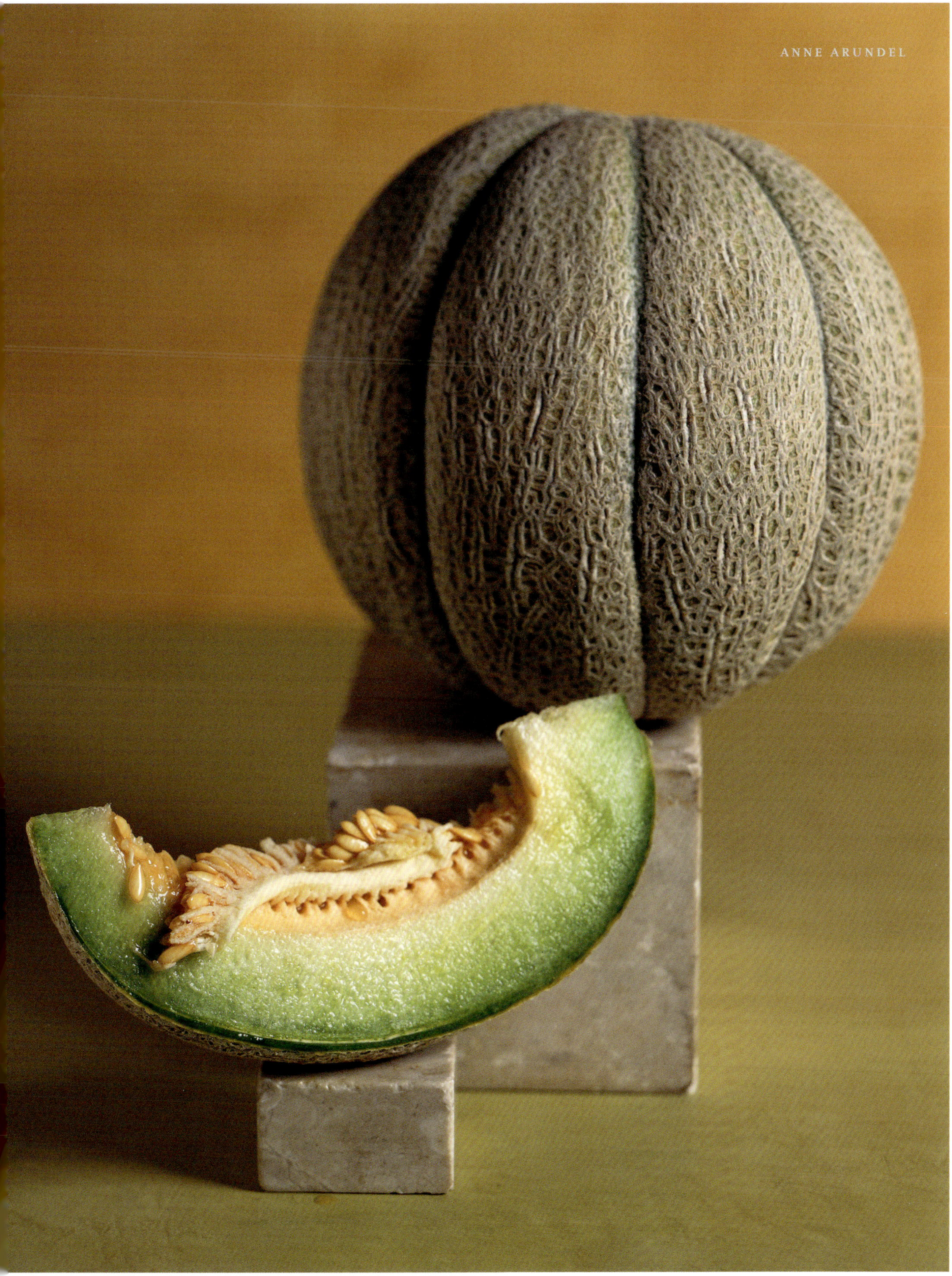
ANNE ARUNDEL

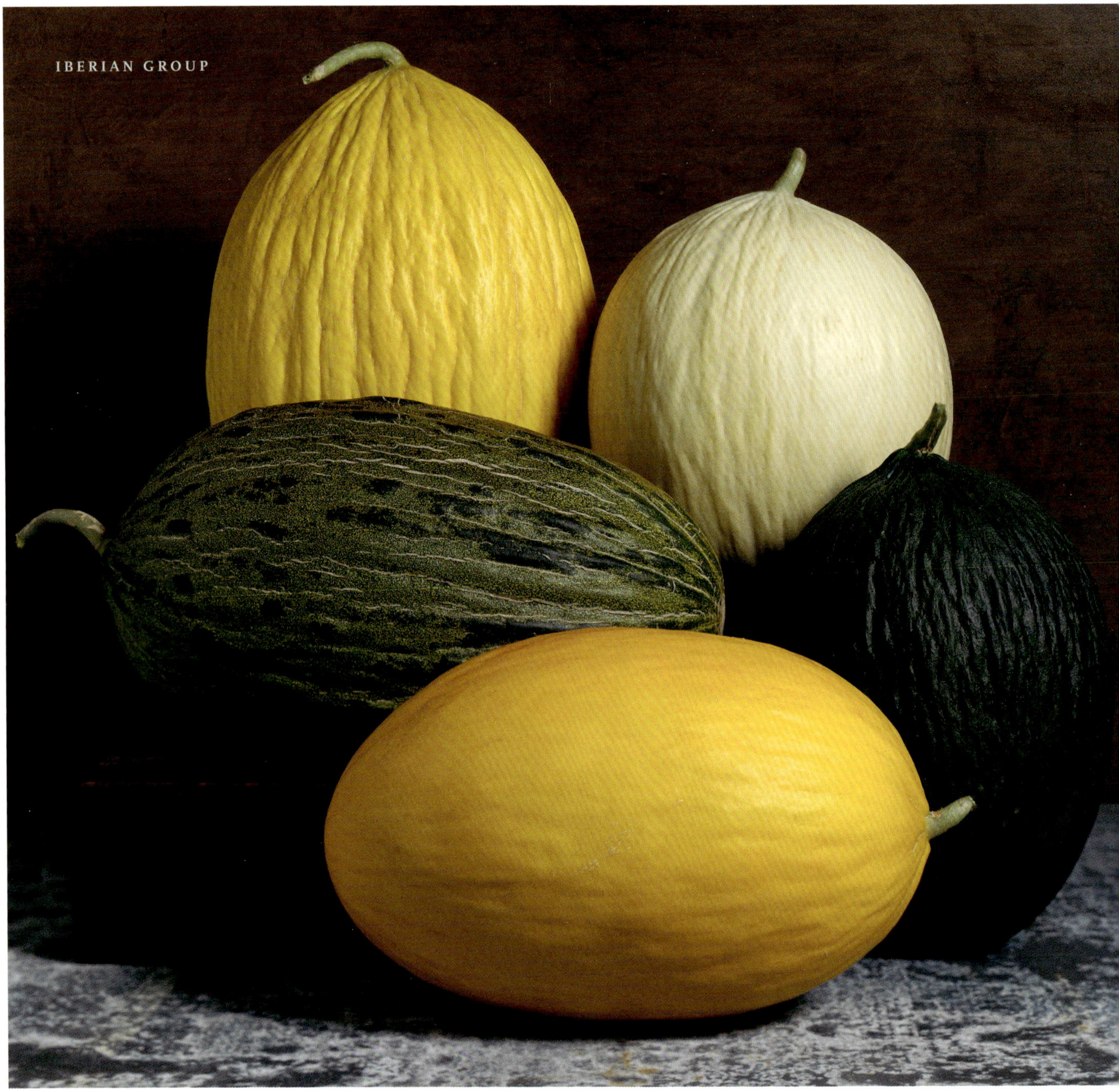
IBERIAN GROUP

Inodorus Group

WINTER MELON Inodorus melons comprise a very large cultivar group. They are commonly known as "casabas" or winter melons. Spanish types are prominently represented—including Amarillo, Piel de Sapo, Rochet, and Tendral—as are melons from Turkey, including Kirkagac or Santa Claus, all the way to Turkmenistan. (Not the same as *Benincasa hispida*, which also is known as winter melon.)

Casabas are considered to be the sweetest melons in the world. But in order to accumulate enough sucrose and achieve maximum expression of sweetness, they require long, hot, rainless summers with high insolation and optimal growing conditions. Sugary melons were initially selected by farmers in one such place: Khorasan, a broad area in Central Asia encompassing modern Turkmenistan, Uzbekistan, Afghanistan, Tajikistan, and northeastern Iran. It is to this region that casaba melons can be traced as far back as the ninth or tenth century.

According to Paris, Amar, and Lev's account (2012a), Muhammad Abu al-Qasim ibn Hawqal, a Kurdish native of what is now southeastern Turkey, wrote in his travelogue in Arabic, in 955 CE, that elongated melons that were *qabih al-mandar*, of ugly appearance, but *ghaya fi al-halawa*, infinitely sweet, were very good to eat. They were grown in the village of Ardahar (Persia), present-day northwestern Iran bordering Armenia, Azerbaijan, and Turkey. These melons, he added, were comparable in quality to the renowned melons of Khorasan.

Casaba melons, bigger than most, and ugly to some due to their wrinkles and acutely tapered ends, take longer to grow, require sufficient warmth and sun, and often cost more in the market. But, when well grown, they're worth it. What they lack in fragrance and rich flavor, they more than make up for in sweetness. Their juicy, often crisp flesh is usually white or green, but can even be tricolored like Amarillo Oro's. The thick skin is usually inconspicuously lobed or non-lobed and may be smooth or wrinkled. Skins come in a variety of colors and patterns. Shapes vary from round to oval to elongate; many are acutely tapered at the stem end. Winter melons are generally nonclimacteric and have a shelf life of a month or more. Necky thin-skinned casabas or crenshaws such as Bidwell, Crane, and Sweet Freckles are traditionally included: they are climacteric and the most perishable of the lot. Knowing how to judge a ripe winter melon is an art.

Opposite, clockwise from top: Rugoso di Cosenza Giallo, Branco, Tendral Verde, Amarillo Oro, and Piel de Sapo. All but Branco are Spanish casabas. Branco is of Portuguese origin.

Bidwell Casaba

The first time I saw a Bidwell Casaba, it nearly took my breath away. As melons go, this one is big. Its tawny, leathery coat has the look and feel of pigskin or perhaps the carapace of a turtle. What's inside tastes like heavenly orange sherbet. With soft, wet flesh, it seems much sweeter than an objective measure of its sugar content would indicate.

Like so many things from Chico, California, the melon derives its name from General John Bidwell (1819–1900). This Paul Revere of the Gold Rush trumpeted the news of the mother lode from Sutter's Fort to San Francisco, and he struck it rich himself. He was a general in the Civil War and a U.S. senator, but for our purposes, at least, he is most fondly remembered as the father of the Bidwell Casaba.

Using seeds he had procured from the Department of Agriculture in 1869, Bidwell planted Casaba on his ranch in the hot, arid Sacramento Valley. The original melon was said to be similar to Green Persian and Bay View, not to the wrinkled casabas from Spain, Portugal, and Turkey. It has changed in two ways: The occasional netting is sparse (in the furrows) and the flesh is orange (rather than recessive green). No other grower who was tendered these seeds by the USDA succeeded like the General. The melon was his darling, but not everyone concurred. One houseguest, General William Tecumseh Sherman, was exasperated with the menu: "My God, casabas for breakfast, casabas for lunch, casabas for dinner!" What guests didn't eat, Bidwell's hogs did: they were turned into the melon fields to glean.

Robert B. Miller has been guardian of the seed for years, growing the Bidwell to enormous proportions with sky-high Brix readings. He is the latest link in a seed chain that leads directly back to Chico: If not for local melon farmers like Zea Sonnabend and Clara Fay and Lowell A. Stringfellow, there would be no more Bidwell Casaba in the home garden.

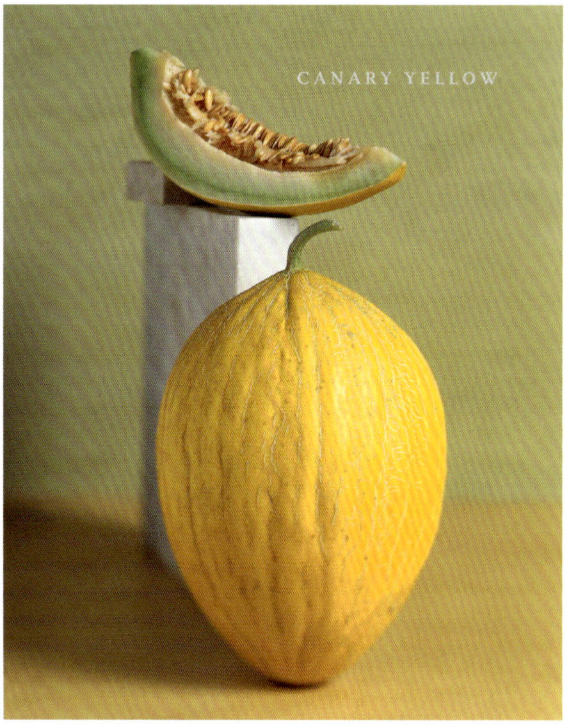

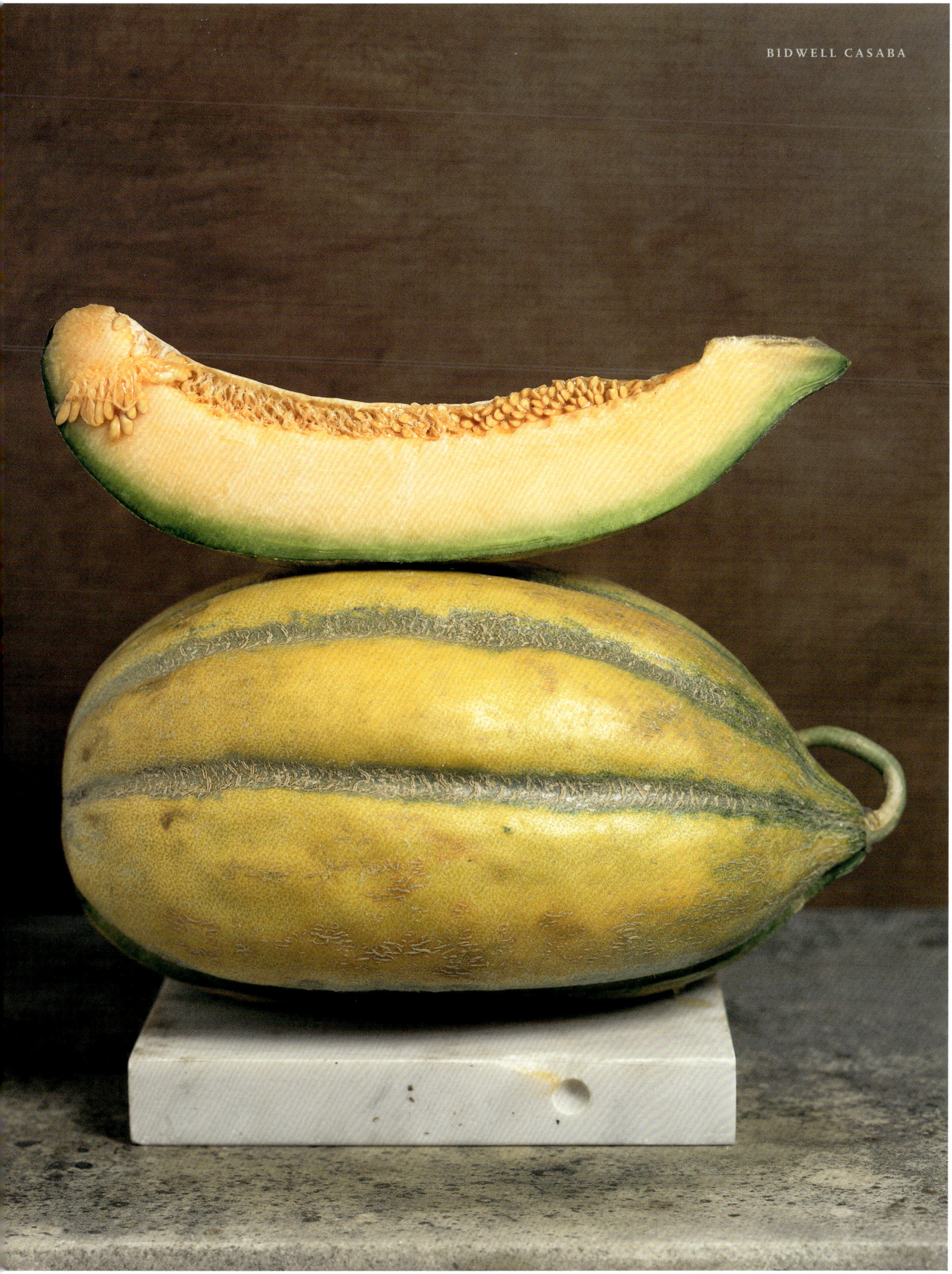

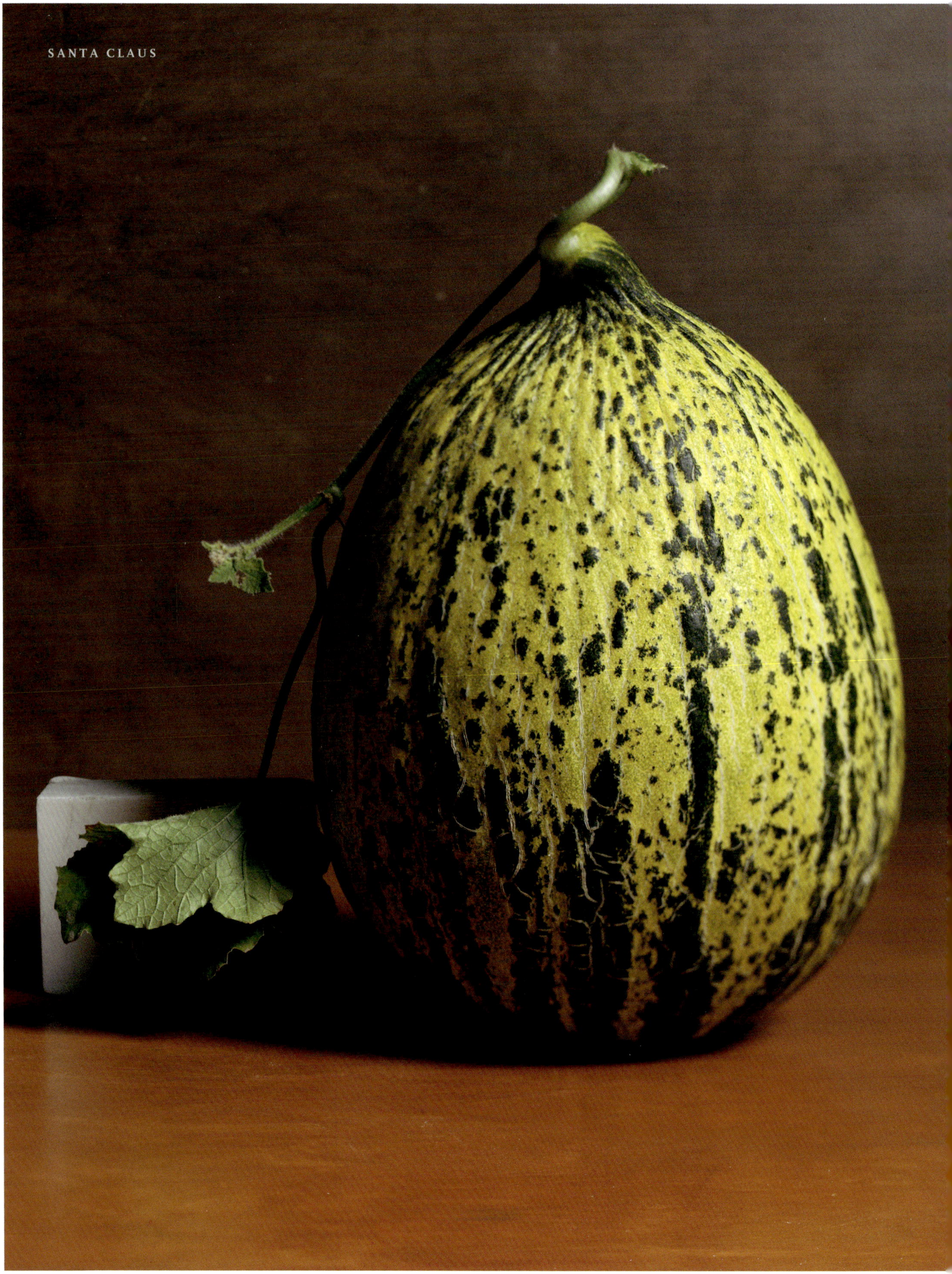

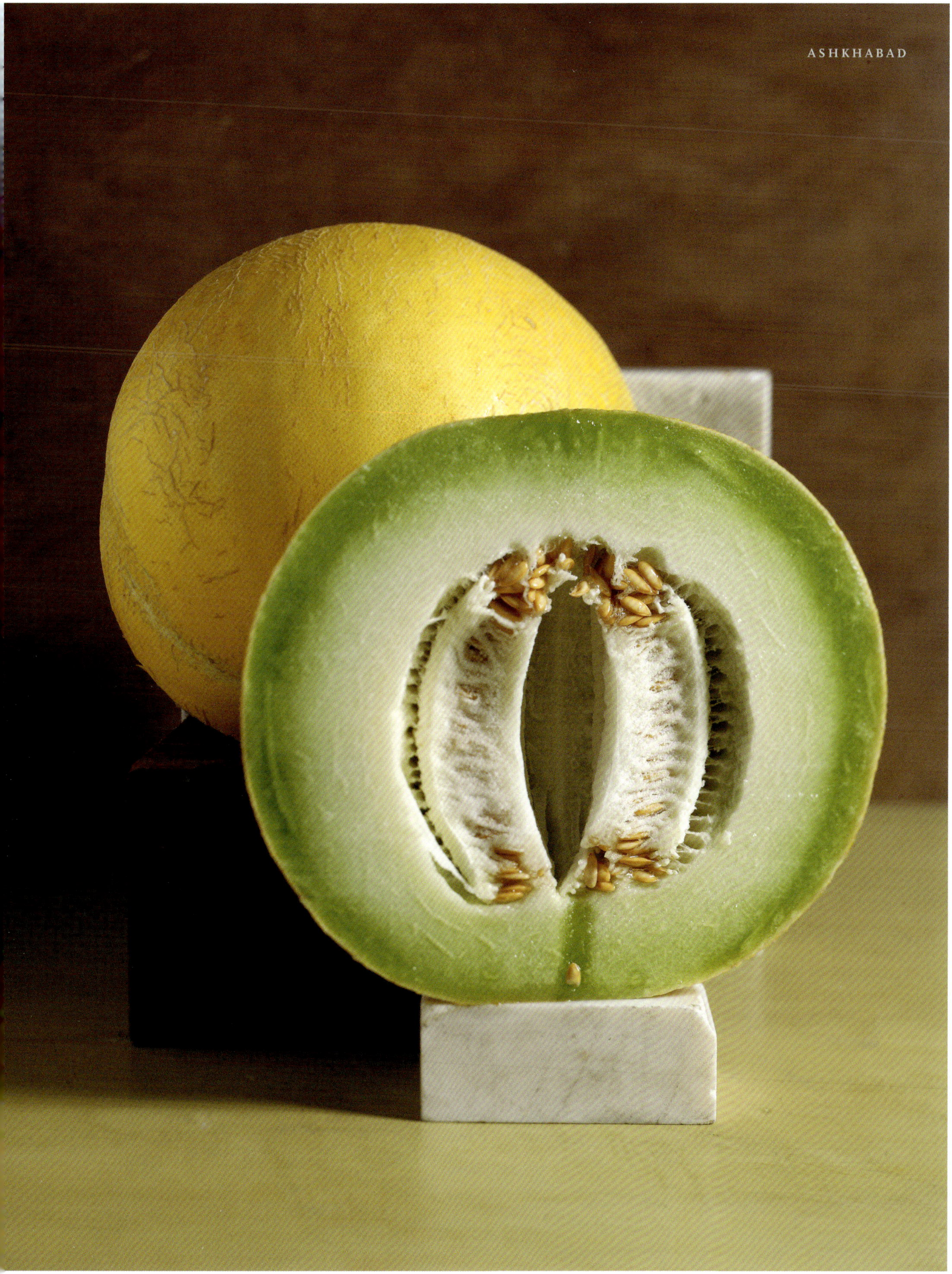

Amarillo Oro

There are lots of good melons; Amarillo Oro is a great one. I always hesitate a little before cutting it open—the anticipation of biting into its succulent, tricolor flesh, with gradations of color from green to white to orange at the seed cavity, is almost as enjoyable as actually doing it. Almost. Tasting it transports me to Spain.

Spain is a land blessed with casaba melons in abundance. They are grown along the Mediterranean, northeast of Gibraltar, for export and local markets. Whether the melon is dusted with powdered ginger, wrapped with Serrano ham or salami, or washed down with Manzanilla sherry, the Amarillo Oro is a melon the Spanish take to heart. Amarillo casabas have yellow skin—by definition—and are wrinkled or smooth and generally without lobes. The sugary flesh is usually white and crisp. Tipo Amarillo is well represented in this book. They do better in my garden than the Tendral and Piel de Sapo types. As Dr. Jose "Pepe" Esquinas-Alcazar, for many years the Secretary of the Intergovernmental Commission on Genetic Resources at the Food and Agriculture Organization of the United Nations, wrote in his 1977 dissertation, "The distinctive characteristics of Spanish melons suggest a common origin for all of them." The results of his study pointed toward India as their possible remote source. Harry Paris later traced the remote source to tenth-century Khorasan in Central Asia. Pepe's finding that Spanish melons have little in common genetically with Italian or French cultivars favored invasions by the Moors (rather than Roman occupation) as their more proximate source.

According to Paris, Amar, and Lev (2012a), "The first description of sugary melons in Europe is from the south-western end of the continent, Andalusia, which has a semi-arid climate, and was written in the late eleventh century by Abu al-Khayr of Seville." Those sugary melons, with narrow, acute stem ends, fit the description of casaba—which was probably the first type of sweet melon grown in Andalusia. The authors believe it most likely that casaba would have reached Andalusia with the Moors during their tenure (711–1492). The Moors also brought sugar, rice, oranges, and saffron, transforming the landscape and the scope of Spanish cuisine.

Honeydew

Who's the sweetest of them all? Many say Honeydew. (Others say Piel de Sapo or . . .) Originally known as Melon d'Antibes Blanc d'Hiver (White Antibes Winter), this melon has delighted connoisseurs for over a century. It was grown for decades in the Midi of France near Antibes and in Algeria for export. None was more agreeable or pleasing to the senses among the dozens cited in an anthology by Vilmorin-Andrieux (1891). Modern-day measures of sugar content confirm—with refractometer readings of 16 degrees Brix—that Honeydew has few peers. And it has the ability to retain its goodness for weeks or months.

Americans knew little about winter or Inodorus melons when renowned horticulturalist Liberty Hyde Bailey grew the French White Antibes in upstate New

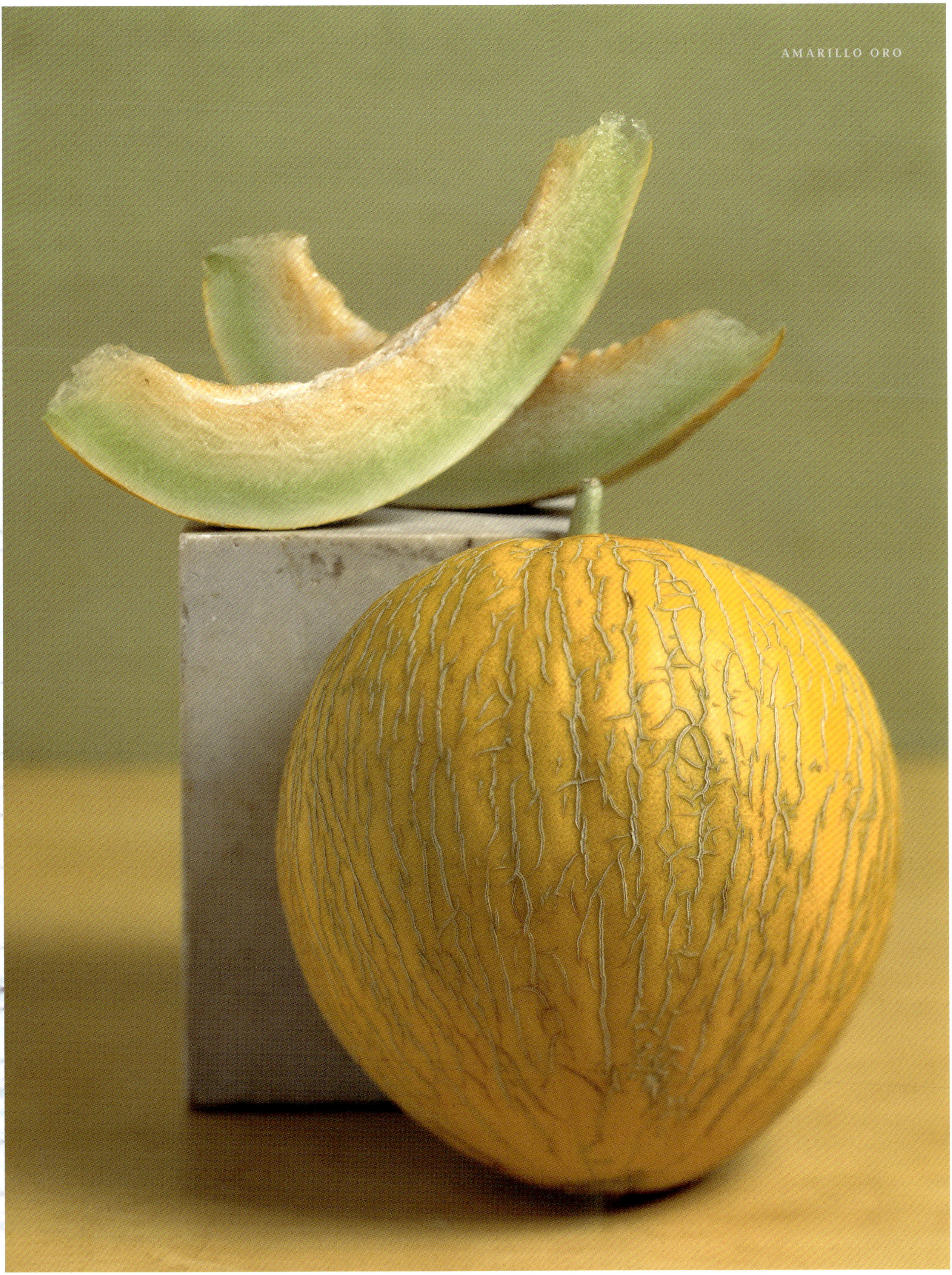

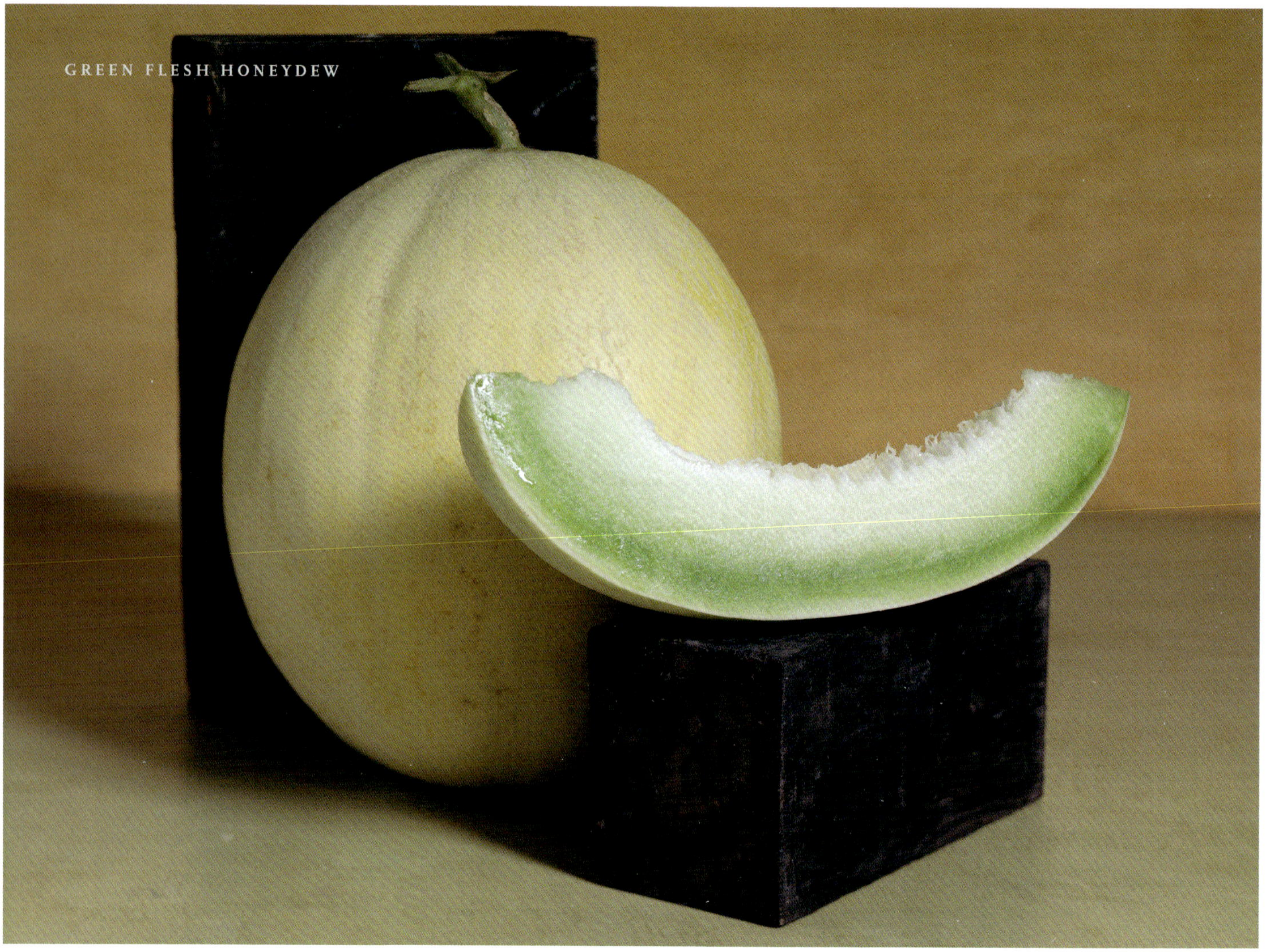
GREEN FLESH HONEYDEW

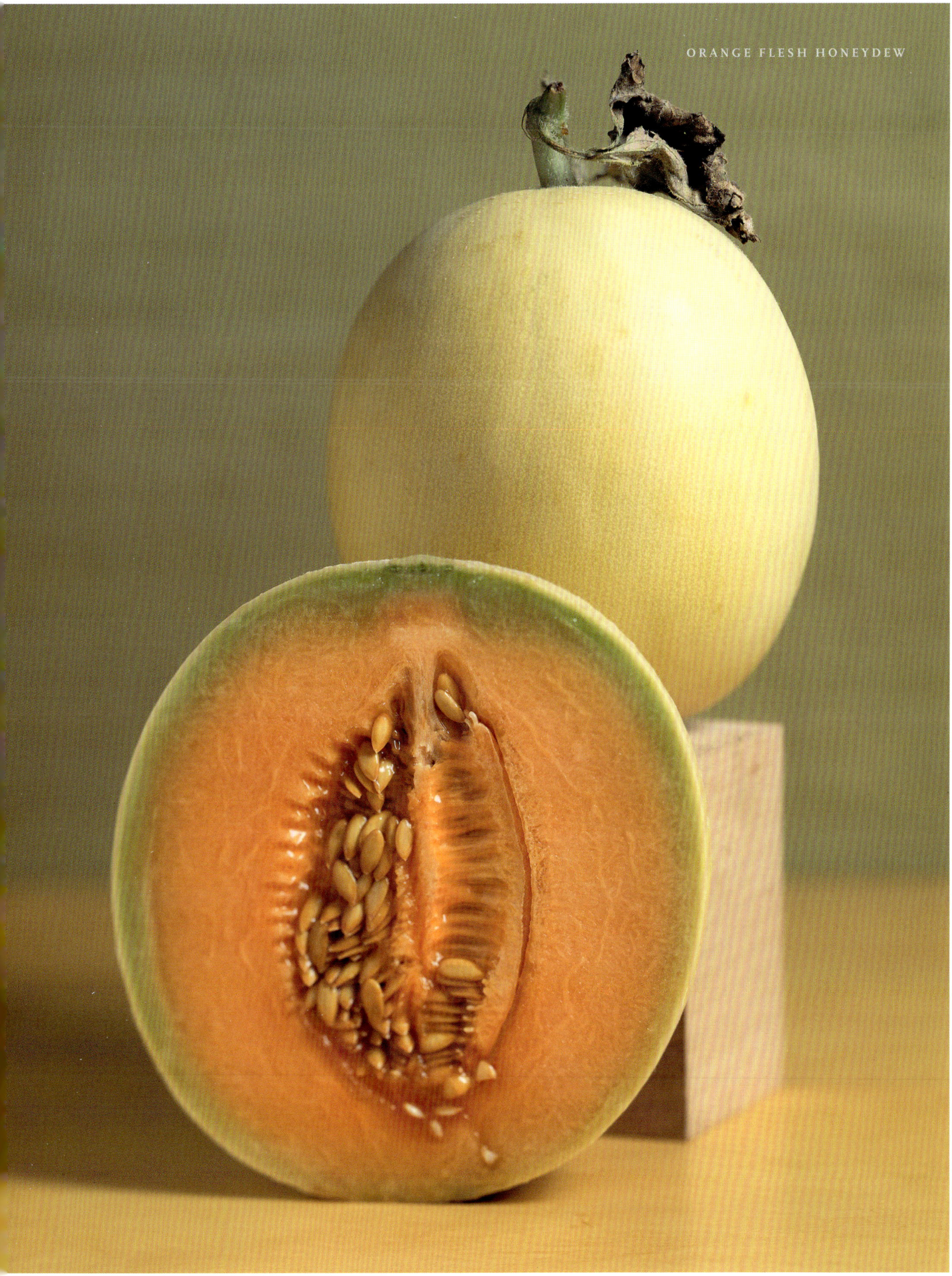

York at the Cornell University Agricultural Experiment Station (1895). He found White Antibes to be a very long keeper and worth growing for home use—yet not as good in quality as the familiar summer muskmelons. He advised harvesting it before frost when fruits are not yet fully ripe and storing them in a cool and dry fruit-room.

We pick up the story about fifteen years later when a White Antibes melon somehow found its way onto the menu in 1911 at a fashionable New York City hotel. A dinner guest was so besotted that he saved the seeds and sent them to Melodew breeder John E. Gauger in Swink, Colorado, to grow. The melon was identified as White Antibes Winter by Dr. Shoemaker of the USDA, and in 1915 Mr. Gauger aptly renamed it Honeydew.

As written by Mr. Gauger in 1917 in the H. D. Garwood and C. S. Woodside catalogue, "The original seed of this new melon was secured from a melon shipped from Africa to New York City. This seed was planted at Rocky Ford, Colo., and crossed with the genuine, original 'Rocky Ford' cantaloupe by pollenization. The following season this hybrid seed was again crossed with the Improved Hybrid Casaba. . . . I named the melon Honey Dew for several reasons: It is as sweet as honey and has the mingling of several flavors—pineapple, banana, and vanilla predominating."

Honeydew quickly became America's most popular winter melon. The original Honeydew is green fleshed, but the Orange's star is rising. We've all tasted Honeydew that did not live up to its potential. For a fruit of this caliber, that's a shame. The fault lies not with the melon but with where and how it is grown and handled in storage. When cultivated by the home gardener and left on the vine to catch the last surge of sucrose in a good season, harvested when the skin has turned from light-green to creamy, then the melon can truly earn its name.

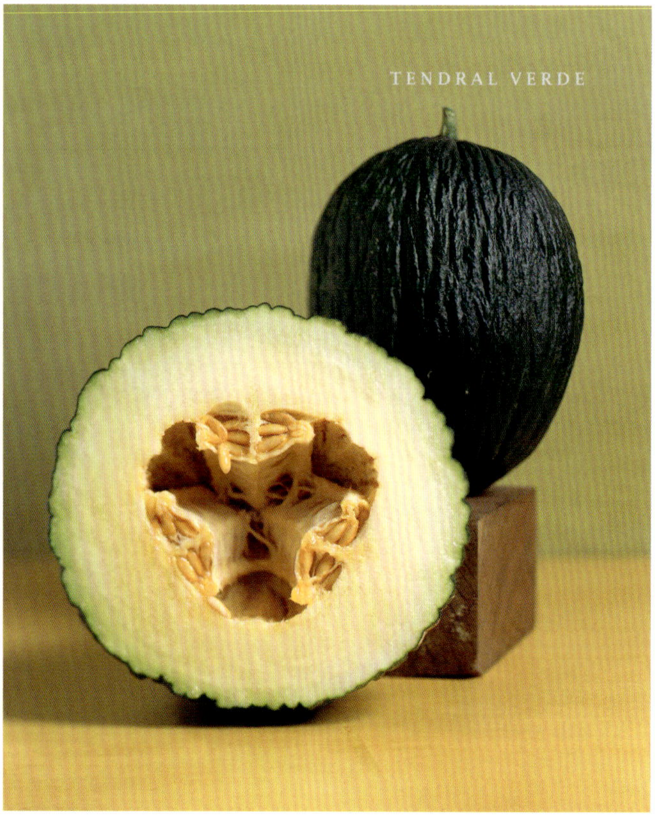

TENDRAL VERDE

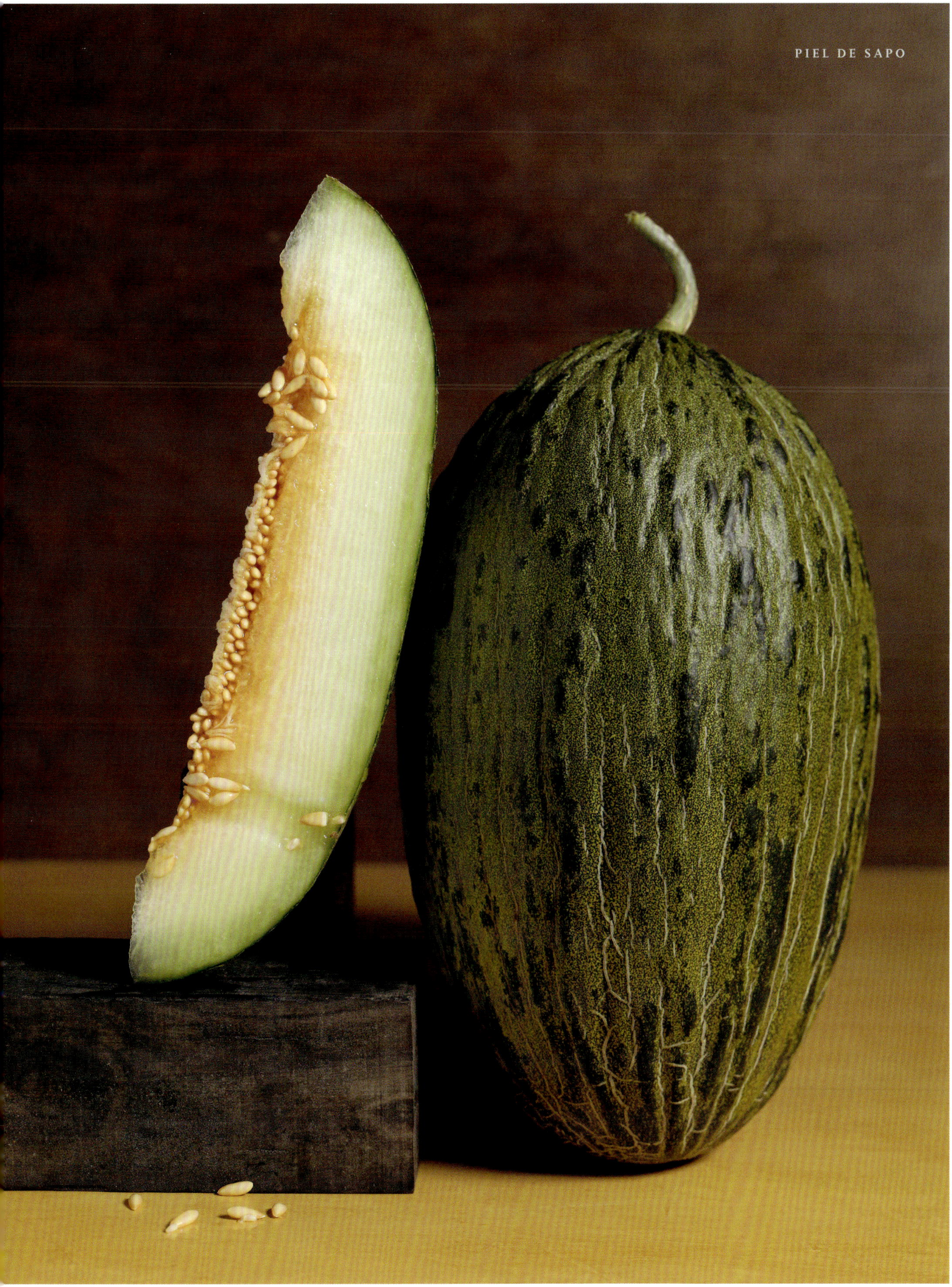

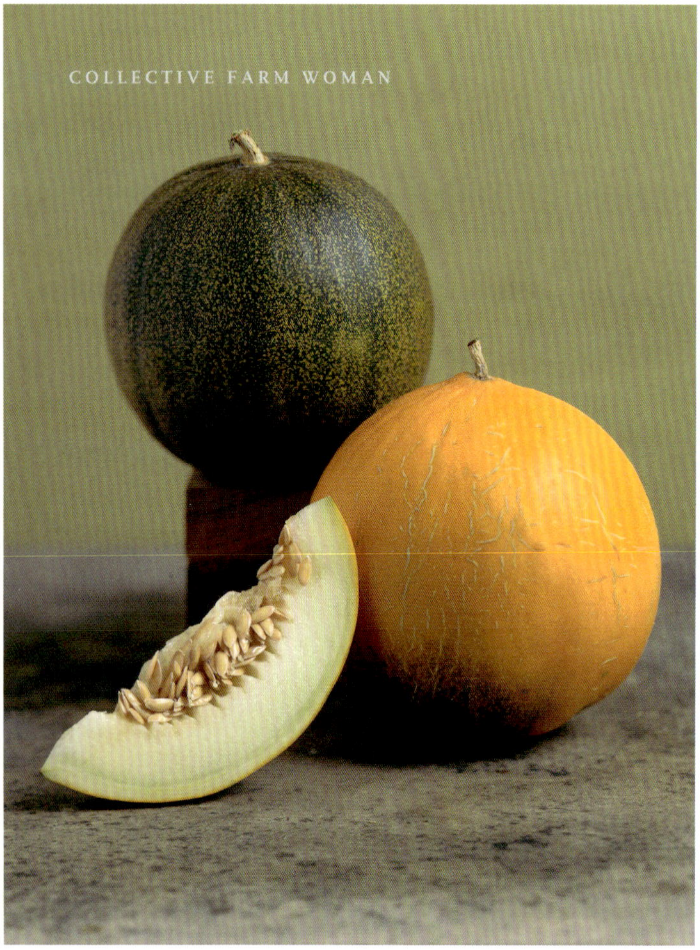
COLLECTIVE FARM WOMAN

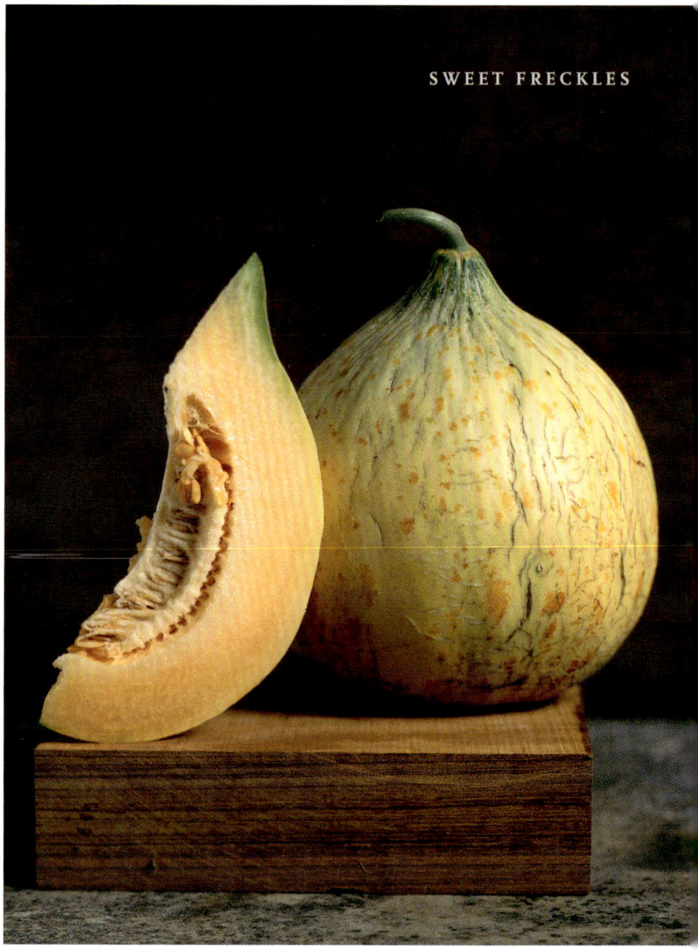
SWEET FRECKLES

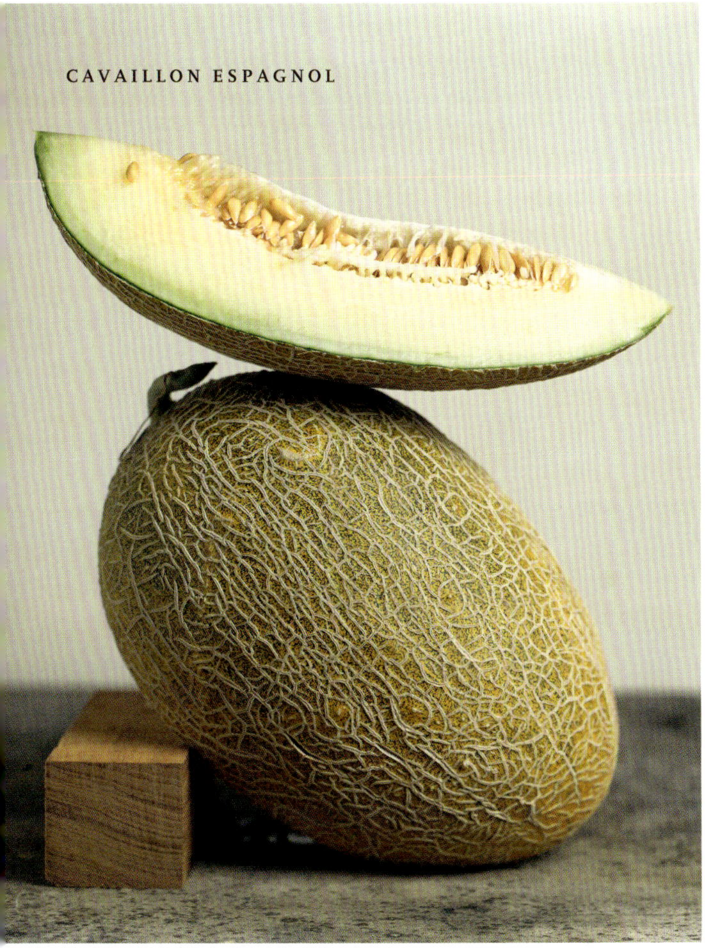
CAVAILLON ESPAGNOL

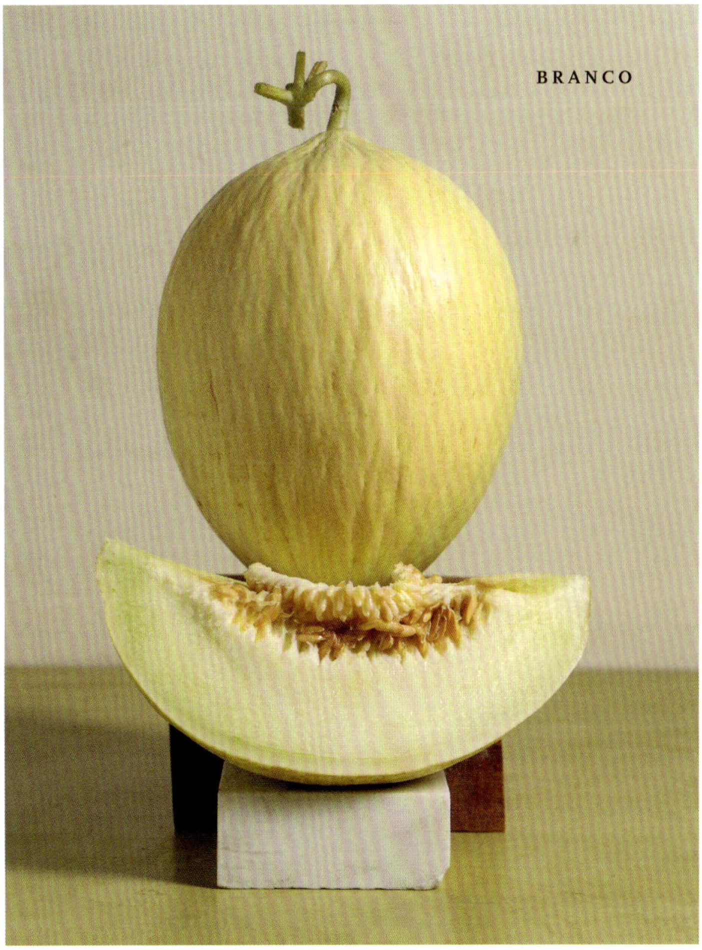
BRANCO

Crane

The Crane melon speaks volumes about the land and people who fostered it. For over one hundred fifty years, six generations of Cranes have farmed the black adobe soil of Santa Rosa, California. Oliver Crane developed the melon in the 1920s; his descendants are dedicated to growing and preserving this heirloom. The family has resisted development pressures on what is now prime Sonoma County real estate and set aside one hundred acres for growing the melon. The Crane family apparently agree with their old friend and neighbor, Luther Burbank, who said, "This is the chosen spot of all this earth as far as nature is concerned."

Oliver Crane hoped to produce "the perfect melon," according to his great-grandson Richard, by crossing a number of favorite varieties. Although Oliver left no records, one of the varieties might well have been the Japanese melon. Popular in the local markets of California at the time, this melon was similar in many respects to the Crane, and was itself an apparent cross of Persian and Golden Beauty Casaba. Written on the face of the Crane are the netting of the Persian, the blotches of the Japanese, and the neckiness of the Casaba.

The Crane is a rare commodity in a world filled with colorless, odorless, and tasteless melon. It is everything the mass market will not allow. Fragile and short-lived, it's bred to taste good. "The sad part today is, most people don't know how good a melon can be," Mr. Crane says. The locals in Sonoma know. They come from miles around to shop at the Crane Melon Barn on Petaluma Hill Road. We should all be so lucky.

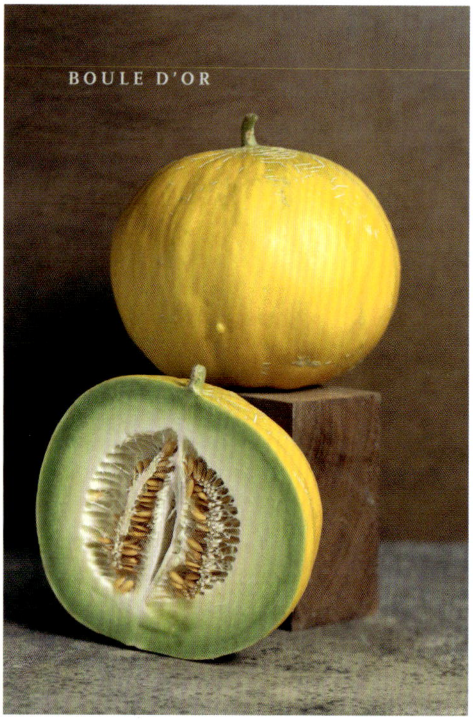

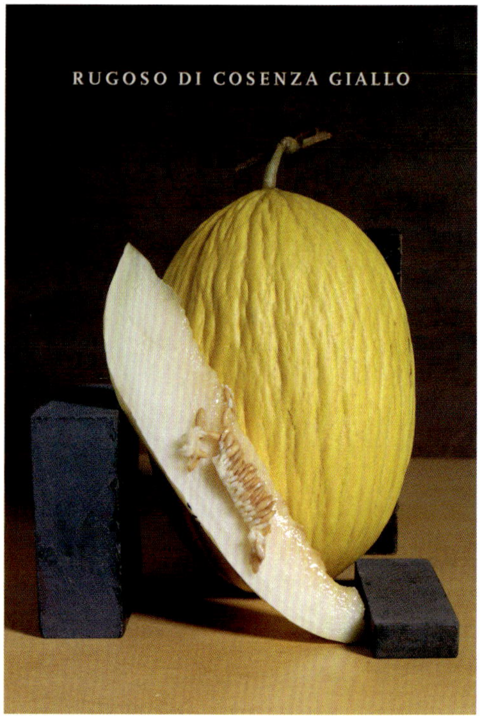

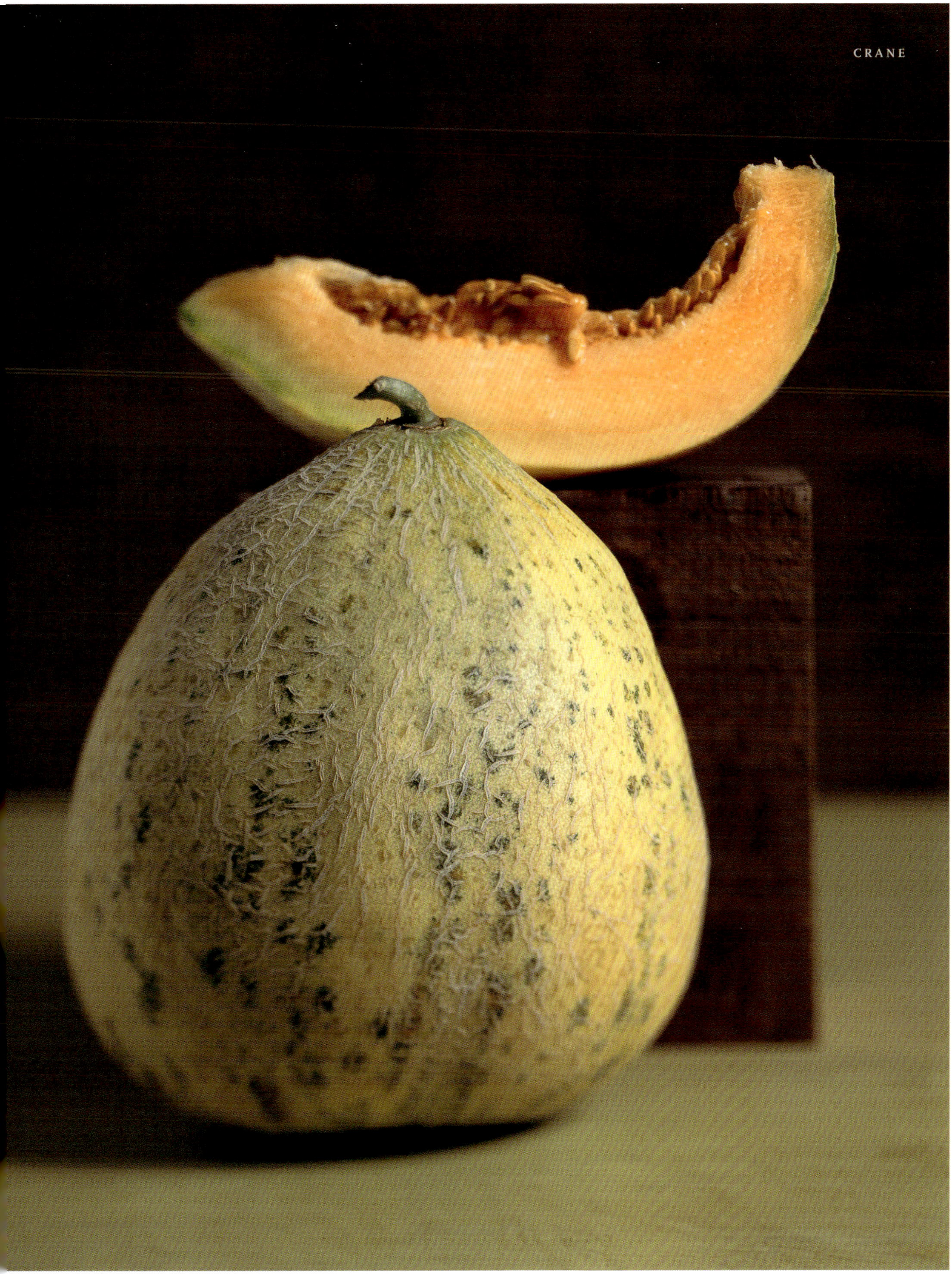

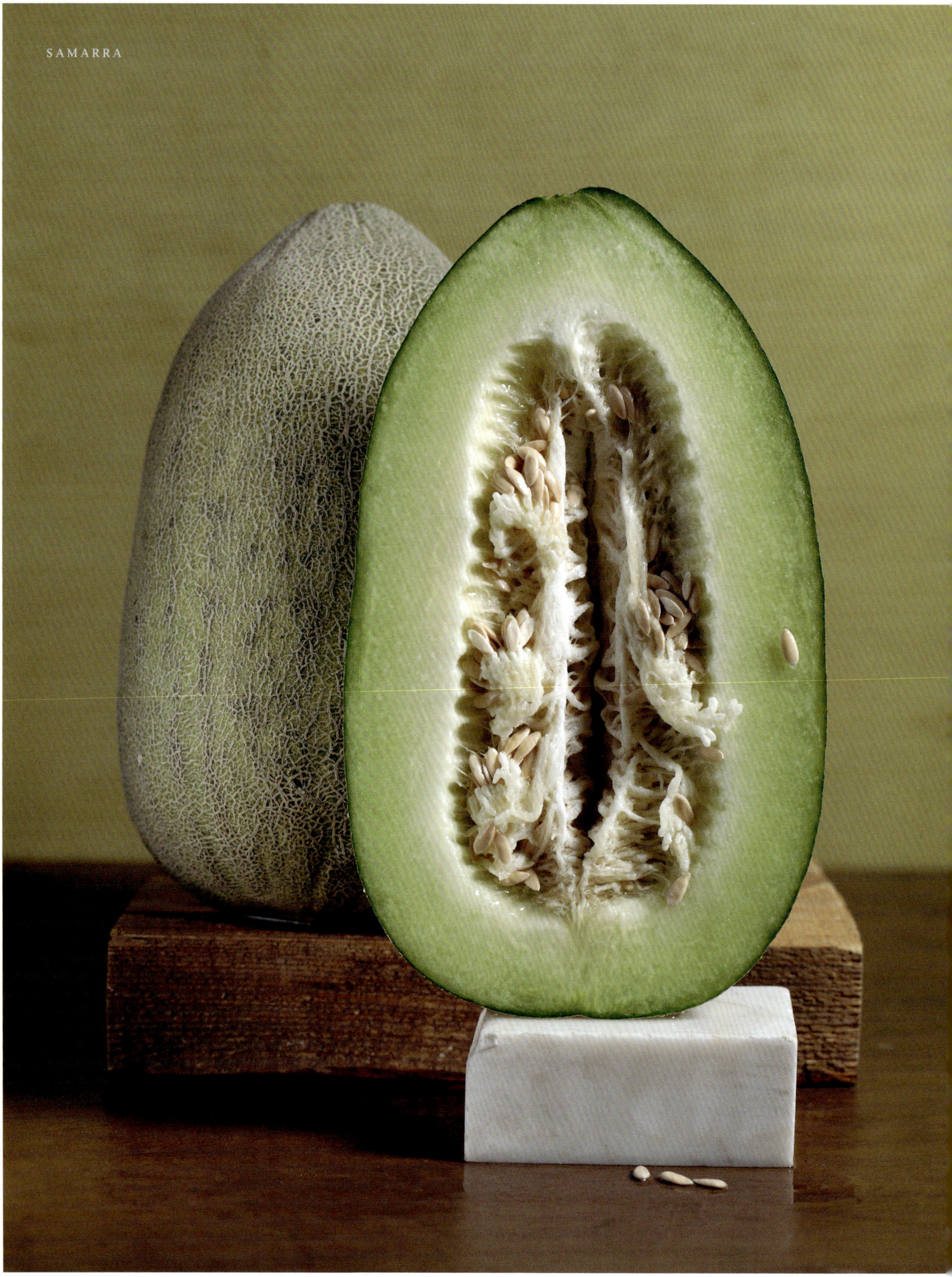

Ameri Group

AMERI MELON Ameri melons thrive in farmers' fields in fabled Khorasan—the cradle in which sweet melons were born, and home to the greatest sweet melon diversity today. (A visit is definitely on my wish list.) But they are not cultivated much in the United States. The next best thing to being there is immersing oneself in the illustrated pages of *Melons of Uzbekistan*. According to Ravza Mavlyanova, of the Uzbek Research Institute of Plant Industry, and her co-authors, Ameri melons account for about half of the diversity featured in that book. On display are twenty-six different Ameri types, making for an otherworldly parade of large, elongated oval to spindle-shaped melons variously adorned with speckles, spots, streaks, stripes, and "ribbons" (decorative vein tracts). Turkmen and Uzbek sorts are very similar—the countries are, after all, neighbors.

Ameri melons are adapted to long, hot, dry summers, and fare poorly in temperate climates with cooler, wetter weather. That would rule out most places in the contiguous forty-eight states. Jim McCreight, who works with the USDA–ARS Crop Improvement and Protection Research Unit in Salinas, California, believes that Ameri and Inodorus melons from Turkmenistan hold promise for local and regional farmers in California and Arizona. He has identified potential candidates from among the many Turkmen farmer- and breeder-selected melons his group collected in 2008 as part of a collaborative Turkmenistan–United States germplasm expedition.

Back at the farm, in upstate New York (Zone 5b), I have succeeded in growing five Ameri varieties: Ananas, Altaiskaya, Ananas Yoqne'am, Hami, and Samarra. Reduced irrigation seems to have helped. But when it rains on their parade during ripening, those melons—with the exception of Ananas—become sloshy or "juggy" and unappetizing. Four Ameri sub-groups have recently been proposed based upon skin color and presence of vein tracts (Pitrat 2017). Three of the varieties featured here fit well within the new Ananas sub-group since they sport light-orange, regularly and finely netted skin without vein tracts. They are plain vanilla (unadorned)—but that's my favorite flavor. Melting and sweet in hot, dry summers. Samarra and Hami are best placed in the Maculati sub-group. The other proposed sub-groups apparently account for the remaining variability: Bargi and Mashhadi.

Opposite: Perhaps centuries old, this green-fleshed melon landrace is dryland farmed in the vicinity of the ancient city of Samarra, Iraq. Samarra resembles many Ameri melons grown in Turkmenistan and Uzbekistan today.

Ananas Yoqne'am

The story of Ananas Yoqne'am, as relayed by Harry Paris, Ron Cohen, and their fellow researchers (2013), is a prime example of how melons evolved under cultivation. The first traces go as far back as the thirteenth and fourteenth centuries, to the sweet, yellow melons of Akko and Nablus (now a town in western Galilee, and a city in the territory of the Palestinian Authority, respectively), known in Arabic writings as *battikh asfar*. Fast-forward to the twentieth century and to the likely direct descendant: a melon landrace grown by Arab farmers and known by the names *shammam*, *battikh asfar*, or else Baladi (Of the Land). It is traditionally dryland farmed by seeding in early spring in prepared fallows made moist from winter rains, and grown without additional irrigation.

Enter Natan Rabinovich (1906–1990), a Jewish immigrant to the land of Israel (native to Romania), who set out to improve upon success. He was a nurseryman schooled in agriculture, with a deep love for the fruits of the vine. His raw material was the local *shammam* or melon, which was highly variable in almost every respect. (The original *shammam* can occasionally still be found today. It is often used when small and immature, like a cucumber, and in the same ways.) In short, Rabinovich "cleaned up" the local landrace over time by mass selection for better quality and uniformity. By around 1950, he was producing the new cultivar at his home, Yoqne'am, a *moshava* or private farming community in the Yizre'el (Jezreel) Valley, and shipping to local markets. What would eventually become known as Ananas Yoqne'am was immediately met with rave reviews—and clamoring for more. The flesh was described as "cream-colored, juicy and very sweet, with superb taste and aroma." Natan had to zealously guard his seed stock from pilfering.

He formed a beautiful partnership with the neighboring collective farm community, kibbutz Hazorea, that would enable large-scale dryland melon production, and endure for twenty-five years. Zevi Dor (nicknamed Dola) coordinated melon growing and provided commercial and marketing savvy. Natan took care of the professional side, including the actual growing of the crop to maximize yield and fruit quality and produce uniform seed stock for future plantings. Yohanan Ma'oz, who briefly helped run melon production later on, says Dola and Natan were the dream team. And Natan had a wry sense of humor. Yohanan recalls they even placed a special sticker on their melons symbolizing the two spies sent by Moses, but bearing a melon instead of a cluster of grapes.

The Ananas Yoqne'am is the direct ancestor of perhaps all of the widely commercialized Ananas-type melons grown today. Breeders continue to improve the melon's table quality and disease resistance. The slew of Ananas Yoqne'am descendants comprise an established international market type. Some of them are drop-dead sensational.

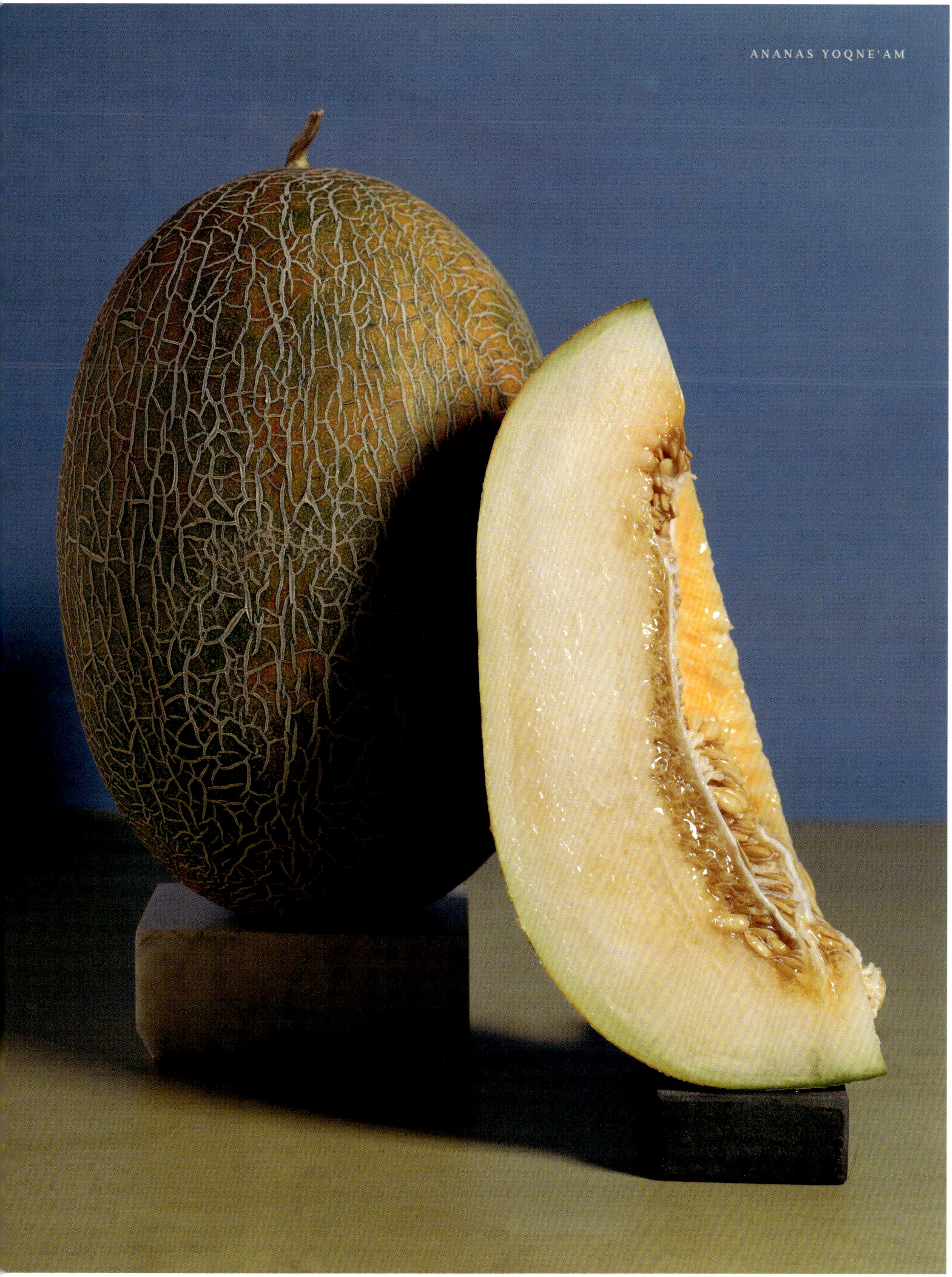

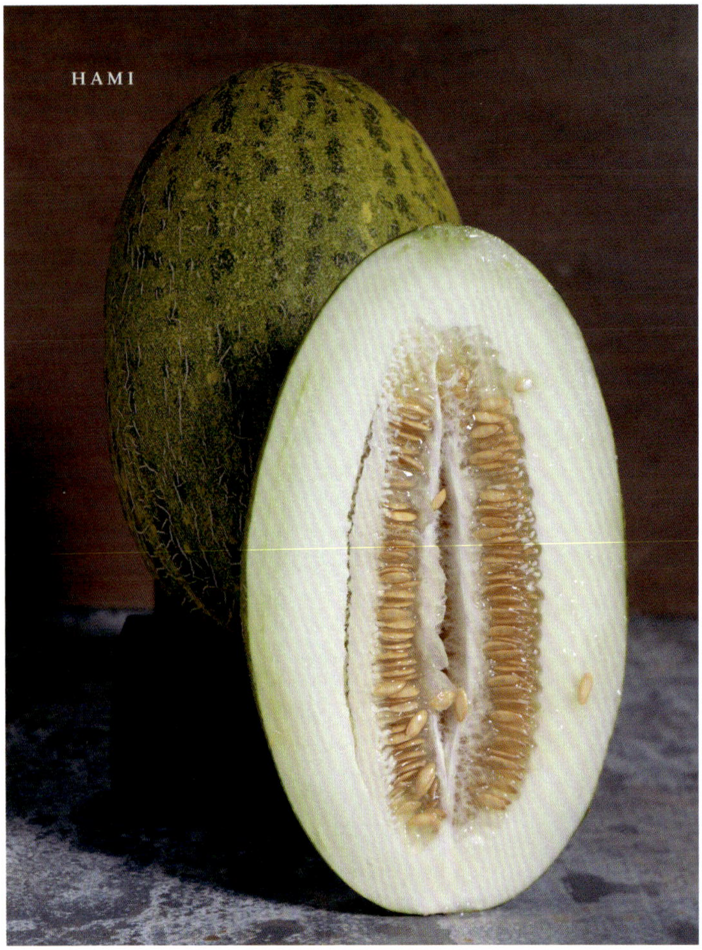
HAMI

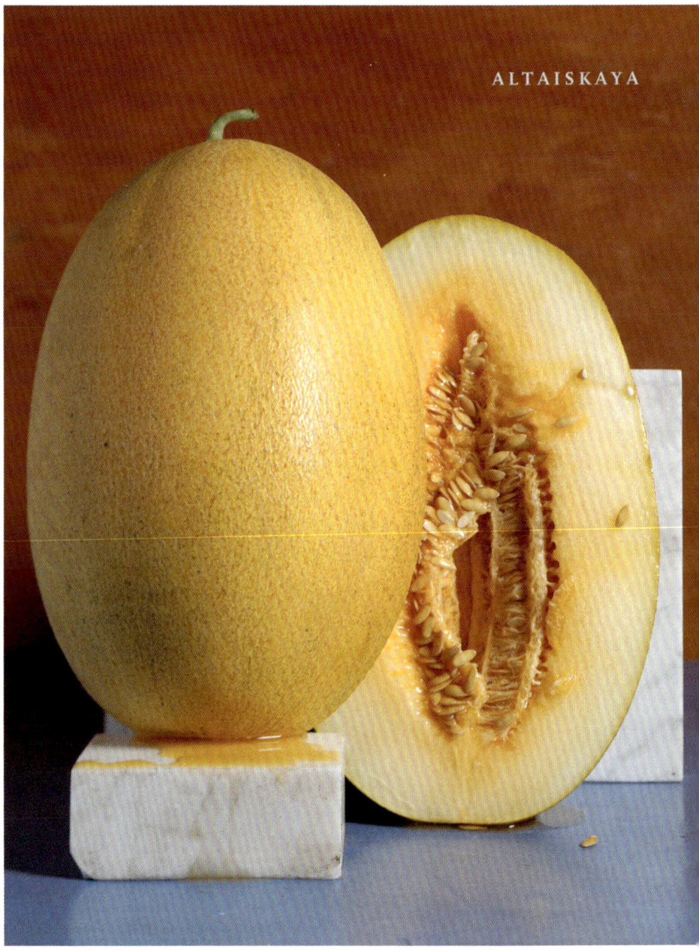
ALTAISKAYA

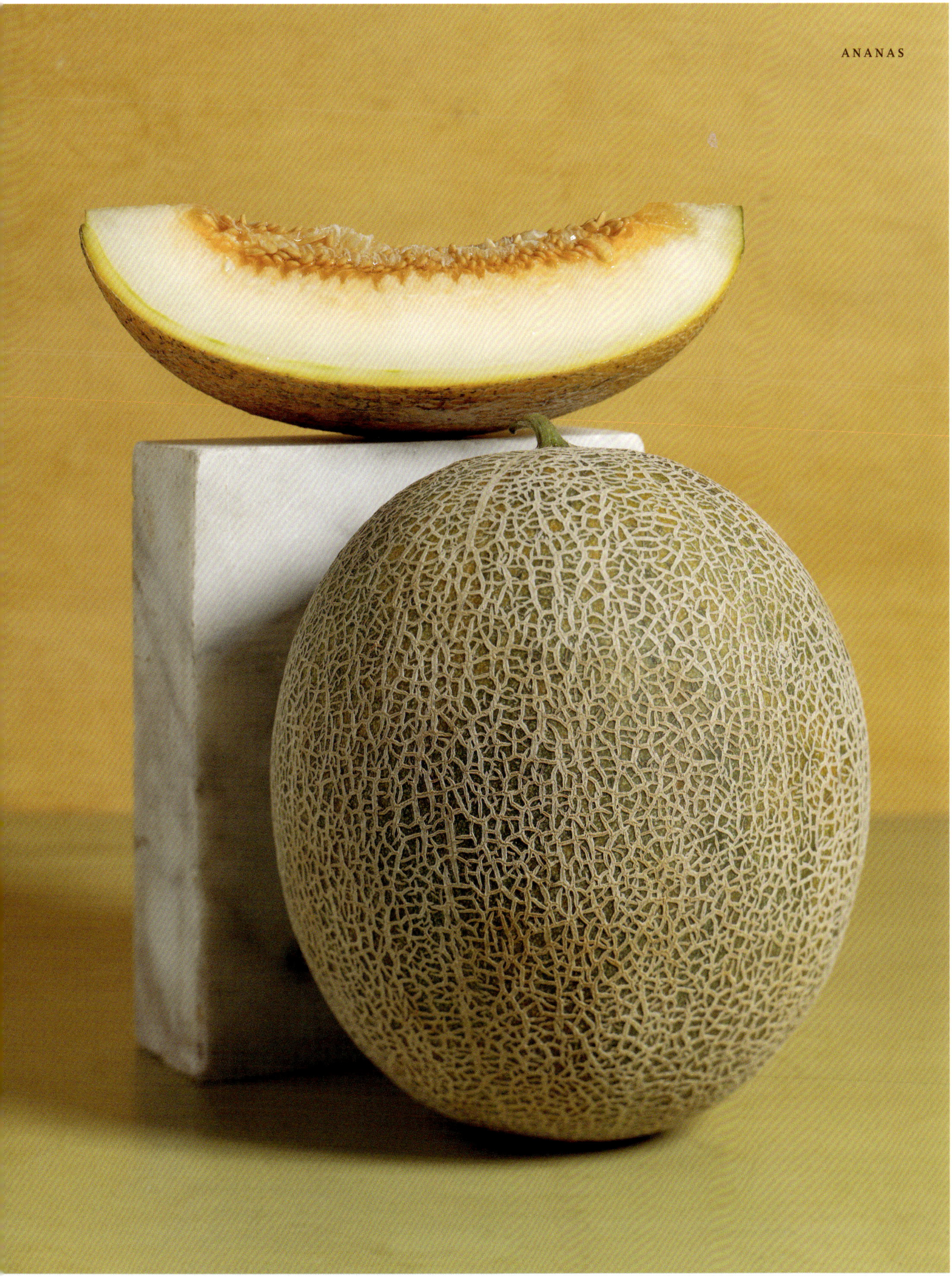

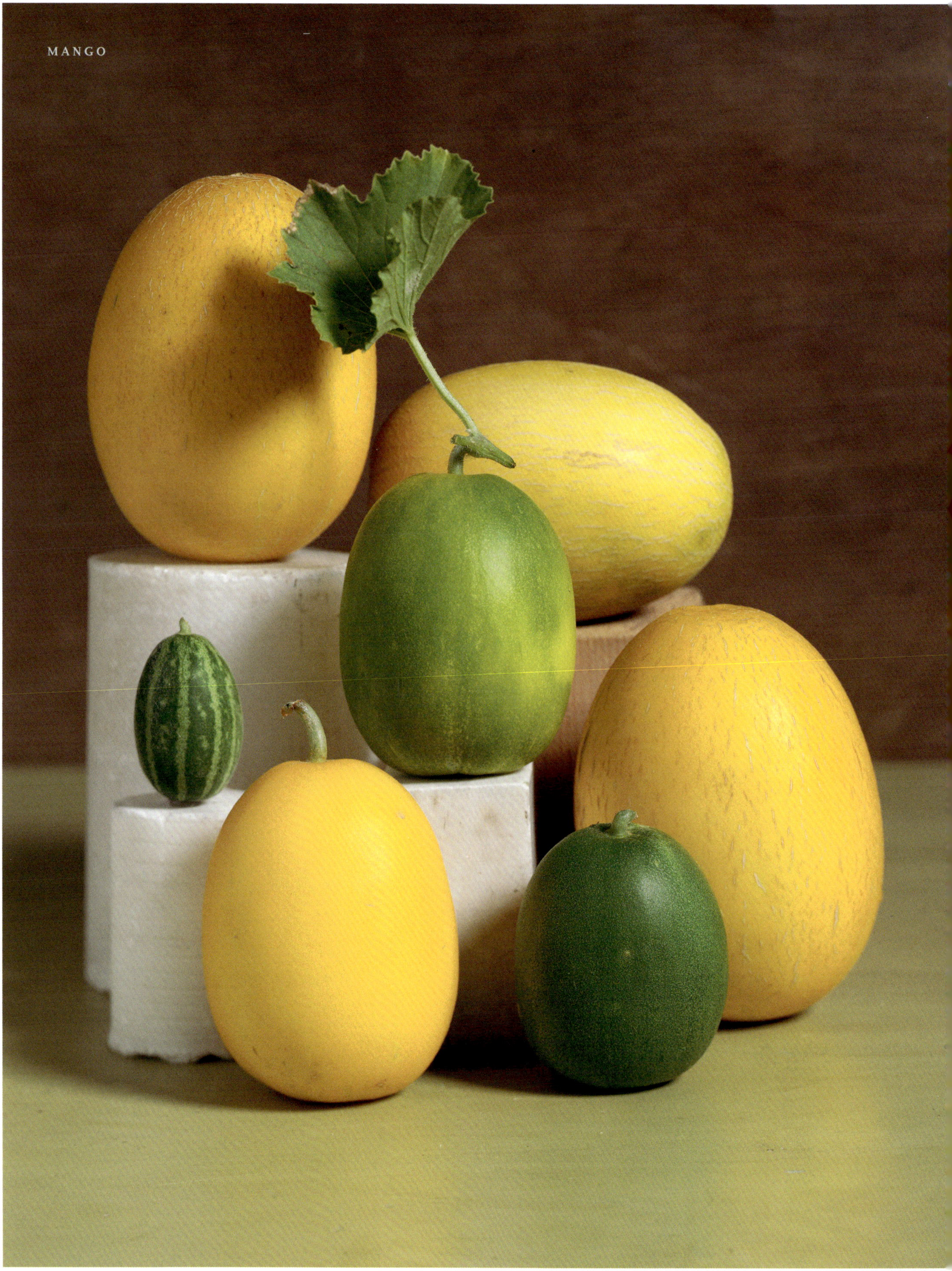

Chito Group

CHITO OR MANGO MELON Best known in the trade as Mango melon, Chito's many synonyms are a testament to its popularity, which came and went in the latter half of the nineteenth century: Garden Lemon, Melon Apple, Vine Peach, Glass Melon, Lemon Cucumber, Melon Peach, Melon de Quito, Melon de Grenade, Chito de la Havane, Orange Melon, Vine Orange, and Vegetable Orange.

The Chito melon's biggest and best fan was Charles Morren, a celebrated Belgian botanist (1807–1858), director of the Botanical Garden at the University of Liège, and producer of *La Belgique Horticole*. In September 1848, M. Dutrieu de Terdonck, senator from Malines, steered Morren to the horticultural novelty he had grown at an agricultural exposition in Brussels. Dutrieu had obtained Chito seeds from M. Hauwers, Malines town councilor, who got them originally from his son in Havana, Cuba. (Cubans at the time apparently enjoyed the melon, which was pronounced "kito," "cito," or "tchito," served with boiled meats and sometimes as a dessert fruit with sugar.) At the exposition, Morren was smitten by the fruit's delicate, mouthwatering perfume—which smells like flowers and cinnamon to me. After sampling Chito at a dinner party, he proclaimed it exquisite—when doctored a bit. The best way to eat the melon? Lop the top off (as one would a soft-boiled egg), remove seeds but retain juice (a tricky maneuver since seeds are encased in a gelatinous sheath), macerate the pulp with sugar and champagne, then savor it with a spoon.

In 1849, Morren described and named the species, Cucumis Chito, in *Annales de la société royale d'agriculture et de botanique de Gand*. Two years later he sang the praises of Le Chito de la Havane in *La Belgique Horticole*. He donated free seed grown by his son, Prosper Morren, to thousands of subscribers, exhorting them to spread the wealth. The Chito's popularity rose. The excitement caught on in the United States, but for more sober purposes than in Belgium—the making of sweet "mango pickles," pies, preserves, and ornaments for home decoration.

MANGO MELON
Chito melons are small, round to oval or short oblong in shape. Bland, not bitter, when young, they are climacteric, mildly sweet, and pleasantly fragrant at maturity. Their skin color changes from green to yellow, with or without netting or green striping. Flesh is firm and can be light green, white, or cream. Chito is prolific and slips from the vine when ripe. Although cultivated in some places, including my garden, for its ripe as well as unripe fruit, this melon is generally considered to be feral. It can be found free living in the Caribbean and Central America. Morphological and molecular data show it is not the same as the wild melon populations found in North America (sometimes known as Texanus group).

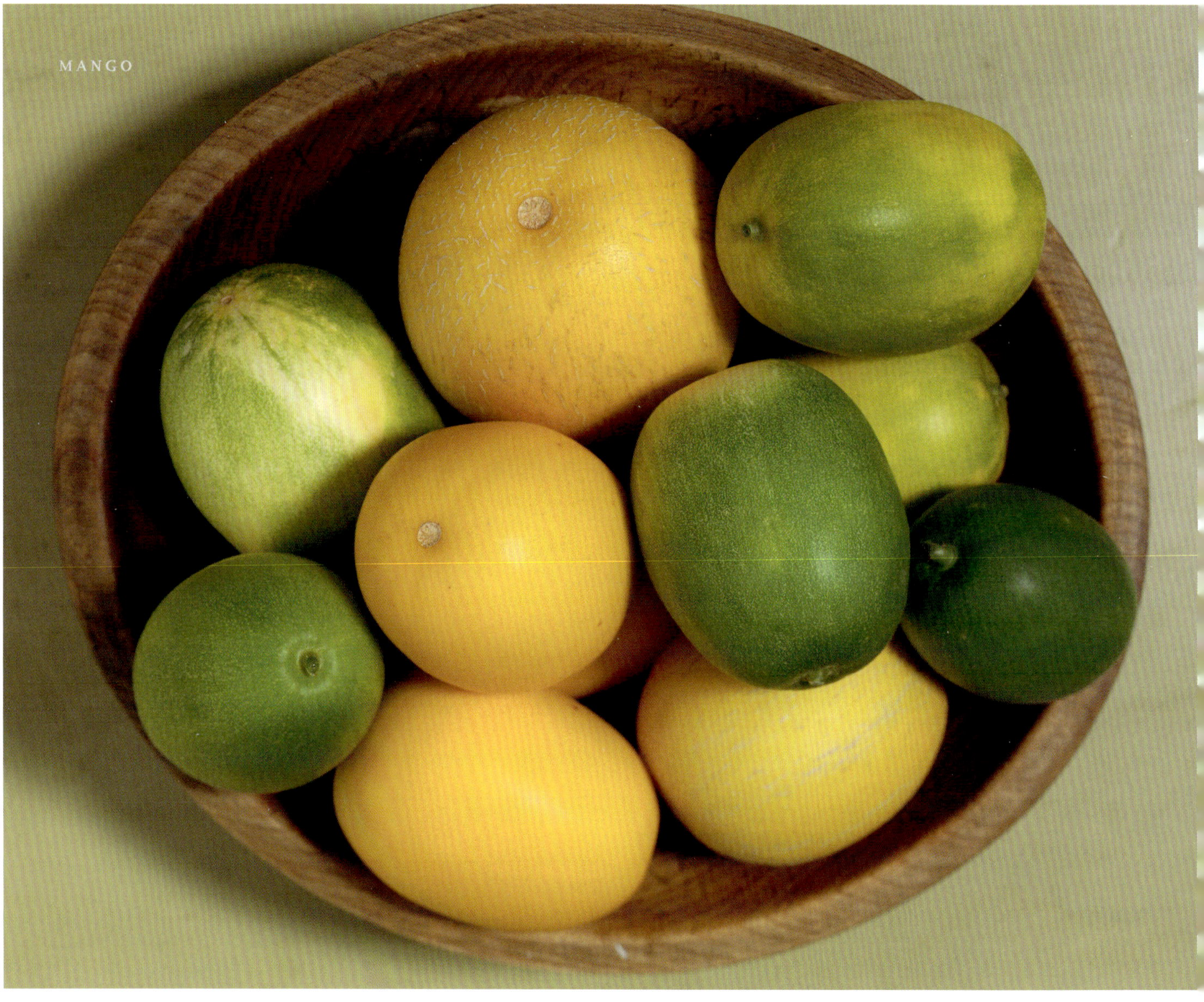
MANGO

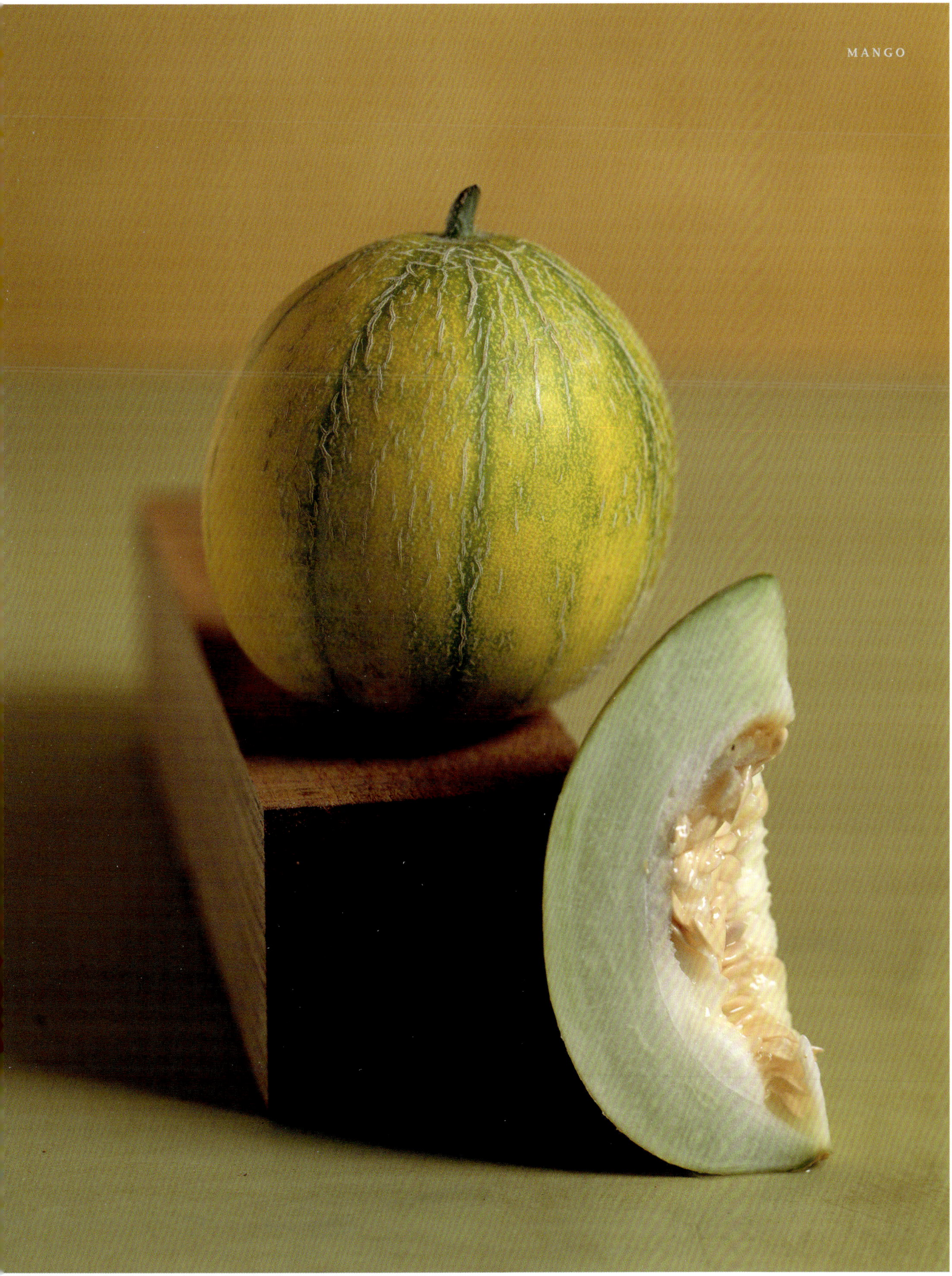

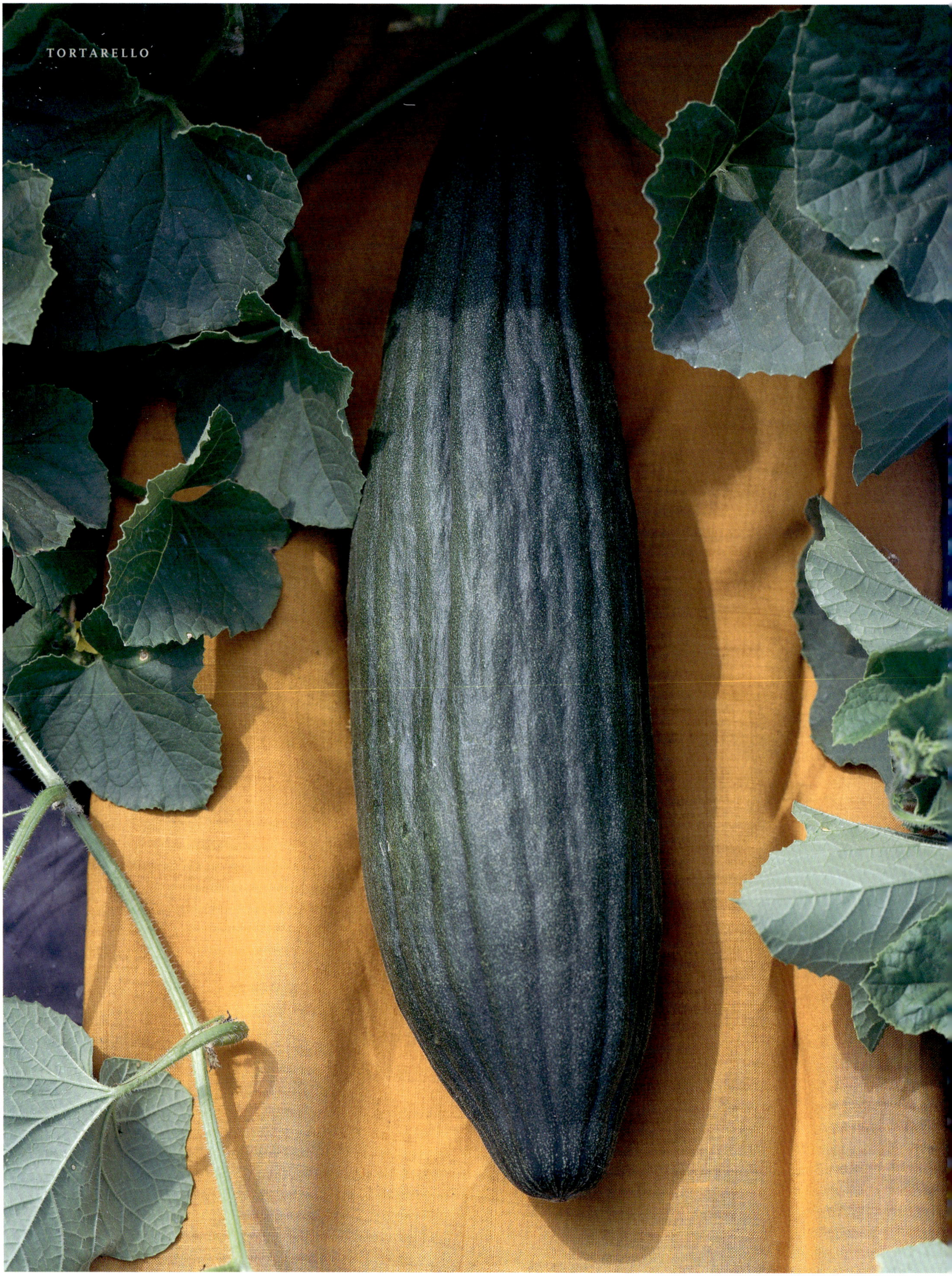

Flexuosus Group

SNAKE MELON The snake melon is often mistaken for a cucumber—and for good reason. It looks like one and tastes like one, but genetically it is, in fact, a melon. It crosses only with other melons and not cucumbers. Like other melons, Snakes are hairy, rather than spiny like cucumbers.

Flexuosus group melons are the longest fruited of all: Their length to greatest width ratio generally falls between 4:1 and 6:1. They have been cultivated and selected by farmers over the centuries in Mediterranean lands and India for long-fruitedness. Designed to be eaten immature when crisp, these non-sweet "vegetable melons" become insipid or sour at maturity. Snakes are longitudinally furrowed, wrinkled, or smooth. They come in three basic skin colors: light green, dark green, and striped light and dark green. They are well adapted to the Old World tropics and subtropics. Plants are monoecious, early, and prolific.

Snake melons were widely grown and appreciated in the Mediterranean region from antiquity through classical times. They are depicted in a Roman-era fresco at Herculanum (destroyed by the eruption of Vesuvius in 79 CE and later reclaimed), Roman mosaics in Tunisia, and early Byzantine mosaics in Israel and Lebanon.

Familiar during the medieval period, snake melons were eventually supplanted in Europe by cucumbers originally introduced after the Roman period, which are better adapted to temperate zones. In Arabic-speaking countries snake melons are still commonly found in markets and known as *faqqous*. They are largely unknown and under-appreciated in America except in home gardens. The next time you feel the urge to serve that childhood favorite, cucumber stuffed or topped with cream cheese, try making it with a snake melon instead.

Harry Paris, in his studies of crop history, has examined Semitic-language records and iconographic evidence, including images from ancient Egyptian wall paintings and sculptures, and mosaics from the Roman Empire. He confirmed that the five vegetables the Children of Israel longed for during their exodus from Egypt (Numbers 11:5) were snake melon, watermelon, leeks, onions, and garlics, in that order. Harry reports, "The sikyos of Greek, the cucumis of Latin, and the qishu'im of Hebrew, thought by many to be cucumber, Cucumis sativus, have now been identified as snake melons."

PIQQUS OF THE FAQQOUS

Like the Children of Israel of yore, I long for the qishu'im (snake melons), too. They have none of the bitterness or burping that cucumbers can impart; the flesh is drier, so melon pickles are crunchier. And when snake melons are cut in cross-section, the raised lobes confer beautiful scallops, making them even more desirable. I practice what Harry Paris preaches: the ancient art of piqqus (Mishnaic Hebrew for soft hairs or their removal) of the faqqous: rubbing off the soft hairs on the fruit surface of young snake melons before eating. This is best accomplished with a scrub sponge—no need to peel the skin.

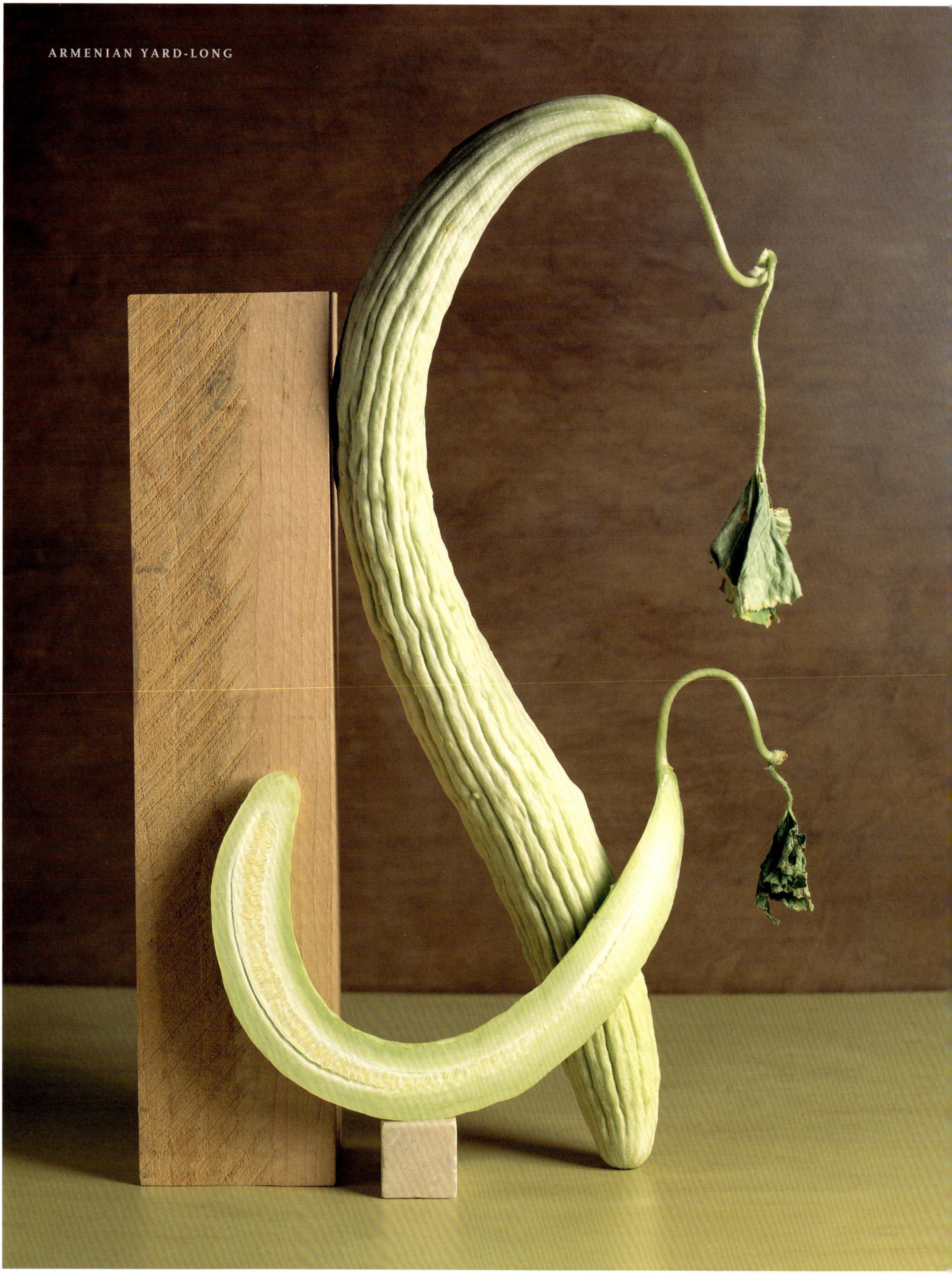

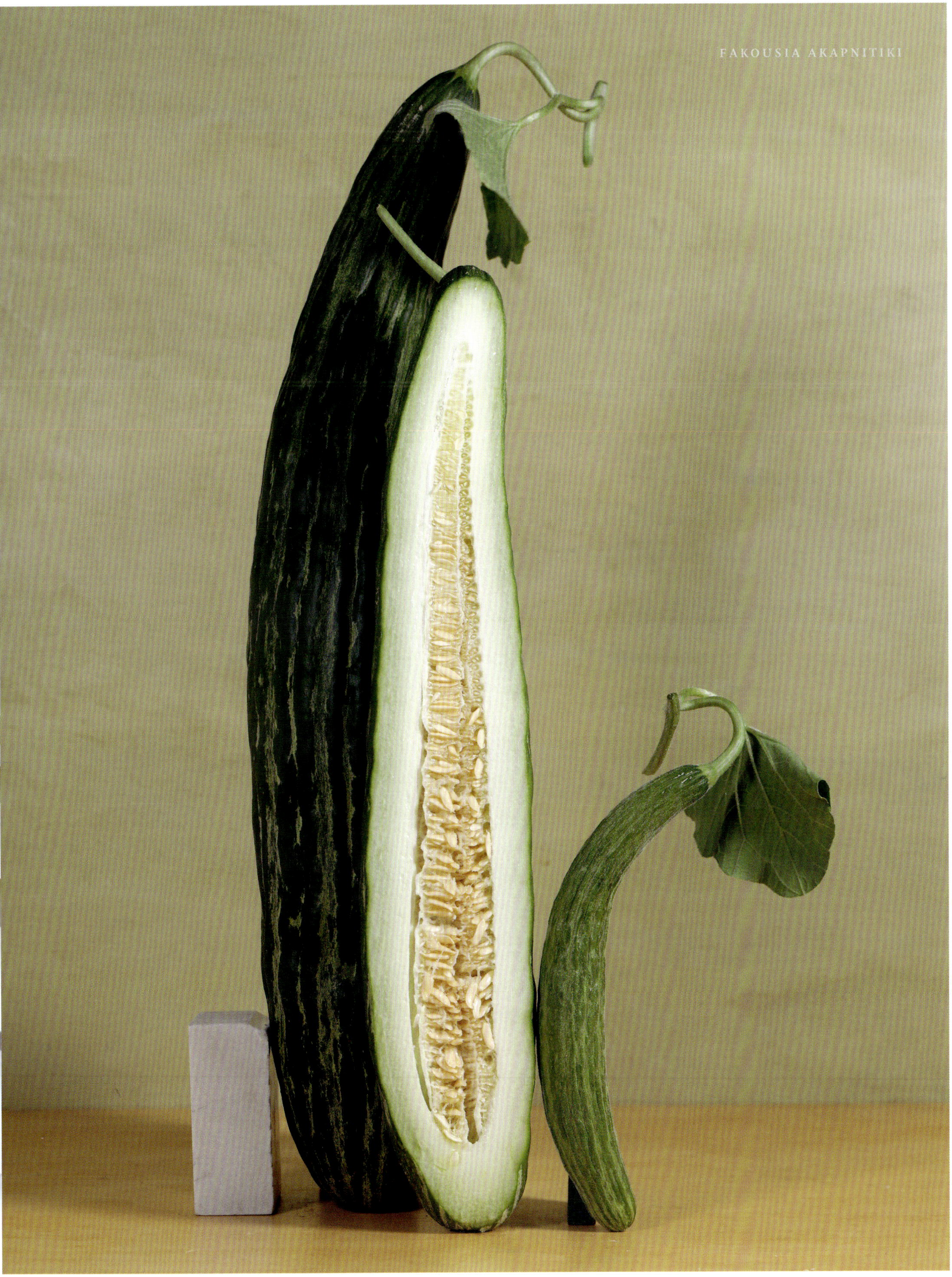
FAKOUSIA AKAPNITIKI

STRIPED ARMENIAN

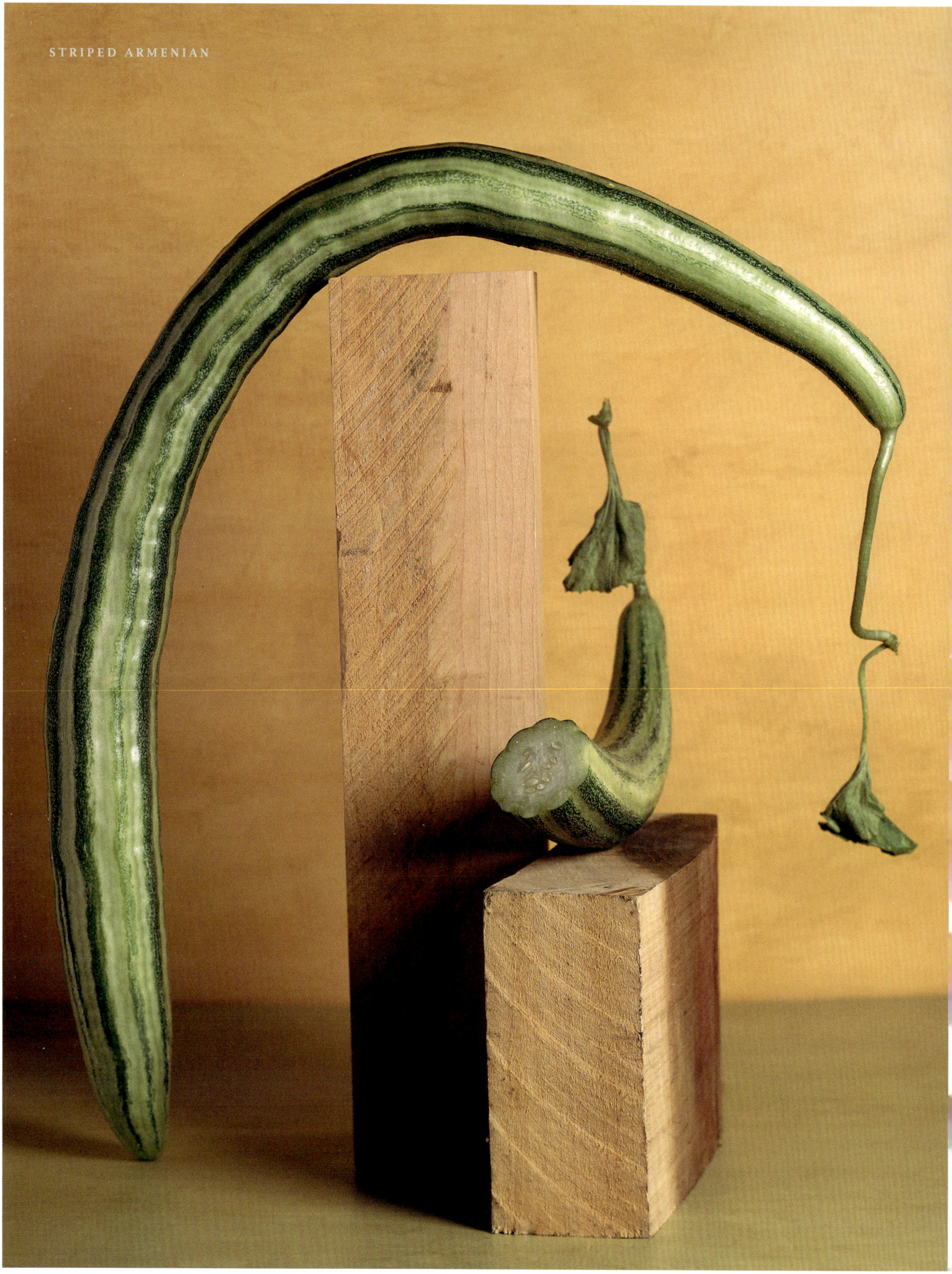

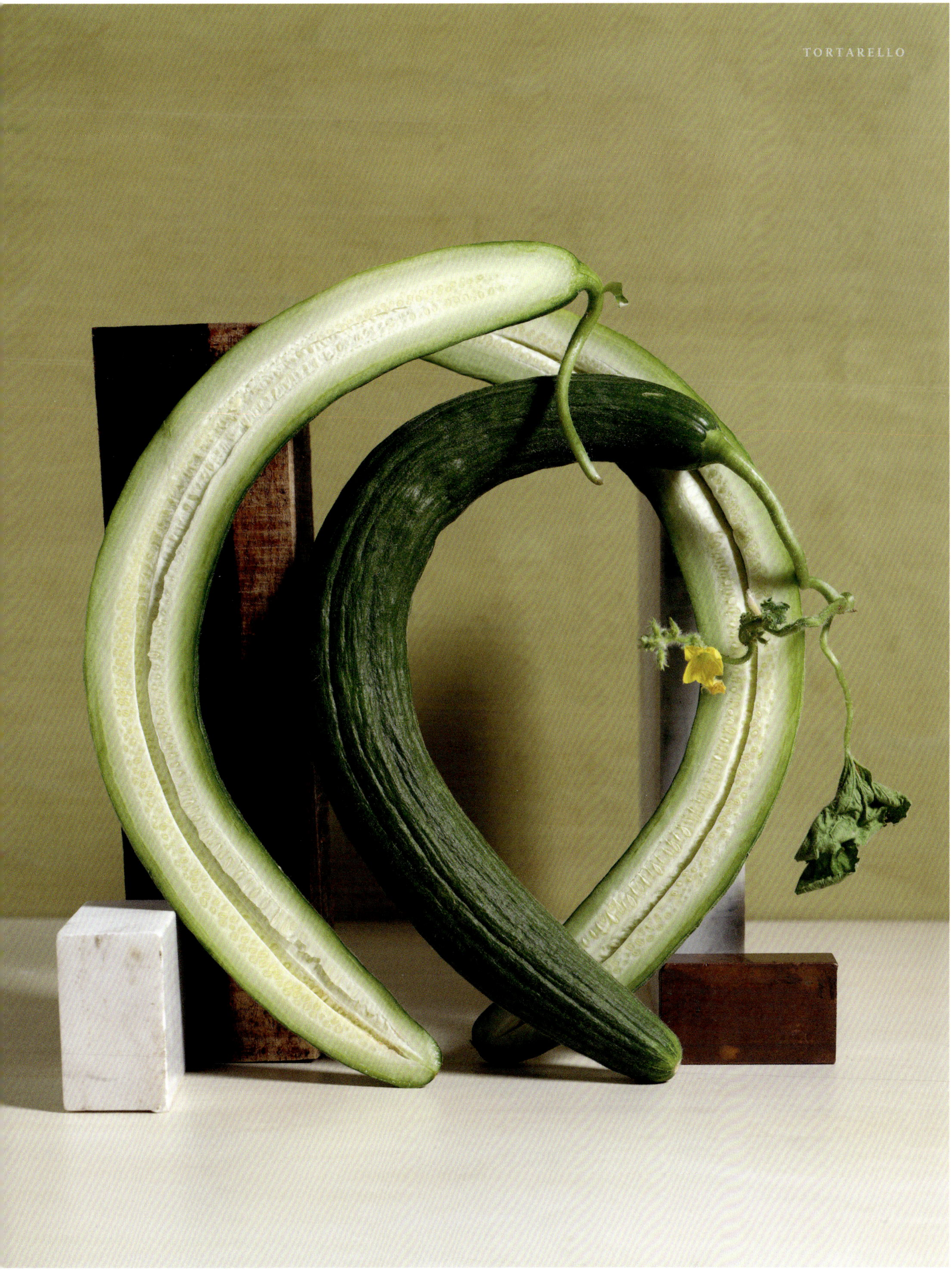

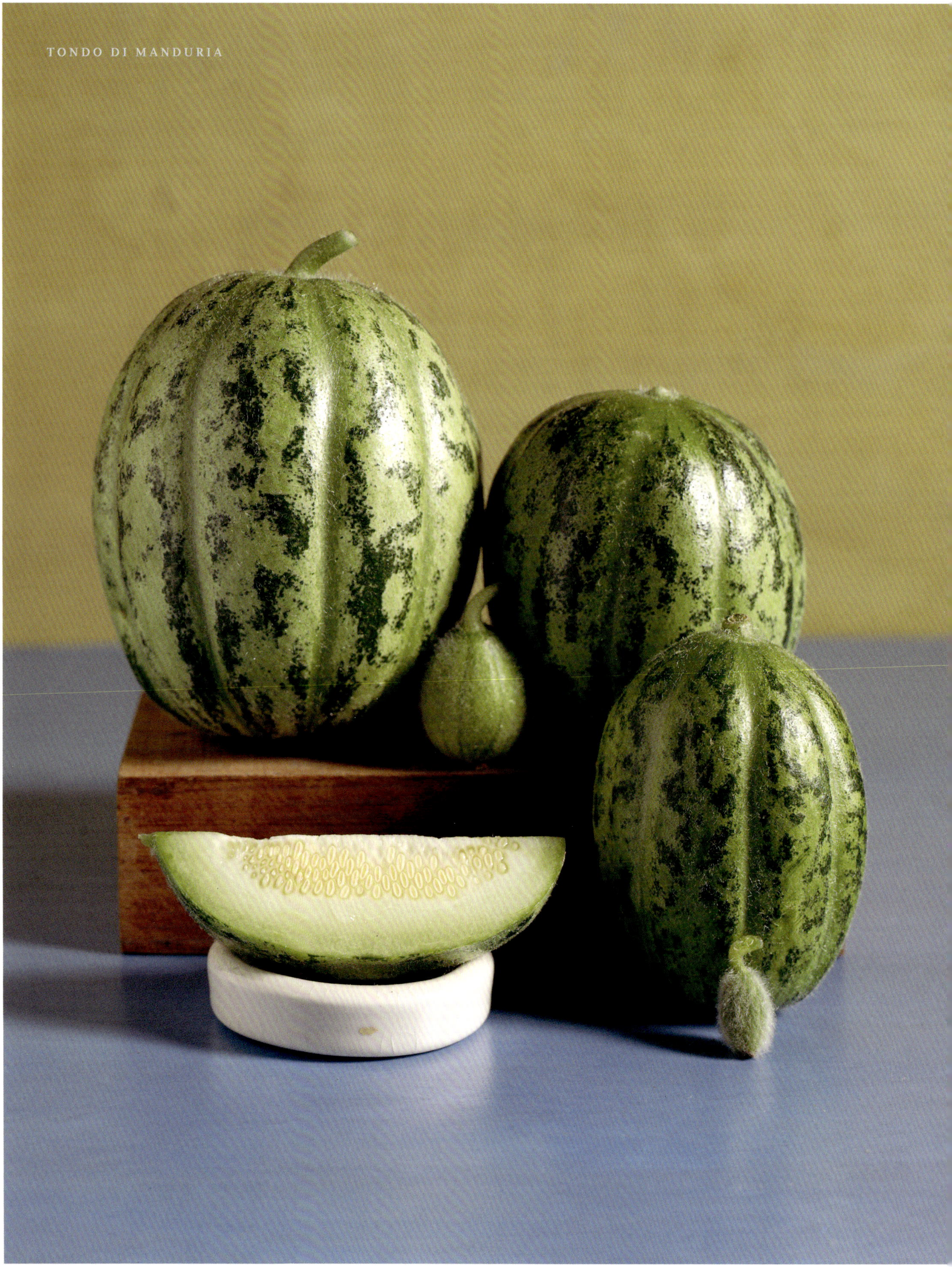

Adzhur Group

CAROSELLO MELON I came to the Adzhur or Chate group later in life—and have since been making up for lost time. Adzhur has been grown for ages in Mediterranean lands and appreciated for its immature bland fruit. Melons like these were illustrated in Egyptian wall paintings over three thousand years ago and depicted in a Byzantine floor mosaic from Israel (circa 410–600 CE). Beautiful, original, and highly detailed illustrations of Adzhur and Flexuosus vegetable melon plants and fruit appear in a number of manuscripts of Italian origin (including what is known as the Manfred Manuscript and the Carrara herbal), dating to late Medieval times (circa 1330–1400 CE). By then, cucumber (*Cucumis sativus*) had made major inroads, and would eventually displace vegetable melons in most of Europe. Six hundred years later, Adzhur melon, in the form of Carosello landraces (local cultivars), survives as a relict crop in sunny and dry southern Italy, in and around Puglia—the heel of the boot. In recognition of its long history of cultivation and contribution to Italian cuisine, Slow Food Italy recently honored Carosello by boarding it onto its Ark of Taste.

Carosello is generally elongate (sometimes oval or almost round) with a length-to-greatest-width ratio of less than 4:1. (You might think of it as a shorter and stubbier Flexuosus melon.) Most often the green skin is distinctly lobed and furrowed rather than smooth. Turn to Carosello if you are looking for a cucumber substitute in or out of the garden and are attracted to novelties. The varieties featured here have a "birdnest" growth habit or phenotype. Early in the season or growth cycle, birdnest types produce an abundance of female blossoms, and bear a large concentrated fruit set, on vines with shorter spaces between nodes (they have the *si-2* gene for short internodes). You may want to sow successive crops or let the plants produce a second wave on vines with longer internodes. In my experience, the first immature carosellos are ready to eat sixty days from seeding (a month from transplant). Some varieties and landraces reportedly have a high degree of salt tolerance.

FUZZY WUZZY WOOLY

If you grow Carosello yourself—and I hope you do, since it is scarce in the marketplace—you are sure to be amazed by the fuzzy-wuzzy-wooly long and soft hairs that cover the ovary and primordial (youngest) fruits. Harvest fruit immature, about ten days after flowering, and enjoy it as a raw vegetable. Any remaining light fuzz can be gently rubbed off with a damp scrub sponge.

Carosello's light green or white flesh tastes like a firm and agreeable cucumber. When grown to maturity, the fruit flesh becomes soft and mealy.

A note about etymology: The word Adzhur derives from the Arabic 'ajjour, which refers to short vegetable melons, and I prefer it to the more commonly used word Chate (derived from the Hebrew qishu'im and the Arabic qatta or qitha), which encompasses both short and long vegetable melons.

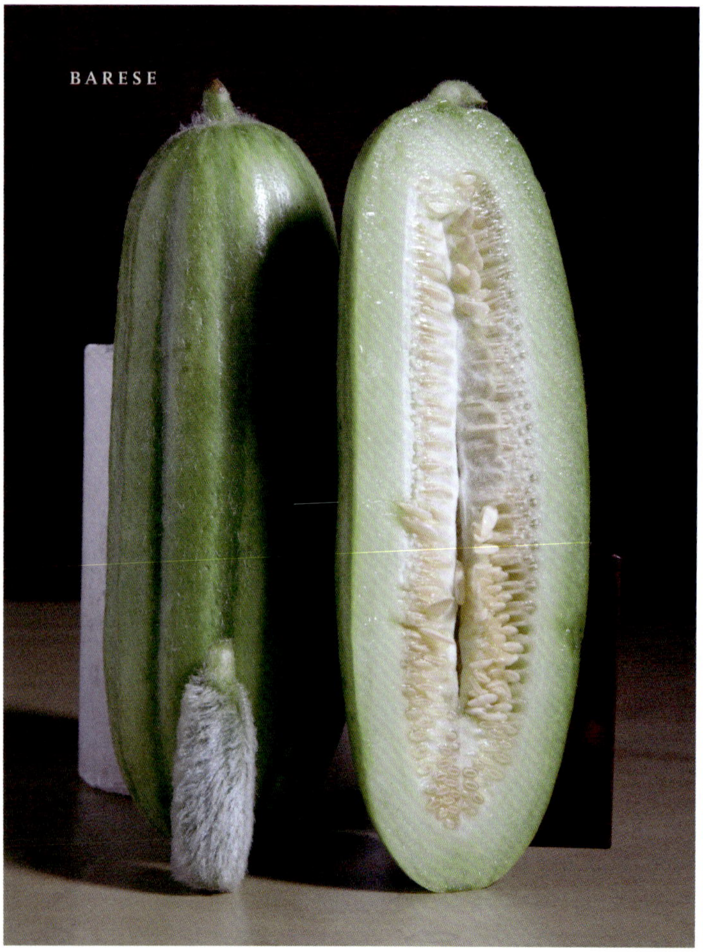
BARESE

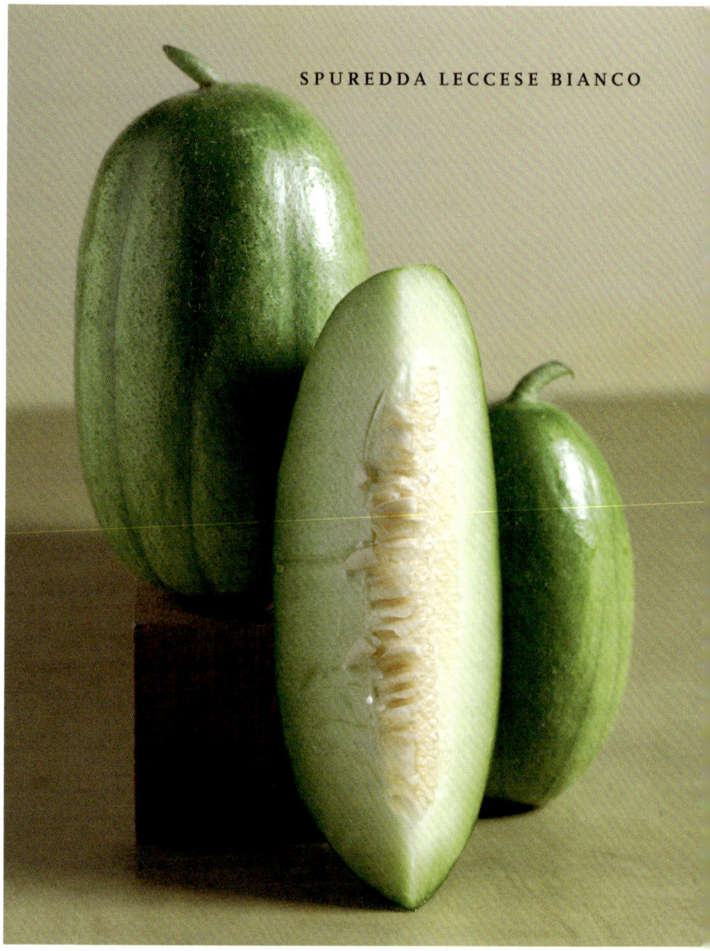
SPUREDDA LECCESE BIANCO

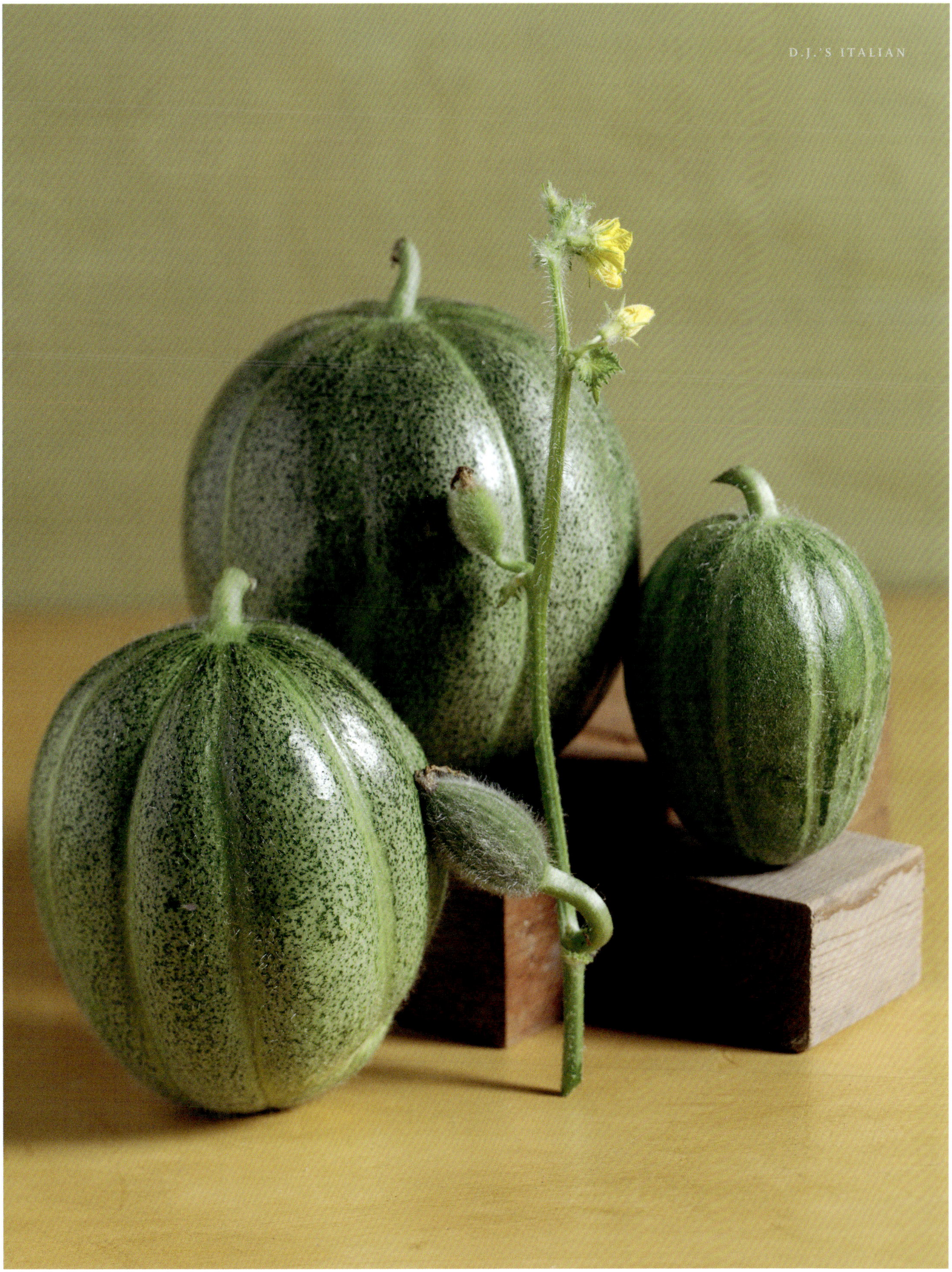

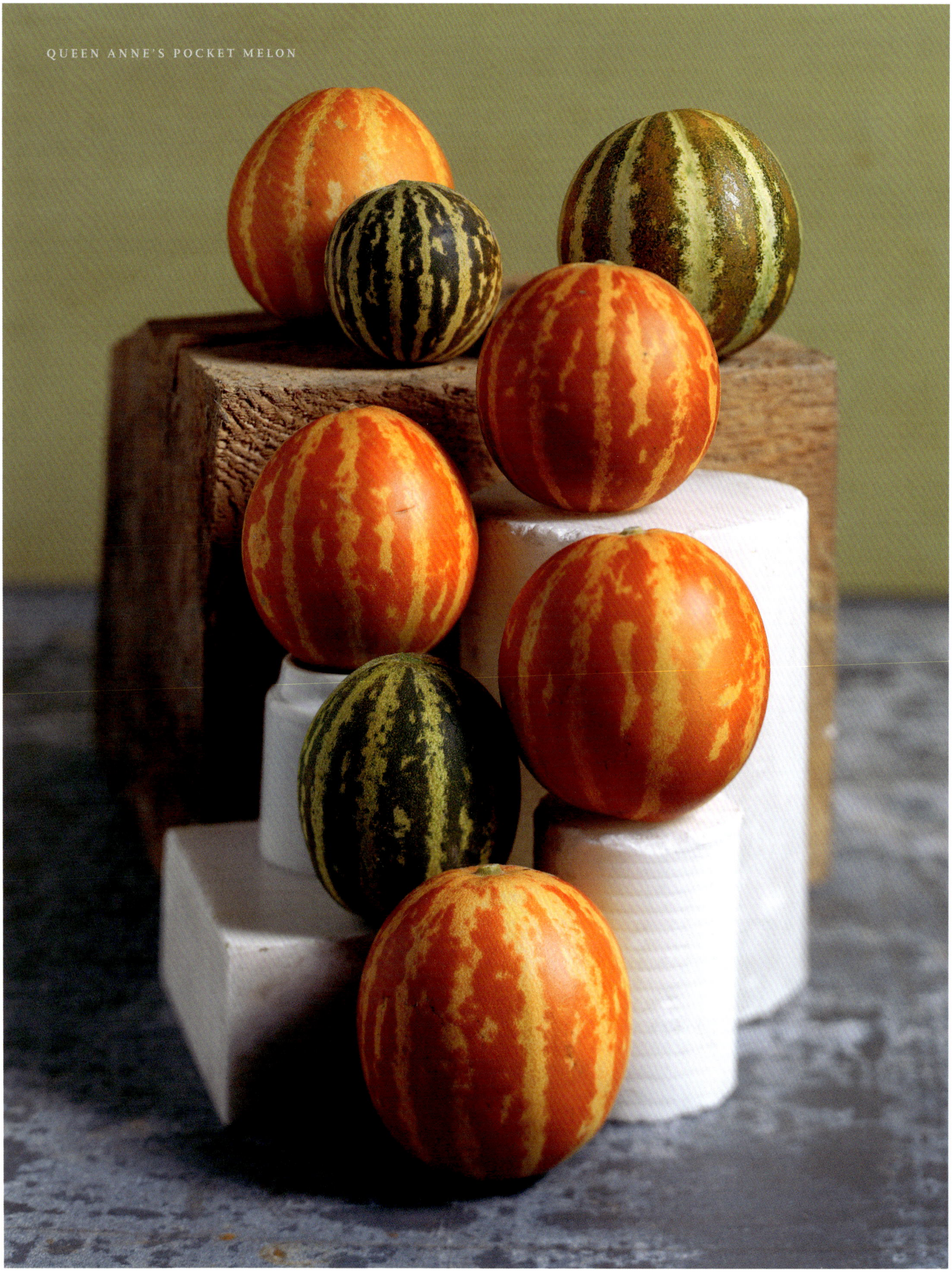

Dudaim Group

DUDAIM MELON The Dudaim (also known as duda'im) has enchanted melon fanciers for centuries, and it's easy to understand why. The exotic maroon or dark orange and yellow striped melon is not only beautiful to behold, but it gives off an unforgettable perfume. Its lush fragrance comes courtesy of volatile compounds—many unique—concentrated in the skin. Take a Dudaim in your hand and bring it to your nose, and you'll know why it is revered. It's better than a tussie-mussie. Flavor is not its strong suit, however: the thin white flesh is insipid and its placental cavity is juicy, filled with hundreds of seeds. In many ways it's more like a wild melon than a cultivated one, including in appearance (small, spherical, striped, smooth, velvety). The Dudaim has recently captured the imagination of American home gardeners; however, for some farmers in Texas and Louisiana, the feral plants have become a pesky weed. It matures one month post-anthesis, abscises from the vine, and produces a prolific crop.

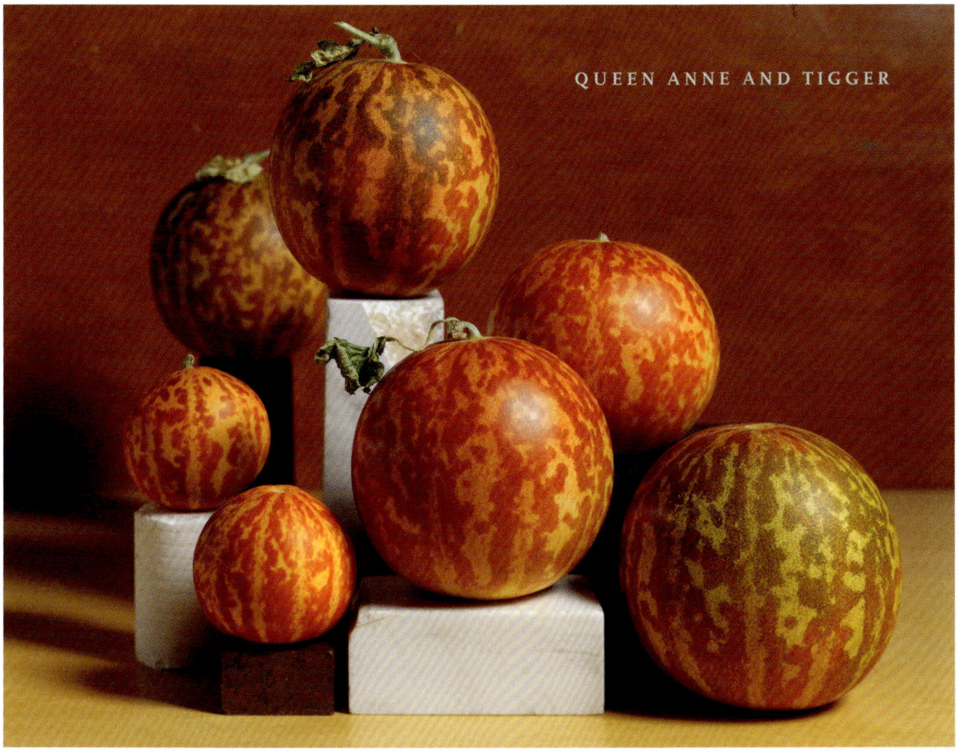

QUEEN ANNE AND TIGGER

Queen Anne's Pocket Melon

The pocket melon has an illustrious history, although it is not known which Queen Anne lent it her name. Well before the reign of Queen Anne of England (1702–1714), it was mentioned in Egypt but with the Persian epithet *destbouyeh* (surely indicative of where it actually came from!) and described by Al Tamimi (circa 975) as small, round, with red and yellow stripes. Linné (1753) attributed this melon to Egypt as well as Arabia, and used the Hebrew word *duda'im* to classify it. The name duda'im derives from the Biblical Hebrew (Genesis 30:14) for mandrake (*Mandragora* L. spp.), an herbaceous plant native to Israel and surrounds, which bears fragrant round orange-yellow fruits.

By examining historical records from various geographical areas in several languages, Harry Paris has documented the earliest known descriptions of the dud'aim melon, making it possible to trace its history to the mid-ninth century in southwestern Iran or Baghdad. According to Paris, "Sabur ibn Sahl, a Nestorian physician and pharmacologist who worked in southwestern Iran and, later, Baghdad, wrote (in Arabic) a book of prescriptions called *Al-Aqrabadhin* (*The Dispensatory*) (circa 850). One prescription, no. 184, calls for the use of the skin of 20 *dastabouya*." Harry further reveals that *dastanbouya* is a compound Persian word for duda'im melons that literally means "hand perfume."

Hand perfume—if only we could bottle it. That would be much more convenient than carrying the melon around in your pocket—precisely what women used to do when daily bathing was not de rigueur and aerosol deodorants were undreamed of. Wily fruit mongers have also been known to hide them in crates of inferior fruit to impart their perfume.

If not Queen Anne of Austria (1601–1666), then certainly her son the Sun King, Louis XIV (1638–1715), must have been acquainted with this melon. It is illustrated in color on vellum (as *Colocynthis fructu rotundo variegato*) in the *Florilegium of Prince Eugen of Savoy*, now known as the Codex Miniatus 53, along with other specimens from the royal gardens. Jean-Baptiste de La Quintinie, the king's renowned chief gardener, grew melons there surrounded by espaliered figs—Louis's favorite fruit. Something for the lord to eat out of hand, and something for the lady's pocket.

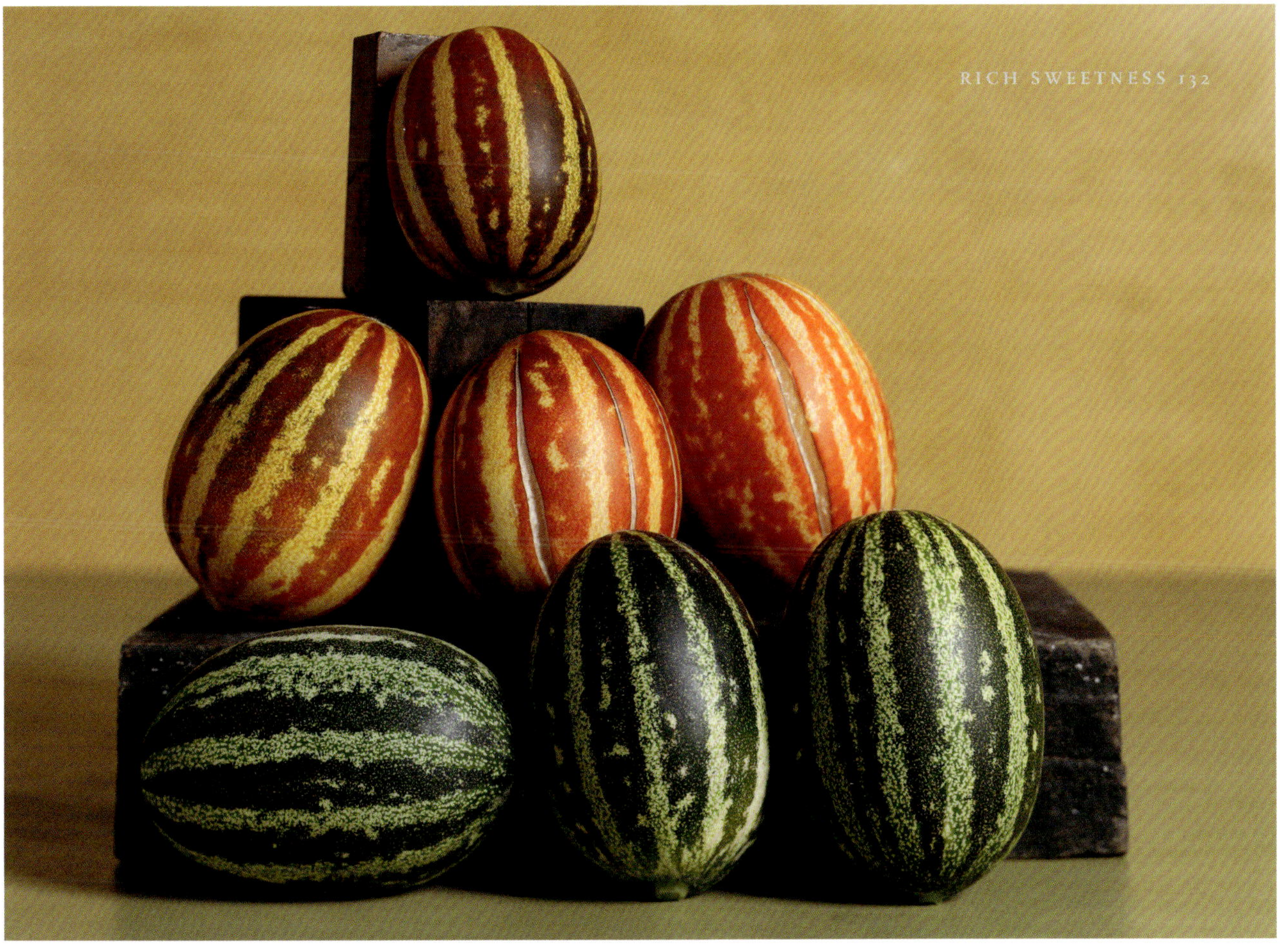
RICH SWEETNESS 132

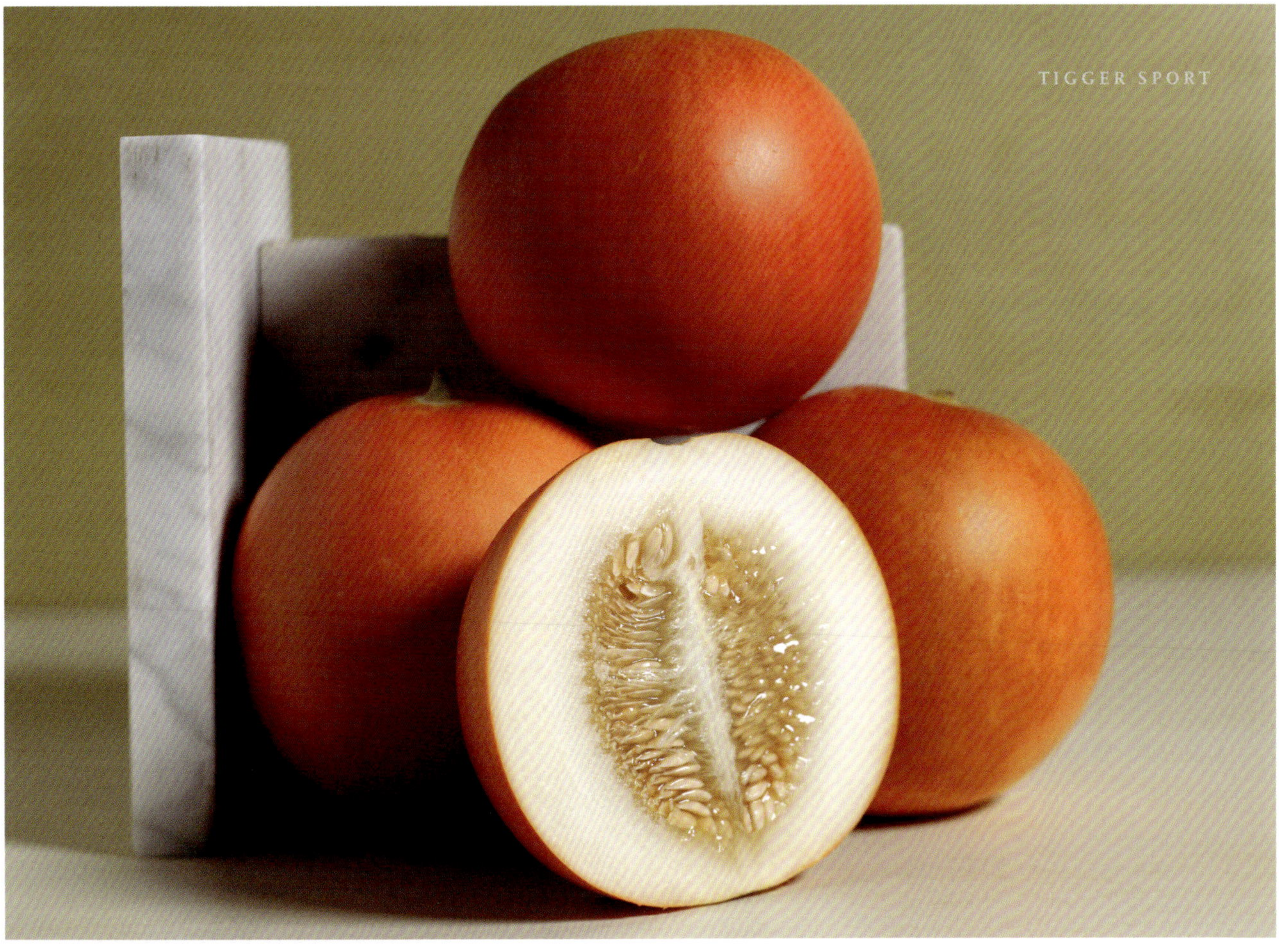
TIGGER SPORT

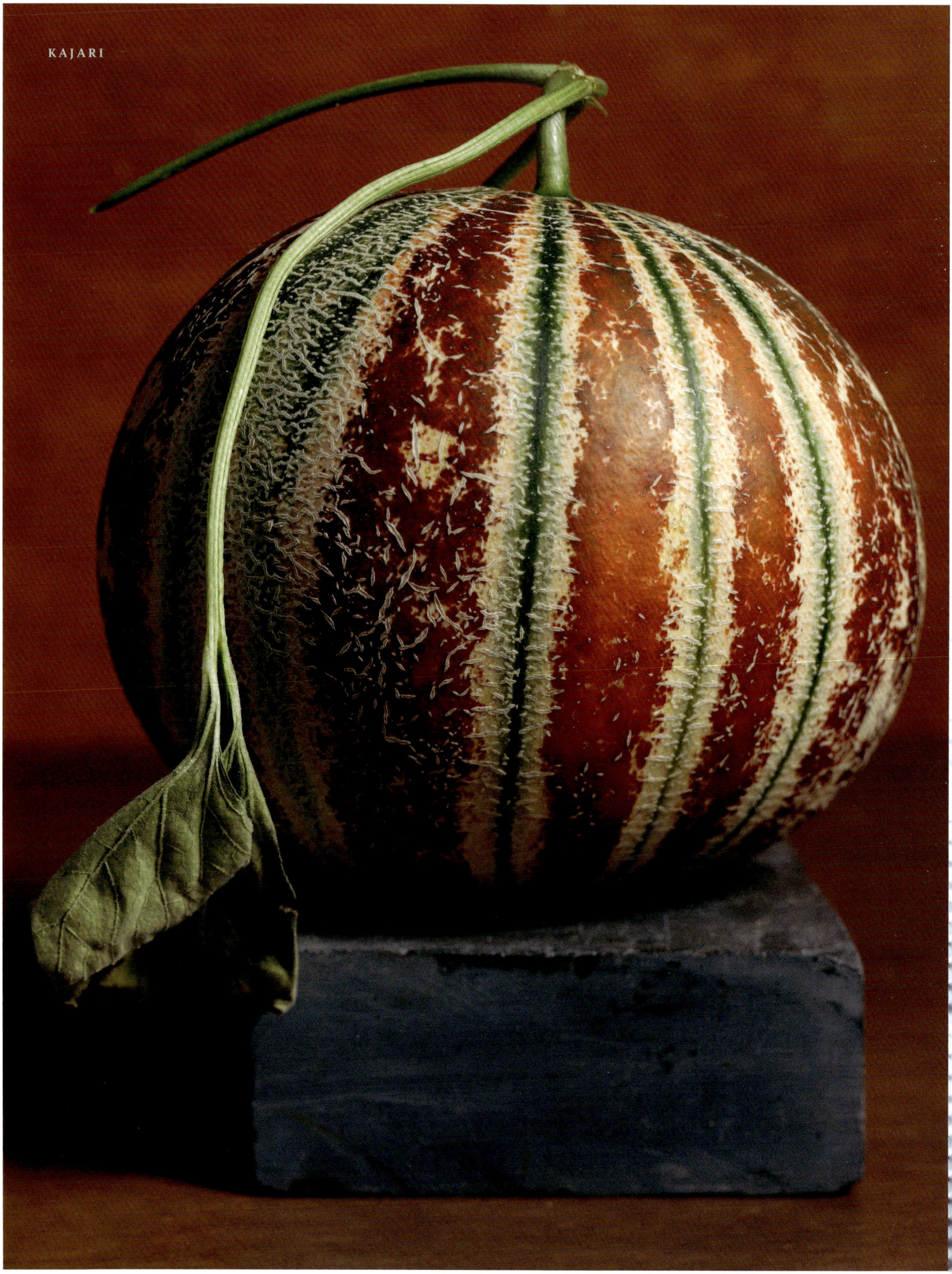

Khandalak Group

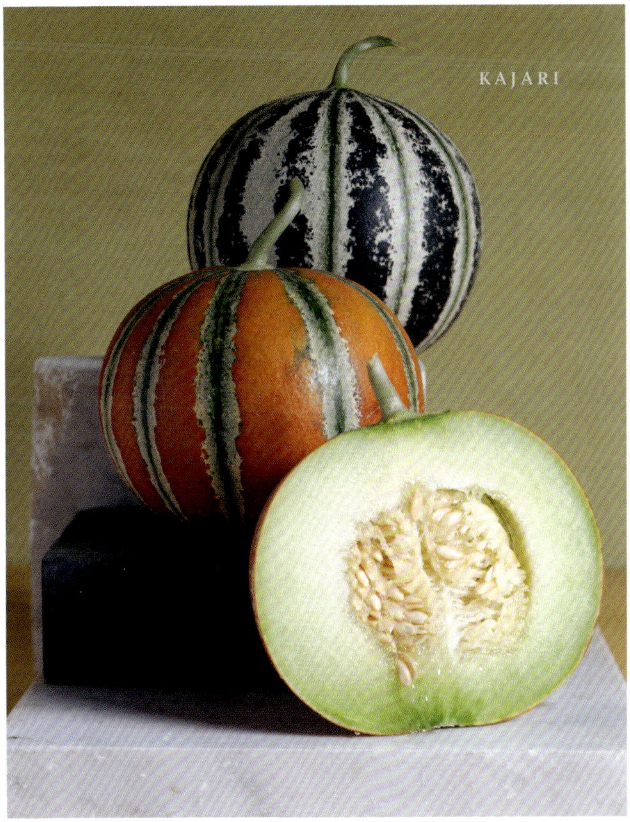

KHANDALAK MELON Khandalak melons (aka Chandalak or Handalak) are virtually unknown to gardeners in America. I've been growing melons for close to fifty years and until Kajari came along in 2015, introduced by Baker Creek Heirloom Seeds, I had no clue what I was missing. My sample size is as small as can be—one cultivar grown twice, in successive summers—but it was enough to get bitten by the Khandalak bug. Later, while flipping through the pages of *Melons of Uzbekistan* and the *Turkmen Melons Atlas*, I was excited to find Kajari look-alikes and many other dreamy sorts just waiting for me.

Khandalak cultivars are primarily grown in Central Asia and India. These are sweet dessert melons, climacteric, that slip from the vine when ripe. Their shelf life is shorter than most. They are either round or oblate, sometimes netted, but not prominently lobed, wrinkled, or warted. Their skin ranges in color from cream to maroon to orange brown and is sometimes speckled; green vein tracts are usually in evidence. Khandalaks ordinarily have green flesh, but some have white or orange.

Opposite and above: Kajari is a dreamy sort, adorned with decorative "ribbons" or vein tracts.

Pitrat (2017) recently proposed four sub-groups: Zami, Tachmi, Garma, and Bucharici. Kajari qualifies as a Zami. The boldness of its maroon stripes and veinal tracts, intense green at their center, is a study in contrasts—and the melon's most remarkable feature. I flipped my lid when I first saw it on the vine.

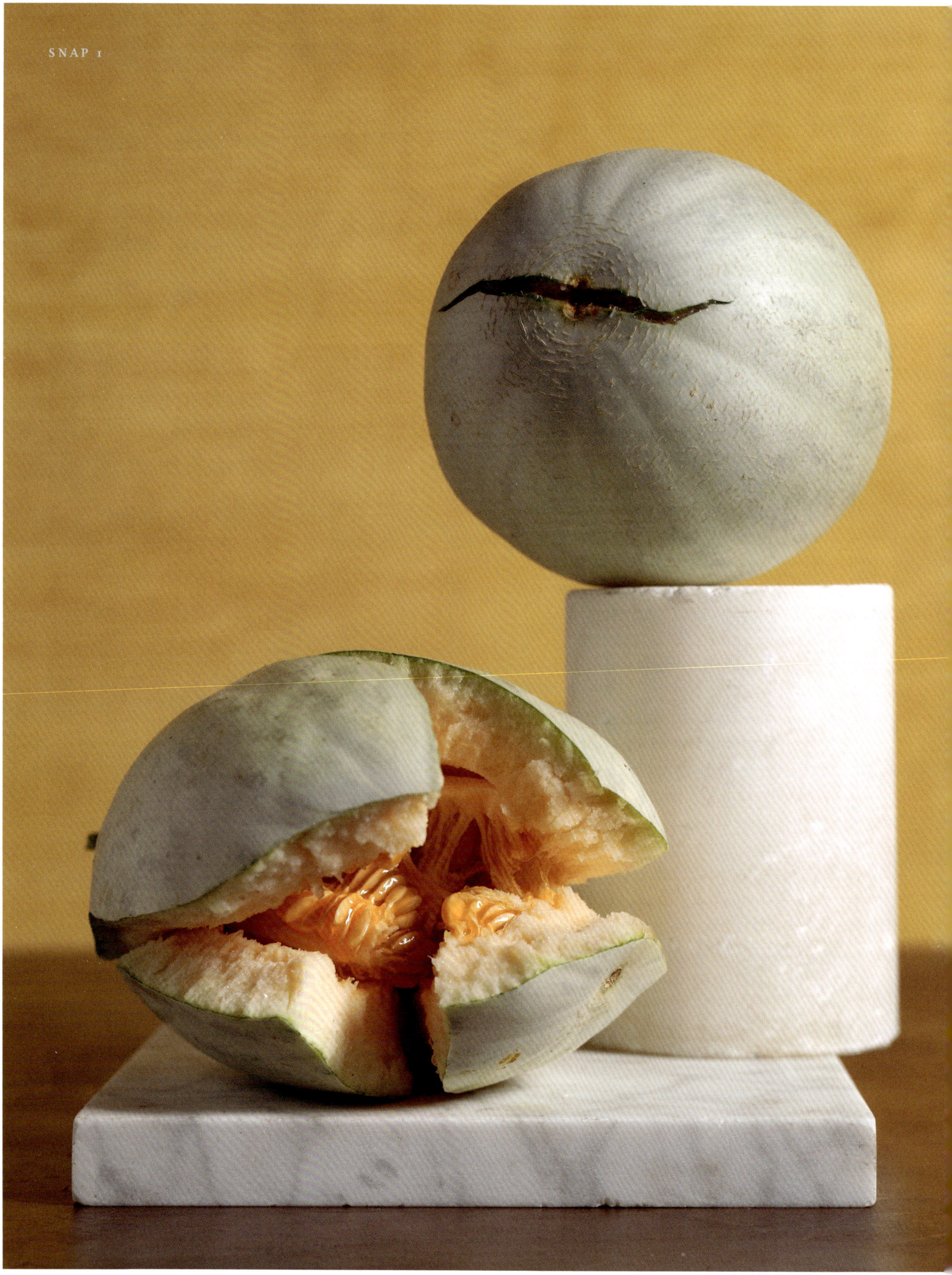

Momordica Group

SNAP OR PHUT MELON I am fascinated by the way these exotic melons behave at maturity. Known as phut, which means "to split," in their native India, snap melons (or snapmelons) almost always crack, snap, or peel open. I saw this happen at close range after gingerly cradling and ferrying an intact yellowing melon, harvested the previous day, from the kitchen to the barn studio to be immortalized by Victor's camera (see Snap 3, page 124). As he placed the specimen on a pedestal, setting the stage for a still life photograph, the melon began to snap, and then slowly unraveled, the peel in pieces like leather. I wish we'd had a video camera.

Given their thin skin, tendency to self-destruct, and very short shelf life, Momordica melons would be easy to underestimate. Don't. Snap melon plants are heat-tolerant and drought-hardy. They thrive and produce at very low levels of irrigation. There is tremendous untapped diversity in Indian snap melon landraces. Their genes hold great promise in breeding more resilient "climate smart" melons worldwide, according to snap melon advocate Narinder Dhillon, who leads the Cucurbit Breeding Unit at the World Vegetable Center. Small farmers in India develop and cultivate snap melons in semi-arid areas during the rainy season, providing local people with a valued foodstuff including at least one melon selected for resistance to cracking and hence longer shelf life.

In a recent survey, middle-aged housewives in northwestern Rajasthan reported using unripe snap melons known as *phoot kakari* or *kakadia* in a curd-based preparation with vegetables, or dehydrating them for off-season use. They used ripe fruit flesh—which is characteristically mealy, low-sugared, and often light orange—in jam, salad, dessert, and milkshakes. Seeds processed commercially served as an ingredient in baked goods and *thandai*, a traditional drink. Indian housewives also employed snap melon medicinally as a treatment for various stomach ailments. Momordica melon is not solely confined to India—it is also grown in Southeast Asian countries such as Myanmar and Vietnam. And there is a long history of cultivation in Japan: Their seeds have been found in archeological sites dating to the Nara and Heian periods. More recently, snap melons have been enjoyed there as sugar-dressed flesh.

YEOMAN SERVICE

If you live in the United States, chances are you've never tasted Momordica melon, but you have probably benefited from them. Snap melon genes have been bred into the pedigrees of many modern American melon varieties to improve resistance to fungal and viral diseases, insects, and nematodes—and also to increase tolerance to salinity, high temperatures, and drought. As Narinder Dhillon says, snap melons have done "yeoman service"—starting in the United States with the development of PMR 45 (Powdery Mildew Resistant Cantaloupe No. 45), introduced in 1935. Novel alleles (variant forms of genes) from the donor parent, the historic Indian landrace known as California 525 (a self-pollinated increase of the Indian landrace Big Round), conferred heat and salt tolerance as well as powdery mildew resistance.

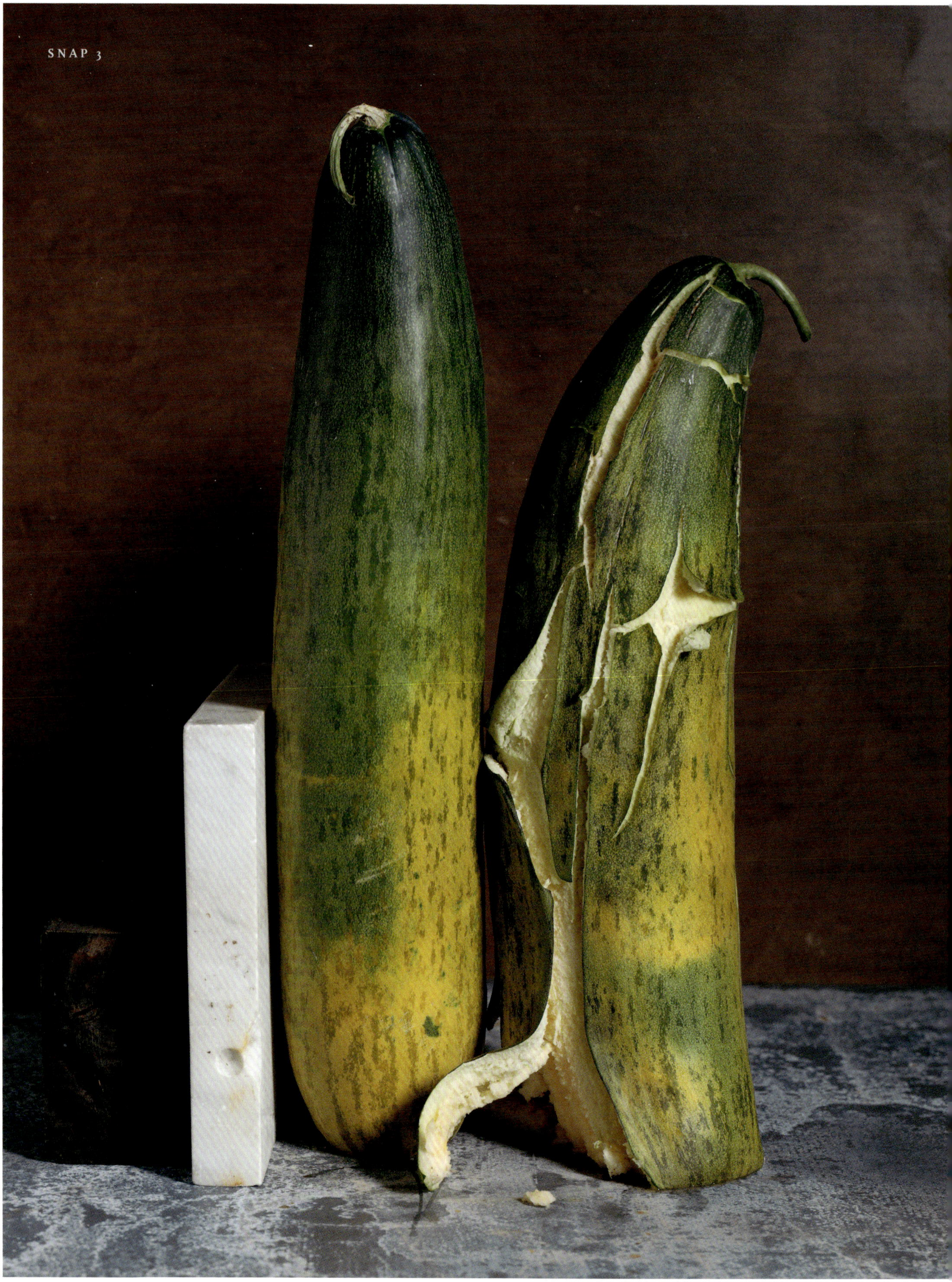

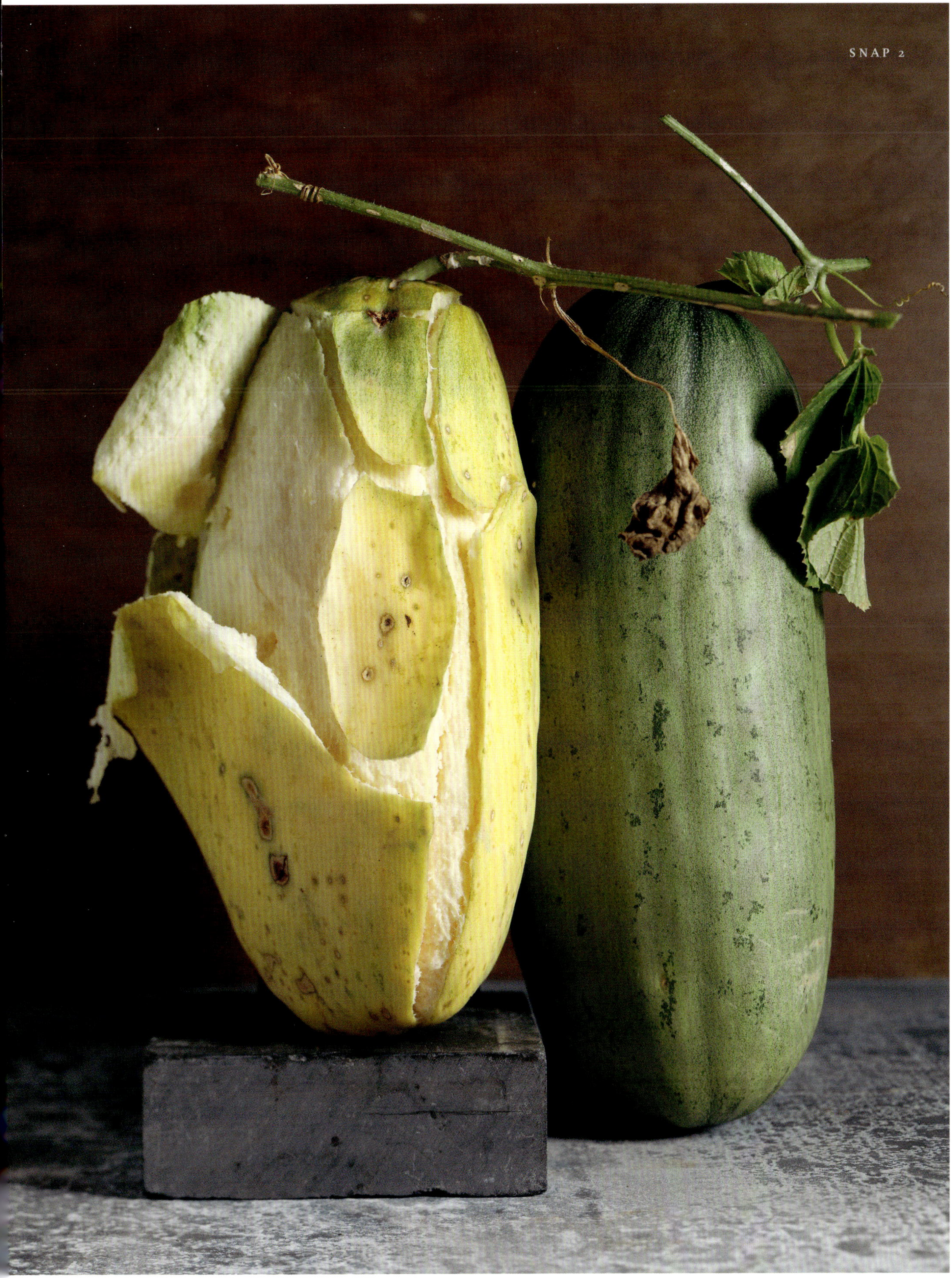

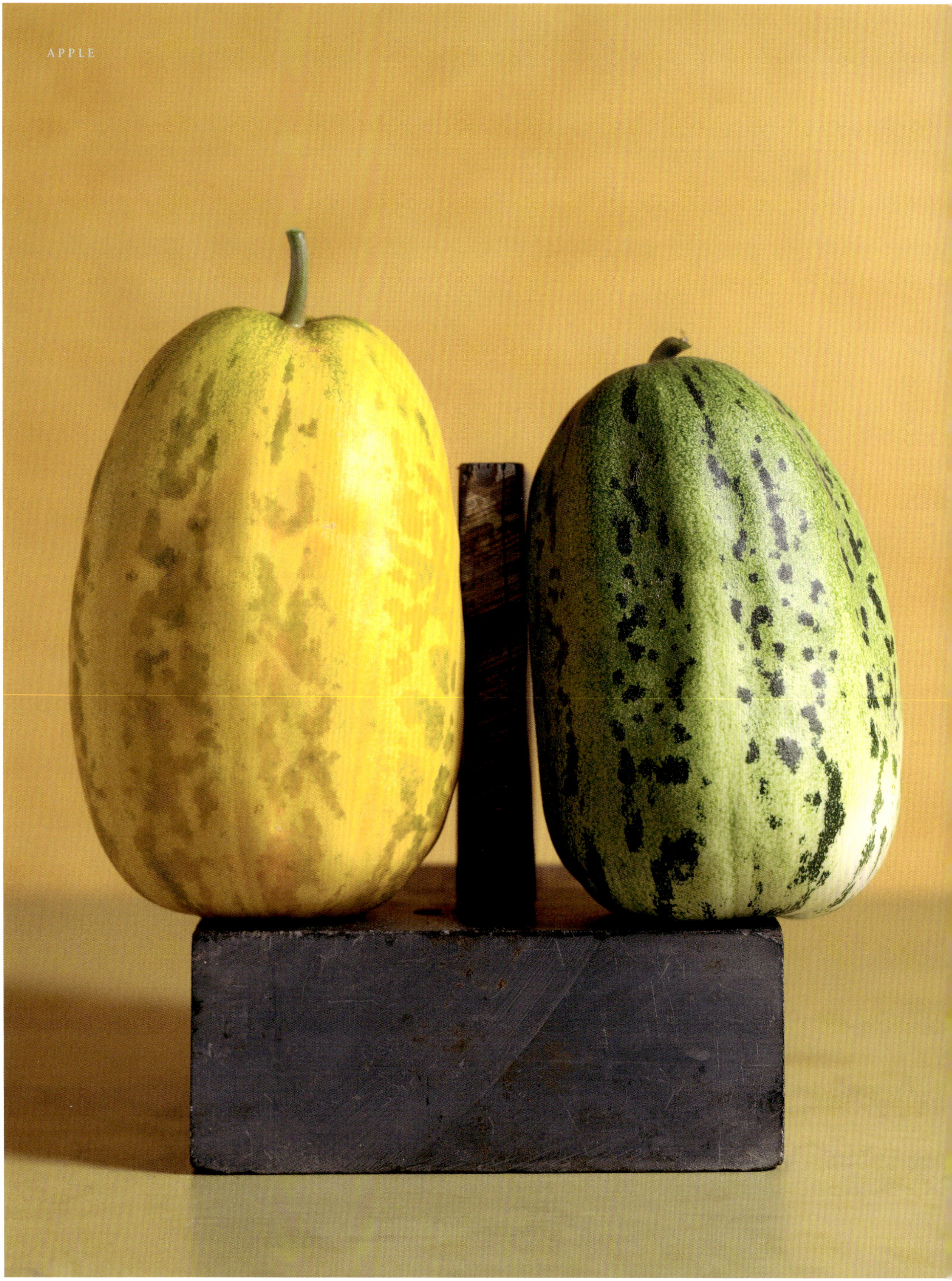

Makuwa Group

ORIENTAL SWEET MELON Makuwa group melons are new on the American scene, but in East Asia they've been known for thousands of years. They were most likely cultivated in China as early as 2000 BCE. According to Japanese scholar Noriyuki Fujishita, they probably originated in the Manchuria area of China and traveled to Japan through Korea. *Makuwa uri* first appeared in Japan in the Kyushu area near Korea; excavations from the Yayoi period have unearthed 2,100-year-old Makuwa seeds. Minoru Kanda, of the seed company that bears his name, believes that the melon was named for a small village in Aichi prefecture, where farmers grew the finest *uri* fruits. *Useful Plants of Japan* (Dai Nihon Nokai 1895) illustrates and describes *makuwa uri* with these remarks: "It is produced much in the village *Makuwa* in the province *Mino*, whence the name is derived. . . . They are oval-shaped, about five inches long, and of a yellow colour, with longitudinal stripes. They are eaten one or two days after having been collected, and are very sweet and delicious. There are several varieties of different colours and forms."

Oriental sweet melons differ from the sweet dessert melons well known to Westerners. They are petite by comparison—one might say "personal sized." Their smooth edible skin comes in off-white, light green, and yellow. I find the easiest way to remove the skin is with a peeler rather than a knife. Some cultivars have longitudinal vein tracts in silvery white or green. Shape varies from oblate to oval to pyriform. Flesh is thin, firm to crisp, and generally white, but can be greenish or pale peach. Immature fruits can be bitter; sugar content of mature melons ranges from low to very high. Mature climacteric melons have a pleasant external aroma that many people miss.

Long suited to the wet climate in Japan, and treasured over the centuries, open-pollinated *makuwa uri* are not as popular there anymore. They've lost market share in recent decades to high-quality Japanese F1 hybrid netted cultivars with green flesh, cultured in glasshouses. (Many of those cultivars have Earl's Favourite—an English heirloom melon imported in 1925—in their pedigree).

ORIENTAL SWEETS

In Korea, Oriental sweet melons are known as *chamoe* (which translates as high-quality cucumber). The Chamoe or Korean melon remains very popular in Korea and among Korean-Americans. It's delicious and refreshing eaten sliced in half, slivered, or cubed, with or without the skin. In the manner of many Koreans, I've learned to prize and eat what most of us usually discard: the sweet and nutritious small seeds and gel. Melon seed coats are thin and soft and can be easily consumed.

Makuwa melons are a gold mine for plant breeders searching for resistance to various pests and diseases. Pitrat (2017) has proposed six new sub-groups based on skin color, presence of vein tracts, and seed shape: Ogon, Nashi-uri, Yuki, Kanro, Ginmakuwa, and Seikan. In the melon patch, Makuwa are among the earliest, most vigorous, and prolific.

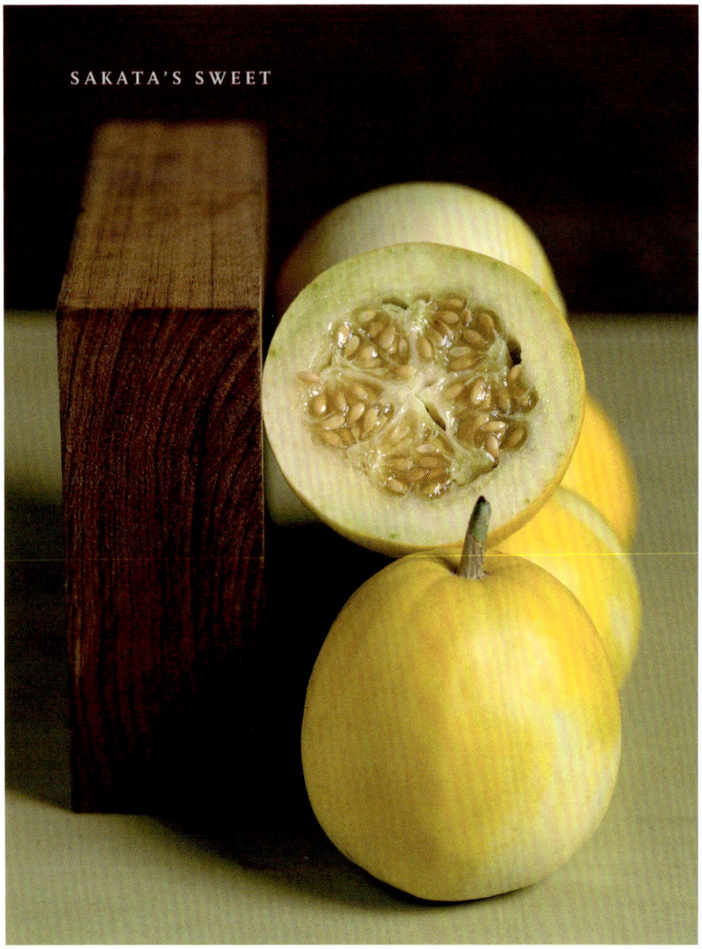
SAKATA'S SWEET

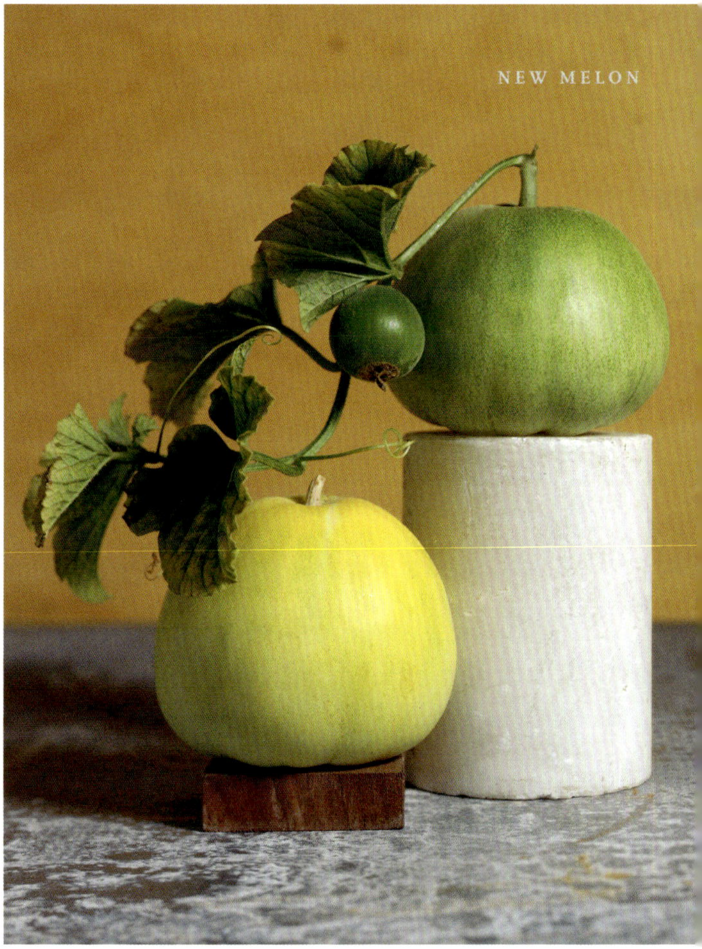
NEW MELON

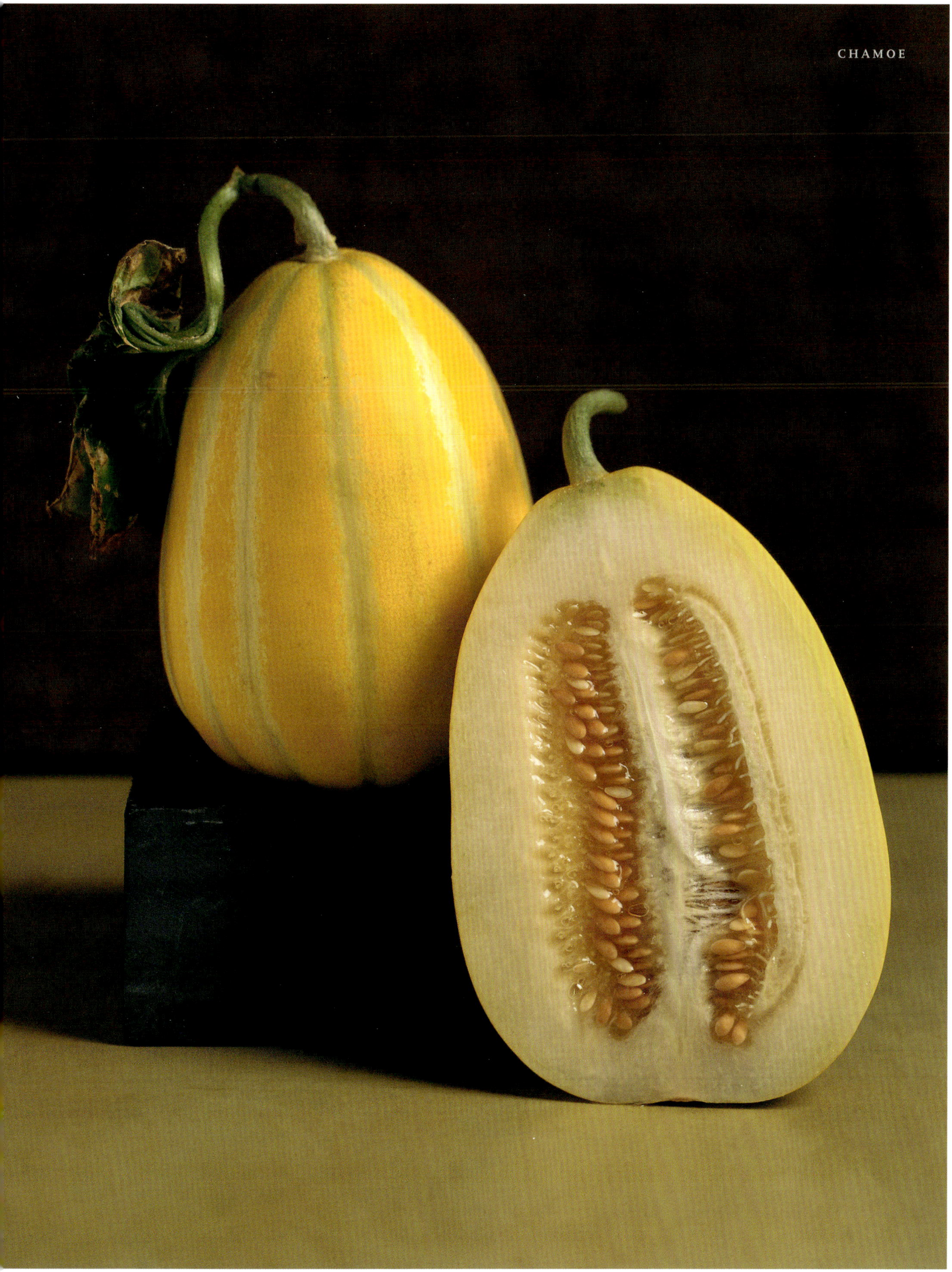

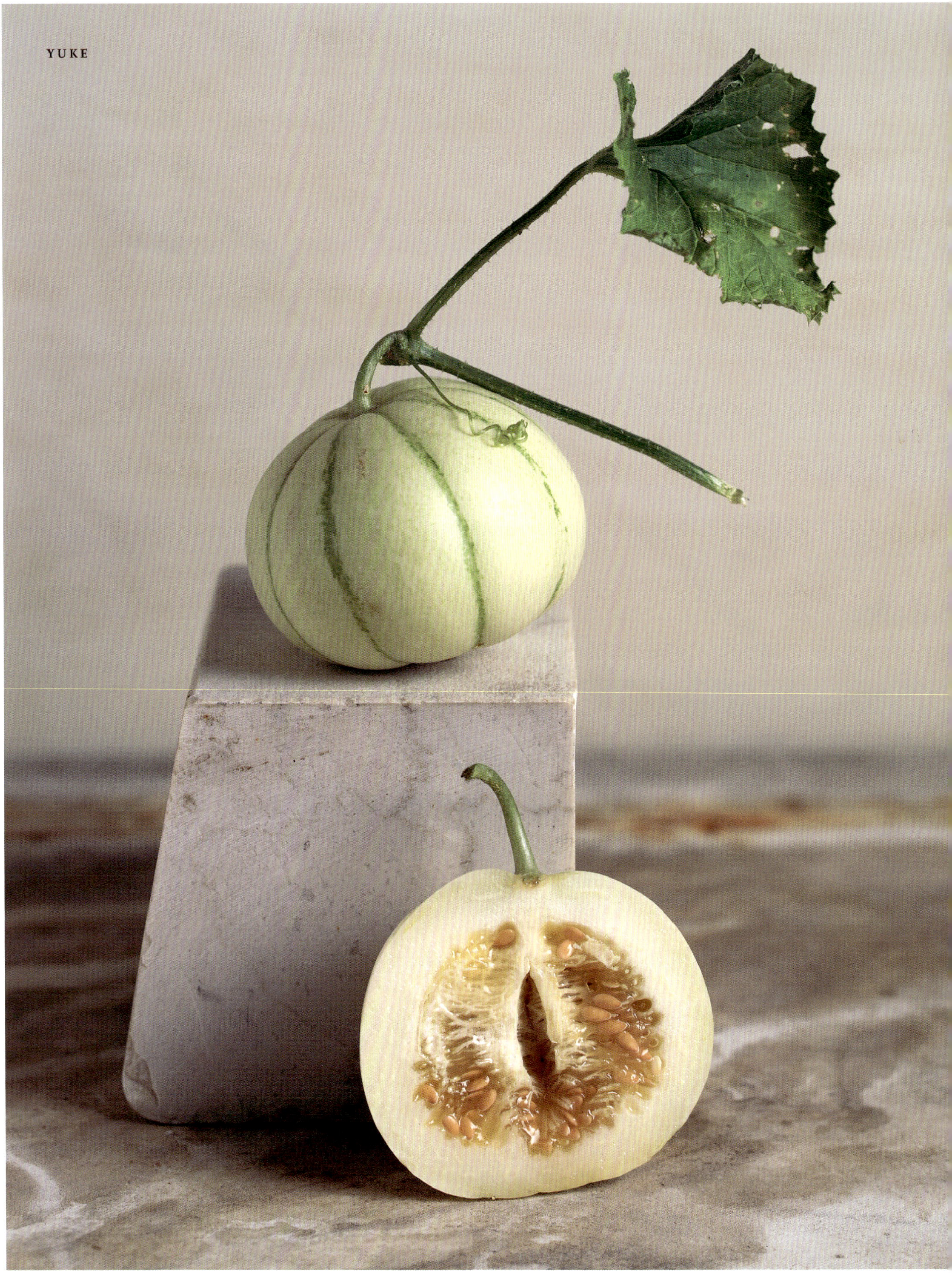

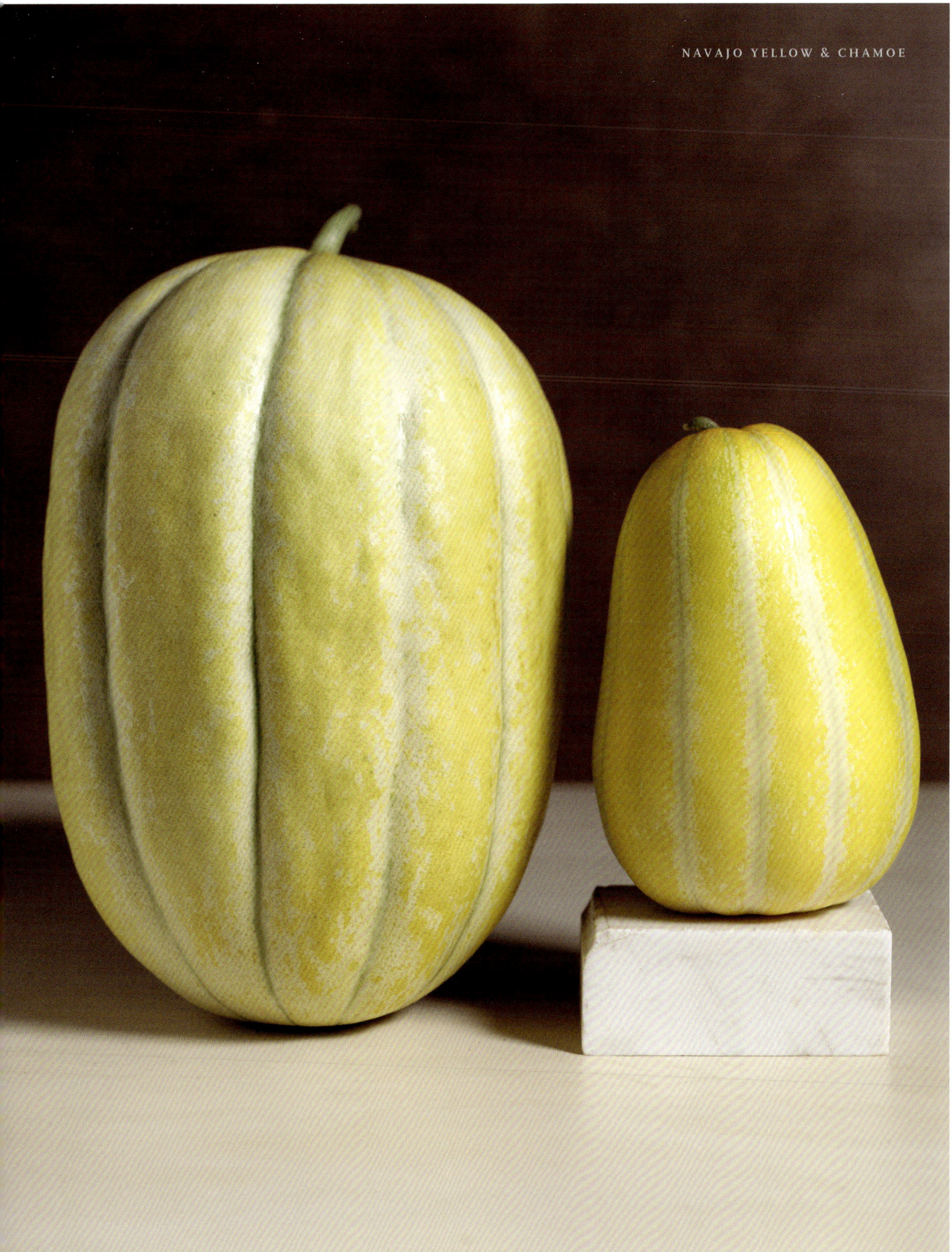

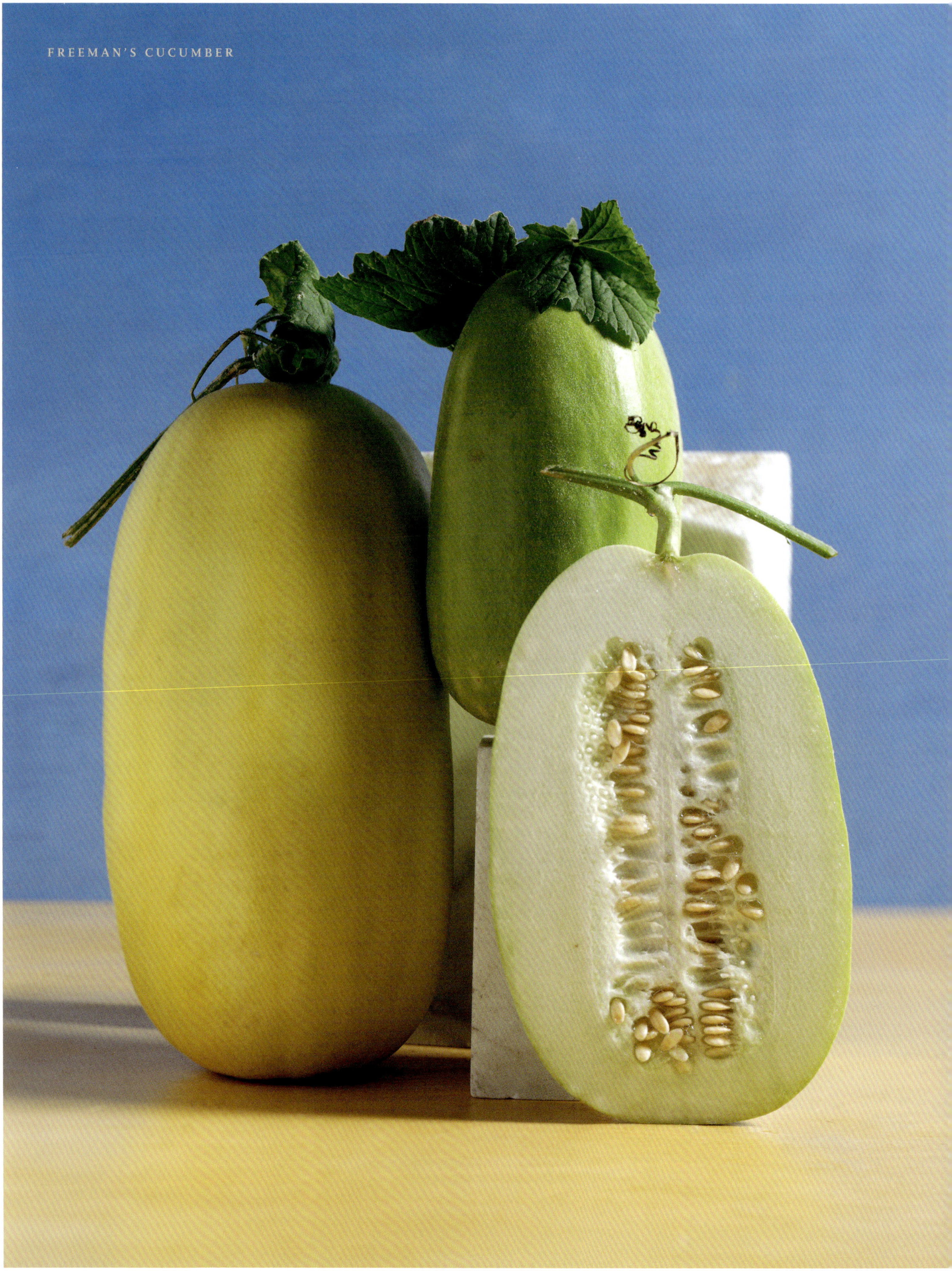

Conomon Group

ORIENTAL PICKLING MELON Not all melons are designed to be sweet. For American palates, elongated vegetable melons are a novelty. If you love cucumbers or summer squash, as I do, then consider the Oriental Pickling melons known as *tsuke uri*. Fruits are smooth, generally oblong, but can be oval or club-shaped, and come in white, green, yellow, and striped skin. When harvested full-sized yet immature, most cultivars have crisp pale flesh that is pleasantly mild. They're more typically eaten stuffed and baked or stir-fried than raw or in salad in Japan today. The immature fruits have been used for centuries in Japan for salt pickles and *kazu-zuke* or *nara-zuke* (sake lees pickles, and mirin and sake lees pickles respectively). Pickling of Conomon melons in this way is almost a lost art. The current interest in pickling and fermentation may be the melon's salvation. Could *tsuke-uri* be the next kimchi?

Oriental picklers are easy to grow. The ones featured in this book bear an early concentrated fruit set with a birdnest growth habit and produce a second crop on longer vines. You can eat the skin along with the flesh in most cases, though I prefer not to.

This type of melon probably originated in China and Japan. It was first noted in Chinese literature in 560 CE. The Japanese know the melon as *uri* and the pickles as *tsukemono* or *koko*. The name Conomon is likely a play on the words *koko no mono*, which translates as "material for making *koko*." Today these ancient melons are grown mainly in China, Korea, and Japan. Cucumbers seem to be taking over the world.

Opposite: Freeman's Cucumber, botanically a melon, was used by Zvi Karchi of Israel's Agricultural Research Organization to introgress resistance to Cucumber mosaic virus (CMV) into Galia-type melons, a widely popular international market type.

Katsura Giant

Katsura is the name of a tree native to Japan and an imperial villa in Kyoto. Katsura is also a melon unlike most ever encountered by Americans. Paillieux and Bois (1892), who made a point of cultivating exotic comestibles in France, didn't quite know what to make of this sort of melon, either. So they cut it in rondelles, fried it, and made beignets—some would say fritters.

Katsura Giant was born to be pickled. For centuries, the Japanese have pickled everything in sight: from roots to shoots, and flowers to seeds. Visit any Tokyo food emporium or small-town covered market and you'll smell and see the pungent and dazzling *tsukemono*. Pickles increase the appetite and amuse the palate; they provide fiber as well as lactobacilli and friendly flora to enhance digestion.

To make *koko*, don't use vinegar—*uri* is traditionally salt-pressed. After the melon is deseeded and sliced, salt it well and place it in a pickling pot. Crown it with a wooden pressing lid with heavy stones on top, and let the melon form its own brine.

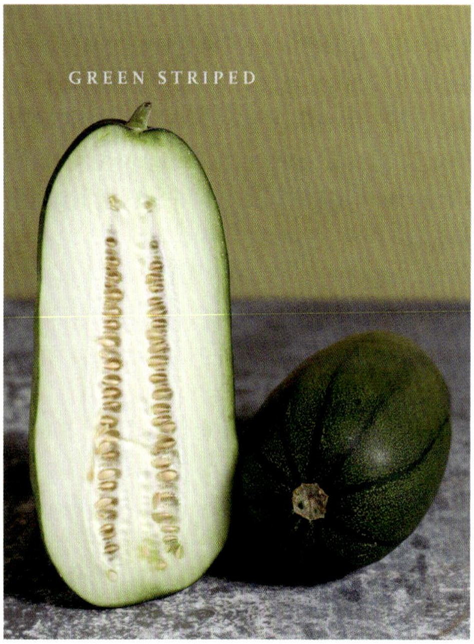

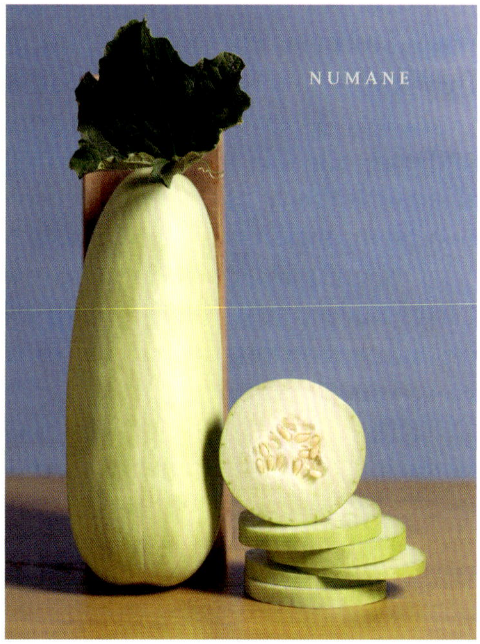

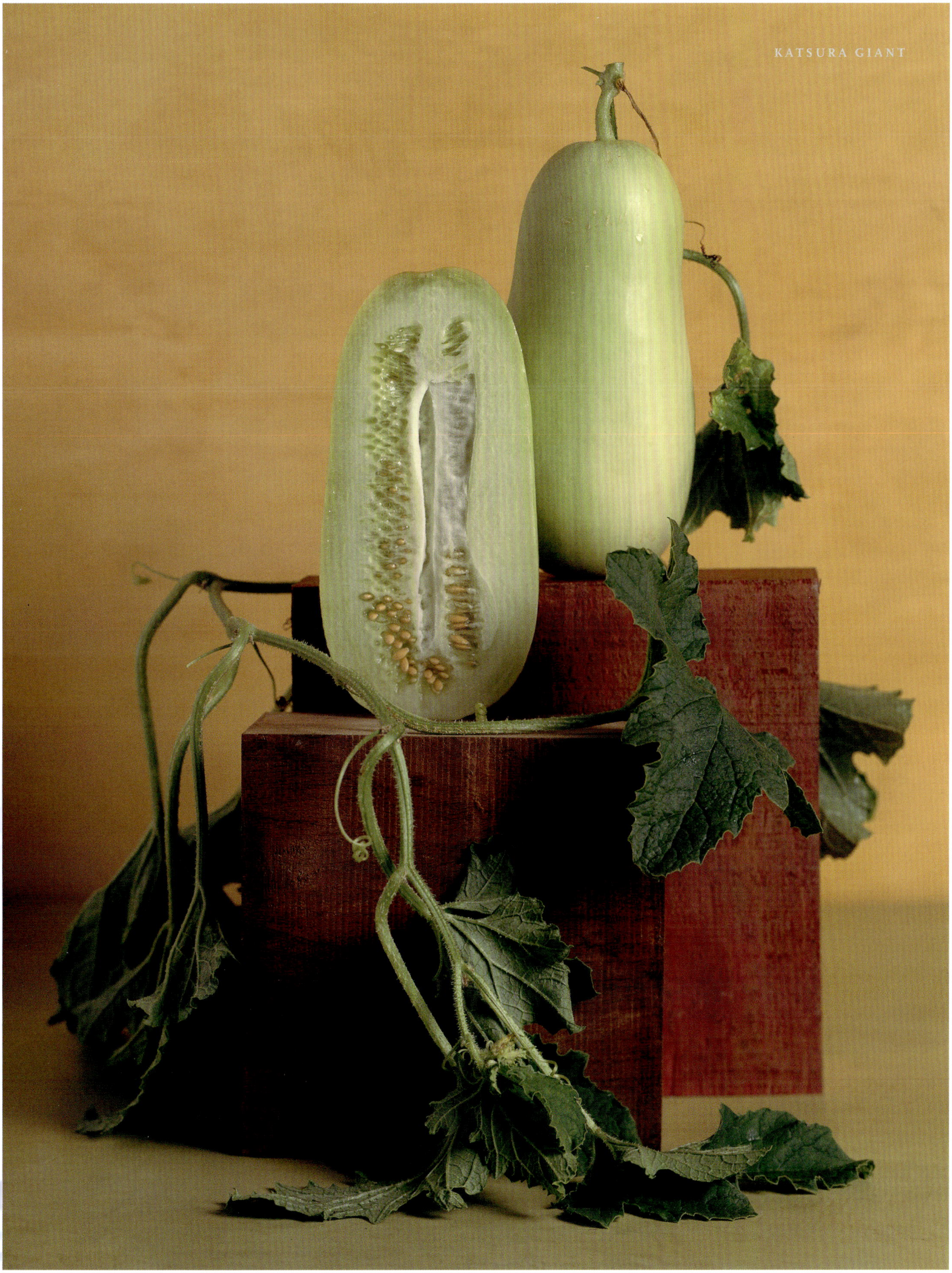

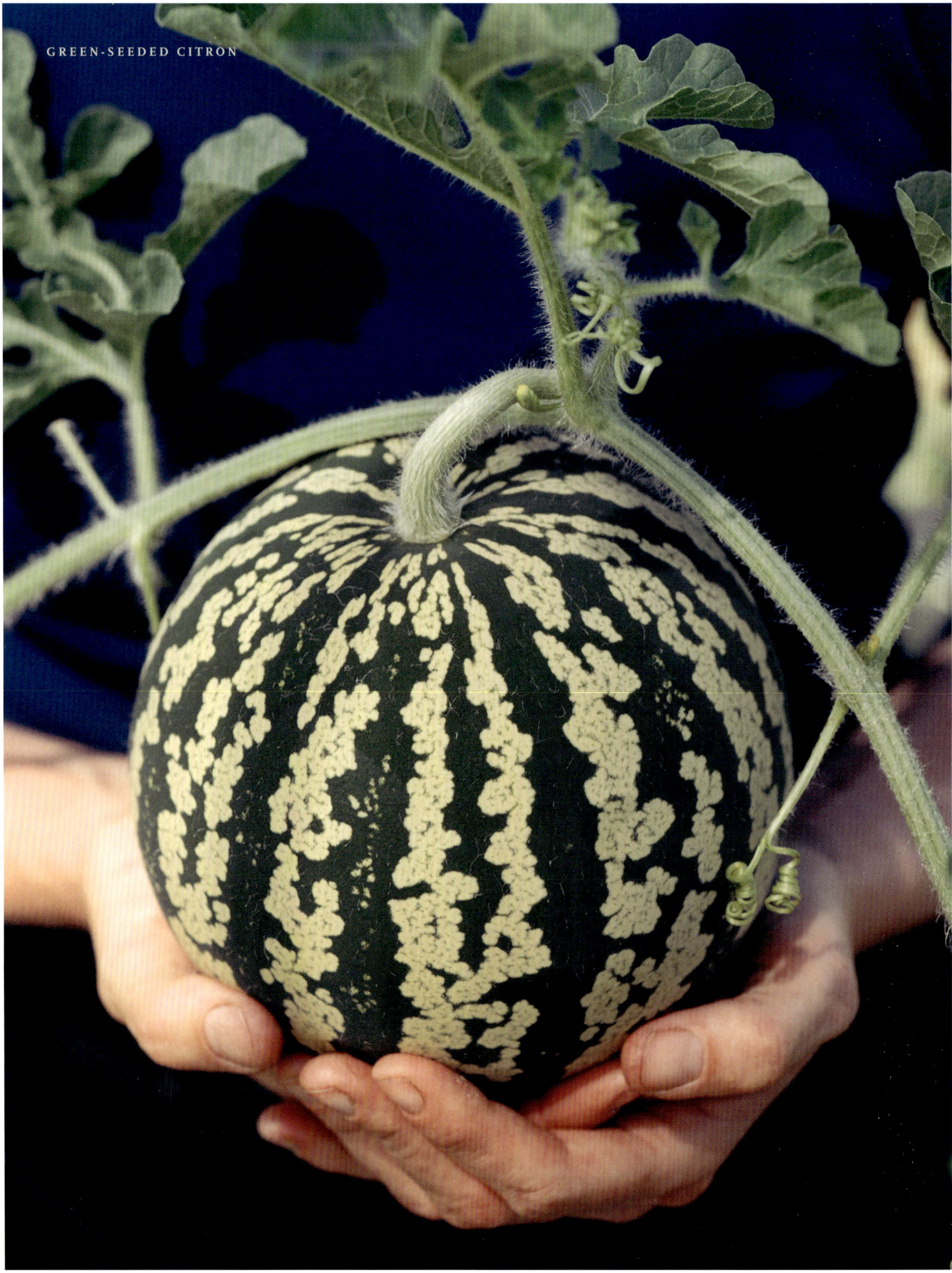
GREEN-SEEDED CITRON

CHAPTER SEVEN

Watermelon Portraits

Today's cultivated sweet dessert watermelons (*Citrullus lanatus*), with their tender, highly colored flesh, have come a long way in terms of both flavor and geography. Paris (2015) attests to that with considerable evidence from the archeological, iconographic, and literary records. Dessert watermelons originated in northeastern Africa (Sudan and neighboring countries) and were domesticated in that region by 2000 BCE. Seeds and illustrations in Egyptian tombs of nobility indicate that watermelon was valued in ancient times—even King Tut's tomb contained seeds (1323 BCE). One particular image of a large, long, boldly striped fruit, served up on a platter, in a wall painting in a tomb at Neir, Old Kingdom, more than four thousand years old, reads as a cultivated watermelon. It reminds me of a modern-day Georgia Rattlesnake (see page 149).

Amnon Levi, of the Agricultural Research Service of the USDA in South Carolina, and his colleagues (2017) have discovered that despite the great phenotypic variation exhibited by today's dessert watermelons, most in fact have a narrow genetic base. So how did our beloved sweet dessert watermelons evolve, what did their ancestors look like, and where did they come from? Harry Paris of Israel's Agricultural Research Organization thinks he knows the answers. He has learned that small, round, and striped wild and primitive dessert watermelons (*C. lanatus* subspecies *cordophanus*)—used for food, fodder, and as "botanical canteens" or as a source of clean water in the dry season—can still be found growing in Sudan and neighboring countries. (These are not the same as the colocynth [*C. colocynthis*], which is extremely

CITRULLUS SPECIES

Watermelon, which comprises the genus *Citrullus*, was recently divided into seven species based on a landmark molecular phylogenetic study done by Guillaume Chomicki and Susanne S. Renner (2015). Accordingly there are three widely cultivated species: the dessert watermelon (*C. lanatus*), the citron watermelon (*C. amarus*), and the egusi watermelon (*C. mucosospermus*), the last characterized by its soft, mucilaginous seed coats. A fourth species, the colocynth (*C. colocynthis*), has been sparingly cultivated for medicinal purposes. There are three species of crop wild relatives: *C. ecirrhosus*, *C. rehmii*, and *C. naudinianus*.

Opposite: A proud bearer of a Green-Seeded Citron watermelon still attached to the vine. Notice the branched tendrils and pinnatifid leaves.

bitter and small, and tends to desiccate rather quickly, or the citron watermelon [*C. amarus*], which is occasionally bitter but always hard and watery.) The flesh of the local non-sweet dessert watermelons is watery, pale in color, greenish white or white, and either bitter or insipid. *Gurum* is their Sudanese name. They are called *gurma* in neighboring Egypt. Paris explains that these are "living representatives of the ancestor of the sweet dessert watermelon."

The domestication of watermelon for water and food, and subsequent selection by ancient farmers in Mediterranean lands for sweet, tender, and well-colored flesh, happened over many centuries. The Children of Israel longed for the *avattihim* (watermelons) of Egypt in the Sinai Desert during the time of the Exodus (Numbers 11:5)—but it would have been for their cooling effect, to quench their thirst, and not to sate their sweet tooth. Paris believes that the longed-for watermelons were not sweet, because they were mentioned with *qishu'im* (vegetable melons), leeks, onions, and garlics and were used in the same manner: eaten raw, pickled, or cooked. Paris concludes, "The *pepon* of the Greek physicians, the *pepo* of the Latin encyclopaedist Pliny and the *avattiah* [singular] of the Hebrew scholars were one and the same large, watery, fruit—the watermelon."

"THE DOMESTICATION OF WATERMELON FOR WATER AND FOOD, AND SUBSEQUENT SELECTION BY ANCIENT FARMERS IN MEDITERRANEAN LANDS FOR SWEET, TENDER, AND WELL-COLORED FLESH, HAPPENED OVER MANY CENTURIES."

By around 200 CE, watermelon had been selected for sweetness. Non-bitter, sweet, and tender watermelon had emerged in Israel and possibly Greece: it was listed in the tractate on tithing in the Jewish religious codex known as the *Mishna* (compiled by Rabbi Yehuda HaNasi, in northern Israel in the second century CE) as part of a common class of sweet, juicy fruits eaten raw, including figs, table grapes, and pomegranates. Dessert watermelon subsequently found its way to Mediterranean Europe: to Italy, and perhaps France, by the beginning of the sixth century CE, and to Spain by the tenth century at the latest.

When did the watermelon acquire its colored flesh? We don't know. A Byzantine floor mosaic, found at Mamshit in southern Israel (circa 425–600 CE), seems to portray two halves of the same round watermelon cut equatorially, revealing yellow-orange flesh; the stones representing the seeds are pale-colored. (The flesh and seed coat colors exactly match those of the modern Orange Flesh Tendersweet.) We do know that red-fleshed watermelon had emerged at the latest by 1395 in northern Italy: a full-page rendition in a *Tacuinum Sanitatis* manuscript (referenced as Vienna 2644, and likely executed in Giovannino de Grassi's workshop in Lombardy) vividly illustrates a man in a gray tunic, knife in hand, harvesting a long, oval, green-striped watermelon from the vine. The garden is green and lush, brimming with a bumper watermelon crop; a basket at the man's feet includes two red-fleshed harvested specimens.

A hundred years later (1492) Columbus arrived in the New World. The ensuing transfer of flora and fauna between the Old and New Worlds would become known as the Columbian Exchange. Spanish colonists and later Africans introduced watermelons to North America. Colonial records of Spanish Florida show that watermelon was grown there as early as the late sixteenth century. Court documents from 1576 quote farmer Juan Serrana, who testified that Santa Elena Island was fertile ground for maize, pumpkins, and watermelons (*sandias*). Sweet watermelon quickly diffused from Spanish settlements to the far reaches of the continent. Native Americans were early adopters of watermelon—as were the Spanish of New World pumpkins and maize. As early as the end of the seventeenth century, Amerindians were growing watermelons in the American Southwest, the Northeast, and the Great Lakes region. Later historical records indicate that some tribes (including the Huron and Hidatsa) cultivated watermelon in the same way they did squash (*Cucurbita pepo*). This makes sense—and would have speeded adaptation—since both are cucurbit crops that grow on long, trailing vines. In 1763, Charlevoix described how some native people propagated seed using the heat of a fire (Blake 1981): "Sun-Flowers, Water-Melons, and Pomkins are set by themselves; and before they sow the Seed, they make it shoot [germinate] in Smoke, in light and black Earth."

The sweet dessert watermelon in America today is a many-splendored thing. Although it originated in northeastern Africa, it has become over the centuries a part of the American vernacular, ingrained in our culture and traditions. For such bounty, we owe a debt of gratitude to early agrarians, immigrants, amateurs, professional breeders and, ultimately, seed savers.

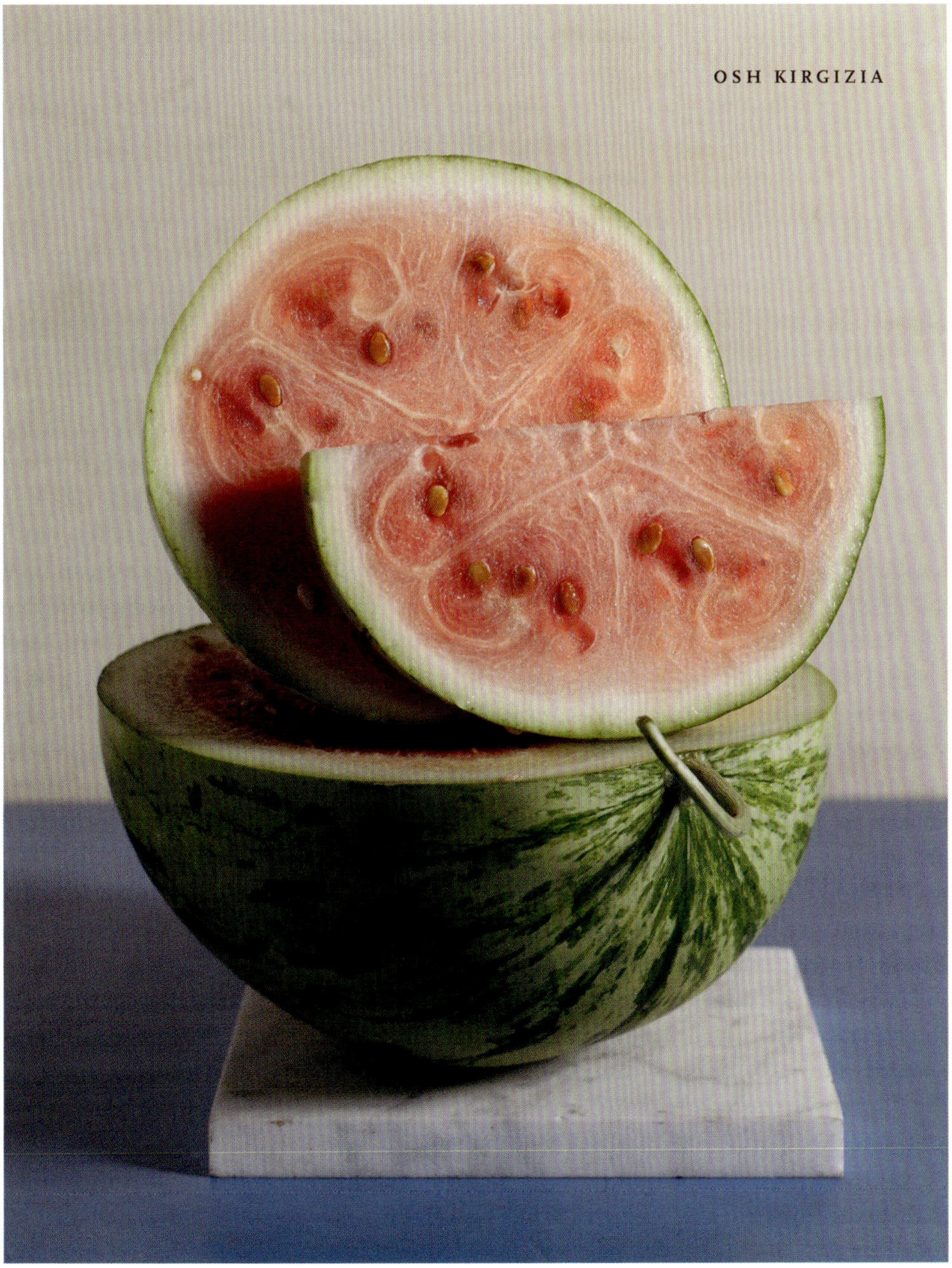

OSH KIRGIZIA

In the gallery of watermelon that appears in this book, grown by yours truly in the Hudson Valley of New York, there's a cultivar for almost every place, occasion, family, and inclination. Watermelon comes in many sizes (mini, icebox, small, medium, large, and exhibition) and shapes (round, oval, blocky, elongate, and pyriform) and can be classified into market types based on those criteria. There are no cultivar groups as there are in melon (*Cucumis melo*). Watermelon rind surface is usually smooth and waxy (some varieties are furrowed) and lacks certain topographical features common to melon. Flesh may be red, pink, orange, yellow, white, and rarely bicolor or green. Unlike other cultivated cucurbits, watermelon exhibits a wide variety of seed coat colors (black, brown, tan, green, red, white) and patterns (white tan-tipped, white pink-tipped, black banded, dotted or stippled black).

What appeals to me most about watermelons—apart from devouring a sweet and crisp one—are their coats of many stripes and patterns. Rinds generally remain green (light, medium, or dark) or "gray" (speckled light green) at maturity, with or without darker green jagged stripes (narrow, medium, wide). There are some with pale orange rinds. I love the quirky patterns: spotted like Moon and Stars, intermittently striped like Osh Kirgizia, marbled like Scaly Bark, or penciled like Golden Midget. Commercial heirlooms such as Crimson Sweet and Allsweet are among my favorites.

Nearly all watermelons featured in this book are cultivated for their sugary flesh. Others are valued primarily for their rinds or seeds. Many carry names that reflect their origin or fruit attributes, such as Orangeglo and Luscious Golden. Choose old-fashioned big mamas, like Odell's Large White, with its thick rind, for making rind pickles. Check out Killy Edible Seeded or Malali if you enjoy snacking on nutritious watermelon seeds. Carolina Cross, Cobb Gem, and Black Diamond are the ones to grow for contests and exhibitions.

I've included a few citron watermelons of a different species (*Citrullus amarus*) with red or green seeds. Citron watermelons are a far cry from the modern sweet dessert watermelons (*Citrullus lanatus*) most of us know and love. Their flesh is typically hard but nonetheless watery, insipid (although sometimes bitter), and weakly colored (white, light green, or yellow) at maturity. But that doesn't mean they are good for nothing. Citrons were independently domesticated in southern Africa (the Namib-Kalahari region) thousands of years ago as a source of uncontaminated water, food, and animal fodder. A recent molecular phylogenetic study of the genus *Citrullus* reveals that citron watermelon is not the progenitor of the sweet dessert watermelon as was once believed (Chomicki and Renner 2015). Known locally as tsamma or cow melon, it remains an important crop—used for fodder and often cooked or pickled in areas of southern Africa where water and other resources are scarce.

In Europe and the United States, cultivated citron watermelon is a minor or relict crop, used as a sweetmeat prepared with sugar and commonly known as Preserving Melon. It arrived in Mediterranean lands during or prior to the Roman era, according to Harry Paris. It was as common in Europe as sweet dessert watermelon in the early Renaissance—probably because it was more hardy, better adapted to cool (then cooler!) Europe. Dessert watermelons could be grown successfully only in Mediterranean Europe. The citron was later grown by Americans in colonial times and remained popular throughout the nineteenth century.

In my own garden, I've never seen a citron watermelon felled by anything other than frost—they are incredibly vigorous and prolific. These same assets can also be liabilities and, to many farmers, citron is a demon: The feral watermelon escapes into

Desert Watermelon

Wild in the deserts
Of Egypt and Sudan,
Grows hard and bitter fruit
Called gurma in the land.

Harvested and hoarded
Somewhere in the shade,
It holds a fount of water
In green flesh that it made.

Water for dry seasons,
Water kept in storage,
Water for a Pharoah's Ba
On his celestial voyage.

The fibrous fruit was pounded,
And the water bound would flow.
A gift to desert dwellers
Five millennia ago.

From one gene only dominant,
Its bitter taste was made,
So when recessive flowers met
The bitterness did fade.

Melons bearing yellow flesh,
By Common Era's time,
Rabbis grouped with grapes and figs
As sweet within the rind.

The gene for sugar links with red,
Though DNA was not yet spelled,
Medieval farmers bred
A fruit fit for angels.

Ruby slabs of watermelon
Decorate my table,
While in the wild deserts
Its ancestral stock is stable.

Civilization could collapse,
There could be Armageddon.
But in five thousand years,
Our kin could
—Again—
Have watermelon.

—*Kate Rauner*

the wild in the southern United States, overruns cotton and sorghum fields, and readily mates with other watermelons, contaminating the seed stock, although not the fruit on the vine.

The citron watermelon's assets far outweigh its liabilities—as elaborated by research geneticist Amnon Levi at the USDA–ARS in Charleston, South Carolina, and his collaborators, in a 2017 review of watermelon genetic resources. Sweet dessert watermelon cultivars (*Citrullus lanatus*) notoriously share a narrow genetic base, leaving them susceptible to pests and diseases and abiotic stressors. Various genebank accessions of the far more genetically diverse citron (*Citrullus amarus*), collected in southern Africa, contain genes or alleles that are proving useful in watermelon breeding programs as sources of resistance to root-knot nematodes, fusarium wilt race 2, gummy stem blight, anthracnose (races 1, 2, 3), powdery mildew (PM), and potyviruses. And with its more efficient and vigorous root system, it may serve as rootstock or hold the genetic key to higher yields, enhanced heat and drought tolerance, and water use efficiency in sweet dessert watermelon.

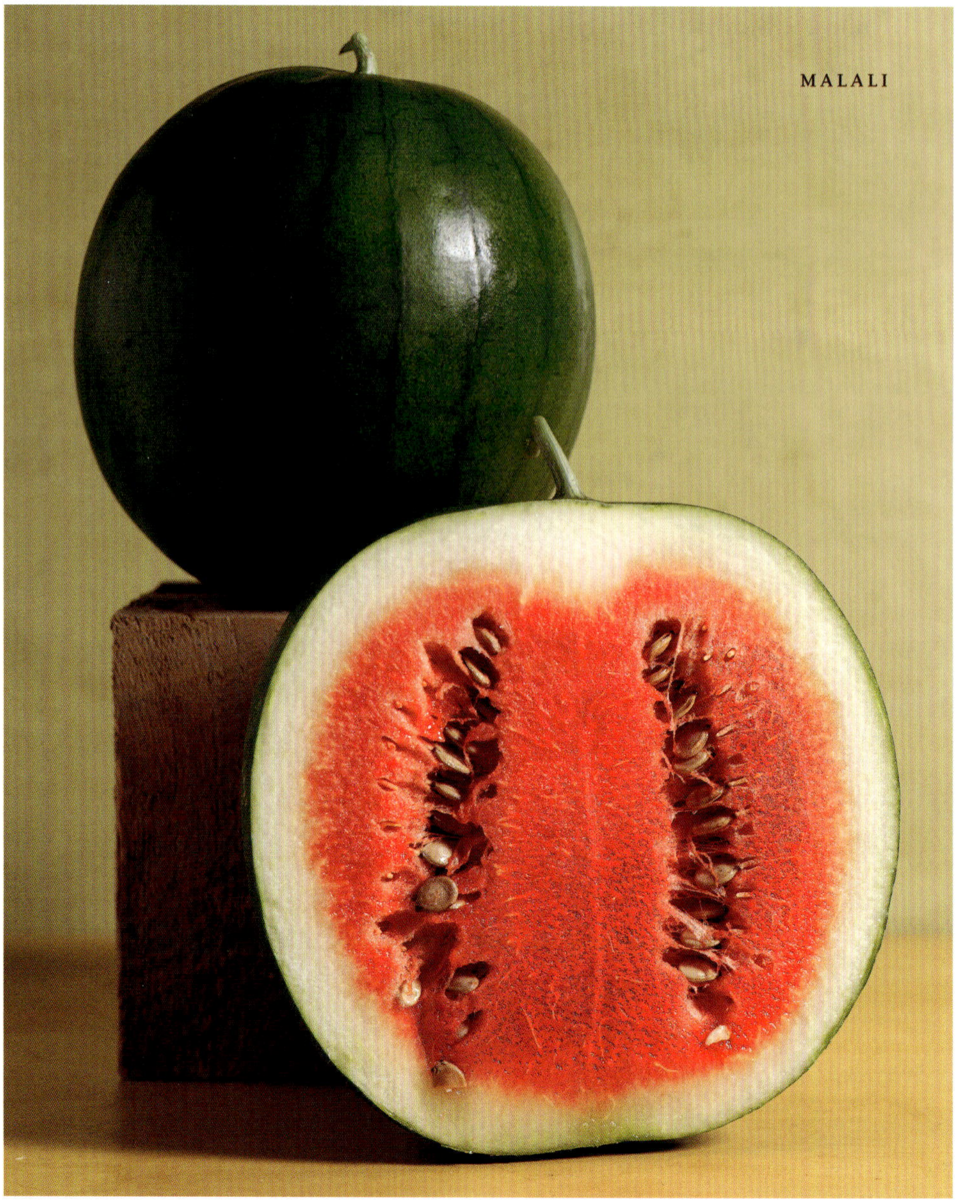

Malali is a "confectionary" watermelon traditionally grown in Israel for seed consumption. But it has a dual purpose: the vivid red flesh is good enough to eat. Opposite: Irish Grey is a prolific producer of luscious pink-fleshed, pale-green oblong watermelons. Practically impervious to sunscald.

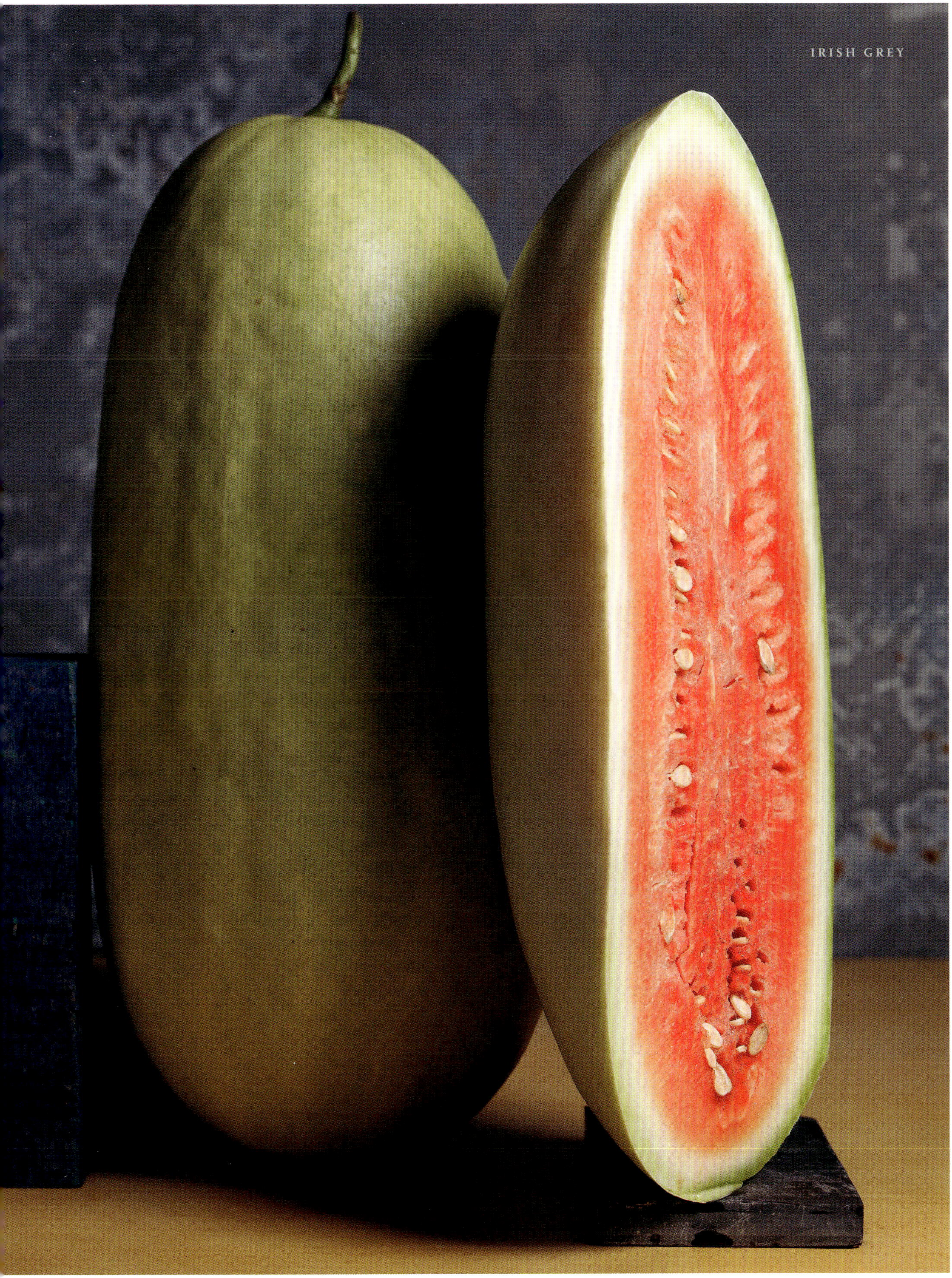
IRISH GREY

Odell's Large White

In September of 2016, I chanced to meet Rodger Winn and his wife, Karen Shealy Winn, at Monticello's Heritage Harvest Festival in Charlottesville, Virginia. Naturally we got to talking about watermelons . . . and some that were presumed dead and gone. I showed them a photo of the long-lost Jumbo Triumph in the copy of Lloyd Bright's pamphlet "World Record Giant Watermelons 1917–1987" that I'd been carrying around in my bookbag. That's when Rodger shyly mentioned that Karen's family, the Metzes, had been keeping a watermelon named Stoney Mountain for over a hundred years in what is known as the "Heart of the Dutch Fork" of central South Carolina. Their friend James Everett Kibler had identified the melon as the missing Odell's Large White a year earlier. I grew curiouser and curiouser.

Back at home again a few days later, I received in the mail a couple of photographs of the Metze family heirloom and seeds of it to trial the following summer. The variety had been grown by Karen's great-grandfather, John Quincy Metze, of Little Mountain, South Carolina, around 1900, and by his father before him. The family maintained the watermelon in Little Mountain (which is very stoney) in isolation until the 1980s and then nearly lost it when freezers thawed on two occasions and seeds were thrown out along with the other contents. Luckily the Winns were able to obtain viable seed from Glenn Long, Karen's mother's first cousin, who lived nearby. Glenn died a week after turning over the seed.

I've grown the variety in my garden from Rodger and Karen's seed and can vouch for its quality: its sweet, juicy, somewhat coarse pink flesh is better than most. It has the shape, size, and furrowing of a Black Diamond, but the rind is gray rather than intensely dark green. The plant is a prolific producer of big lunkers that are a target of crows, who like to dive-bomb. Netting is the only thing that will dissuade them. They know a good melon, too.

Odell's Large White is described in Fearing Burr's *The Field and Garden Vegetables of America* (1863 and 1865) and in Burr's *Garden Vegetables* (1866), and except for the change in seed color to white, the description fits perfectly: "Size very large, sometimes weighing sixty pounds; form round; skin gray, with fine green network spread over its uneven surface; rind nearly three-fourths of an inch in thickness; seeds large, grayish-black, and not numerous; flesh pale red; flavor fine; quality very good. Productiveness said to exceed that of most other kinds. This remarkably large melon originated with a negro man on the property of Col. A. G. Sumner [*sic*], of South Carolina. Its large size, and long-keeping quality after being separated from the vine, will recommend the variety, especially for the market."

I recently had an enchanting telephone conversation with Dr. Kibler, who lives about twenty-five miles from Little Mountain. "It's been an interesting ride," he told me. When visiting the Winns on August 10, 2015, he'd seen some very large, round, whitish watermelons growing in their perennial borders and immediately thought "Oh my God, could this be?" He grew sixteen hills of the variety the following summer from seed of a fifty-pounder grown by Rodger, and harvested a bumper crop that bred true to Burr's description (except for the seed coat color). Kibler is an expert on Adam Summer (1818–1866) of Ravenscroft, the plantation where Odell's Large White was born, and his brother William Summer (1815–1878). They operated

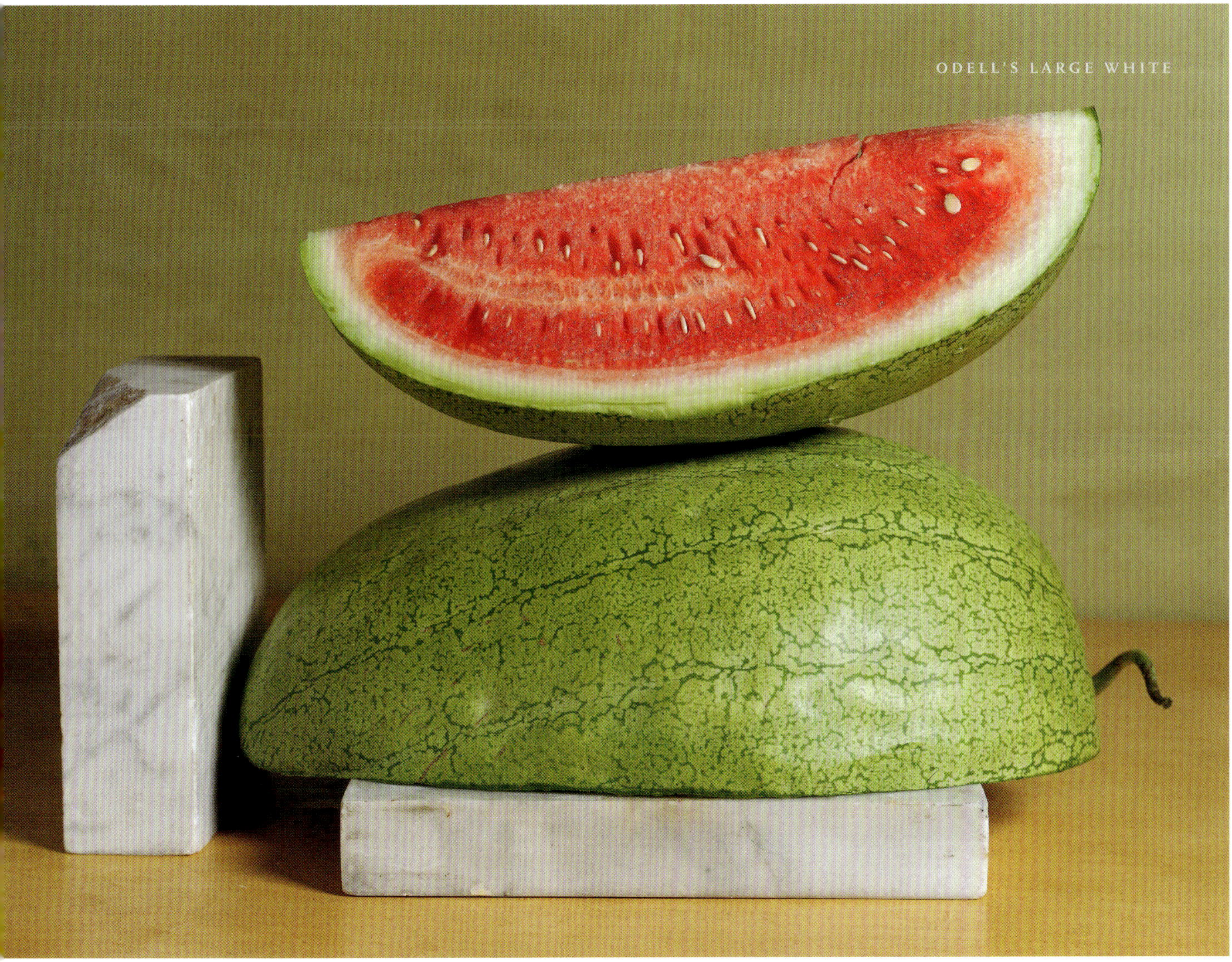
ODELL'S LARGE WHITE

Pomaria Nursery, which was one of the most prominent nurseries of the antebellum South (founded by William in 1840 and run by his heir until about 1890). Both were also distinguished journalists, editors, and nature writers. Adam grew a diversity of crops at Ravenscroft and experimented with plant breeding there; the plantation was also the site of the nursery's tree stock farm. The distance between Ravenscroft plantation and the Metze farmstead (neighbor to the Shealy-Winn farm)? Less than six miles as the crow (or raven) flies.

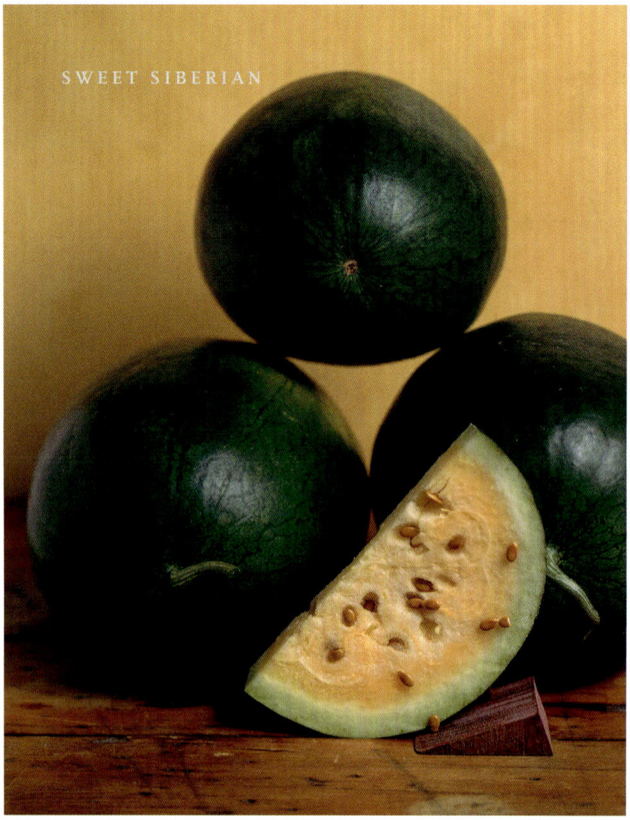

Kibler is editing the Summer brothers' garden calendar and carefully sifting through letters, ledgers, and tangential materials—much of which he saved from ruin or obscurity—that have already yielded correspondence between the brothers, Dr. William Draper Brinckle (Brincklé) (1798–1862), and Fearing Burr (1815–1897). Brinckle was a prominent seedsman and horticulturist, one of the founders of the American Pomological Society, and a friend of fellow-member William's. In the 1850s, the Summer brothers exchanged grafting stock with Brinckle and sent him new fruits that they had bred. Brinckle wrote a piece on Odell's Large White that appeared in *Horticulture* in 1857, and he was Burr's informant on Odell's as well as Downing's (1857). The watermelon carries the name of Milton Odell (1826–1895), a farmer and good friend of Adam Summer's. Odell sent a prize specimen for exhibition (probably to Augusta, Georgia) in October 1853 that weighed fifty-four pounds and received wide acclaim.

The Ravenscroft watermelon, described by Brinckle and featured in Burr, also got its start on the property of Adam Summer. Whereas the Odell's Large White was apparently the result of a chance seedling, according to Dr. Kibler, the Ravenscroft was a product of Adam's breeding program (probably between 1846 and 1850). And now Kibler has apparently found the missing Ravenscroft watermelon, kept by Mr. Carold Wicker, an elderly gentleman farmer and a friend of Kibler's, who calls the watermelon Leitzsey (his mother's family name). The Leitzsey homeplace is located about fourteen miles from Ravenscroft plantation, but they were once neighbors to Ravenscroft. The Kibler and Wicker families have attended the same church, located about six miles from Pomaria plantation, for about 250 years, and are related in a vast cousinage. Kibler describes Mr. Carold thus: "Quite a character and legendary locally as a curmudgeon, who in comparison would make Wendell Berry look like a cross between a Monsanto executive and Elton John."

Siberian Yellow is an eye-catching watermelon with Persian orange flesh. Adapted to the far north. Introduced in 1899 by Peter Henderson & Company. Opposite: Early Jumbo is a champ—I'm wowed by its sweet, crisp-textured, deep pink flesh. Needs more devotees.

To paraphrase Dr. Jim, "Finding both watermelons less than fifteen miles from Ravenscroft, where the melons originated over a century-and-a-half ago, is testimony both to the strong farm traditions of the area and to the stamina of the melons themselves." I know we'll hear more from him about the Odell's and the Ravenscroft. I can't help wondering: Who was the African American man associated with the remarkably large Odell's melon? Kibler says, "There were several free black families associated with the Summer family living nearby, all of whom were intermarried, so the man Brinckle describes was not necessarily one of Adam's slaves."

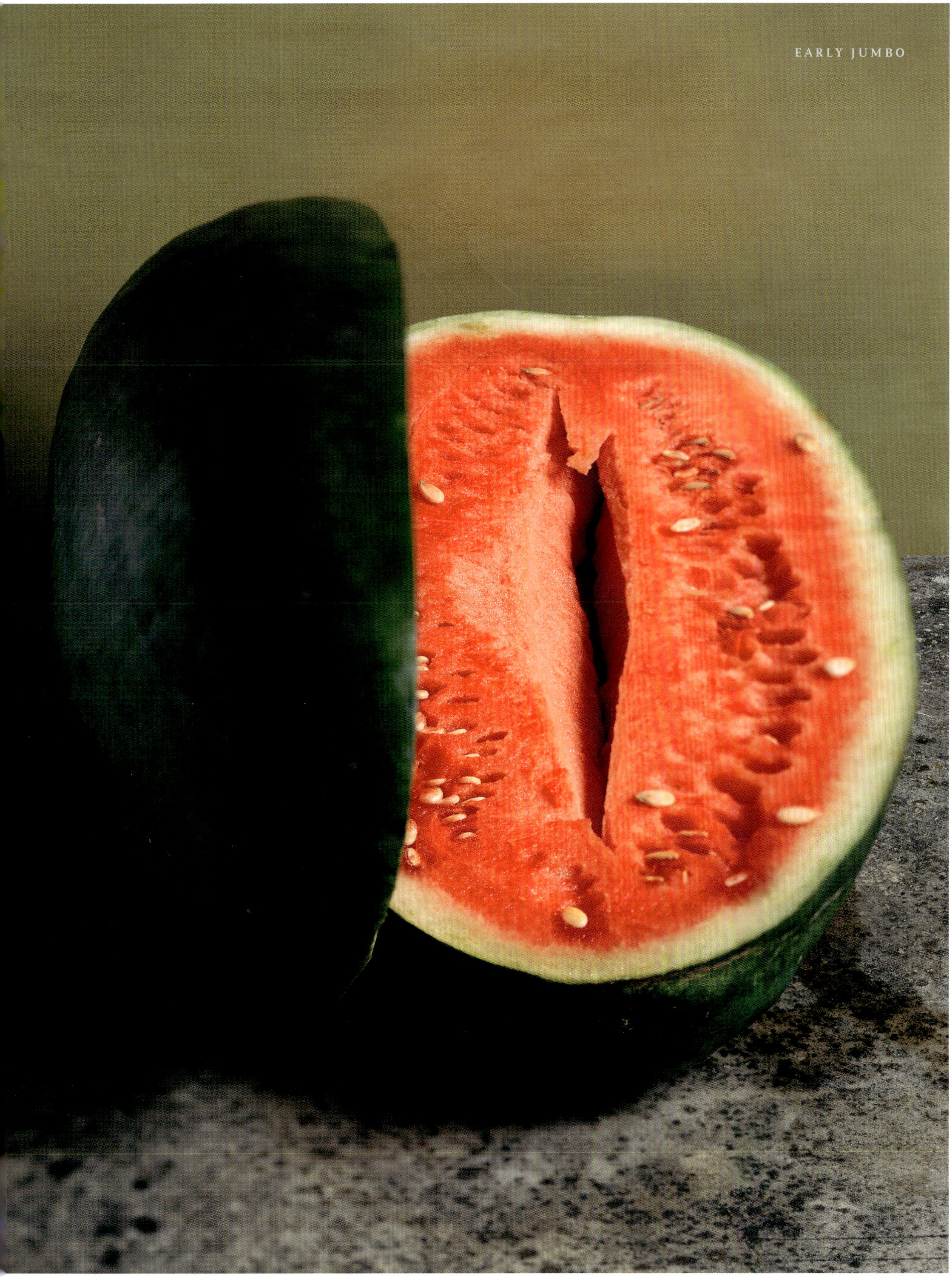

EARLY JUMBO

Georgia Rattlesnake

One of my favorite watermelons. Statuesque, extra-large and heavy, thick-rinded and sturdy, uniformly cylindrical, with jagged dark "rattlesnake" stripes of medium breadth. It holds a world of goodness inside: designed to quench our thirst, satisfy our sweet tooth, and bring joy. A melon for the ages, no? Rattlesnake resembles a long, boldly striped fruit served on a tray depicted in a wall painting in an Egyptian tomb over 4,000 years ago (Paris 2015). Makes me think how far the sweet dessert watermelon has come from ancient times, when watermelon flesh was neither sweet by modern standards nor highly colored. Georgia Rattlesnake has the distinction of being one of two parents of the world's heaviest variety: the Carolina Cross.

Georgia Rattlesnake is known as "the watermelon that made Georgia famous," the Augusta section in particular, as fertile ground and point of origin for watermelons shipped North by rail beginning in the 1860s. The earliest mention I've seen of the Rattlesnake is in an article titled "Cultivation of the Watermelon at the South," by Hawes H. Coleman of Cachemasso, Arkansas, contained in the *Report of the Commissioner of Patents for the year 1855*. Coleman offered suggestions that resonate: "Melons need all the ground they occupy. The smallest weeds affect them. They require all the sunshine that falls on the plat in which they are planted, and every breeze that sweeps over them, and that the soil should be light and well worked." He listed the Coleman, the Rattlesnake, and the Bough watermelons. Of the Rattlesnake he wrote: "It should be planted in hills twelve feet or more from center to center [*sic*], on land properly prepared. It takes its name from its resemblance to the rattlesnake. It is moderately large, weighing from twenty to forty pounds. The seed is almost white, the rind thin and brittle, and the meat red. It is larger and more beautiful than [his own] Coleman, but inferior in flavor."

Dr. David S. Shields, Carolina Distinguished Professor at the University of South Carolina and an expert in Southern foodways, provided me with his file description of the Rattlesnake. It contained a reference to an article written by P. J. Berckmans in the May 1866 issue of the *Southern Agriculturist*. I had a heckuva time locating this reference—there was no journal by that title in print in the year 1866—but succeeded with the help of three talented archivists who work in special collections: Stephen Sinon of the New York Botanical Garden, Jennie Oldfield of the Cherokee Garden Library, and Carolyn S. Jackson at Texas A&M University Medical Sciences Library. The article appears in the May 1866 issue of *Southern Cultivator*. The author, P. J. Berckmans (1830–1910), is best known as the father of peach culture in the South for "his work in originating, introducing, improving, or disseminating the South's most famous varieties of peaches" (Range 1952). He described the Rattle Snake watermelon as "Large to very large; long, with alternate green and white stripes of equal width, and dark green spots; flesh red, sweet, crisp and much water, very good; rind thick; seeds white, with black edge, and two very small black specks at the base. This variety grows to a very large size, but bears sparingly." Berckmans counted it among the best varieties in cultivation. "For market purposes, the Mountain Sweet, Sheephead, Cuba, Lawson, Rattle Snake and Strawberry, are the best. For garden culture, where quality is preferred to quantity, the Mountain Sweet, Orange and Imperial, will be found very valuable."

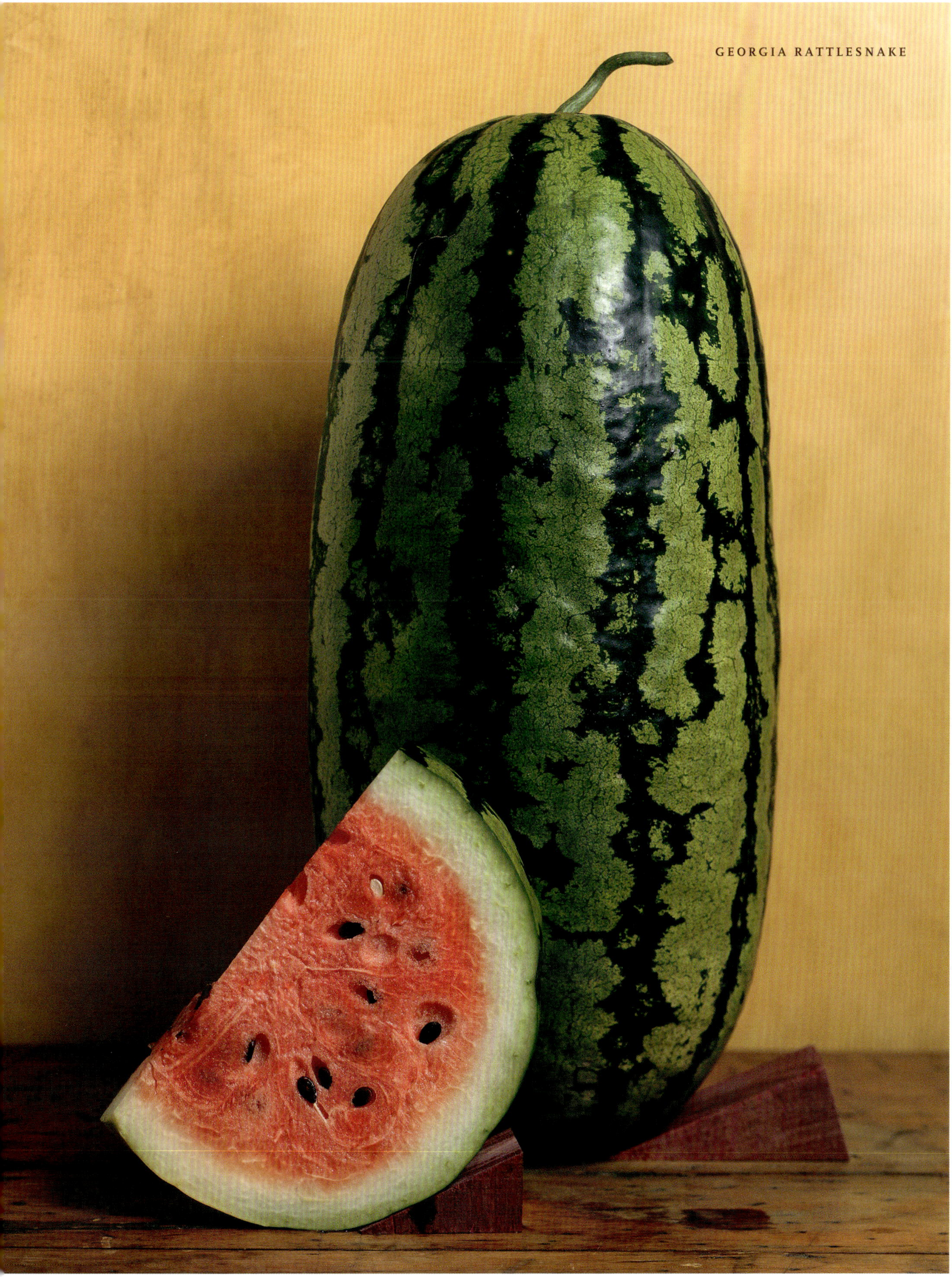
GEORGIA RATTLESNAKE

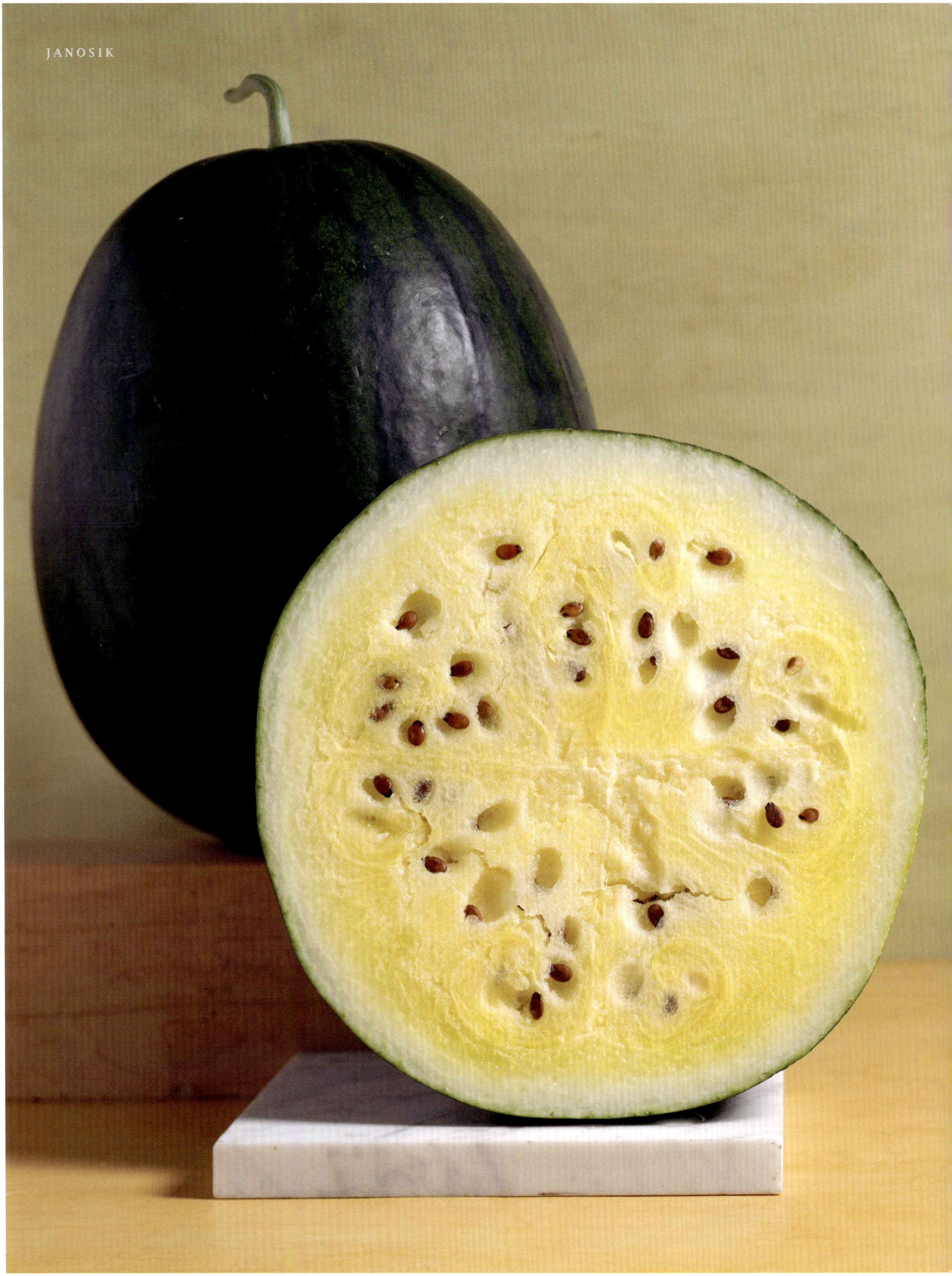
JANOSIK

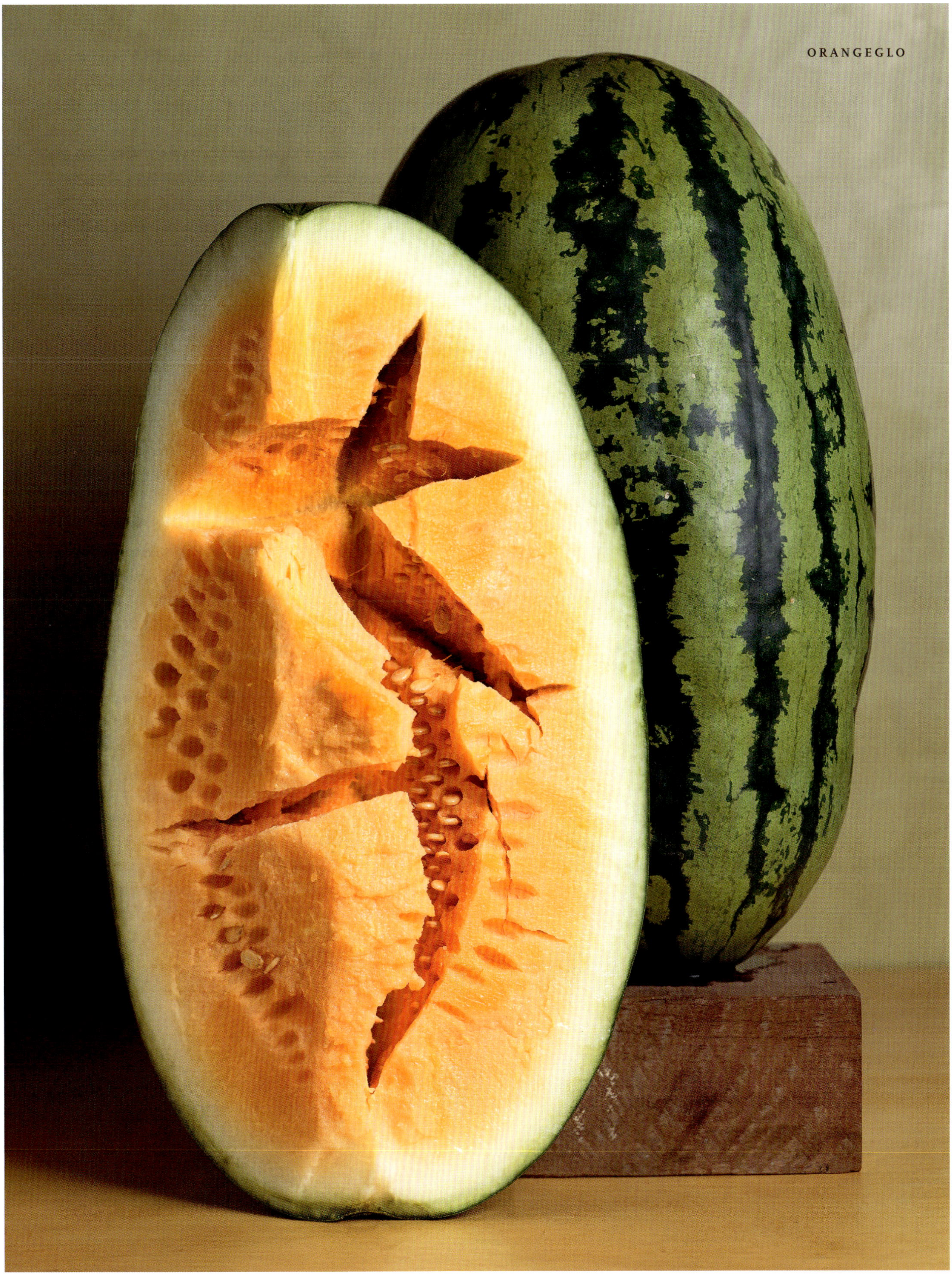

Dr. Shields believes the Rattlesnake to be the progeny of a cross between two of those varieties: Lawson and Mountain Sweet. He says the first to offer seed for sale were Plumb and Leitner, seedsmen and pharmacists of Augusta. The earliest commercial listing I've found is Rattle Snake in Thorburn's 1871 *Catalogue of Seeds* (New York City's own)—earlier than Berckmans's Fruitland Nurseries. The Georgia Rattlesnake held its ground through the early part of the twentieth century, then gave way to new arrivals. And Fruitlands, P. J. Berckmans' home and nursery, "the Southland horticulture Mecca for all tree and plant lovers," became the domain of the Augusta National Golf Course.

Moon and Stars

The Moon and Stars watermelon is the poster child of the heirloom seed movement. Its buttery moons and constellations of stars are awe-inspiring. Unique among watermelons, both fruit and foliage are bedecked with heavenly bodies. This miraculous trait is the handiwork of a higher power: the *Spotted* (*Sp*) gene.

Peter Henderson & Company of New York introduced the novelty in their 1924 catalogue as Sun, Moon and Stars, and illustrated it with an out-of-this-world engraving. The catalogue copy reads: "We often hear the old saying quoted, to the effect that 'there is nothing new under the sun,' but even if the essential truth of the old saw is conceded, it must be admitted that there are innumerable variations of things that already exist. This new watermelon, named by its originator 'Sun, Moon and Stars,' is a conspicuous instance. The circles and stars imprinted on its rind are certainly an extraordinary variation from the usual type, and they have persisted for so long a period as to really make Sun, Moon and Stars a new variety. Seeds were sent to us originally by a practiced melon grower of Rocky Ford, Colorado, who fixed the type, and named it. The rind, which is very thin, is medium green in color, with an obscure dark stripe; and conspicuously marked with cream-colored circles and star-shaped spots, both large and small. This gives the melon a very distinct appearance, and has suggested its name, Sun, Moon and Stars. We do not, however, offer seeds of this melon to our friends because of its appearance, but for the reason that its 'meat' has such a delicious taste, and because this variety can be successfully grown in most parts of the United States. It is of medium size and just about right for private use. Those of our friends who have not hitherto grown Watermelons, should give this variety a trial. Price, 25c. pkt., 5 pkts. for $1.00."

But no matter how stellar the watermelon, the Sun, Moon and Stars almost disappeared into a black hole. Thirteen years after its introduction, the variety was still in circulation, according to the J. D. Robinson Seed Company, which illustrated the fruit in 1937 and described it as novel but not without merit (of about average table quality). For unexplained reasons it was dropped from commercial production (Henderson discontinued it in 1932). For decades it was nowhere to be found.

Its loss and recovery is a cautionary tale. Kent Whealy (1946–2018), cofounder of Seed Savers Exchange (SSE), along with a number of members, set out to find the fabled variety. After four years of fruitless searching, Kent hit pay dirt about a hundred miles southeast of his Princeton, Missouri, home. In the spring of 1981,

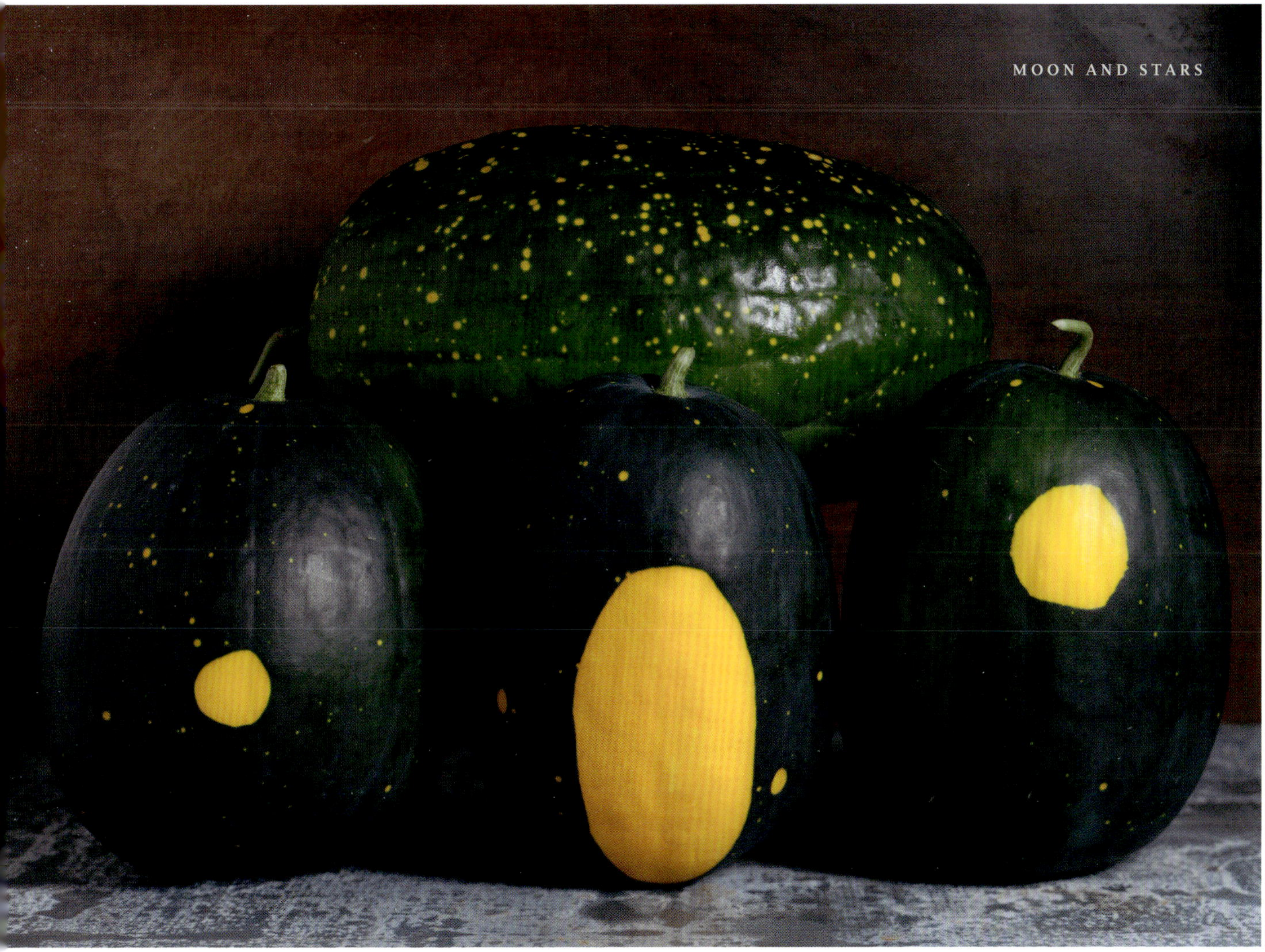
MOON AND STARS

Kent gave an interview on local television station KTVO in Kirksville, in which he broadcast a plea for help in the search. After the show aired, a letter arrived. Merle Van Doren wrote, "My dad always grew Moon and Stars in Tennessee and I have grown the melon on my farm near Macon for as long as I can remember. . . . The melon is like the Black Diamond [see page 172] but more disease-resistant. The fruit usually matures in about 90 days. I will save some seed for you."

In *Gathering: Memoir of a Seed Saver*, Diane Ott Whealy, cofounder of Seed Savers Exchange, vividly describes the visit to the Van Doren farm in the fall of 1981 with Kent, their children, and Larry Hollar, Jeanne Malmgren, and a photographer from *Mother Earth News*: "The melon patch was behind the Van Dorens' farmhouse, and a mound of about ten melons had been placed on the lawn in front of the garden. The bright yellow moons and stars glowed against the grass, with a background of yellow marigolds and a few red velvet cockscomb peeking out for contrast. . . . The afternoon ended around the Van Dorens' kitchen table with a modest lunch of fresh tomatoes and open-faced pimento loaf sandwiches—and of course slices of Moon and Stars watermelon. Then Kent and I loaded the car with as many watermelons as we could cram in, and drove back to our homestead feeling as if we had found a pot of gold."

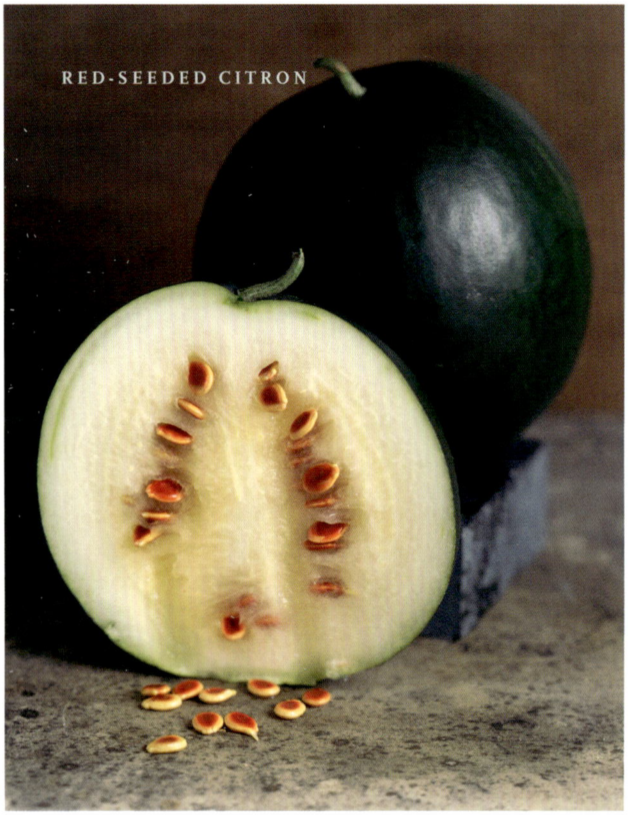

Red-Seeded Citron watermelon produces seeds that retain their gorgeous red color for years. Opposite: Ledmon is a sublime North Carolina heirloom: sugary, juicy, firm, habit-forming.

Merle famously said before the Whealys departed that day, "I'm sure some folk would give a pretty penny to get their hands on this melon seed." Kent recalled, "I told him I was afraid I was just going to have to give it away to as many people as I could. He just grinned at me." And Kent grinned back. SSE proceeded to distribute seed to approximately 125 members of its 1982 Growers Network. It wasn't long before Moon and Stars was reintroduced commercially: Seeds Blüm of Boise, Idaho, was the first company to list the rediscovered watermelon, in its 1985/6 catalogue.

The watermelons that Kent and Diane carted home in 1981 from Macon, Missouri—a "founder population"—were variable in shape and size. But overall they closely resembled the original, slightly furrowed, Black Diamond–type Sun, Moon and Stars introduced by Peter Henderson in 1924 (oblong, pink-fleshed, brown-seeded). Today these are known simply as Moon and Stars or Van Doren strain. After "the discovery," other variants emerged and, over the years, growers selected for novel flesh or seed color and improved horticultural traits. And there are always happy accidents, including my own Moon and Stars and Stripes (see page 176).

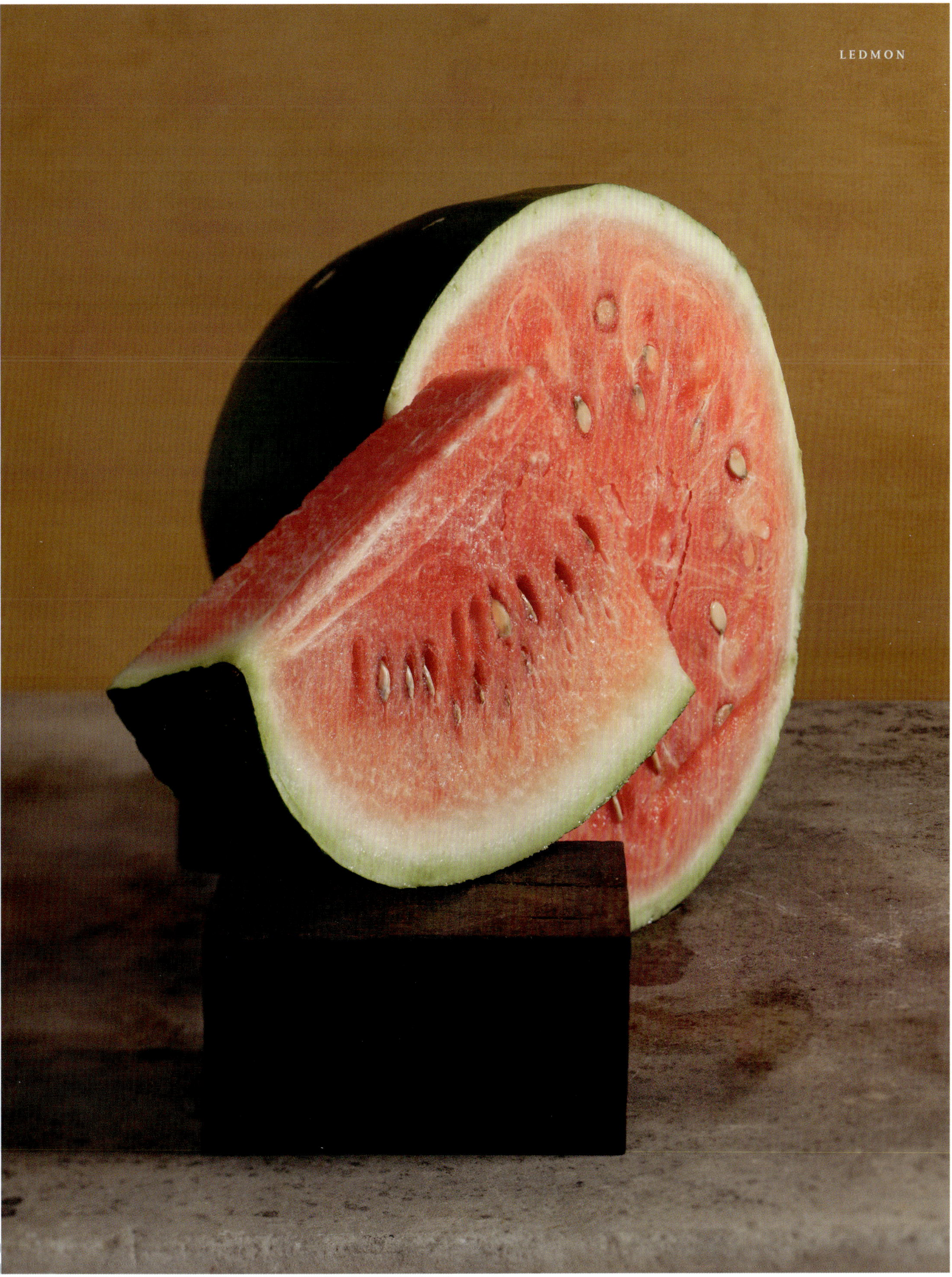

Tom Watson

The Tom Watson is a good old-fashioned watermelon. It's the kind Allen Lacy (1984) might call the food of holy communion. Born and raised in Texas, Lacy "came from a long, proud line of watermelon eaters, watermelon experts, and watermelon thieves." He knew watermelon was meant to be shared by as many people as possible, so only a mammoth melon from Texas, Florida, or Georgia, sweet and in season, would make the cut.

This good old boy is known to produce giants weighing over a hundred pounds. It was one of the record-setting Hope watermelons (including Triumph and the Triumph-Watson hybrids) grown by the Laseter clan in Arkansas in the 1920s. Size matters, but taste—known in the trade as "cutting quality"—is the bottom line. The Tom Watson is no slouch here, either. The vivid flesh "cuts red" (has no hollow or white heart) and is crisp and scrumptious. Tom's superior flavor and thick, hard rind made it the most popular shipping watermelon during the Great Depression. Archival photographs show these family-sized beauties stacked vertically being transported to railroad loading spurs on Model T trucks and horse-drawn wagons.

The Tom Watson watermelon was named for Thomas Edward Watson (1856–1922), of Thomson, Georgia, a noted trial lawyer and prolific author, U.S. congressman, and later a United States senator. He is lauded as the architect of the Rural Free Delivery mail system. Though his campaigns for the presidency were unsuccessful in 1904 and 1908, the watermelon was—and remains—an uncontested winner.

According to Tad Brown of the Watson-Brown Foundation, a nonprofit organization that focuses on historic preservation and owns and operates three Watson historical sites in Georgia, "The Watson watermelon seems to have originated around 1900, when a Florida farmer forwarded to Tom Watson an envelope of melon seeds he had named in Watson's honor. Watson summarily passed the seeds along to his brother, Top, a local merchant and struggling farmer who had experience growing produce. After a few growing seasons, Top Watson discovered the unusually large melons grew a very thick rind, making them ideal for shipping long distances. . . . They were branded "Tom Watsons" and quickly became a commercial success. Soon brother Top was cultivating more than 50 acres of melons exclusively for their seeds."

William Augustin "Top" Watson (1859–1937) is credited with introducing the Tom Watson watermelon to Southern growers prior to 1906. Top is described by the Foundation as a "seedsmith" and his production of a "grander scale and refinement" than that of his nephew, J. D. Watson, Tom's own son, who once grew seed produced on his father's farm, and employed "stomping children" to mash ripe watermelons. By the 1920s, Top and his son, Truman, were devoting more than 350 acres to their endeavors. The family firm, William A. Watson & Son, and its successor, Wm. A. Watson's Sons, maintained the original strain with the utmost professionalism for decades.

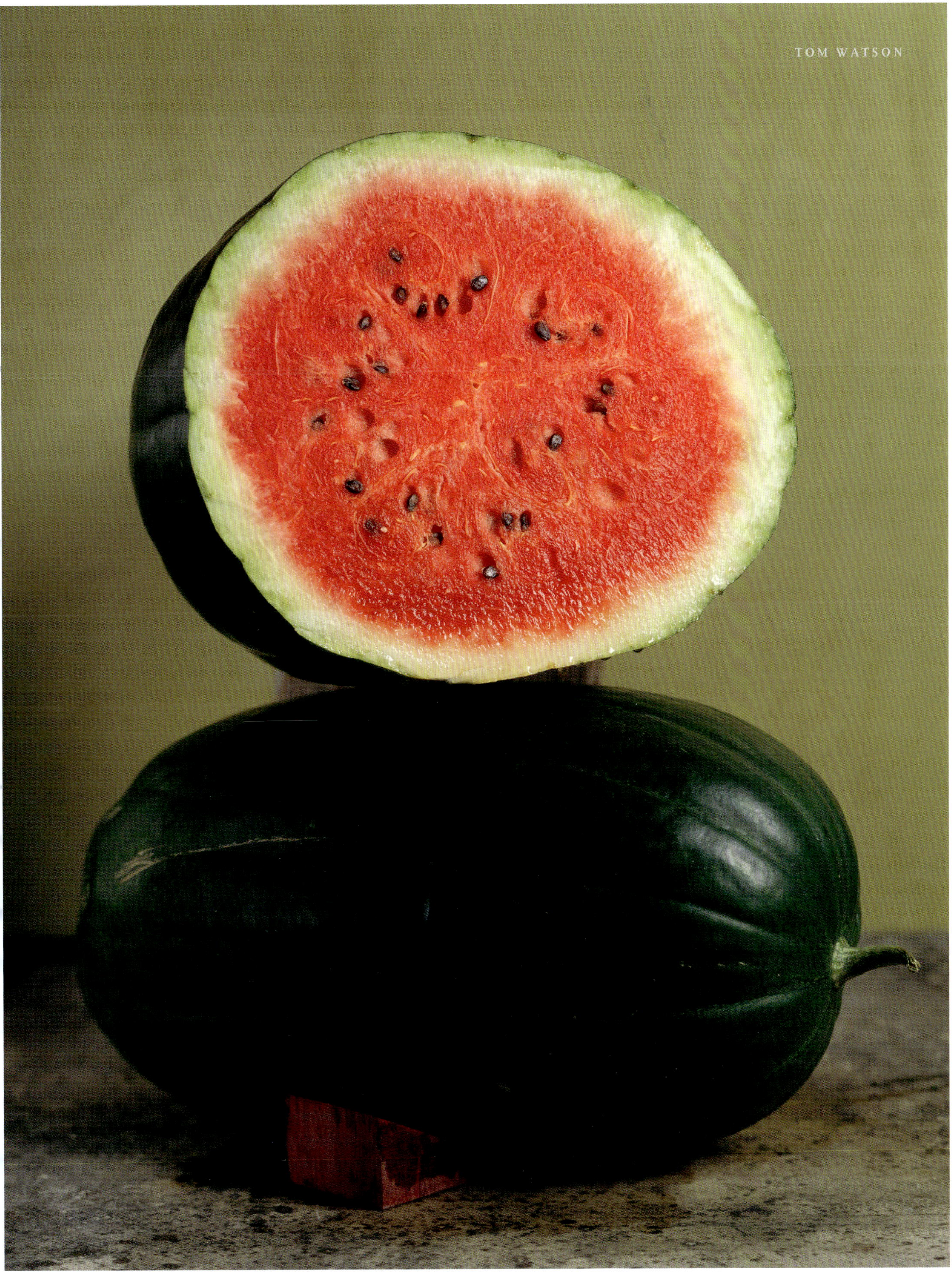
TOM WATSON

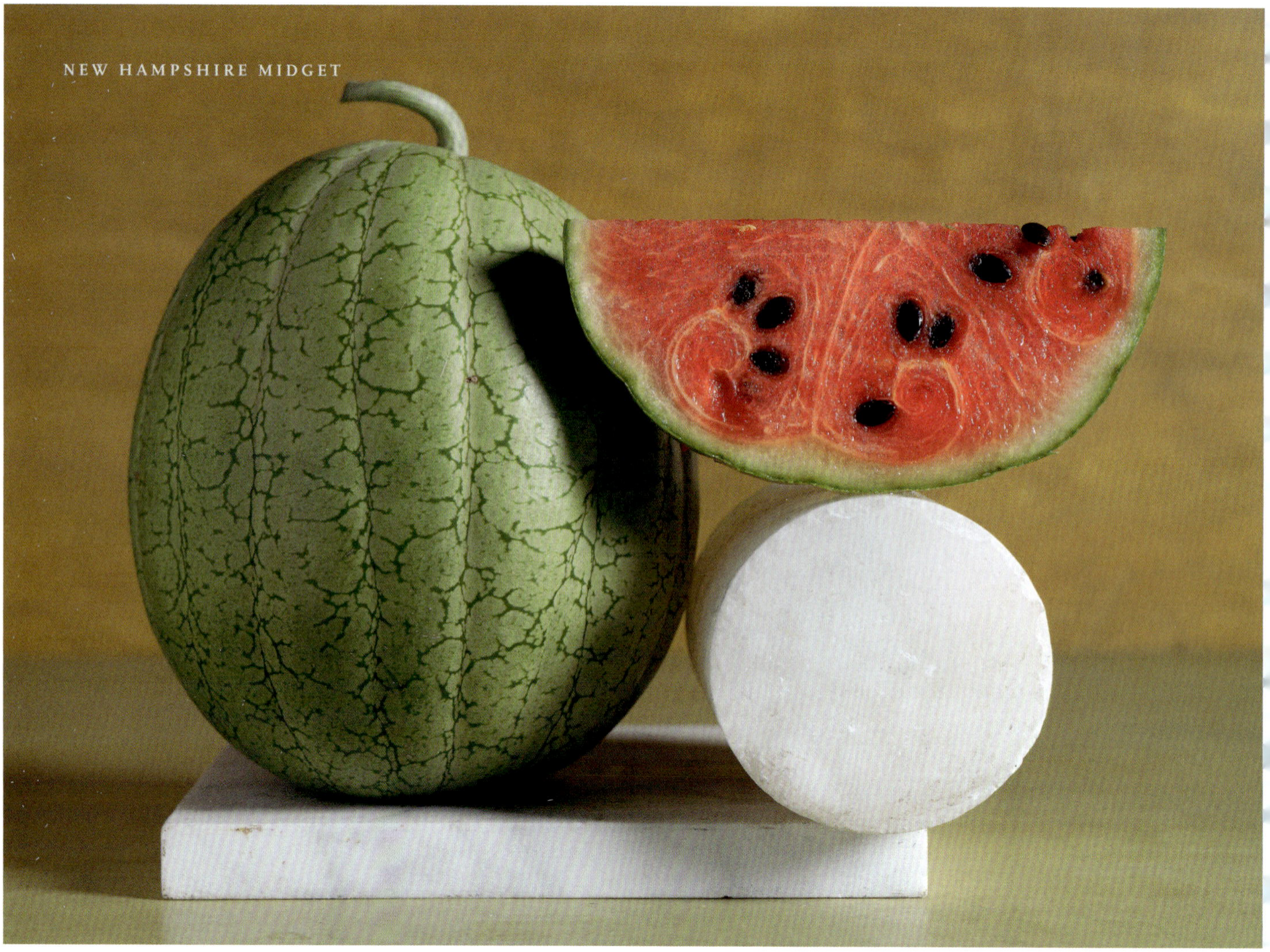
NEW HAMPSHIRE MIDGET

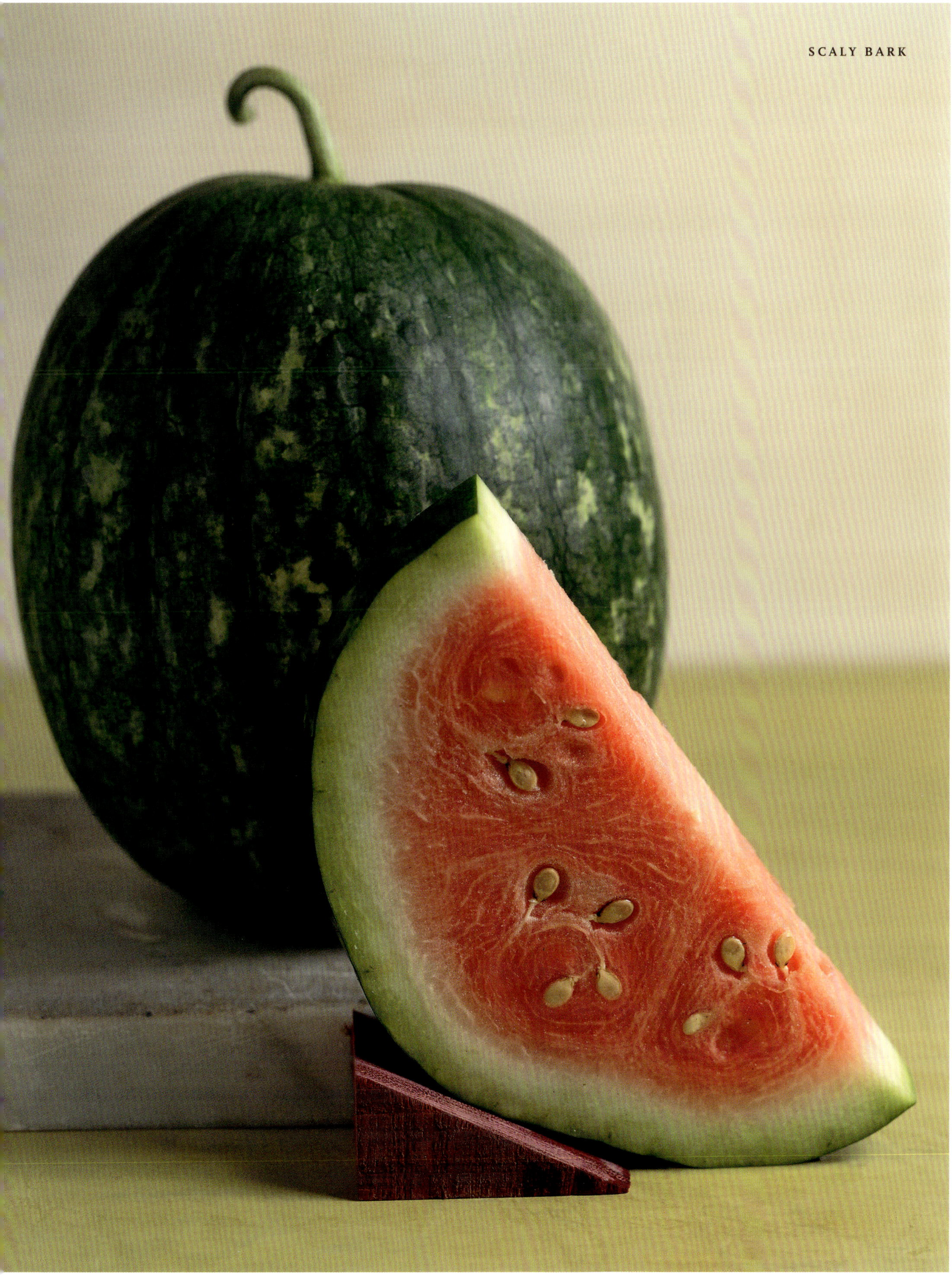

Killy Edible-Seeded

This is one seedy watermelon: a confectionary type grown for its large edible seeds. Best known as Wanli in China, where such watermelons have been grown and enjoyed for a very long time. Peter Valder, author of *The Garden Plants of China* (2005), cited A. H. Crook, who claimed in 1935 that the eating of watermelon seeds by the Chinese is mentioned in a letter written around 1080. Crook also observed that many Chinese had grooves in their incisors formed by years of opening watermelon seeds with their teeth. Ernest Henry "Chinese" Wilson, better known as E. H. Wilson, the famous English plant collector and explorer, wrote in 1913 that "The seeds of the water-melon are esteemed a great delicacy. They are slightly roasted, and are consumed in enormous quantities; no banquet is complete without them, and over their gossip in tea-shops or restaurants, scholars and coolies alike regale themselves with these delicious morsels." (Pardon the impolitic language but you get the drift.)

Killy was offered by Gleckler's Seedmen as early as 1972 (as Edible Seeded, then Kily Edible) and by Peace Seeds in 1985 and 1986 (as Killy Edible Seed or Seeded, then supplanted by Wanli Edible Seed F2 in later years). It eventually made its way into the Seed Savers Exchange network. Large-seeded watermelons grown for confectionary purposes are not currently available from commercial seed houses in the United States and Canada. We're simply not a culture of watermelon seed eaters—especially not when we can munch out on native plant seeds like pumpkin and sunflower.

The big black-rimmed seeds are gorgeous—I was flummoxed the first time I saw them. They reminded me of the large tan seeds of Sucrin de Tours pumpkin, which have prominent rims around the periphery. Wanli seeds are big and fat and require a little finesse in cracking the thick, hard seed coats. Eating them is a pleasant occupation. You can find roasted seeds in local Chinese dry goods stores (or order them from China) with licorice, plum, and other flavors. Why not try cultivating Wanli in your own garden? It's a simple matter to extract the many seeds embedded in the fruit (two paper plates full per melon)—they pop out at the touch of a finger. May be easier than shelling peas.

Crimson Sweet

Crimson Sweet cries out to be eaten and enjoyed. Those gorgeous broad dark stripes make this watermelon recognizable by striping pattern alone in home gardens and farmers' fields everywhere. It's a world-beater—literally. Today, Crimson Sweet may be one of the most popular watermelons on earth. Does an heirloom watermelon need to be rare? Not at all.

Crimson Sweet exemplifies the value of land-grant university plant breeding programs, and the breeders working magic behind the scenes. This sweetie was developed by Dr. Charles V. Hall at Kansas State University, in Manhattan, Kansas, over a ten-year period. Seed stock was made available for increase in 1963, with enough supply to allow full commercial release for 1966. Crimson Sweet has a good

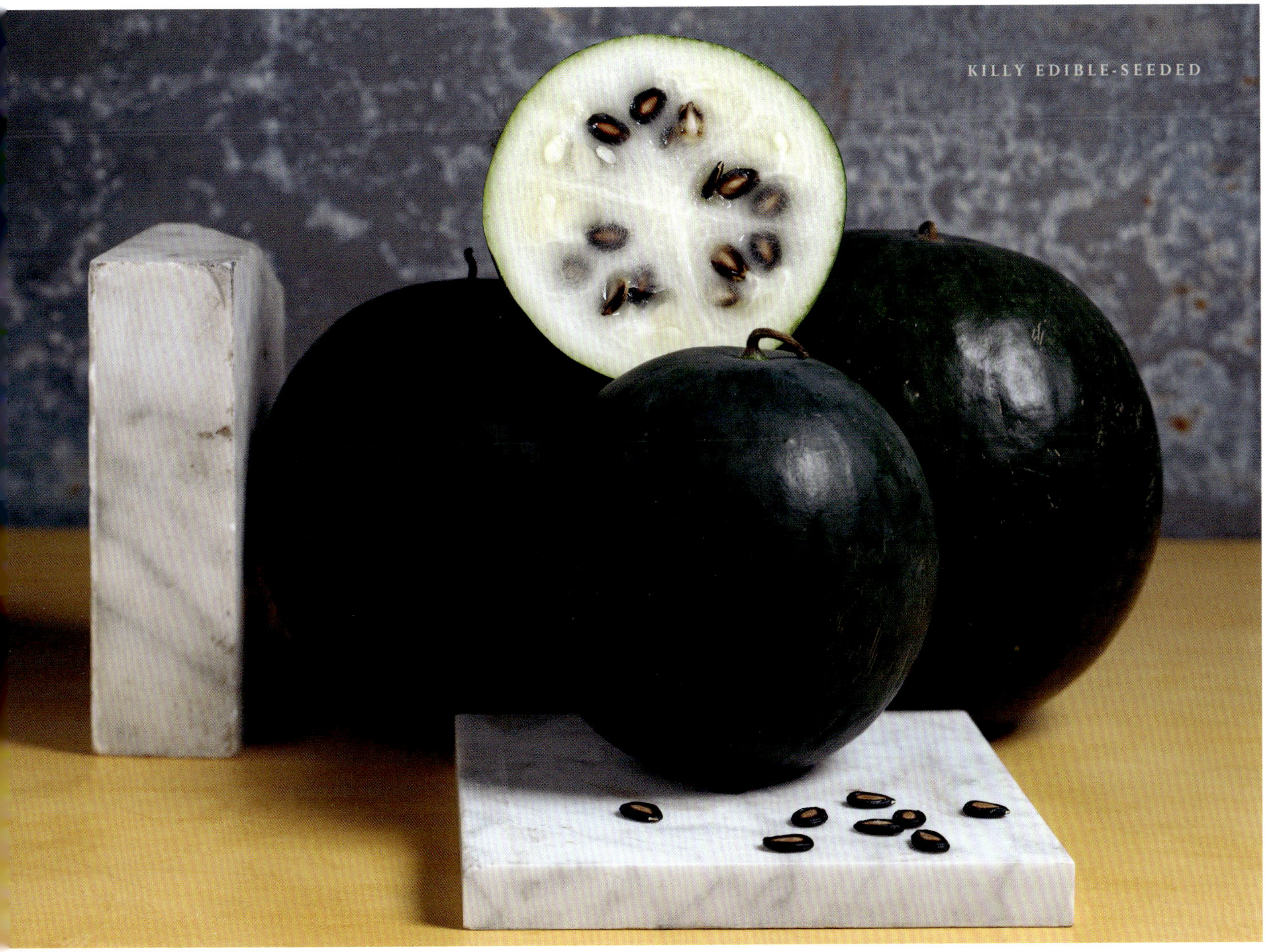
KILLY EDIBLE-SEEDED

pedigree: (Miles x Peacock) x Charleston Gray. And it has an extremely good "sister" in Allsweet, selected from the same breeding line (see page 175).

I can't claim bragging rights but I do have a personal connection, through my husband, Cary Fowler, to the Miles watermelon, one of Crimson Sweet's parents. Noted for its very high sugar content and resistance to fusarium wilt, Miles was developed in the 1940s at the West Tennessee Agricultural Station in Jackson, Tennessee. Cary's maternal great-grandfather, Martin Van Buren Exum (1839–1918), lauded as a modern farmer in his day, played a major role in the establishment of that station. His daughter, Allie (Al) Florence Exum (1882–1973), was Cary's beloved grandmother. Cary couldn't have known his great-grandfather, but he came alive in the stories Al told Cary as they sat on the porch swing of the family farmstead in Jackson.

One Christmas morning, Martin gave young Al a shotgun, which didn't go over well; she would have preferred a doll like the ones he gave to her older sisters. In time she came to appreciate the compliment, and it was Al who took over the farm when her father died. A dark portrait of Martin hung over the fireplace in the dining room—which scared the daylights out of Cary and his sister, Jo, when they were little. Years later, Al insisted that Cary take the painting because she had a feeling it wouldn't be there the next time he visited. Cary says, "She was right. Several months later, the house, built with bricks made from clay from the farm, home to multiple generations of my family, burned to the ground." That portrait, restored and cleaned, now hangs on a wall in our 1788 farmhouse in the Hudson Valley. (And I can see where Cary got his red hair from.)

Wilson Sweet

I love this watermelon. It may not be the sweetest in the world (it's good enough) when grown in my garden, but it's one of the most striking. It combines a marbled, scaly-bark rind pattern, vivid red or crimson flesh that makes me blink—the likes of which I've rarely seen in heirlooms—and attractive seed coat patterning when fresh. The rind is quite thin, so you get more for your melon.

This cultivar has only recently been reintroduced by Glenn Drowns of Sand Hill Preservation Center. He wrote in his 2006 catalogue: "A uniquely colored melon that we are pleased to offer again to the public after many years of being out of circulation. We were thrilled when a customer sent us a sample of this variety which I'd only seen a picture of in a 1961 seed catalogue. The rind is somewhat mottled in appearance giving it excellent sunburn resistance. The 10 to 20 pound melons have solid, firm, super sweet red flesh."

Glenn recently forwarded to me a scan of that 1961 Hollar Seeds clipping from his archives: "The Wilson Sweet is another small family size melon weighing about 20 pounds, with Cletex marked rind [marbled like Spotted Watson] that is very tough and thin, will ship or haul anywhere. The shape is round, exactly as pictured; flesh beautifully red, solid and firm, fine textured; cutting qualities perfect. Maturity approximately 85 days. They have yellow bellies when ripe; medium size seed that are white with black tips."

I reached out to Bruce Carle, a plant breeder at Hollar Seeds in Rocky Ford,

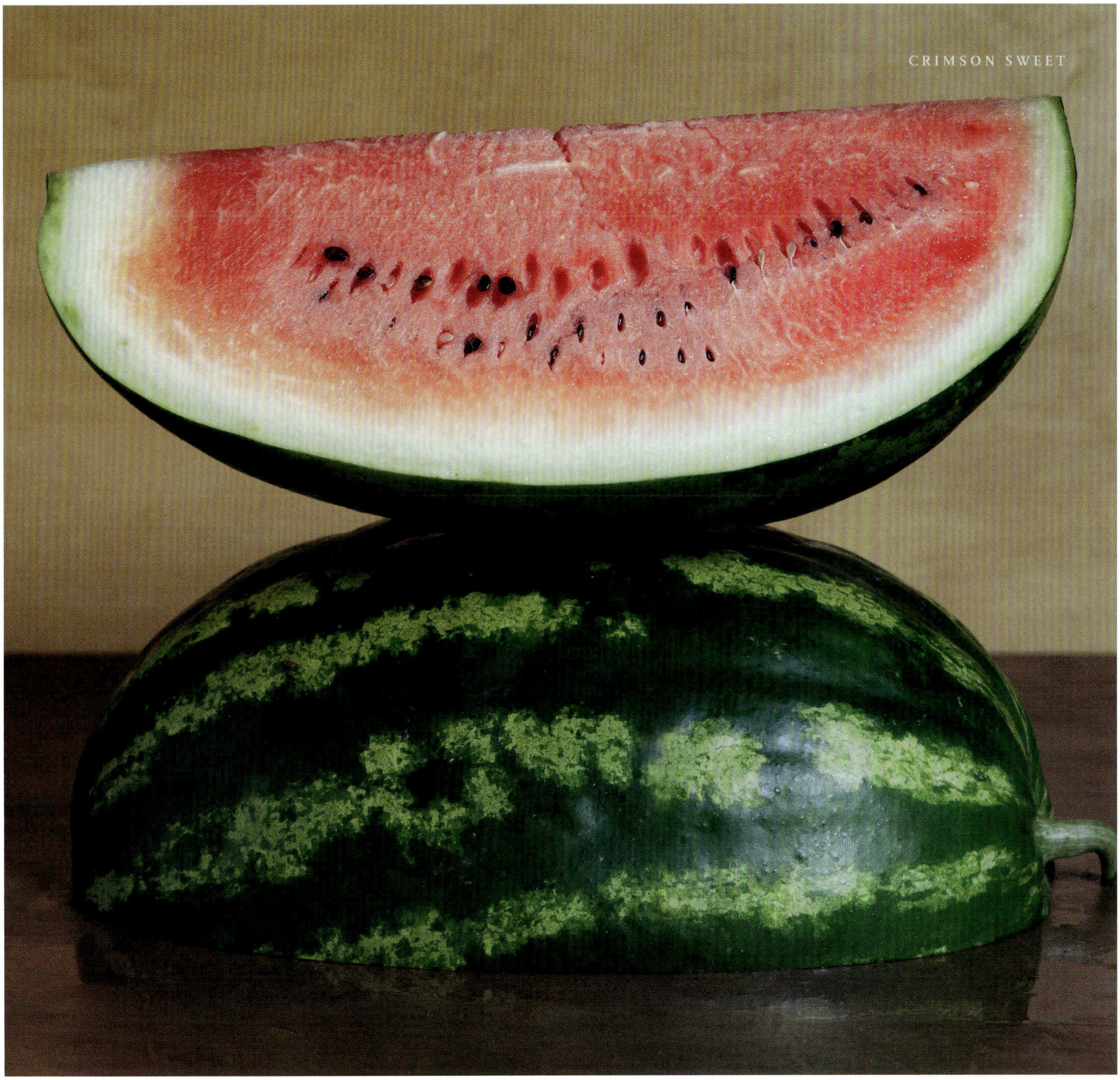
CRIMSON SWEET

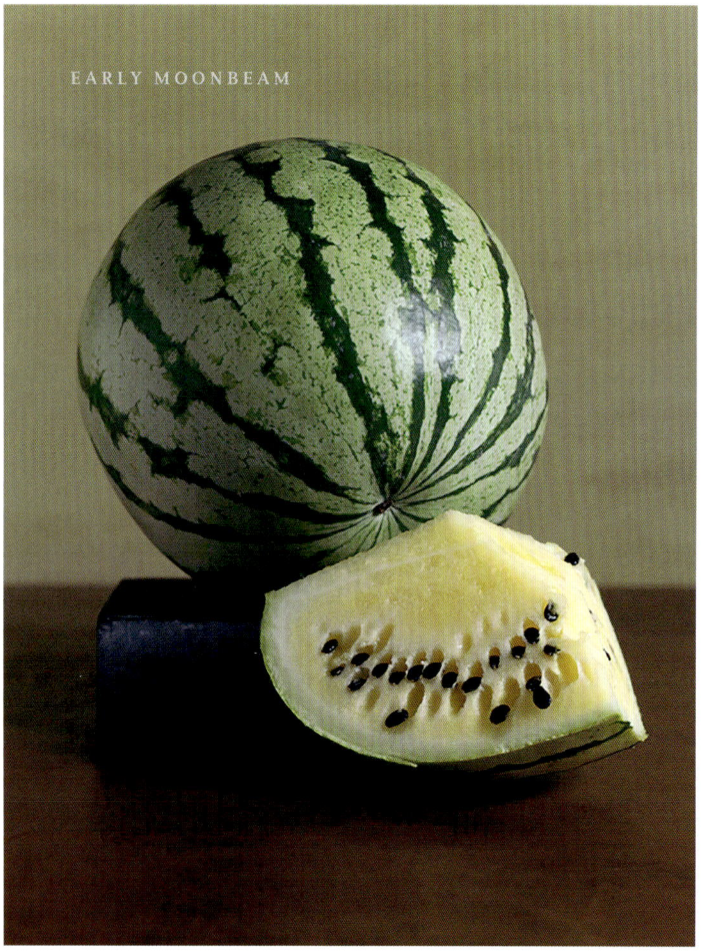
EARLY MOONBEAM

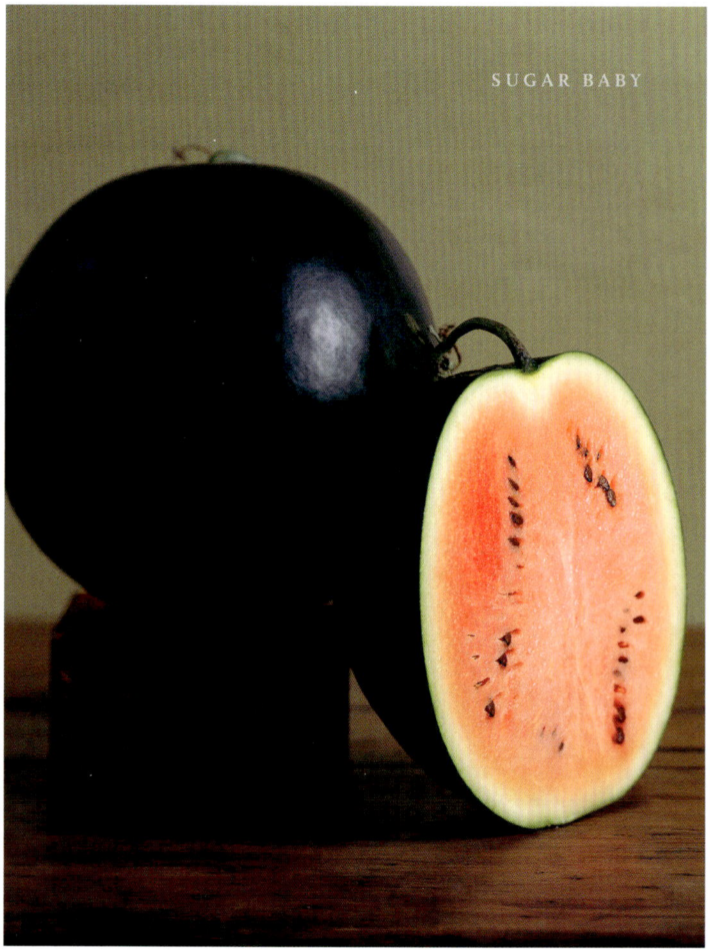
SUGAR BABY

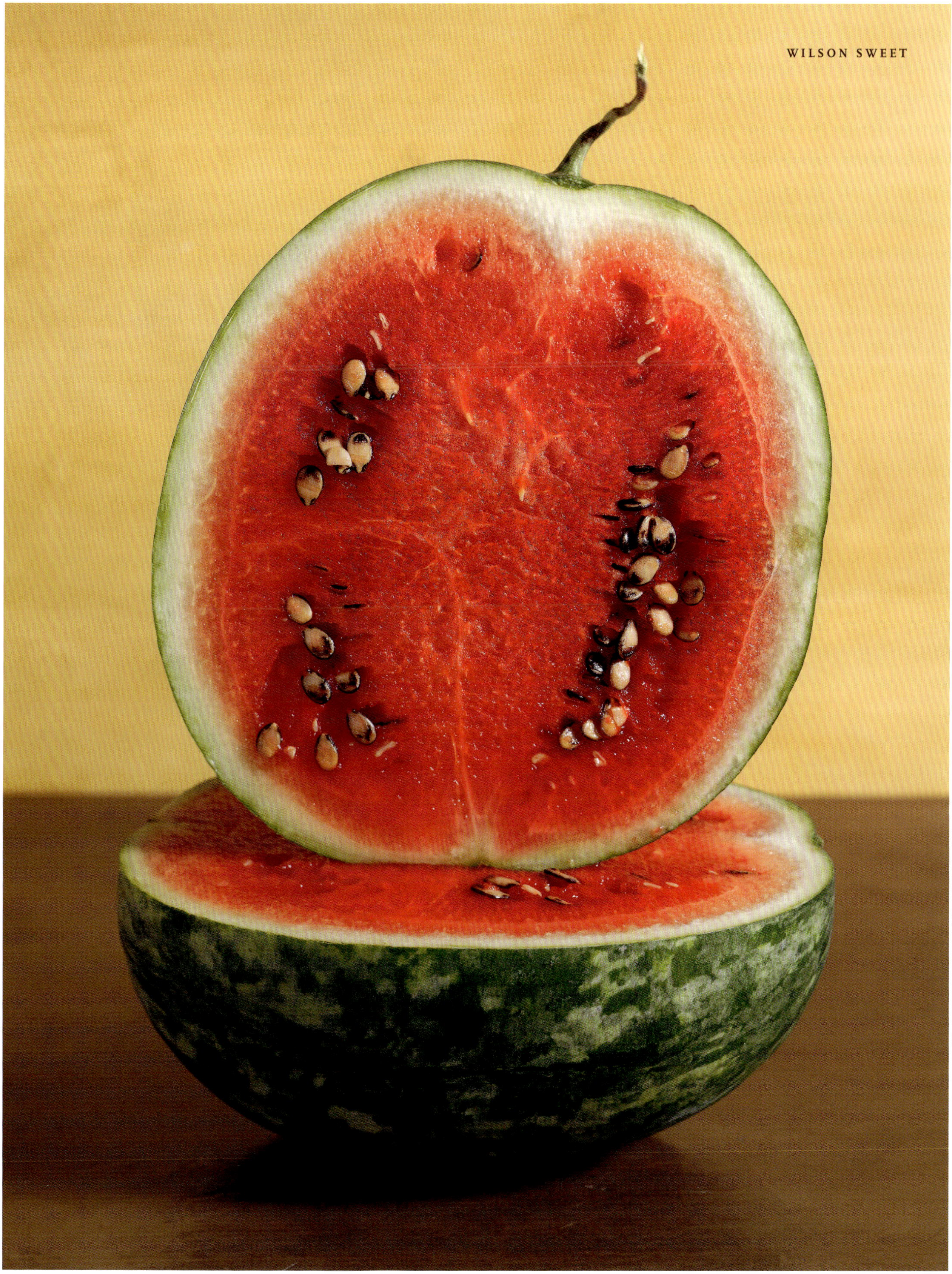

Colorado. He checked his database and talked to Bob Nelson, their retired production man. The upshot is that Wilson Sweet may have been around as early as 1916, and originated in Lee County, Virginia. Bruce explained, "At that time the fledgling seed trade was acquiring popular local watermelon varieties from growers, producing them, and offering them for sale. The primary source for watermelon varieties was the southeast. Those varieties that proved successful endured in the trade but were eventually eclipsed and replaced by USDA and University developed varieties. Bob said Wilson Sweet had a following because it was early. Its problem, however, was that it had a fragile rind and did not ship." They carried the variety until about 1969, but sales were limited to the pack trade in the southeast.

Hollar did not introduce Wilson Sweet (it predates 1950, the year Hollar went into business). Bruce says that Vic Hollar, founder of the company that bears his name, was "instrumental in establishing a certified seed program in Colorado for watermelon. Prior to that, most of the commercial watermelon seed was produced in Florida and Georgia and had off-types due to out-crossing with the citron watermelon. Consequently, he was the go-to guy, adding varieties to feed the demand. Rather quickly, all the major seed producers moved their watermelon (and melon) production to Colorado and then to California. Getting away from the citron problem and taking advantage of the dry climate and irrigation to produce disease-free seed."

Vic is now 102. His son, Larry, who joined the Whealys' expedition to Merle Van Doren's house in Missouri to retrieve Moon and Stars watermelon, is enjoying retirement. Which leads me to paraphrase my own father, Sol: "If you eat watermelon for a hundred years, you'll live a long time."

Arikara

A fascinating watermelon from the Arikara tribe of North Dakota. One of the earliest to mature in my garden. Said to descend from watermelons introduced by the Spanish centuries ago. Not a dessert watermelon by modern standards (its flesh is pale pink with a white heart, coarse, juicy, and bland). It strikes me as an exceptional seed store. When I sat down to count the dried ones I'd extracted a couple of months earlier from the watermelon pictured in this book, I was astonished by the bottom line: 1,176 small black seeds. I don't know how the Arikara used the watermelon in traditional cookery, but I've sent some feelers out to tribe members. I imagine they ate it raw as a refreshment and saved the nutritious seeds for other purposes (besides regrowing). Some tribes used watermelon seeds medicinally (whether mashed, chewed, or infused) to treat various kidney ailments: water retention, bedwetting, blood in the urine, and kidney stones.

The Arikara watermelon came to notice—and was almost certainly salvaged—by Oscar H. Will and Company, of Bismarck, North Dakota. From the company's inception in 1884, Oscar Will's mission was to introduce and perpetuate seed stock obtained from the *great* Plains agriculturists: the Mandan, Hidatsa, and Arikara Nations. The three affiliated tribes, now known as MHA Nation, are located at the Fort Berthold Indian Reservation in central North Dakota. Will introduced the Arikara watermelon in 1925: "Up until two years ago we believed our Will's Sugar

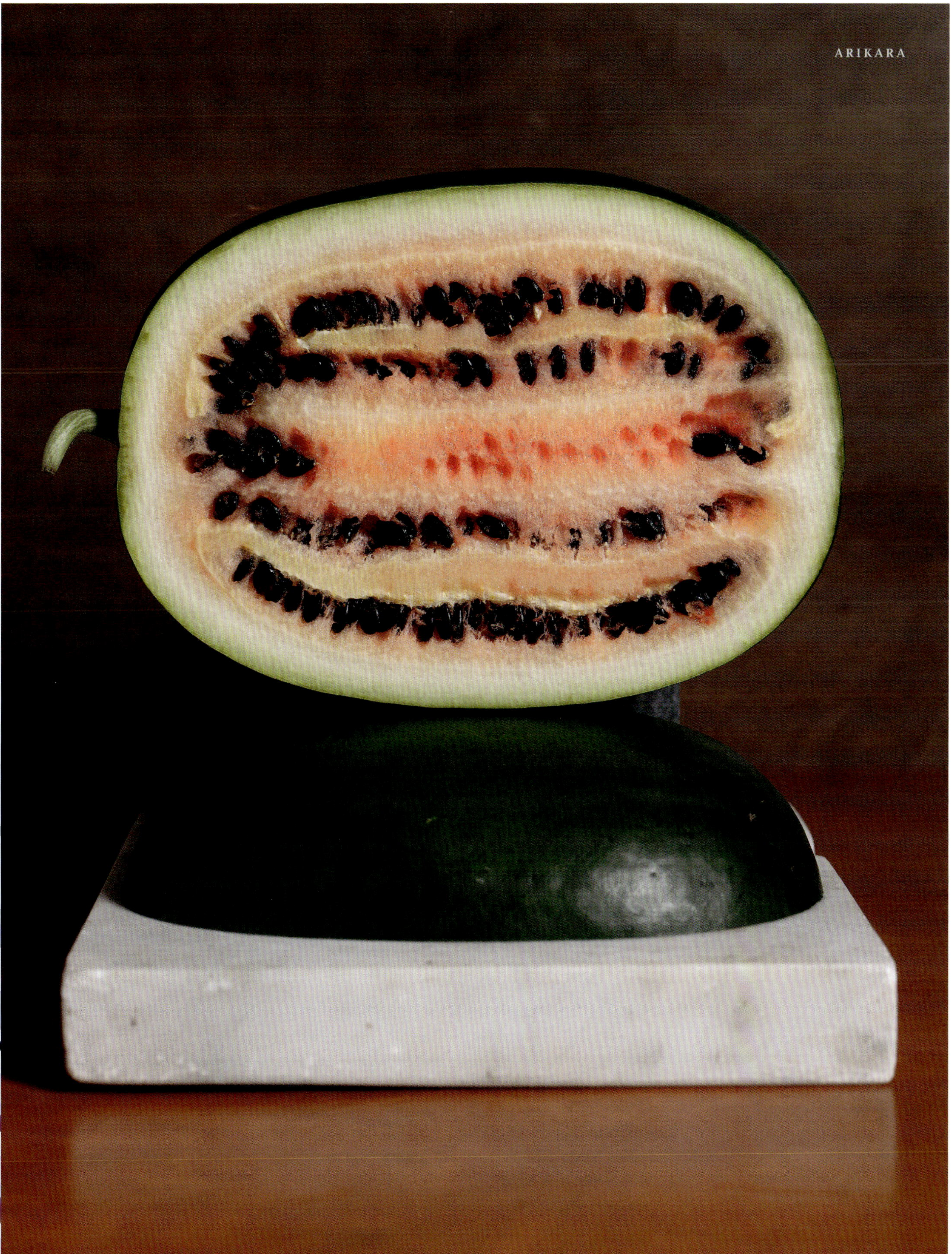

Melon to be the earliest watermelon in the world. We have now decided after several years of growing and observation that we have an earlier one. It is the water melon which Lewis & Clark found growing among the Arikara, Mandan, and Hidatsa Indian tribes in North Dakota [on October 10, 1804] when they made their famous journey. We had given up hope of getting hold of this melon after a number of years search, when Dr. Gilmore discovered a few seeds among some of the Arikara tribe. [Melvin Randolph Gilmore was an ethnobotanist and author of many works on Indian agriculture] Like the other Indian vegetables this little melon is exceedingly hardy, cold resistant and early. It was the only water melon of which we ripened a full crop this last season. Very dwarf, diameter from 6 to 12 in., almost black in color with pink flesh and black seeds, with an occasional melon having yellow flesh and brown seeds, flesh sweet and pleasant, quality somewhat coarse. Supply limited." The Arikara watermelon that has come down through the years is little changed from the one introduced and illustrated in a photograph by Will in 1925—if anything, it's purer, with no off-types that I've seen. That's thanks to seed stock given by Oscar Will's son, George F. Will, to good stewards at the Museum of the Fur Trade, in Chadron, Nebraska, and obtained by Seed Savers Exchange and others.

Oscar Will introduced a number of other notable Arikara varieties or offshoots including the Arikara Yellow bean, Arikara squash (which I adore), and Gilmore squash (a cross between Winnebago and Arikara). Will's Golden Anniversary watermelon and Ranger watermelon both resulted from crosses between Arikara and Kleckley Sweet. Taken together with the many Mandan and Hidatsa corn, bean, and squash varieties that Oscar Will introduced, they formed the backbone of his catalogues—and were the lifeblood of the MHA Nation. Rowen White, Coordinator of the Indigenous Seed Keepers Network (a project of the Native American Food Sovereignty Alliance, or NAFSA), is nourishing and assisting the seed sovereignty movement within tribal communities across Turtle Island (North America), and returning native seeds to Native farmers.

Cream of Saskatchewan

White-fleshed dessert watermelons are rare. And this one is excellent—sweet and lemony. The rind can even be eaten raw. It's early, prolific, and reliable in northern gardens or where growing seasons are short. The only drawback is the explosive rind gene, *e*, which causes the thin, tender rind to burst open at the glint of a knife.

Cream of Saskatchewan seems to have appeared out of the blue in Seed Savers Exchange's 1984 Winter Yearbook: Watermelon maven Curtis Sylvester Showell of Bishopsville, Maryland, was the first to list and offer seeds to members. (It appears that 1984 was the first and only year he officially participated in the Exchange.) Three years later, Seeds Blüm of Boise, Idaho, catalogued it for the first time, and it quickly gained fans. John Navazio, plant breeder, then of Bar Harbor, Maine, wrote in SSE's 1989 Winter Yearbook: "Of all the cool-climate melons I've grown in Oregon and Maine this takes the cake; not just sweet, but real character and flavor; small ice box-type if you can get one from the field to the icebox without splitting open; white to light yellow flesh with only 60-70% of the seeds, fully mature when perfectly ripe for

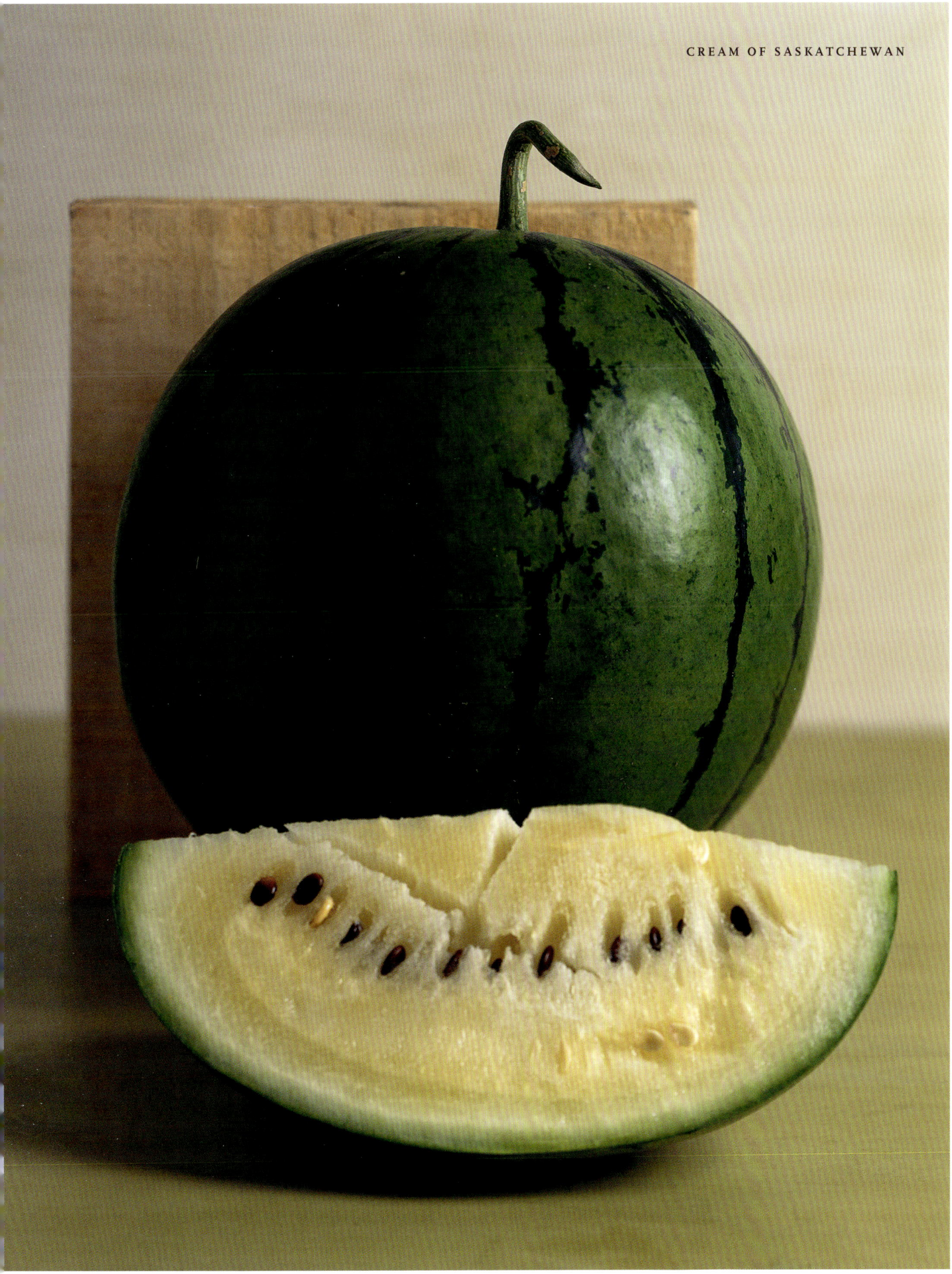

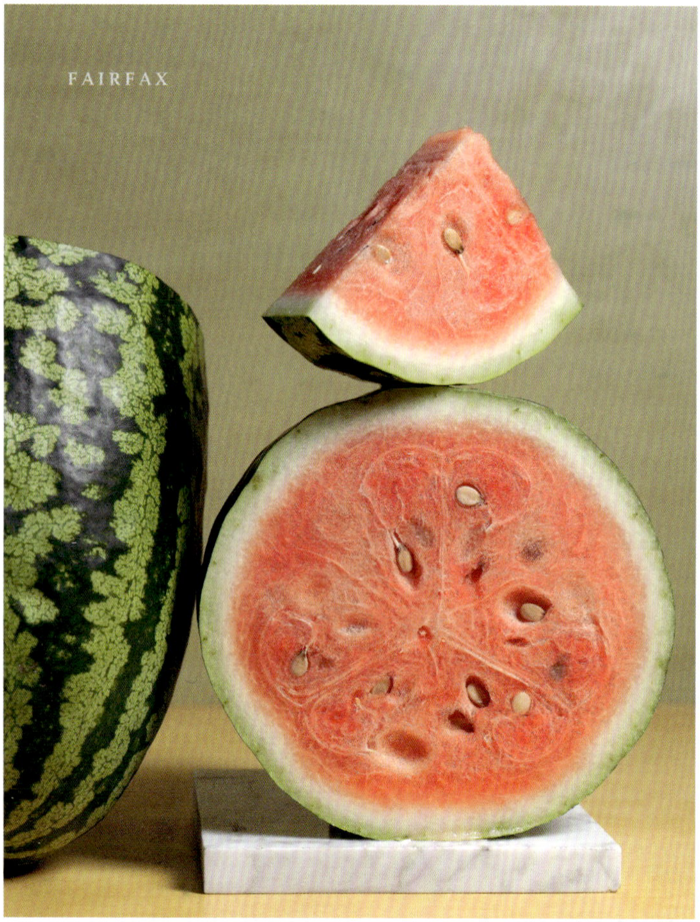
FAIRFAX

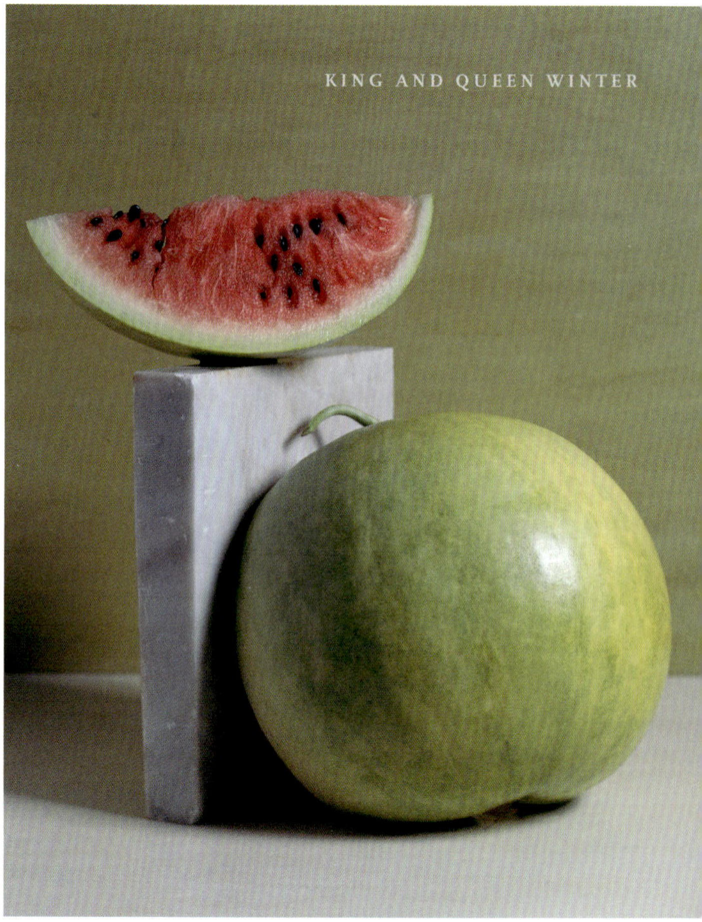
KING AND QUEEN WINTER

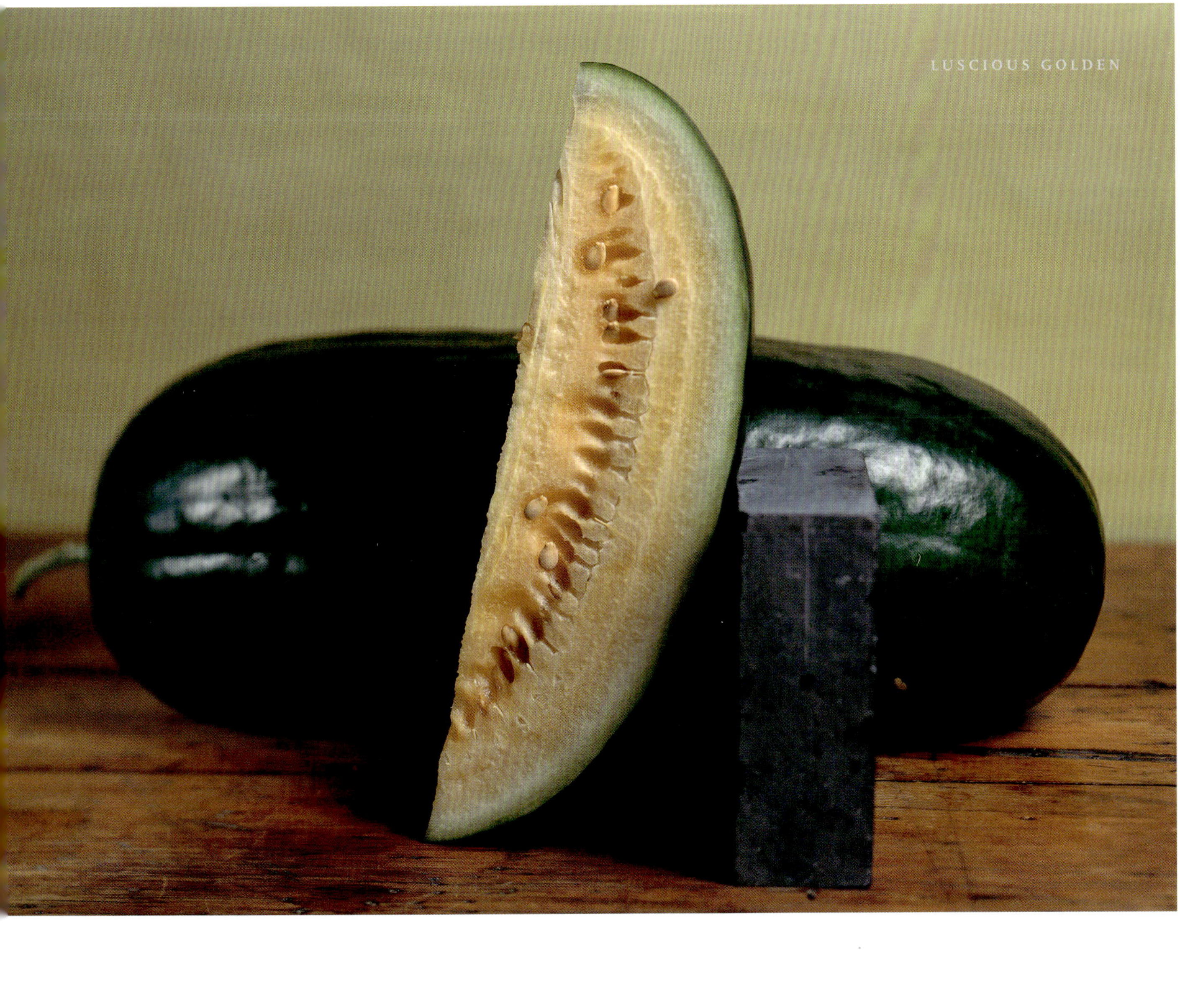
LUSCIOUS GOLDEN

eating." In 2011, another breeder-member of SSE, Ken Ettlinger, weighed in: "Early white (very light yellow) melon, looks like the popular hybrid outside, Yellow Doll, and it has the same tendency to split as you cut into it. When I sink my teeth into a cold slice of 'Saskatchewan,' I can't help but think I'm eating a fine Italian Lemon Ice."

I've heard stories of a white-fleshed lookalike brought to Saskatchewan with Russian immigrants in the nineteenth century. It's certainly possible. My best guess is that Cream of Saskatchewan is a renamed older variety of more recent vintage: Honey Cream watermelon, introduced by Harris Seeds in either 1946 or 1947. Photographs and catalogue copy match. From the 1947 catalogue: "(New.) Fine for the North - Very Sweet - Yellow Flesh. A new extra early variety of the highest quality having distinctive yellow or cream colored flesh. Fruit medium sized, nearly round and attractively striped like Dixie Queen. Not a shipping type but on roadside stands its excellent sweetness and fine flavor will bring customers back again and again; the yellow flesh gives them assurance of fine quality. Ripens as early as a Bender melon and the yields in Northern sections are excellent."

Whatever its provenance, Cream of Saskatchewan is a keeper. Todd Wehner (2008) recommends that breeders use it for development of white-fleshed cultivars since it's "an adapted accession with the best genetic background."

Black Diamond

A big, old-fashioned, round, intensely green-black melon with red flesh and a white belly when ripe. Comes complete with furrows and a tough rind up to two inches thick. Hugely popular in the mid-1900s as a Southern shipper for northeastern markets. C. F. Andrus, developer of the Charleston Gray watermelon—the first to include resistance to both fusarium wilt and anthracnose—lamented in 1955 the sharp decline of the Black Diamonds.

In 1950, Wm. A. Watson's Sons, of Thomson, Georgia, introduced a red-meated strain of Black Diamond by the name of Superior Black Diamond (also known as Yellow Belly Black Diamond or Yellow Belly Cannon Ball), whose ground spot turns yellow or orange, not white, at maturity. They wrote, "We have developed this superior strain of the Black Diamond beginning seven years ago. We believe that it will eventually be the leading round-type melon on the market. It is smooth of rind and much darker than anything on the market under the Cannon Ball or Black Diamond name. . . . This melon is more resistant to sunburn than other strains. It is dark green, almost blue in color, cuts redder, is sweeter and of a better flavor. We have also found it to be about 5 days earlier. It will pay you to plant a portion of your acreage to this improved strain. We have a limited quantity at the following prices, this being the first year we have offered them." You would have paid four dollars per one to twenty-four pounds of seed delivered, and less for larger amounts.

The Yellow Belly Black Diamond proved to be a mutant of Black Diamond, according to Gusmini and Wehner (2006), who measured the inheritance of novel rind phenotypes. Their test "confirmed the hypothesis of a single gene controlling yellow belly in Black Diamond Yellow Belly. We propose naming this new dominant gene *Yellow belly*, with the symbol *Yb*." And so it came to pass.

Black Diamond sizes up early in my northern garden and keeps on chugging. My

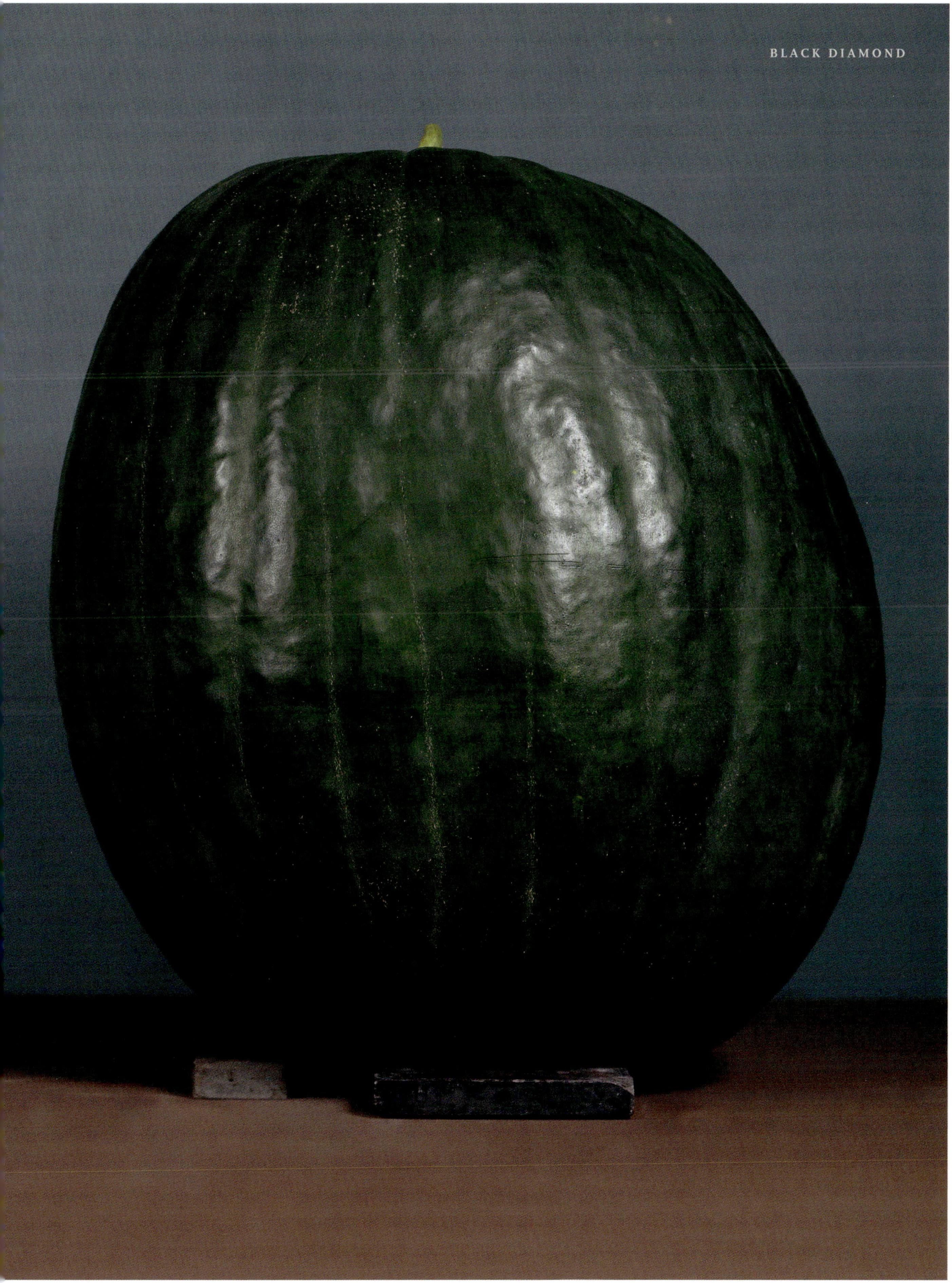

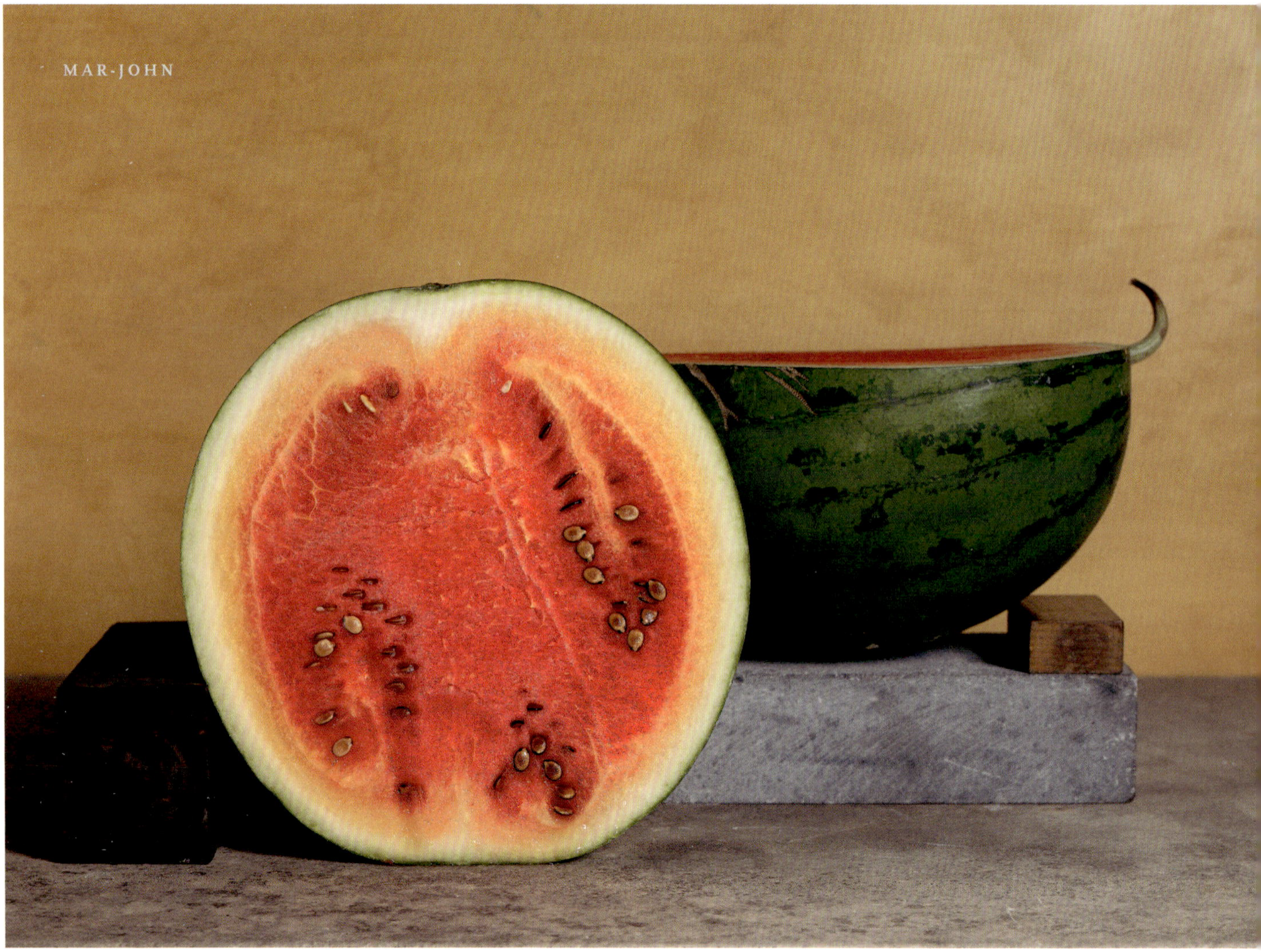
MAR-JOHN

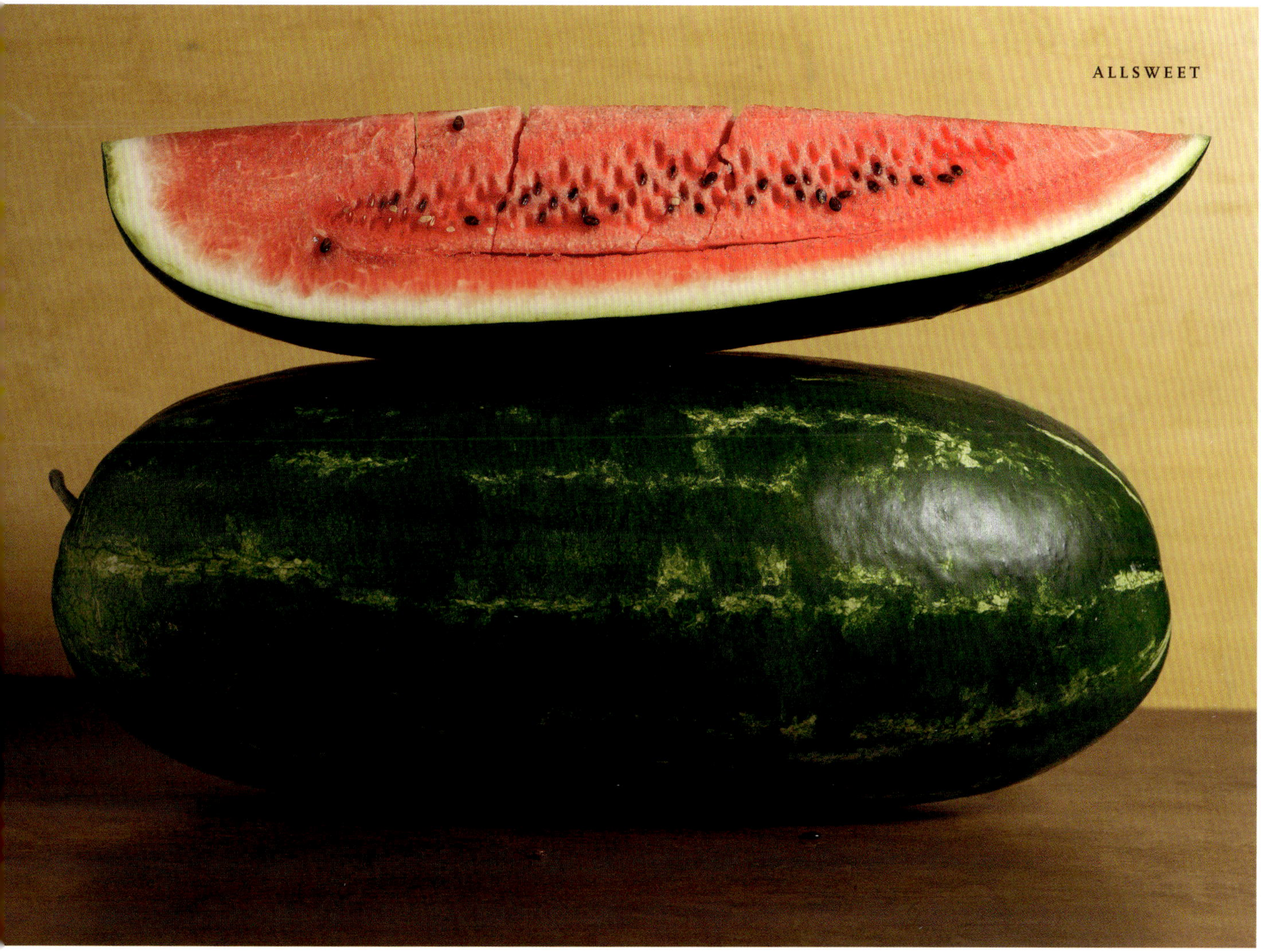

ALLSWEET

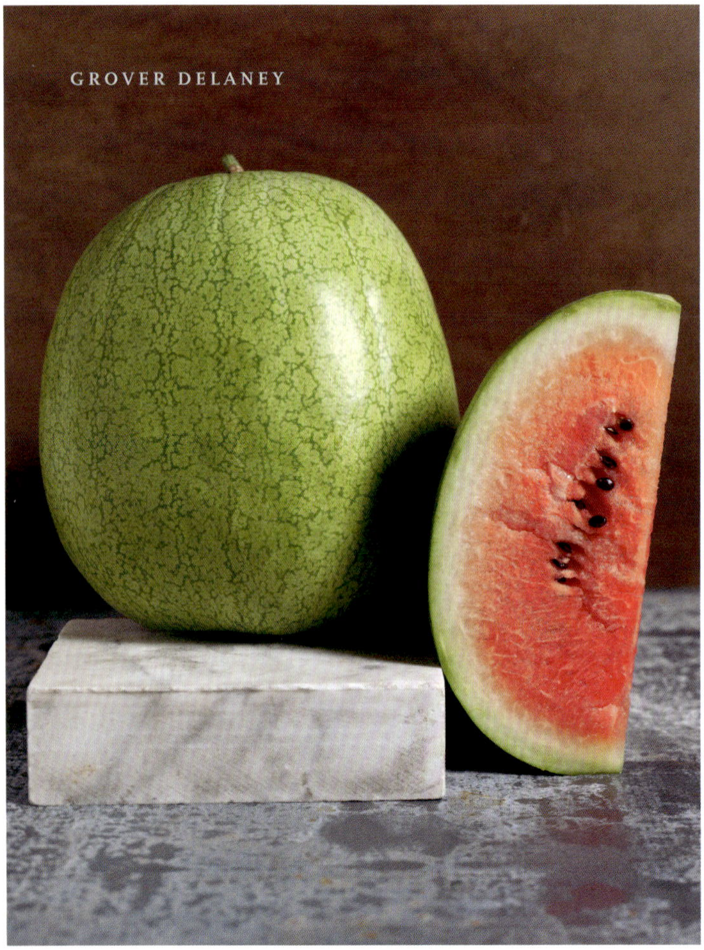
GROVER DELANEY

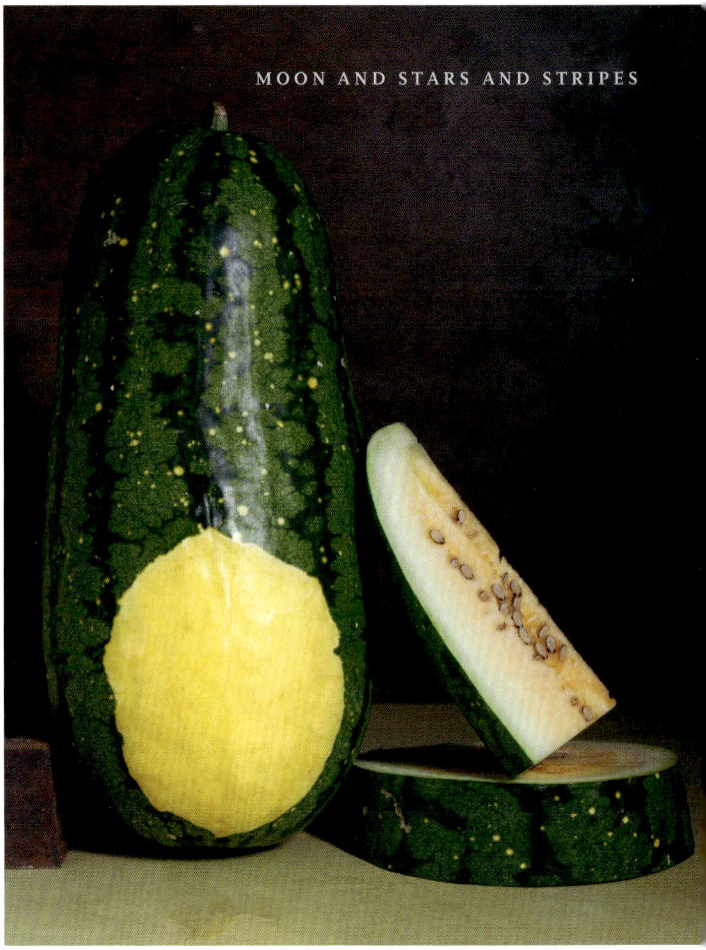
MOON AND STARS AND STRIPES

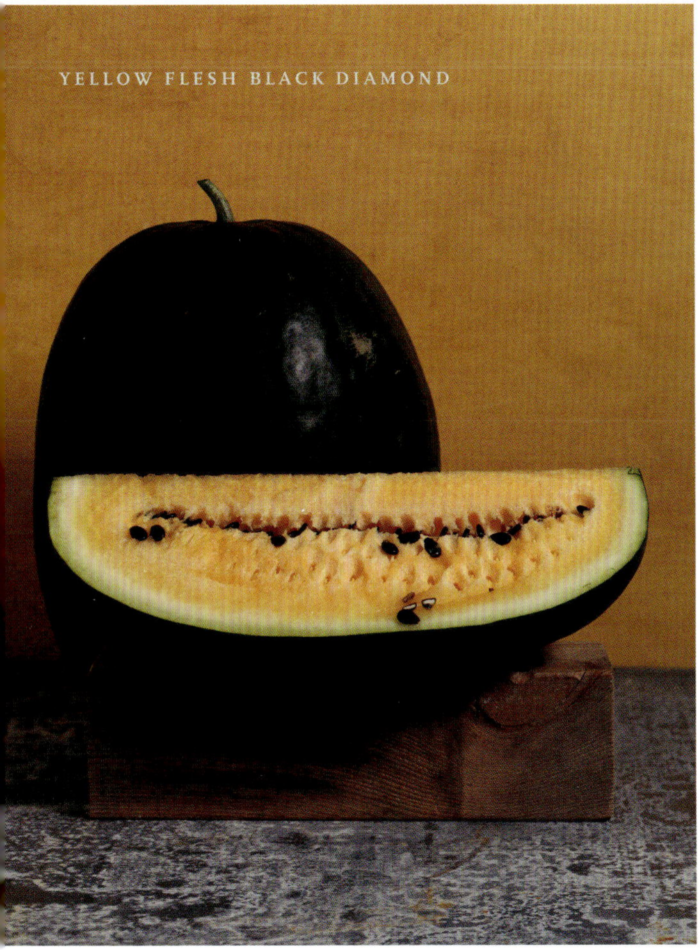
YELLOW FLESH BLACK DIAMOND

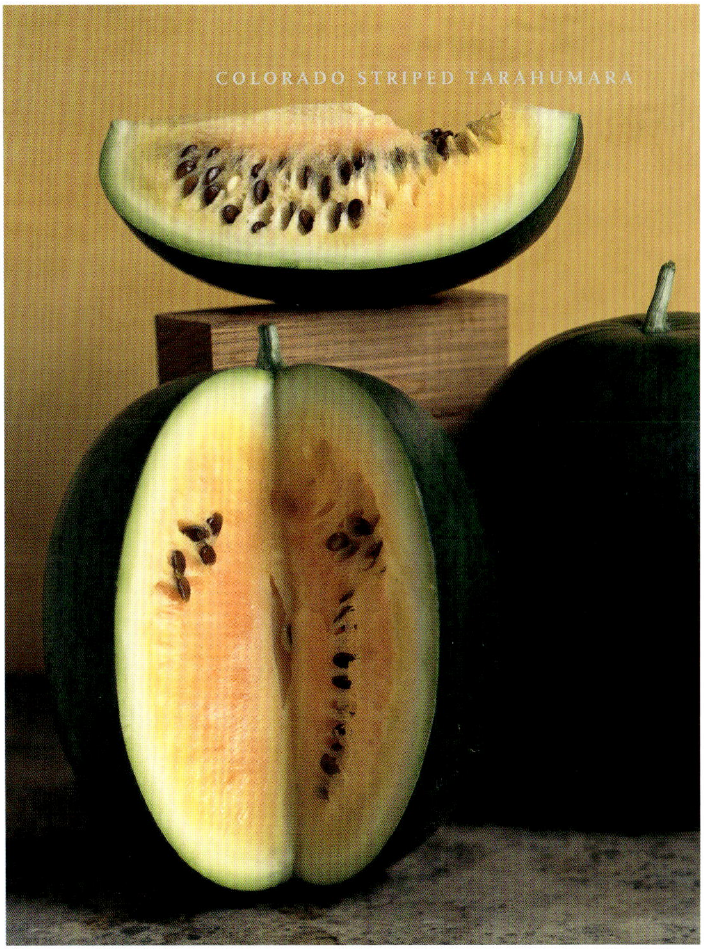
COLORADO STRIPED TARAHUMARA

personal best is the 98-pounder featured in this book. Can't say I love its tough flesh, but it's a hoot to grow. Chris Kent, grower of the world's heaviest watermelon (see Carolina Cross, page 185), told me that he grew Lloyd Bright's Jumbo Black Diamond seed line in 2016 and broke the old dark green watermelon world record from the 1930s with a 205-pounder. I wish C. Fred Andrus had lived to see the day.

Golden Midget

The Golden Midget stands out in a sea of large, green-rinded watermelons. More like a lunchbox type than an icebox type, this oval watermelon in miniature, with its golden-yellow rind penciled with very narrow stripes and its sweet salmon-pink flesh, brings to mind papaya. It's just plain fun to grow and eat. Big, monster watermelons invite a crowd; this one is made for a more select few. Keep it to yourself, or share it with someone special.

Albert Franklin Yeager (1892–1961) and Elwyn Meader (1910–1996) had the golden touch when it came to breeding watermelons. Working together at the University of New Hampshire, they attempted to produce a short-season variety with a compact habit. When Mr. M. Hardin of Geary, Oklahoma, sent them a variety called Pumpkin Rind for trial, it was the answer to the prayers of any watermelon man. Crossed with the New Hampshire Midget (see page 158), a path-breaking 1951 Yeager introduction, it produced the Golden Midget (1959), a watermelon with a built-in ripeness indicator: It turns yellow as it matures.

Success wasn't immediate. The first-generation melons were all green-rinded. The recessive "golden" gene, *go*, which causes older leaves and mature fruit to turn yellow and reduces plant vigor, didn't assert itself until the second generation. Selections were made for table quality and size until Yeager and Meader had a Golden Midget that could ripen anywhere with seventy frost-free days.

Working separately, Yeager and Meader made many plant introductions during their careers. Both were renowned generalists, often favorably compared to Luther Burbank. Yeager was most famous for developing the Buttercup squash. He introduced about forty fruit, vegetable, and flower cultivars including Pixwell gooseberry and Bison tomato. Meader developed everything from blight-resistant chestnuts to raspberries, lilacs, persimmons, and cucumbers. Farmers and gardeners are grateful to both men for leaving their legacy of heirloom varieties in the public domain.

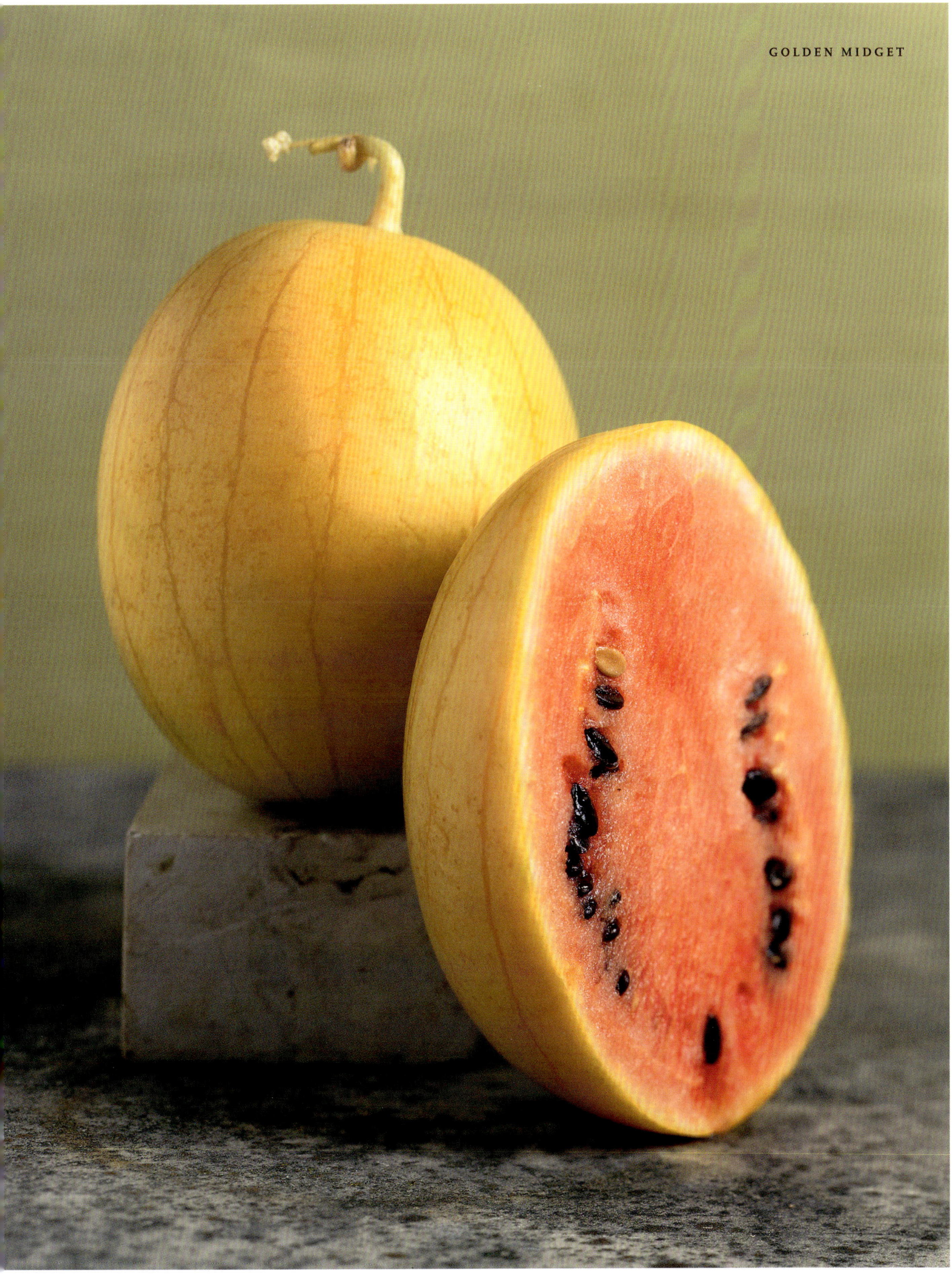

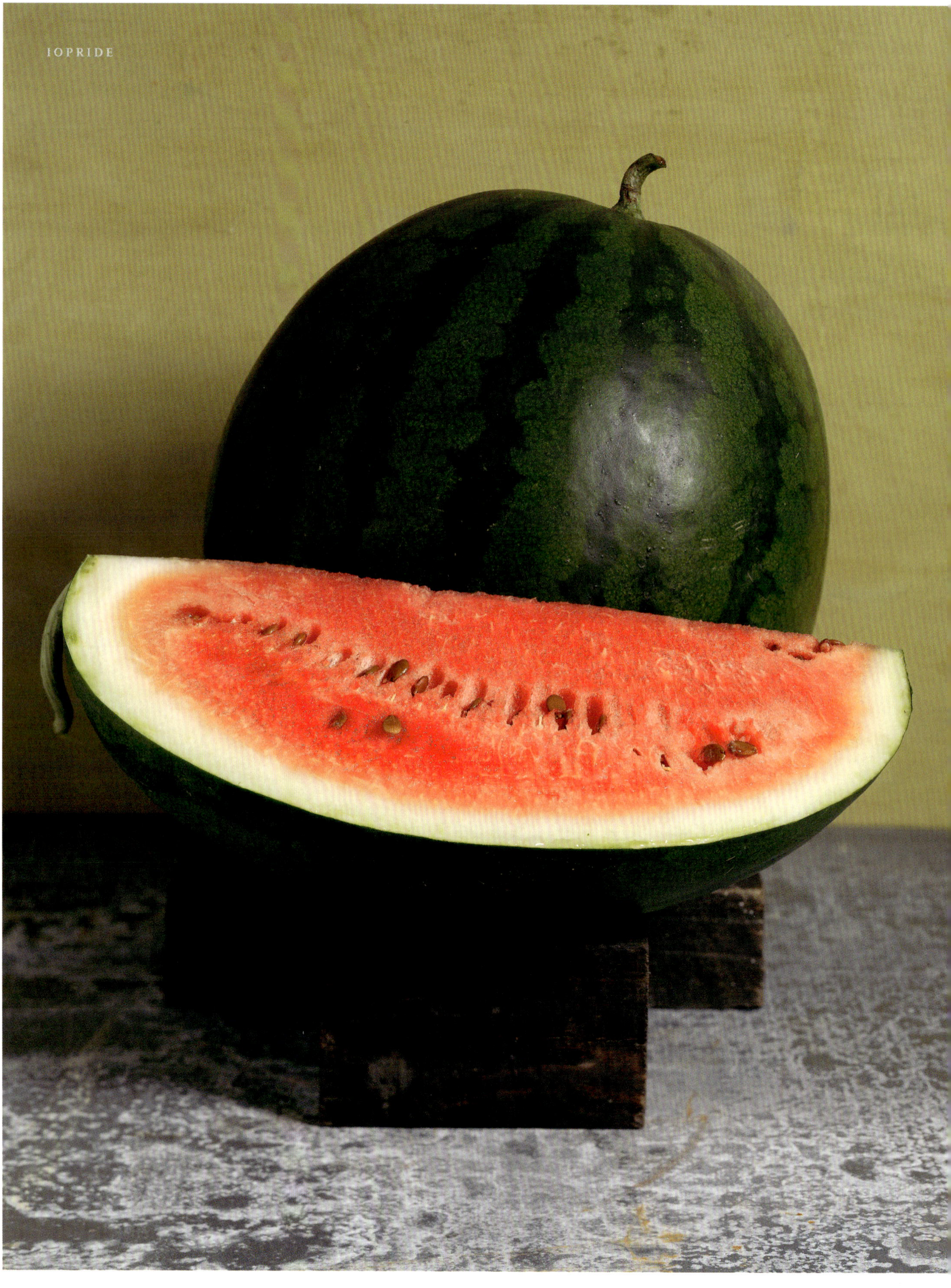

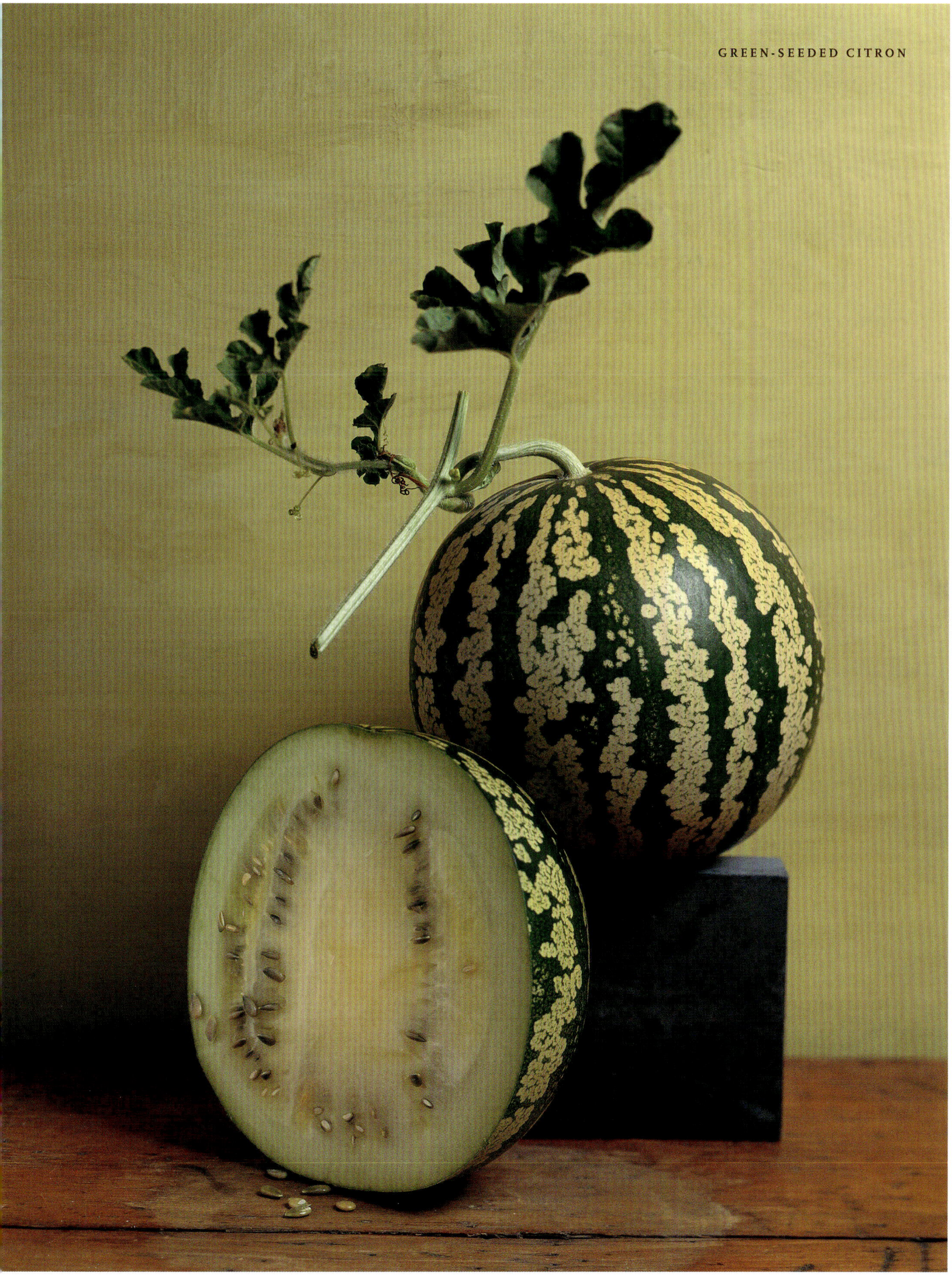

Blacktail Mountain

Blacktail Mountain watermelon is the gold standard by which my daughter, Sara, and I judge all other icebox types. This glossy dark green cannonball with black longitudinal veining is shot full of flavor: Its dense scarlet flesh is sweet, juicy, and crunchy. That it tastes so good is happenstance. Blacktail was bred by Glenn Drowns in the 1970s for early maturity and cold tolerance. Seventeen-year-old Glenn, as precocious as his fruit, was determined to get a watermelon to ripen in his garden in the mountains of northern Idaho. This is not easy in a place where summer nights average 43 degrees Fahrenheit. All he wanted was a watermelon with some red inside it—anyone who's ever cut into a colorless (immature) one understands what drove him.

Inspired by the work of Gregor Mendel, Glenn spent four years crossing heirloom watermelons. One of the known parents was Red Sugar Lump, an icebox type no longer available commercially. He named his creation after the mountain in his backyard, not, as rumored, because the seeds have little black tails.

Glenn has always had a passion for watermelon and for gardening. As a child growing up in a poor family, he savored watermelon just once a year, for his birthday at the end of August. His primers were seed catalogues, and his pastime was winning blue ribbons for vegetables at the county fair. Glenn became a high school science teacher, 4-H judge, and seedsman in Calamus, Iowa. He moved south, weary of the cold and the mixed blessing of melon-pollinating rattlesnakes. Lucky for him—and for us—the Blacktail is just as much at home in hot, humid climes as in the chilly north.

Carolina Cross

Carolina Cross produces the world's weightiest watermelons and is the only cultivar that grows giants weighing 200 pounds or more. The current world record holder is the 350½-pounder grown by Christopher Kent in Sevierville, Tennessee; verified by the Great Pumpkin Commonwealth on October 4, 2013. Chris recently told me via email, "I have always loved growing plants from the time I was a little boy. My grandmother loved flowers and roses and inspired me. As for my seeds in the Carolina Cross line, one of the goals I've had was to improve the weights by doing controlled crossing. Up to that point, most growers were just doing open pollinations and the weights were not increasing. I feel my seed lines are improved because I know the parents and can produce better, heavier melons more consistently. I like to experiment. This year [2018], I have crossed the Carolina Cross with an African melon to try and add some disease resistance and make them more solid."

Carolina Cross didn't appear out of the blue—it stands on the shoulders of giants, those of both men and watermelons. Edward Weeks, founder of the Weeks Seed Company in Greenville, North Carolina, and holder of four Guinness world records for vegetables, laid the foundation by developing the Weeks North Carolina Giant in the 1960s. He told me in a 2001 telephone interview that his Dad gave him some

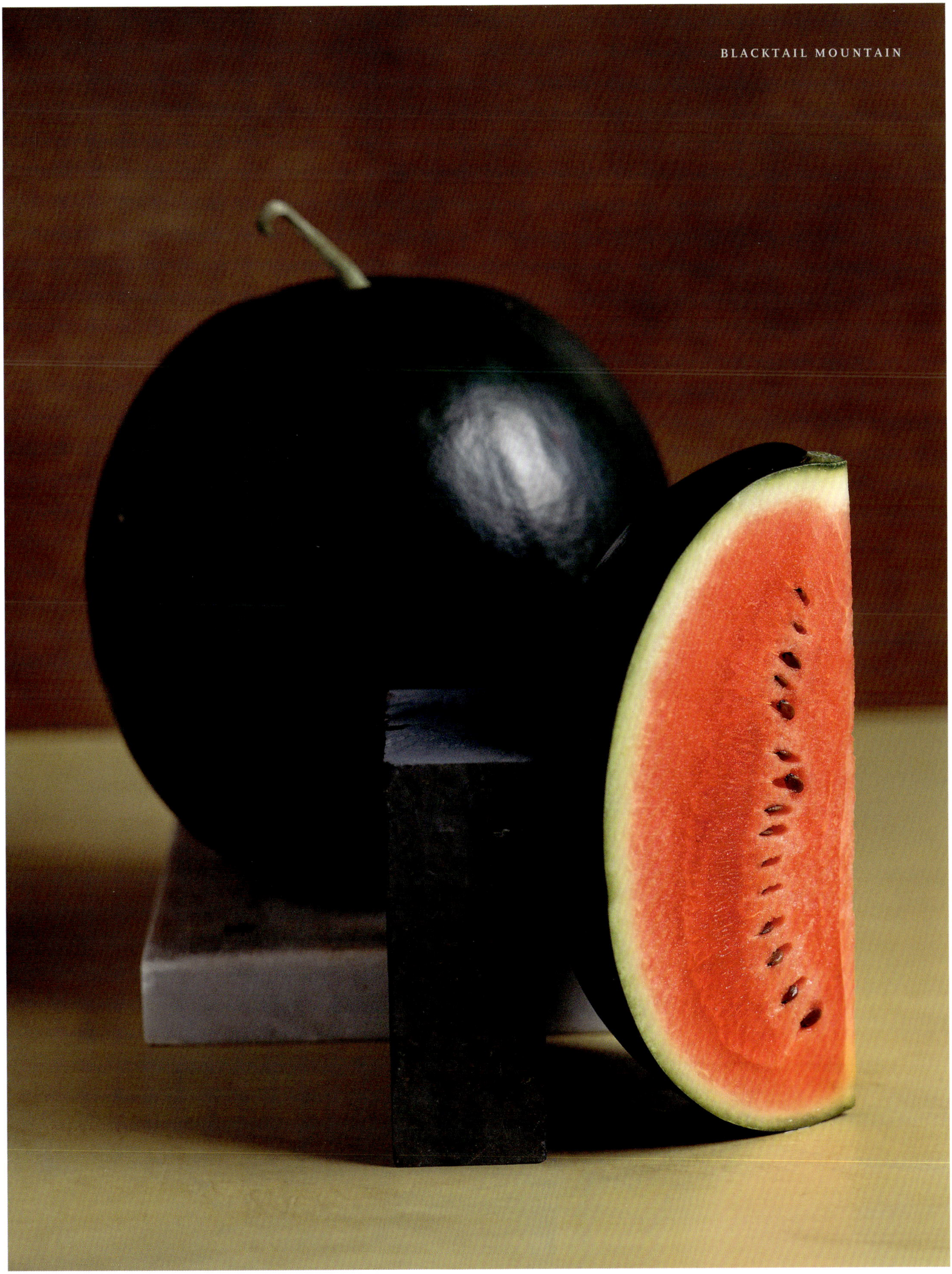
BLACKTAIL MOUNTAIN

(unspecified) giant watermelon seed in 1962 for starters. Edward deliberately grew out the handed-down seed in close proximity to Georgia Rattlesnake and Cobb Gem watermelon plants (see pages 149 and 187, respectively), and perhaps others he wouldn't divulge, and let the bees do a series of crosses (he did no controlled crosses or hand pollinations). He said he got a thousand melons over 100 pounds right away, and by 1966 the new variety was perfected. Weeks eventually produced a 197-pounder in 1975 that won a Guinness world record (thirty-six inches long by twenty inches wide).

The new inbred cultivar was sold through a "very expensive" ad with a photo that Weeks placed in *Progressive Farmer* magazine. It was a stretch and a gamble for this self-described bashful and quiet farm boy. The orders started rolling in, and what started out as a hobby became a business. Lloyd Bright told me in a 2001 phone call that he'd seen Mr. Weeks's ad in the magazine and bought seed from him by mail order. Bright took the Weeks watermelon, crossed it with Cobb Gem around 1972, and marketed the progeny as Carolina Cross. The Bright family produced the world's first 200-pounder in 1979, earning them a Guinness world record—and garnered another in 1985 with a 260-pounder.

In the case of the Carolina Cross, there is no conflict between genetically conferred size and genetically conferred sweetness: this watermelon variety has it all. From seed sourced from Chis Kent, I've grown delicious Carolina Crosses with Brix readings of 11 degrees—consistently. No hollow or white hearts: just pure pleasure. My own personal best is a 129-pound Carolina Cross grown in 2018. It's a good thing I'm only competing with myself these days.

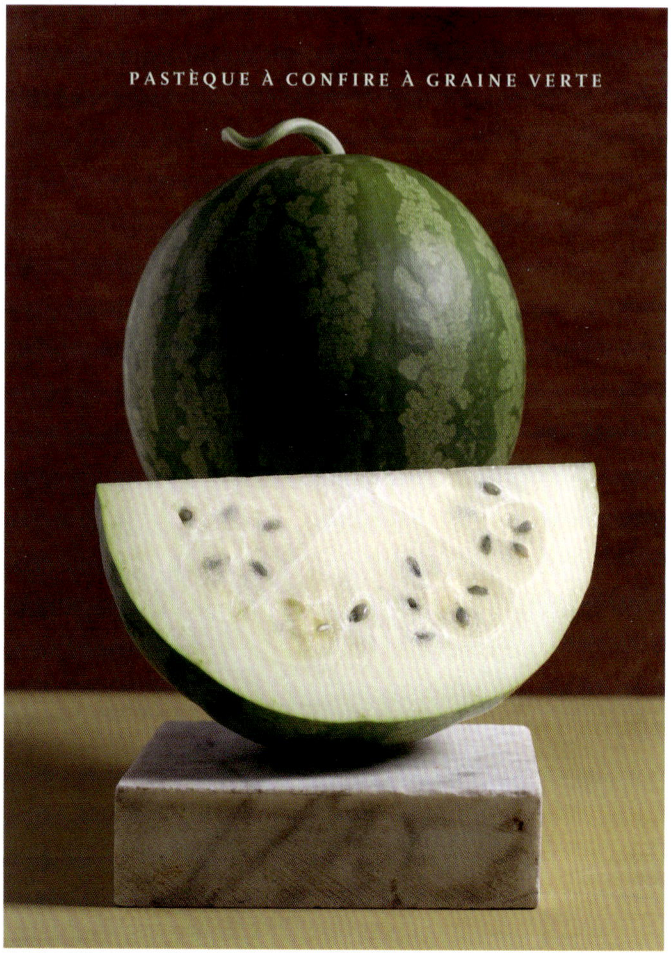

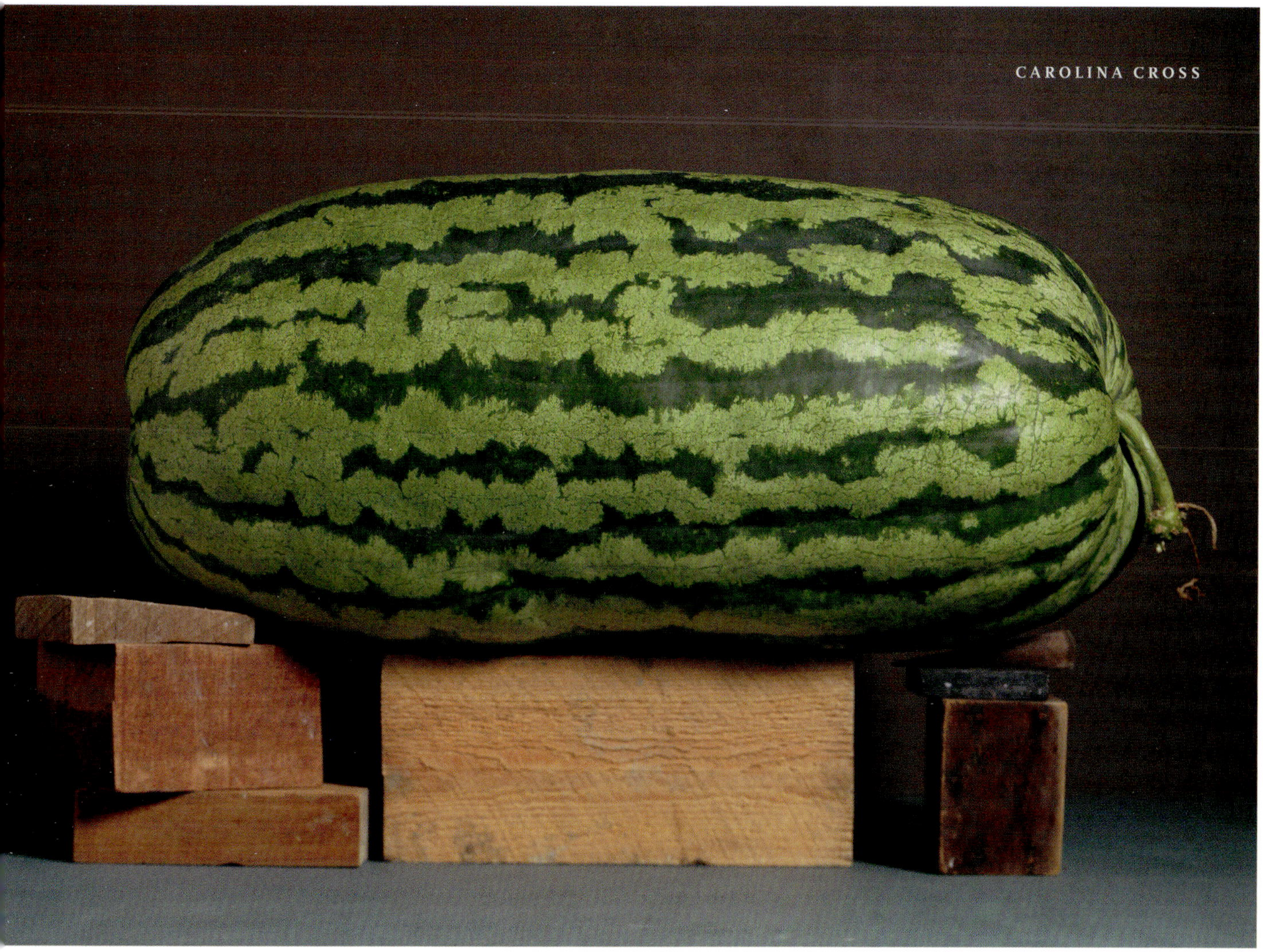
CAROLINA CROSS

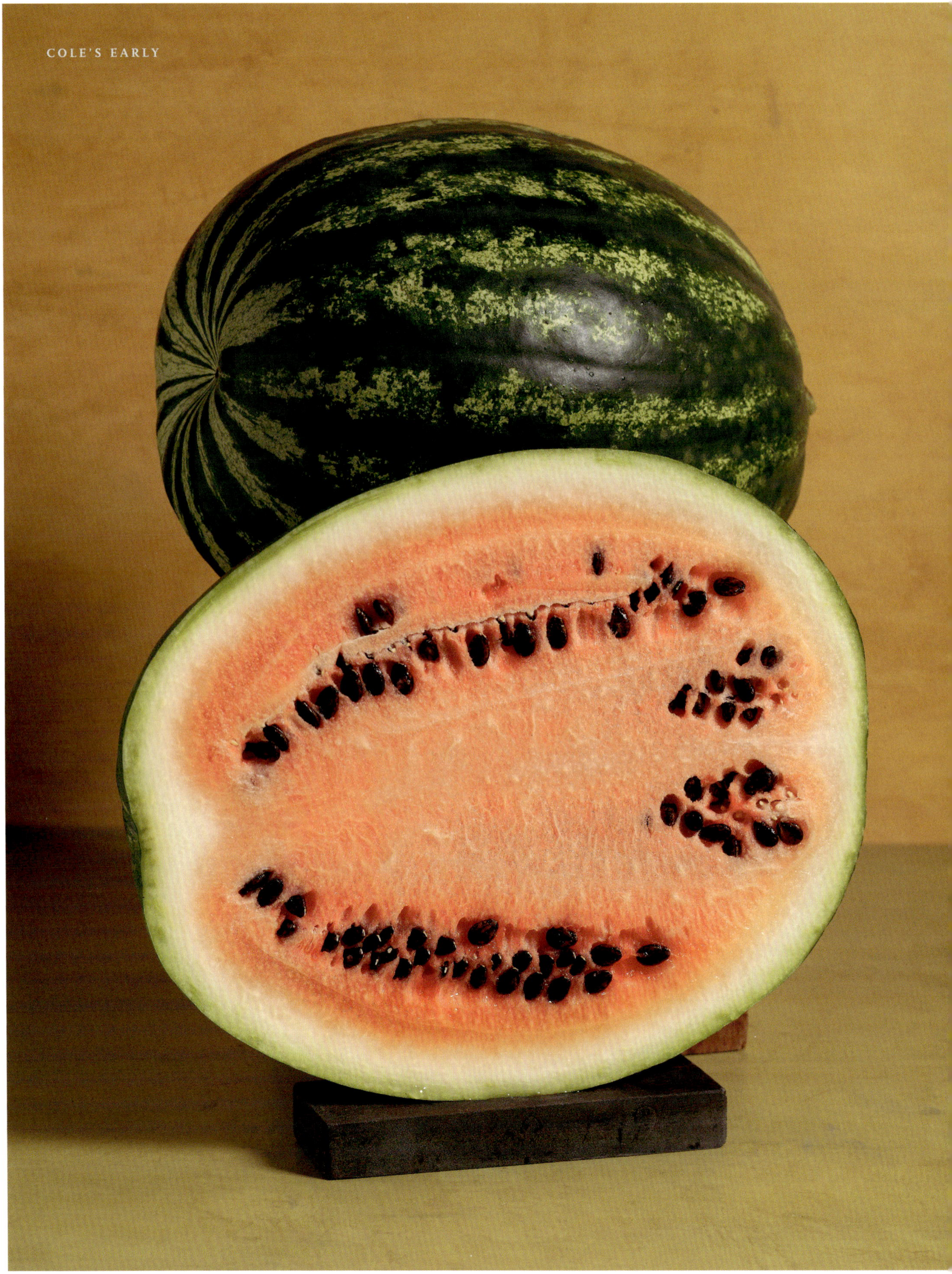
COLE'S EARLY

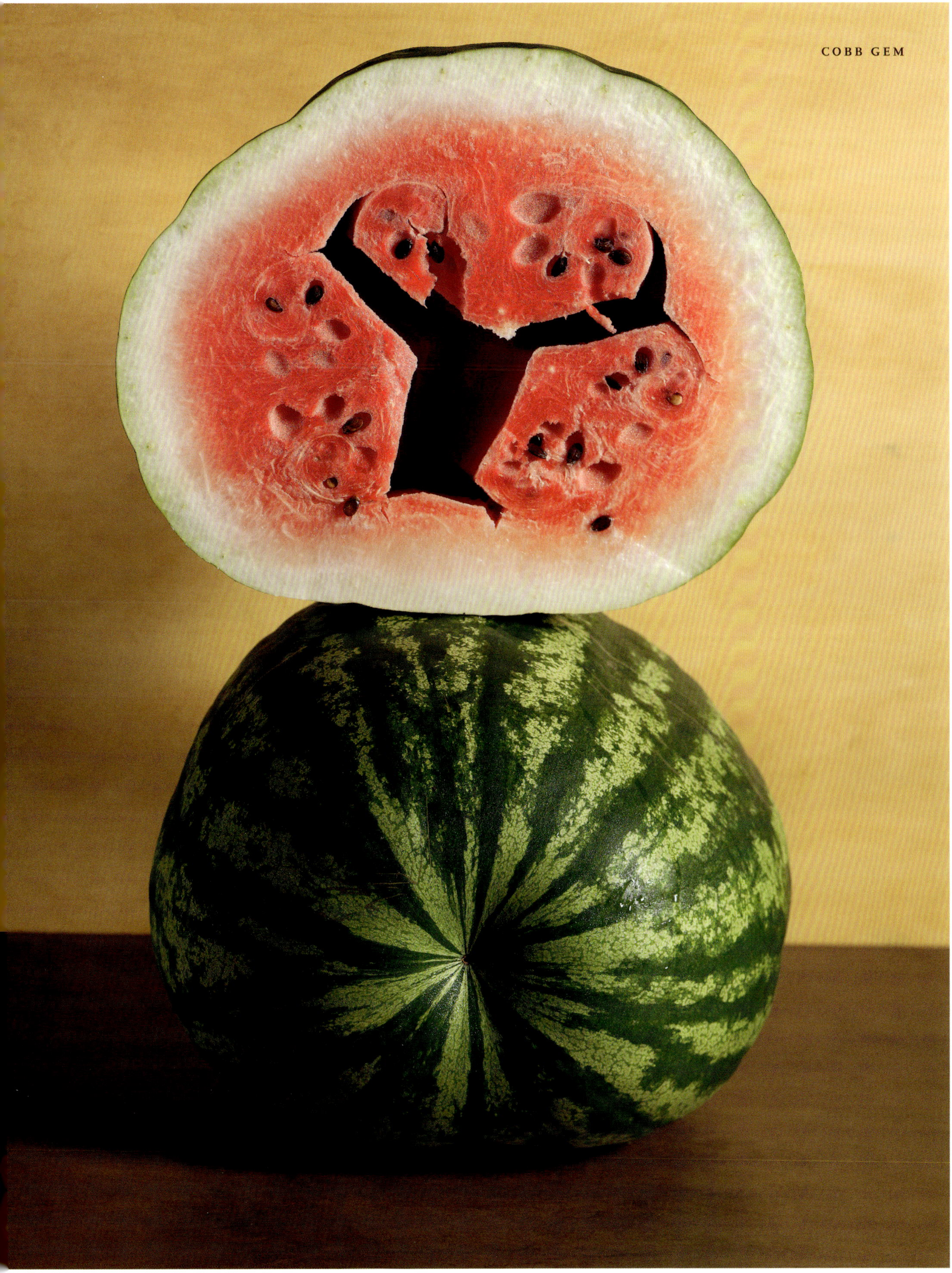

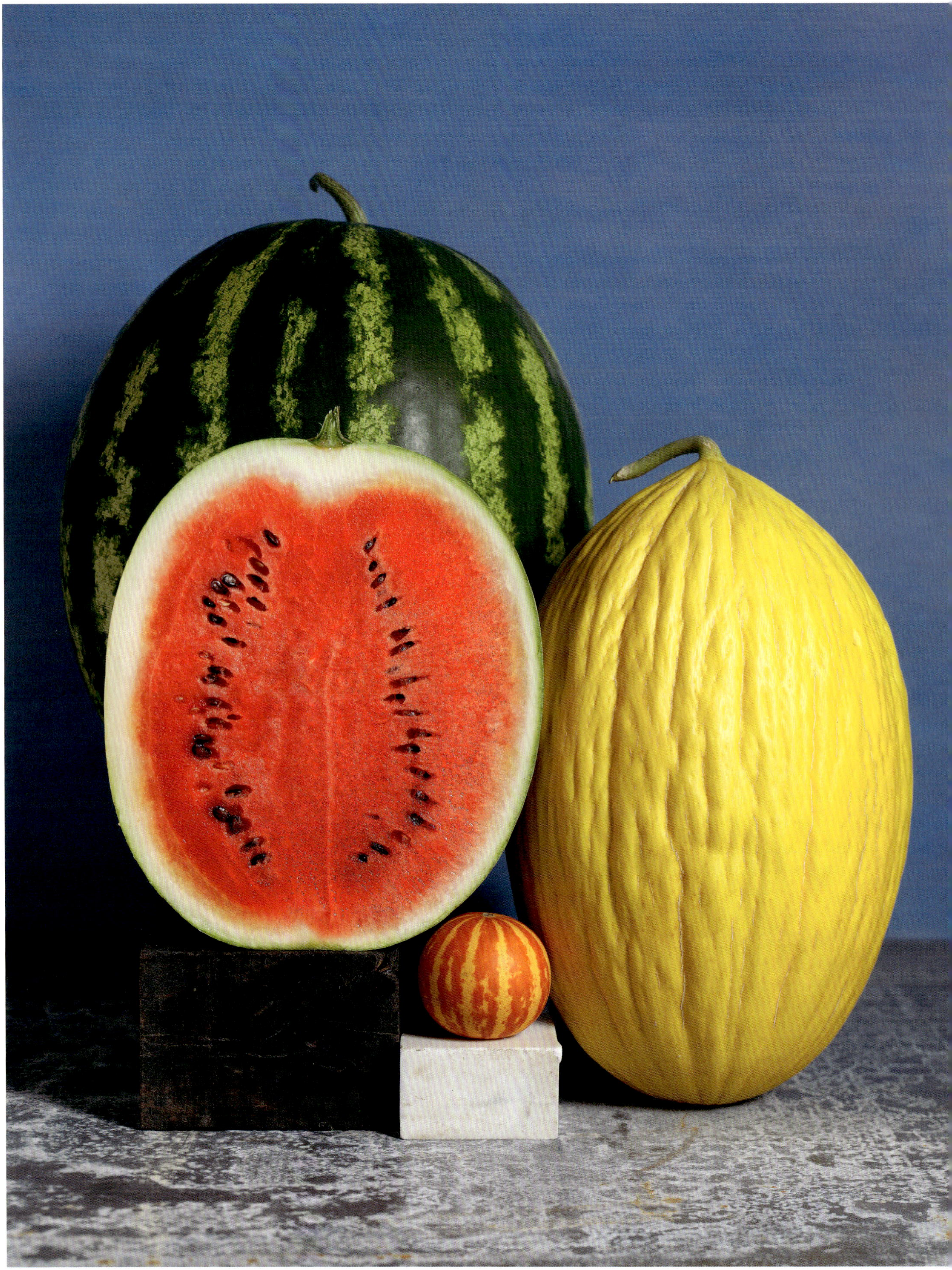

CHAPTER EIGHT

Varietal Descriptions

This is one of the most interesting sections of the book, especially if you eat up lore surrounding heirloom vegetables, want the inside scoop, and crave objective measurements. Here you will find information about all the melons and watermelons illustrated in the book including genus and species, horticultural group (for melon), synonyms, provenance, and telling details. Values for size and weight, sugar content, and days to first edible mature fruit (or first edible immature fruit in the case of the Adzhur, Conomon, and Flexuosus melons) from transplant are based upon personal observations and data derived from my garden over a twenty-year period. Early varieties mature (or are first eaten immature) in 80 days or less; midseason varieties, 81 to 90 days; and late varieties, 90 or more days.

Opposite, clockwise from top: Crimson Sweet, Rugoso di Cosenza Giallo, Queen Anne's Pocket Melon, Mickylee (cut open).

An index of melon and watermelon sweetness is provided, as measured in degrees Brix with a handheld refractometer. Sweet dessert cultivars with a Brix reading of 10 degrees are barely enjoyable; 11 degrees or more will put a smile on your face. The fructose in green-fleshed muskmelons makes them taste sweeter than their orange counterparts, which are high in sucrose, but the difference in taste is not measured by refractometer. Brix values will vary from year to year, place to place, and on fruits from the same plant over the course of a growing season.

Commercial seed sources are provided for most varieties. It's touching to read through that section: In addition to the venerable seed houses (that started out as family businesses) there is so much evidence of handmade, family-focused care. These individuals are literally doing yeoman's work.

Source names are abbreviated under each melon and watermelon entry. For example, if you see ANN listed in the Banana melon description, you'll be able to identify it as Annie's Heirloom Seeds. Sources are extensive but limited to those in the United States and Canada. I want to make it easy for you to find seeds and grow these gems. See page 285 for the source list.

Melon Varietal Descriptions

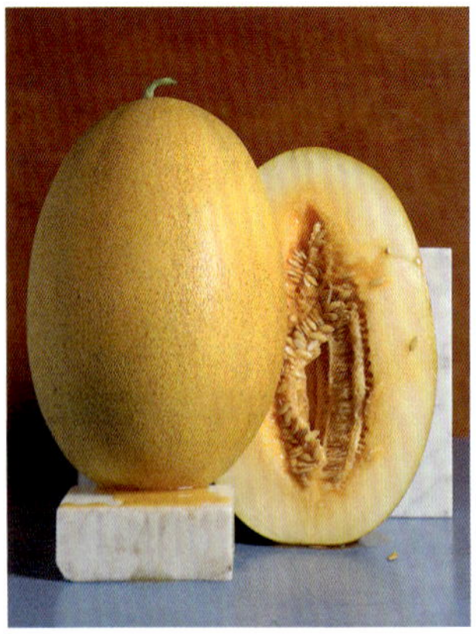

ALTAISKAYA

CUCUMIS MELO **AMERI GROUP (SUB-GROUP ANANAS)**
SIZE: 11" LONG BY 6¼" WIDE; WEIGHT: 7½ POUNDS
SUGAR CONTENT: 7° BRIX
MATURITY: EARLY
SYNONYMS: ALTAI, ALTAI SKAYA, ALSTAJSKAJA, ALTAY
SEED SOURCES: HUD, VICT

Altaiskaya bears its first ripe fruit so quickly—about 60 days from seed—that two crops can be sown (at the beginning and end of June in my garden), if desired. A high yielder of bland melons on short vines as far north as Saskatchewan. Jim Ternier of Prairie Garden Seeds located in Humboldt, Saskatchewan, says Altaiskaya is about as early as the heirloom cultivar Boughem, which itself is seven to ten days earlier than that northern stalwart, Far North.

Early yield is one thing, fruit quality another. Altaiskaya doesn't produce sugary melons unless summers are hot and dry. In the Hudson Valley, my Altai melons split easily on the vine and brim with water. (It's not a pretty picture; see page 100.) Brix numbers are piddling, too.

Dr. Jim McCreight of the USDA told me that about twenty years ago Tom Orton at Rutgers initiated work on adapting Uzbek melons to New Jersey growing conditions for marketing to the large Russian and Central Asian population along the coast. There was no indication of significant progress, and Orton's attention shifted to tomatoes.

Altaiskaya is named for the mountainous Altai region, which sits in the middle of the Siberian taiga, the steppes of Kazakhstan, deserts of China, and high and dry Mongolia. At least on the map, it's a hop, skip, and a jump to Hami, home of the eponymous melon (see page 100).

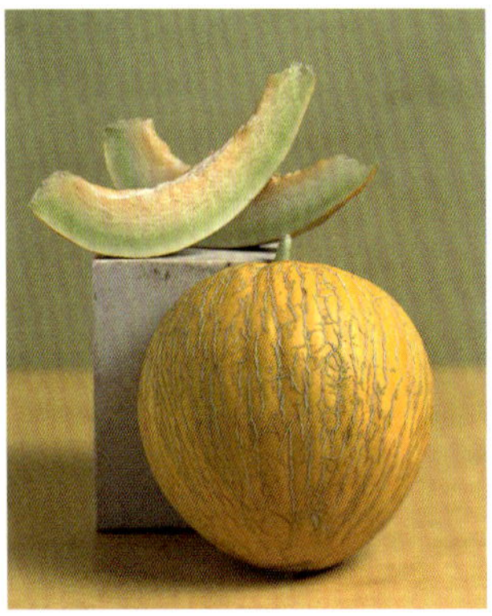

AMARILLO ORO

CUCUMIS MELO **INODORUS GROUP**
SIZE: 8½" LONG BY 6¼" WIDE; WEIGHT: 6½ POUNDS
SUGAR CONTENT: 14° BRIX
MATURITY: LATE
SYNONYMS: AMARELO AURO, AMARILLO ORO DE VALENCIA, CANARY, SPANISH
SEED SOURCES: HEIR, REIM, SAND, SSE, VICT

Amarillo Oro is Spanish for "Yellow Gold." Inodorus from the outside but when the wrinkly deep-yellow skin is cut open a delicate perfume wafts—a prelude to even more earthly delights. (See more on page 86.)

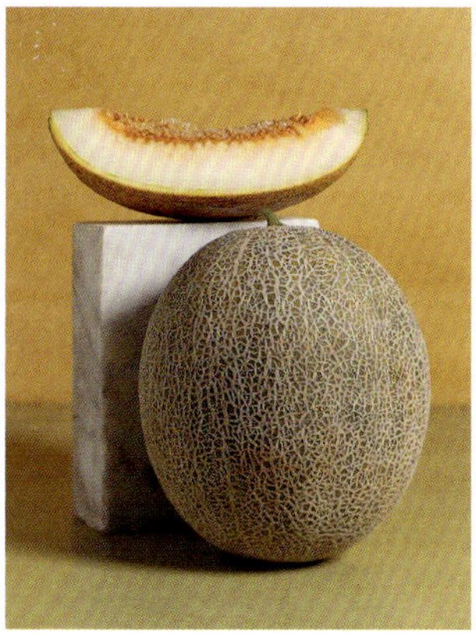

ANANAS

CUCUMIS MELO **AMERI GROUP (SUB-GROUP ANANAS)**
SIZE: 7½" LONG BY 6½" WIDE; WEIGHT: 5 POUNDS
SUGAR CONTENT: 10½° BRIX
MATURITY: MIDSEASON
SYNONYM: PINEAPPLE
SEED SOURCES: EDEN, GOU, HEIR, PINE, REIM, ROHR, SAMP, SAND, SOLA, SSE, TSK, VICT, WHITE, WIL

A cream-fleshed type favored in the Middle East. Harry Paris says the Ananas-type melons sold commercially today probably evolved from melons similar to those grown in Akko and Nablus as reported in Arabic literature from the thirteenth and fourteenth centuries. Ananas Yoqne'am is the more direct ancestor (Paris et al. 2013).

Ananas is adapted to dryland conditions. Still exotic in the United States. Finely and completely netted; lacking lobes and furrows. Sweet perfumed flavor that my daughter describes as heavenly yet earthy. Best to harvest at half-slip when melons are mostly yellow yet mottled with a little green. F1 hybrid versions can be even bigger and sweeter (with Brix readings up to 15 degrees in my garden).

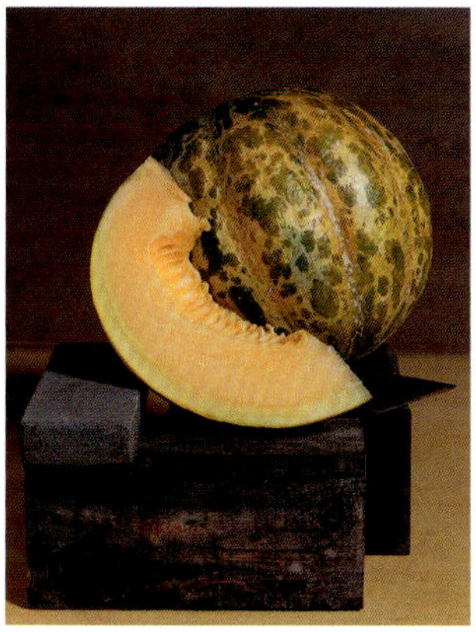

ANANAS À CHAIR ORANGE

CUCUMIS MELO **CANTALUPENSIS GROUP**
SIZE: 5½" LONG BY 6½" WIDE; WEIGHT: 6 POUNDS
SUGAR CONTENT: 9° BRIX
MATURITY: MIDSEASON
SYNONYM: ORANGE ANANAS
SEED SOURCE: NONE IN THE UNITED STATES OR CANADA

Appears to be the same as the Tiger melon popular in England in the 1940s and '50s for growing under cloches; it was catalogued by Bountiful Gardens (California) and Dam Seeds (Ontario) in the 1980s. Tiger is a name applied to several different cultivars that have prominent blotches or stripes on the skin, which are especially noticeable when the melons are unripe. Orange Ananas is different from both the Tiger melon developed by Hill Luz of Bonanza Seeds and the Tiger featured in the first edition of this book (2002). Nor is it the same as Ananas d'Amérique à chair rouge.

It's easier to say what this true cantaloupe isn't than what it is. It shares features with Early Frame Prescott and D'Alger (see pages 202 and 200, respectively), including prominent longitudinal lobes and furrows and fruit skin colors and patterns (pale green with dark green galls that turn light yellow and grayish green, respectively). Should be harvested before it gets overly orange.

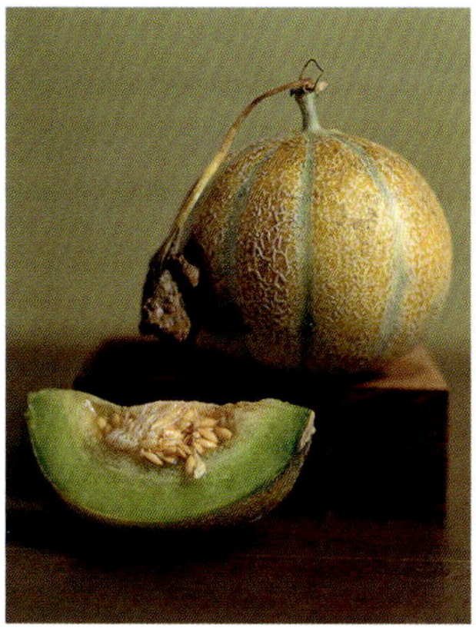

ANANAS D'AMÉRIQUE À CHAIR VERTE

CUCUMIS MELO **RETICULATUS GROUP**
SIZE: 4" LONG BY 5" WIDE WEIGHT: 1½ POUNDS
SUGAR CONTENT: 11° BRIX
MATURITY: MIDSEASON
SYNONYMS: ANANAS D'AMÉRIQUE, ANANAS SANS SOIN, ÉTATS-UNIS, GREEN-FLESHED ANANAS, GREEN-FLESHED PINEAPPLE (OR PINE-APPLE), GREEN NUTMEG, JERSEY GREEN CITRON, MELON CITRON, MELON CITRON VERT, MELON CITRON À CHAIR VERTE, MELON CITRON D'AMÉRIQUE, MELON DE POCHE À CHAIR VERTE, MELON DE LA LOUISIANE, PINE-APPLE, PINEAPPLE
SEED SOURCES: BAKE, SOLA

Bears a remarkable resemblance to Ha'Ogen (see page 43). I wouldn't be surprised if they were related. Larger fruits are rounder (in both cases).

In 1883, Vilmorin-Andrieux suggested that the American pineapple melons, much like Vert Grimpant aka Green Climbing (see page 224), with very long and branching stems, would be well suited for growing on trellises.

It's surprising that this named variety seems to be completely missing from the catalogues and calendars of Bernard M'Mahon (1775–1818), the prominent Philadelphia seedsman. The following melons were listed for decades, with few changes, beginning in 1806: Early Romana, Early Cantaleupe, Early Small Zatte, Early Succado, Black Portugal, Golden Rock, Large Mogul, Minorca, Large African, Rock Canteleupe, Netted Green-flesh, Japan Rock, and Nutmeg. (See more on page 58.)

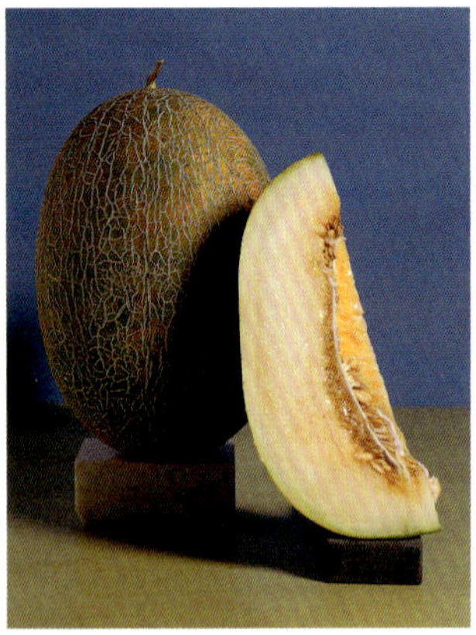

ANANAS YOQNE'AM

CUCUMIS MELO **AMERI GROUP (SUB-GROUP ANANAS)**

SIZE: 10" LONG BY 7½" WIDE; WEIGHT: 9 POUNDS

SUGAR CONTENT: 9½° BRIX

MATURITY: MIDSEASON

SYNONYM: ANANAS YOKNEAM

SEED SOURCE: NONE IN THE UNITED STATES OR CANADA

A very important melon in the scheme of things. Ananas Yoqne'am is the fortuitous ancestor of an important international market type.

I've learned that extra vigilance is required at harvest time because these big, beautiful melons ripen in synchrony overnight (producing one fruit set) and must be promptly picked and brought to the table. Shelf life is very short and fruit quality from my garden disappointing. Perfumed but low sugar, alas. (See more on page 98.)

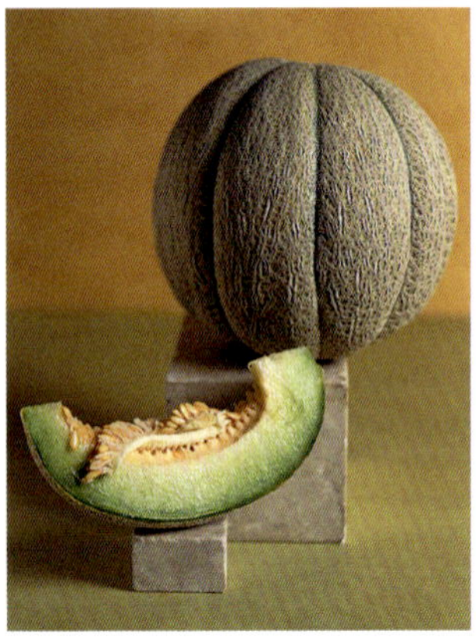

ANNE ARUNDEL

CUCUMIS MELO **RETICULATUS GROUP**

SIZE: 6" LONG BY 6½" WIDE; WEIGHT: 5 POUNDS

SUGAR CONTENT: 10° BRIX

MATURITY: MIDSEASON

SYNONYM: ARUNDEL

SEED SOURCES: SAND, SESE, SSE, THOM

Once a favorite in the Chesapeake Bay area. Believed to have been selected from Acme (1884), which was served in the finest hotels in Philadelphia, Baltimore, and New York. Anne Arundel—the first of the famous Arundel County melons—lost market share soon after its introduction (1890) and was replaced by local upstarts Knight (1908) (aka Sweet Air) and Bottomly (1918) (aka Schramm). According to *The Cucurbits of New York* (1937): "Abe Knight of Pasadena, Anne Arundel County, Maryland, originated the [Knight], possibly as a large fruited selection from Netted Gem." Bottomly, a Netted Gem type, was a selection from Anne Arundel made by Thomas F. Bottomly of Maryland. The Schramm melon was selected from the Bottomly by the Schramm Brothers of Maryland. My rounded Arundel, pictured in this book, more closely resembles Schramm or Bottomly in shape than it does the original Anne Arundel, which is more oval and elongate. I love Arundel for its good looks (symmetry, prominent lobes, and elegant netting) and deliciously sweet, spicy green flesh. (See more on page 78.)

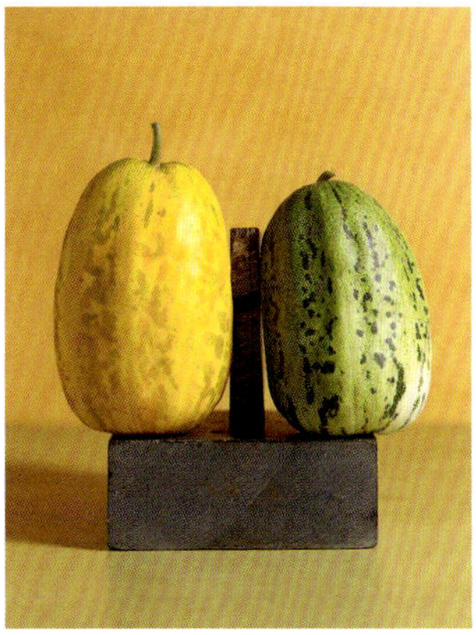

APPLE

CUCUMIS MELO **MAKUWA GROUP**
SIZE: 5¼" LONG BY 3½" WIDE; WEIGHT: 1 POUND
SUGAR CONTENT: 13° BRIX
MATURITY: EARLY
SEED SOURCES: BAKE, SSE

My husband, daughter, and I love Makuwa melons—especially this yellow one speckled green. Apple's pale ripe flesh is very sweet and crisp. When unripe it's mild, not bitter. Has no aroma inside or out. Exhibits a rare characteristic: the fruit's veinal portion is elevated, forming ribs. The skin peels easily nonetheless.

Introduced in 2013 by Baker Creek Heirloom Seeds. Their seed source was the Gingerich family of Kentucky.

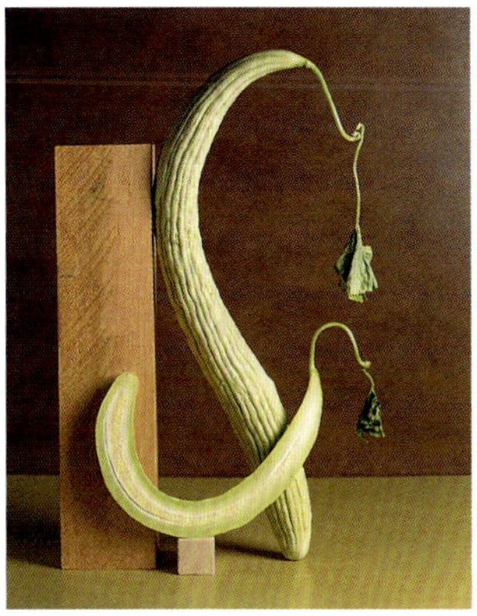

ARMENIAN YARD-LONG

CUCUMIS MELO **FLEXUOSUS GROUP**
SIZE: 24" LONG BY 2½" WIDE; WEIGHT: 3 POUNDS
SUGAR CONTENT: 3° BRIX
MATURITY: EARLY
SYNONYMS: ARMENIAN CUCUMBER OR MELON, ARMENIAN THREE FOOT, DUKE, GREEN SNAKE, GUTA, METKI WHITE, METKI WHITE ARMENIAN, METKI WHITE SERPENT, METKY WHITE TRAILING, SERPENT CUCUMBER, SNAKE CUCUMBER, YARD LONG
SEED SOURCES: ANN, BAKE, BERL, BOT, BUR, BURR, CUC, EDEN, GOOD, GRETA, HEIR, HER, HOPE, HUD, HUME, IRISH, JSKIT, KIT, LEJ, MOR, NAT, NATIVE, NIC, OSC, PEACE, PINE, REDS, REDW, REIM, REN, REST, ROHR, SAND, SHUM, SISK, SOWT, SSE, SSEC, SUS, SWAL, TERRIT, TRADE, TSK, URB, VICT, WHITE, WIL

A very fine and crisp snake melon that crops heavily.

To judge by the number of vendors and synonyms, Armenian Yard-Long is the most popular snake melon in America. It was introduced into the Abundant Life Seeds catalogue in 1976, and later described by them as a "distinctive grey-green variety that may curve. Trellising helpful." The first two Abundant Life Seeds catalogues (1975 and 1976) were handwritten on pieces of paper and then folded into origami shapes. Armenian Yard-Long has proliferated ever since. Some snake melon cultivars apparently can grow two yards long.

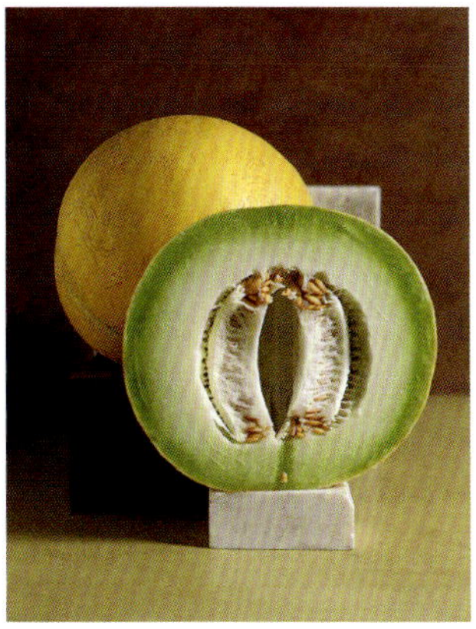

ASHKHABAD

CUCUMIS MELO **INODORUS GROUP**
SIZE: 7" LONG BY 6½" WIDE; WEIGHT: 6½ POUNDS
SUGAR CONTENT: 16° BRIX
MATURITY: LATE
SYNONYM: ASHKAHABAD
SEED SOURCES: BAKE, SAND

When the former Soviet Union collapsed, Turkmenistan was one of the new nations formed. Ashkhabad hails from the Turkmen capital (formerly Ashgabat), in historical Khorasan on the lip of the Kara-Kum Desert. The specimen pictured in *Melons for the Passionate Grower* (2002) caught the measles, a melon disease caused by high relative humidity in which excess salts accumulate where the skin exudes moisture, forming sunken, darkened lesions. This imperfection irked me for years—so we replaced that image with one of a more pristine melon I grew in the summer of 2018. And am I ever glad I grew it that year: Brix readings soared to 16 degrees. Ashkhabad is crisp and sweet beyond measure. Like a great Honeydew (see page 207).

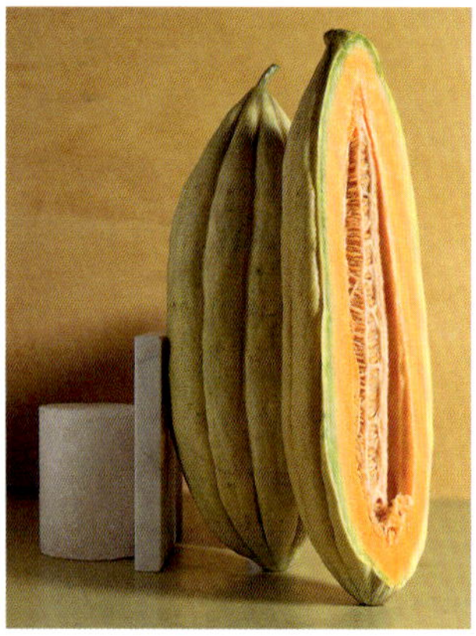

BANANA

CUCUMIS MELO **RETICULATUS GROUP**
SIZE: 14" LONG BY 7½" WIDE; WEIGHT: 9 POUNDS
SUGAR CONTENT: 11° BRIX
MATURITY: MIDSEASON
SYNONYMS: AMERICAN BANANA, BANANA CITRON, CUBAN BANANA, MEXICAN BANANA; PROBABLY THE SAME AS THE BANANALOUPE OFFERED BY PEACE SEEDS IN THE 1980S
SEED SOURCES: ANN, BAKE, CLIF, EDEN, ERS, EVER, HEIR, HER, HVSC, MARY, MOR, PINE, SAMP, SHUM, SOLA, SSE, TRADE, TSK, URB, VICT, WIL

My husband has "plenty of time for this melon." I'm happy to bring in the harvest and share the joy. Bananas are so good, like a sweet fragrant crenshaw, that it's hard to stop eating them. And why should we? Their eye-catching pastel yellow form, longitudinally furrowed (and occasionally netted), is like no other melon I've grown. When cut into bite-sized pieces that are naturally scalloped, the yellow skin, half-inch outer layer of firm green flesh, and inner layer of soft orange make for eye-catching finger food. I'm on the same wavelength as seedsman James J. H. Gregory, who wrote in his 1891 catalogue that morsels cut in the same manner make "a fine and striking contrast when brought to the table." Banana shares features with the Log-of-Wood melon catalogued as new and illustrated by Gregory in 1882 with a tempting engraving of a photograph. The Log was longer (up to three feet) and uniformly netted, but both look more like whales than bananas to me. (See more on page 66.)

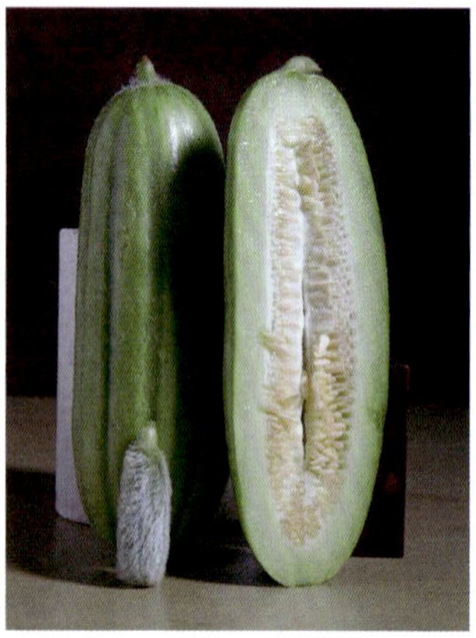

BARESE

CUCUMIS MELO **ADZHUR GROUP**
SIZE: 7½" LONG BY 2¾" WIDE; WEIGHT: 1 POUND
SUGAR CONTENT: 3° BRIX
MATURITY: EARLY
SYNONYMS: MEDIO LUNGO POLIGNANO, MEDIO LUNGO PUGLIESE, MEZZO LUNGO BARESE, MEZZO LUNGO DI POLIGNANO
SEED SOURCES: CUC, HUD, NEWE, SEEDSF, SSE, TSK

I prefer the elongated carosellos to the rounded ones and find them crisper. Judging by the relatively large number of lungo varieties on offer from Italy, I suspect the Italians prefer them as well. Italy is the land where a different cucurbit was selected for long-fruitedness and designed to be eaten immature: Zucchini (*Cucurbita pepo*). Not the same as Tortarello Barese (see Tortarello, page 223).

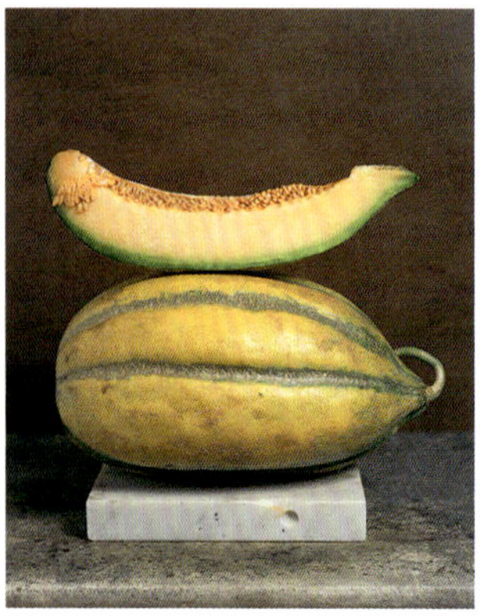

BIDWELL CASABA

CUCUMIS MELO **INODORUS GROUP**
SIZE: 14" LONG BY 9" WIDE; WEIGHT: 14 POUNDS
SUGAR CONTENT: 10° BRIX
MATURITY: LATE
SYNONYMS: BIDWELL, BIDWELL CASSABA, CALIFORNIA CASABA
SEED SOURCES: BAKE, SSE, SSEC, SUS, URB

The Bidwell Casaba may have been the first soft-skinned casaba melon grown in California. Liberty Hyde Bailey wrote in 1927, "Another Cassaba melon was sent out long before [the cream-fleshed, hard-rinded Kassaba sent from Smyrna and grown in California in 1879] by the Patent Office, before the organization of the Department of Agriculture. This received special attention from General Bidwell of Chico, California, and became known as the Bidwell Cassaba. This is a large summer melon, said to grow to one foot in diameter . . . known only locally, although it was grown at Cornell twenty-five years ago."

In a letter to his wife, Annie, in 1876, Bidwell wrote, "Am very fond of fruit. I was the only one that succeeded in raising the Casaba melon, named after the town of Casaba, in Asia. The Agricultural Department imported some of the seeds, and I procured some of them. They were sent all over the U. S. but I am the only one that succeeded in raising them. They would not grow on the coast very well, but in the warm valleys they do better" (Gillis and Magliari 2003). I suspect Annie had heard this story before.

The bigger the melon, the longer to ripen, and the greater the possibility of accumulating high sugars in high sucrose genotypes. Bidwell approaches perfection in Chico: individual melons with thirty pounds of sugary flesh and Brix readings of 17 degrees, grown by the likes of Lowell A. Stringfellow (1914–2011) and his wife Clara Fay. Lowell told me years ago, "You can't beat this melon." Clara Fay pointed to her uncle, Wilber Ganow, as their original seed source. Ganow once operated a farm stand with an honor system on Bidwell Avenue in Chico.

The Bidwell was a trailblazer for that other, quite different, soft-skinned California casaba, the Crane (see page 200), developed in Sonoma in the twentieth century. (See more on page 82.)

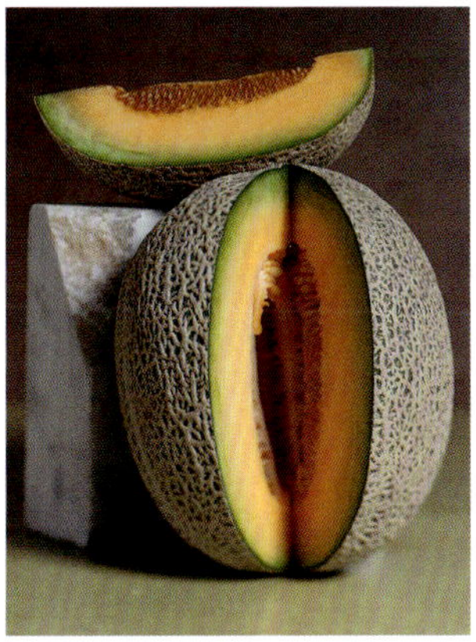

BLENHEIM ORANGE

CUCUMIS MELO **RETICULATUS GROUP**

SIZE: 7" LONG BY 5½" WIDE; WEIGHT: 3¾ POUNDS

SUGAR CONTENT: 9½° BRIX

MATURITY: LATE

SYNONYM: BLENHEIM ORANGE SUPERLATIVE

SEED SOURCES: REDS, SOC, SSE

Blenheim Orange melon was developed in 1881 by Thomas Crump, the Duke of Marlborough's head gardener, at Blenheim Palace in Oxfordshire. Perhaps it was known to Sir Winston Churchill who was born at Blenheim, the family's ancestral home, in 1874. (Blenheim Orange is also the name of an apple that originated near Blenheim, circa 1740.)

Blenheim Orange Superlative was once commonly grown in frames and hothouses in England. When cultivated in my garden, plants become lush with fruits that aren't terribly sweet. Described as extremely perfumed in Vilmorin-Andrieux (1883). Similar to Retato degli Ortolani.

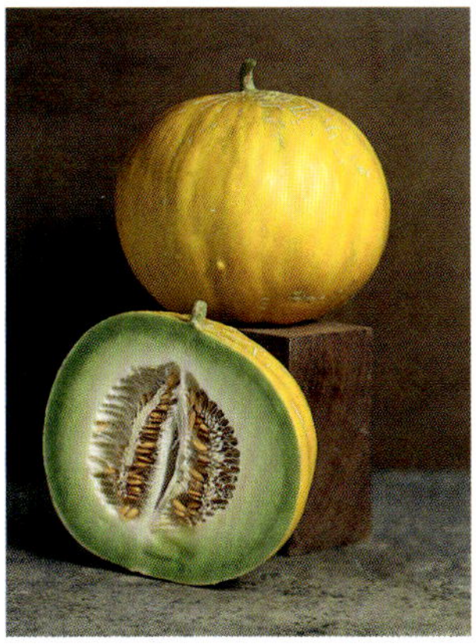

BOULE D'OR

CUCUMIS MELO **INODORUS GROUP**

SIZE: 7" LONG BY 9" WIDE; WEIGHT: 6½ POUNDS

SUGAR CONTENT: 12° BRIX

MATURITY: LATE

SYNONYMS: GOLDEN BALL, GOLDEN PERFECTION

SEED SOURCES: BAKE, HUD, SOLA, SSEC, TRADE

A high-quality Amarillo-type melon with smooth skin, little netting, the occasional wart, and firm, sugary, perfumed green flesh. Aptly named.

Except for its sparser netting, the Boule d'Or of today conforms to the description and engraving of Golden Perfection, later known as Boule d'Or, in Vilmorin-Andrieux's French editions beginning in 1883 and ending in 1925. Appears to be the same as the smooth-skinned Jiley's golden Ball or Boule d'Or de Jiley grown by Naudin (1815–1899) in Paris and described in 1859. Similar to Ashkhabad (see page 193). My original seed source in 1999 was French melon connoisseur Bruno Defay.

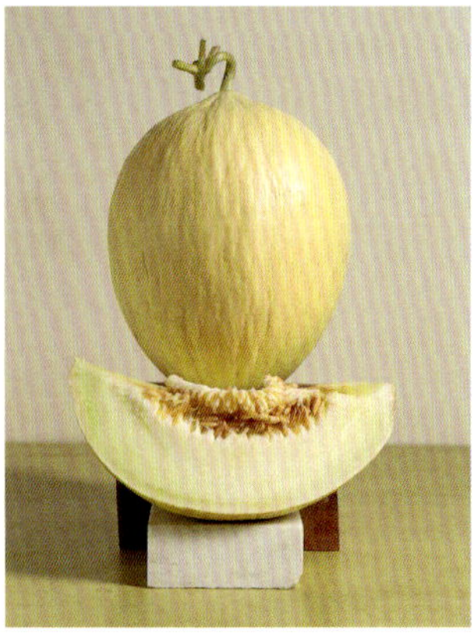

BRANCO

CUCUMIS MELO **INODORUS GROUP**
SIZE: 9½" LONG BY 6½" WIDE; WEIGHT: 8¾ POUNDS
SUGAR CONTENT: 14° BRIX
MATURITY: MIDSEASON
SYNONYMS: BRANCO DE RIBATEJO, BRANCO DO RIBATEJO, BLANCO DE RIBATEJO, DE ALMEIRIM, DE ALMERIM
SEED SOURCE: NONE IN THE UNITED STATES OR CANADA

Branco is the quintessential Portuguese white casaba. Named for its milky-white skin and the region where it is grown. Has wonderfully sweet, firm flesh that can be white, green, orange, bicolored, even tricolored. I once cut open a Branco longitudinally and found one half completely green and the other completely orange!

Can be easily cultivated in home gardens in the United States where summers are hot and dry. A melon resembling Branco was illustrated in northern Italy around 1390 in medieval manuscripts known as the *Tacuinum Sanitatis*; it was labeled, however, as a kind of non-sweet vegetable melon (Paris, Daunay, and Janick 2009). The co-authors surmise, "Probably the climate of northern Italy was not suitable for casaba melons, as they are late to ripen and accumulate sugar, requiring a long, hot growing season."

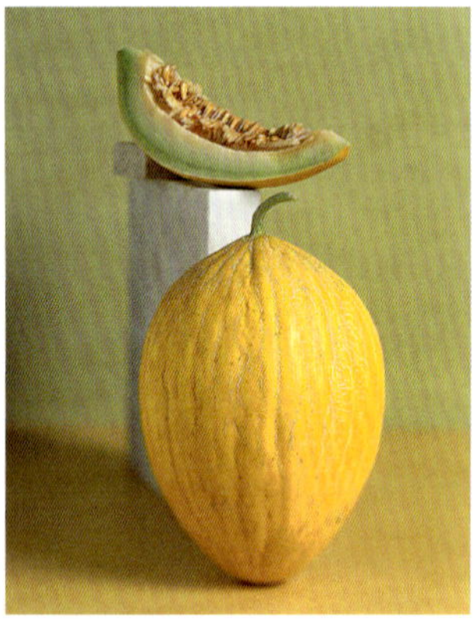

CANARY YELLOW

CUCUMIS MELO **INODORUS GROUP**
SIZE: 7" LONG BY 6" WIDE; WEIGHT: 5 POUNDS
SUGAR CONTENT: 12° BRIX
MATURITY: LATE
SYNONYMS: AMARILLO, CANARY, FRENCH CANARY, ITALIAN YELLOW CANARY, JAUNE DE CANARY [DES CANARIES], YELLOW CANARY
SEED SOURCES: BURR, GOU, HEIR, PEACE, PINE, REDW, REIM, REST, SOWT, SSE, SUS, TRADE, TSK, VICT

A very desirable winter melon that would last months in cold storage if I'd only let it. Yellow Canary satisfies cravings for something sweet, firm fleshed, and healthy. Dr. Marisa Gómez-Guillámon told me years ago that for Spanish melon breeders (such as herself), all yellow winter melons are considered to be canaries or *canarios*. This canary is elliptical in shape with pointed stem and blossom ends, finely wrinkled, and lightly netted. The primordial fruits that form are very hairy—reminding me of those good-luck charms, and sometime keychains, known as rabbits' feet.

Canary melons, which are nonclimacteric, change skin color from light green to intense yellow at maturity independently of ethylene synthesis. According to Yosef Burger and co-authors (2010), preliminary results of their research "suggest that the yellow color of the rind of this variety results from flavonoid synthesis and not from beta-carotene. Therefore, it would seem possible to introduce the combination of longer shelf life and fruit rind color change into muskmelons and cantaloupes, using the Yellow Canary–type casaba melons as a source."

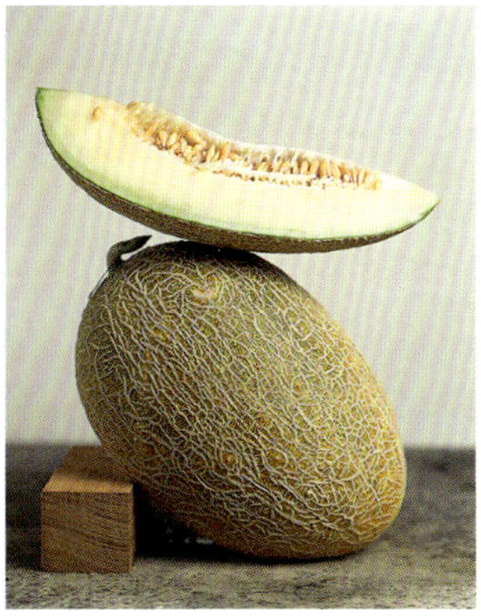

CAVAILLON ESPAGNOL

CUCUMIS MELO **INODORUS GROUP**
SIZE: 9" LONG BY 6" WIDE; WEIGHT: 6 POUNDS
SUGAR CONTENT: 11° BRIX
MATURITY: LATE
SYNONYMS: CAVAILLON À CHAIR ROSE, CAVAILLON ESPAGNOL À CHAIR ROSE, CAVAILLON ESPANOL
SEED SOURCE: SSE

A Spanish Rochet-type melon with thick green skin that becomes yellowish and netted at maturity. Cavaillon Espagnol comes with a prestigious French provenance: Cavaillon is a renowned melon-growing region in southern France. This particular Cavaillon—and there are many—is of excellent quality. Cavaillon Espagnol appears in the *Dictionnaire Vilmorin des Plantes Potagères* in 1946 (with a color reproduction of a seed packet). My translation of the text: a pretty variety with yellow netted skin and rose-tinted flesh (*chair incarnat*). Cavaillon Espagnol served as a replacement for traditional red-fleshed (yellow-orange) varieties like Cavaillon à chair rouge and Cavaillon tranché à chair rouge, both of which were carried by Vilmorin (first listed in 1856 and 1904, respectively).

Cavaillon Espagnol was introduced in the United States by Southern Exposure Seed Exchange in their 1991 catalogue supplement.

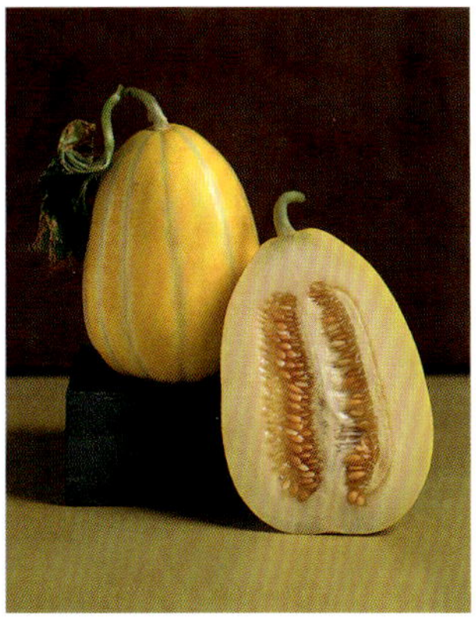

CHAMOE

CUCUMIS MELO **MAKUWA GROUP (SUB-GROUP YUKI)**
SIZE: 6¼" LONG BY 4¾" WIDE; WEIGHT: 2 POUNDS
SUGAR CONTENT: 12° BRIX
MATURITY: EARLY
SYNONYMS: ASIAN, KOREAN, OR ORIENTAL MELON; EARLY SILVER LINE, GINSEN MAKUWA
SEED SOURCES: BUR, EDEN, SSE, SSEC, SUS, TRADE, TSK

A celebrated Korean melon that's finally catching on in the United States. Luminous silvery lines over vein tracts are its signature. Very early and prolific—it's hard to keep up with at harvest time. Excess immature fruits can be pruned off to slow the pace, but it's better to spread the wealth. Chamoe has a pleasant external aroma when brought to the nose. Peel the skin like an apple, slice or dice the fruit, and enjoy the sweet, crisp, white flesh with or without the seeds and gel.

Introduced as Early Silver Line by Burpee in 1989. May be the same as Japan Silver-Striped.

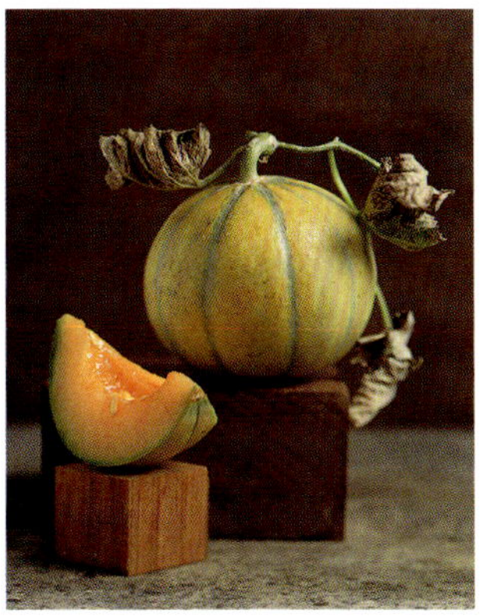

CHARENTAIS

CUCUMIS MELO **CANTALUPENSIS GROUP**
SIZE: 4½" LONG BY 4½" WIDE; WEIGHT: 2 POUNDS
SUGAR CONTENT: 13½° BRIX
MATURITY: MIDSEASON
SYNONYMS: CAVAILLON, MELON CANTALOUP SUCRIN AMÉLIORÉ
SEED SOURCES: AMI, ANN, BAKE, BOT, BUR, EDEN, GOOD, GOU, HEIR, IRISH, JSKIT, KIT, LEJ, MARY, NAT, NIC, PEACE, PINE, REIM, REST, ROHR, SAMP, SAND, SOLA, SOWT, SSEC, SUS, TERR, TERRA, TRADE, TSK, TUR, URB, VICT, WHITE, WIL

Charentais types were painted in Italy (at the Villa Farnesina in Rome) as early as 1513. Five hundred years later, they comprise the most popular "cultigroup" of melons grown in France. Too much of a good (great) thing? Other traditional French cantaloupes (with deep lobes, warts, and galls) have been losing market share since the 1940s. (See more on page 46.)

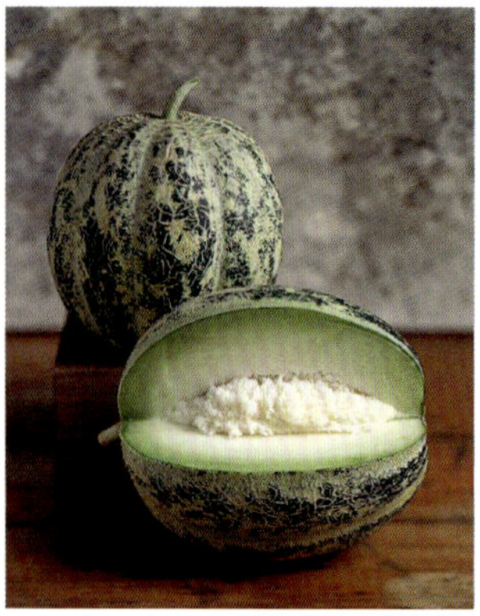

COB

CUCUMIS MELO **RETICULATUS GROUP**
SIZE: 5½" LONG BY 5" WIDE; WEIGHT: 2 POUNDS
SUGAR CONTENT: 10½° BRIX
MATURITY: EARLY
SYNONYMS: APPLE CORE, ICE CREAM, CORN COB
SEED SOURCES: HER, SSE

With seeds clustered in a cob-like fashion at its core, and mealy bland flesh, Cob doesn't fit neatly into any cultivar group. For me it's a curiosity more than anything else. Cob or Ice Cream melon was prized by home gardeners in the nineteenth century in the American West because it tasted like vanilla ice cream to some or even banana to others and was apparently easier to digest than orange-fleshed muskmelon.

Cob Muskmelon was introduced into the Glecklers Seedmen (Metamora, Ohio) catalogue in 1976. The copy reads, "Like corn, the seed on this melon is formed on a cob, and quickly removed. Flesh is creamy white, honey sweet and slightly grainy like home-made ice cream. For a century and more past it was common in old fashioned home gardens throughout the west." Merlin W. Gleckler, Sr., who founded Glecklers in 1943, suffered a severe stroke in 1980, after which his son George L. Gleckler and daughter Martha M. Gleckler took the reins of the business. George reached out to Seed Savers Exchange for Cob seed to restore his father's seed supply, which had apparently been lost during the transition. (See more on page 70.)

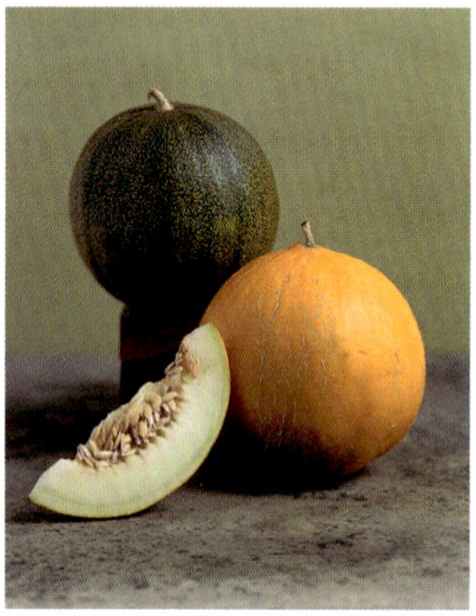

COLLECTIVE FARM WOMAN

CUCUMIS MELO **INODORUS GROUP**
SIZE: 4¾" LONG BY 5" WIDE; WEIGHT: 2 POUNDS
SUGAR CONTENT: 13½° BRIX
MATURITY: MIDSEASON
SYNONYMS: COLLECTIVE FARMWOMAN, KAZAKH, KOLKHOZNITSA 749/753
SEED SOURCES: BAKE, BUR, ERS, GRETA, HEIR, HER, REDS, SOLA, SSE, SUS, TERR, TRADE, TSK

Collective Farm Woman is small but packs a ton of goodness. Add her to your list of must-have melons. The first time I tried one, I had low expectations. The whole fruit is as hard and dense as a baseball—and gives off only the mildest fragrance when ripe. As with a can of ground coffee, the first clue to its goodness comes when you open it and inhale. Collective Farm Woman is floral and crunchy and sweet. I gnawed it right down to the nub.

This melon originated in the arable soil of Ukraine, a soil once bloodied by the purges of Stalin: Millions of farmers and others perished from starvation, mass execution, and exile. When the farms of Ukraine were collectivized, many traditional food crops were lost, making the taste of this melon both bitter and sweet.

Kent Whealy of Seed Savers International (a project of Seed Savers Exchange) obtained seed during one of three collecting trips to the former Soviet Union undertaken between 1991 and 1993. Marina Danilenko, the proprietor of Moscow's first private seed company, supplied the goods. As you may have guessed, her source for the melon was none other than a woman from a collective farm.

Unless you're a devoted member of the Seed Savers Exchange or a crop historian, the following may be too much information. (I eat it up.) Seeds of Collective Farm Woman were first offered in October of 1993 in a brochure mailed to SSE members listing a gift collection of six Russian varieties: Nejinski Cucumber, Small (Shining) Light Watermelon, Collective Farmwoman Melon, Diamond Eggplant, Southern Night Tomato, and Orange Tomato (all seed was grown in Russia). The description of Farmwoman: "Old traditional variety from Ukraine, round 7-10" fruits, smooth yellow-orange skin, white (slightly yellow) flesh, matures early, ripens in central Russia, even matures in Moscow, most popular melon on island of Krim in Black Sea." Nicolas Lindholm of Durham, Maine, was the first SSE member to offer saved seeds (grown on U.S. soil) to other members, via the 1995 annual Yearbook. As was true for Nicolas, my Farmwoman melons don't grow quite as large as those from the old country.

Collective Farm Woman was introduced commercially by Seed Savers Exchange in its 2002 catalogue. SSE member Andrey Baranovski of Byelorussia, Belarus, wrote in the 2018 SSE Yearbook that Collective Farm Woman is synonymous with Kolkhoznitsa 749/753, which was bred before 1945 by Biryuchekutskaya Vegetable Breeding Research Station in Novocherkassk. She is also synonymous with Kazakh, which was carried by Gurney's in the late 1960s according to Glenn Drowns; it first appeared in Abundant Life's catalogue in 1986.

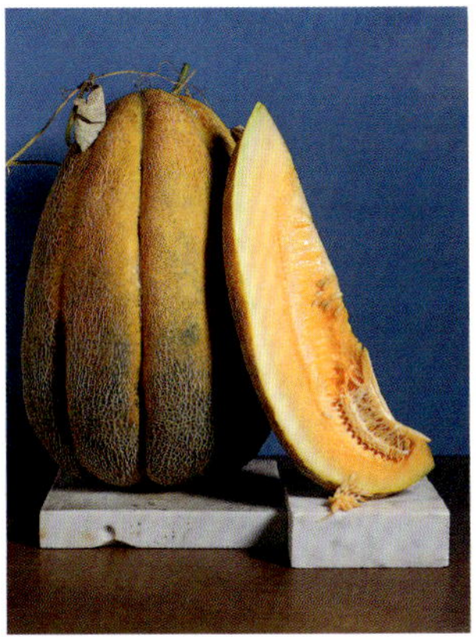

COLOSSAL MUSKMELON

CUCUMIS MELO **RETICULATUS GROUP**
SIZE: 16" LONG BY 14" WIDE; WEIGHT: 33 POUNDS
SUGAR CONTENT: 7° BRIX
MATURITY: MIDSEASON
SYNONYMS: COLOSSAL, COLOSSAL CANTALOUPE, COLOSSAL FURROWED
SEED SOURCES: GIANT, SSE

These are such fun to grow. I can easily pull in thirty-pound-plus Colossals without much fuss—the melon pictured in this book was grown from seed of a 54½-pound Colossal. What if I really tried? Their fruit quality leaves something to be desired: The light orange flesh is abundant with slices up to 2½-inches thick, but watery and low sugar. Happily there are lots of other choices for the table.

Two muskmelon cultivars are known to grow giants: Colossal and North Carolina Giant. The current world record for heaviest muskmelon is held by Danny Vester; his muskmelon clocked in at 65.9 pounds at the State Farmer's Market in Raleigh, North Carolina, on Tuesday, August 14, 2018.

As impressive as that feat is, it surprises me that the world record for heaviest muskmelon hasn't been broken more often—the world record for pumpkin gets broken practically every year. (The previous record holders for muskmelon were Scott and Mardie Robb of Palmer, Alaska; their Colossal weighed in at 64 pounds, 13 ounces, on August 16, 2004, at the Alaska State Fair.) Perhaps the competition isn't as great, the monetary rewards aren't as much, or *Cucumis melo* doesn't have the genetic proclivity for increased size. Things are heating up lately. Another man to watch is Lloyd Bright of Hope, Arkansas. This seedsman, who with his extended family has held many world records for watermelon, came close to setting a new record for muskmelon in 2018. Bright's 66.8-pounder (known as 66.8 Bright 2018) was damaged, and disqualified.

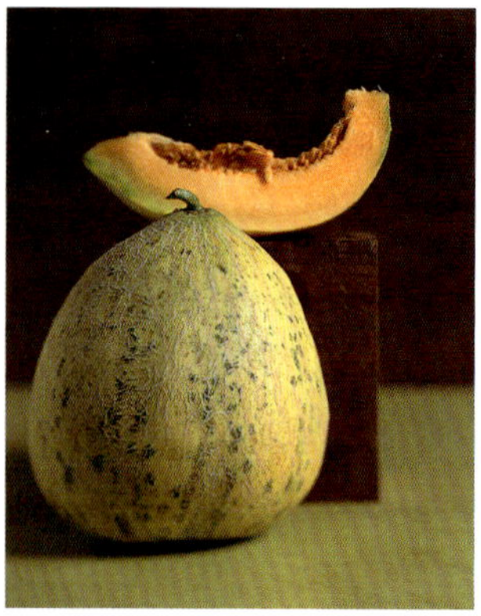

CRANE

CUCUMIS MELO **INODORUS GROUP**

SIZE: 7" LONG BY 6" WIDE; WEIGHT: 3½ POUNDS

SUGAR CONTENT: 10° BRIX

MATURITY: LATE

SYNONYMS: CRENSHAW, CRANSHAW

SEED SOURCES: NIC, REST, SAND, SISK, SSE, TRADE, UPRI

Crenshaws, such as the Crane, are a type of soft thin-skinned casaba melon that became popular for local markets in California in the mid-twentieth century. The Crane melon is an heirloom developed and maintained by the Crane family of Santa Rosa, California. Richard Crane told me in 2001 that the first Cranes were marketed by his forebears around 1926 and sold for sixty cents a dozen. Maintenance of this variety spans generations and is a labor of love. The family's labor became lighter about thirty years ago after they made the switch from short-haul shipping in wooden crates weighing up to eighty-five pounds (holding twelve to fourteen melons) to custom cardboard boxes weighing thirty-five pounds (holding five or six melons) designed to accommodate the melons' pear shape. Richard said the sweet fragrance (non-musky) of tons of fresh Crane melons on the floor of their packing shed in September and October would blow me away.

Of that I am certain. But the Crane melon isn't just for California anymore. Glenn Drowns introduced the Crane into the Sand Hill Preservation Center catalogue in 2003 and described it this way: "Once you've tasted this your taste buds will never want to touch another melon! The texture is smooth. The fragrance is fantastic and the flavor and sweetness will leave you speechless."

The Cranes dryland farm their melons on black adobe ground that retains moisture. Coastal fog helps, too. Crane melons are not partial to intense summer heat and sunburn in places like Chico, according to local farmer Zea Sonnabend. There's fortunately little chance of that happening in my Zone 5b garden, where I'm able to produce lots of scrumptious well-netted fruits on vigorous vines.

There is a precedent for the thin-skinned casaba in California: see Bidwell Casaba (page 194). Crane melon is similar to Eel River, Northern Arizona, and Sweet Freckles (see page 221). Deservedly boarded onto Slow Food USA's Ark of Taste. (See more on page 94.)

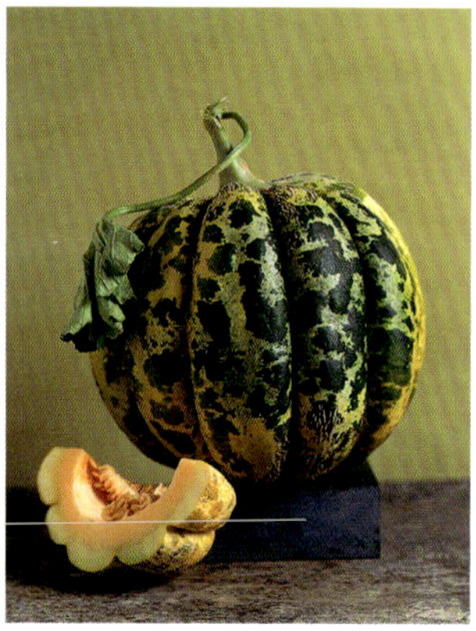

D'ALGER

CUCUMIS MELO **CANTALUPENSIS GROUP**

SIZE: 12" LONG BY 9" WIDE; WEIGHT: 11 POUNDS

SUGAR CONTENT: 6½° BRIX

MATURITY: LATE

SYNONYMS: ALGERIAN CANTALOUPE, ALGIERS CANTALOUPE, CANTALOUP D'ALGER, CANTALOUP DE MAI, ROCK

SEED SOURCE: NONE IN THE UNITED STATES OR CANADA

I'll never forget the excitement I felt the day my first D'Alger ripened on the vine: August 29, 2000. I cut it from the mother plant—its canopy lush and largely spreading—and Victor photographed it that afternoon.

D'Alger is a melon that belongs on a pedestal, the better to view its fantastic form. Is it melon or mirage? Decide for yourself. Its ten very pronounced lobes and deep furrows are reminiscent of pumpkin; its raised spots, or galls, are like clouds. When overripe, D'Alger turns several shades of orange and gives off an intoxicating aroma that can perfume a room after harvest.

I can attest that this cultivar fares poorly in the Hudson Valley: It produces melons with a voluminous seed cavity and a paltry amount of bland orange flesh. It's gotten mixed reviews in France since the nineteenth century. Vilmorin-Andrieux (1856) noted that Cantaloup d'Alger was introduced to French gardeners in 1854 but the company opted not to carry it until later. By 1883, Vilmorin rated it as one of the hardiest summer melons: thick, juicy, perfumed, and always very sweet. Although it was much appreciated in the nineteenth century in the Midi—the south of France, where it's warm and sunny—D'Alger was never popular with market gardeners near Paris. Perhaps the melon's robust vines were too hard to tame in frames; open-air culture was infrequently used in northern France. On the other hand, Charles Naudin (1859) found D'Alger's flesh quality when grown in Paris comparable to the best Prescotts. So much depends on location, climate, seed stock, and gardening skills.

D'Alger is named for the seaport city Algiers. It closely resembles the Cantaloup Gros Orange (aka Cantaloup Grand Hollande) described and depicted in Jacquin (1832). Seedsman James J. H. Gregory of Marblehead, Massachusetts, catalogued it as early as 1881, noting it was allied with Hardy Ridge (see Prescott Fond Blanc, page 215).

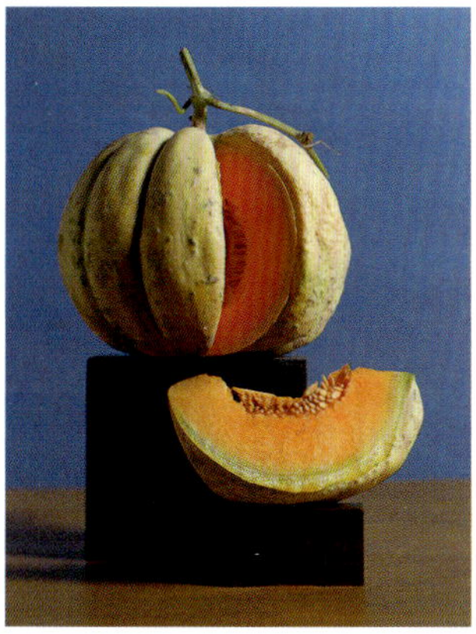

DÉLICE DE LA TABLE

CUCUMIS MELO **CANTALUPENSIS GROUP**
SIZE: 6" LONG BY 7½" WIDE; WEIGHT: 6 POUNDS
SUGAR CONTENT: 10° BRIX
MATURITY: EARLY
SYNONYMS: DÉLICES DE LA TABLE, DELIGHT OF THE TABLE
SEED SOURCES: HUD, SAMP, SOLA, SSE, SSEC

A true rock melon with prominent longitudinal furrowing and a hard, thick skin. Delightful when eaten fully ripe. The first mention I've found of this named variety is in Troop (1911). He listed Delight of the Table as one of four French varieties (including Noir des Carmes [see page 212], Sucrin de Tours [see page 221], and Trévoux) usually grown in frames that also grow out-of-doors under favorable conditions in the United States.

Makes an appearance in Vilmorin-Andrieux's *Les Plantes Potagères* (1925) and is described (and illustrated in living color) in *Dictionnaire Vilmorin Des Plantes Potagères* (1946) as a tasty red-orange–fleshed cantaloupe.

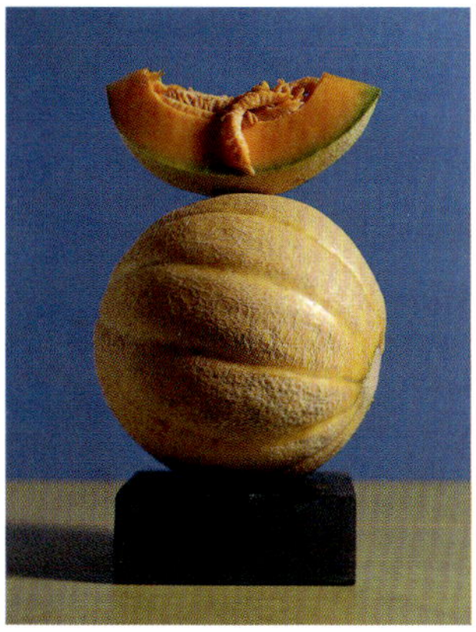

DELICIOUS 51

CUCUMIS MELO **RETICULATUS GROUP**
SIZE: 7" LONG BY 7¼" WIDE; WEIGHT: 6½ POUNDS
SUGAR CONTENT: 12½° BRIX
MATURITY: EARLY
SYNONYMS: CORNELL DELICIOUS, CORNELL DELICIOUS #51, EARLY DELICIOUS 51, GOLDEN DELICIOUS 51
SEED SOURCES: BAKE, BERL, ERS, EVER, FED, GRETA, HEIR, HIGH, HUD, IRISH, JUNG, MOR, NEWE, OSC, PRAI, REIM, ROHR, SESE, SHUM, SISK, SOTE, SSE, SUS, TERRIT, TUR, URB

A "first-early," meaty, delicious, orange-fleshed muskmelon of the highest table quality. Developed by Henry M. Munger at Cornell University and released in 1951. Resistant to fusarium wilt (race 2). Parentage: Iroquois x Delicious. Except for its disease resistance, it is hard to distinguish from Delicious, which was introduced in 1925 as Golden Delicious by Charles J. Lindholm, of Minneapolis, Minnesota. The authors of *The Cucurbits of New York* (1937) described Delicious as "an early strain of the famous Bender from which it was selected and which it resembles in many ways. For the home garden and local trade it has a very definite place in the list of varieties." That applies to Delicious 51, as well as to PMR Delicious 51, which is resistant to powdery mildew.

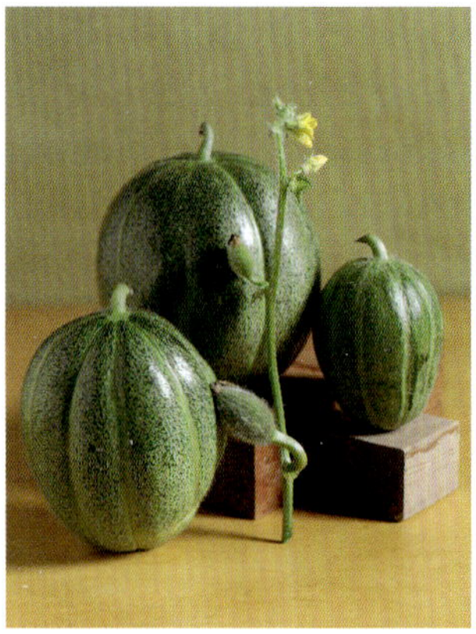

D. J.'S ITALIAN

CUCUMIS MELO **ADZHUR GROUP**
SIZE: 4½" LONG BY 3½" WIDE; WEIGHT: 15 OUNCES
SUGAR CONTENT: 3° BRIX
MATURITY: EARLY
SEED SOURCE: SSE

Crisp and cucumbery. Deeper green, larger, and sweeter at maturity (6½° Brix) than other "tondo" (round) carosellos I've grown. The variety was named for D. J. Luccarelli of Holmdel, New Jersey, whose father brought the variety from Italy to America just after 1900. The seed was donated to Seed Savers Exchange in 1996 by Peter Matey of Collierville, Tennessee. He had worked on D. J.'s farm in the 1970s. Matey wrote in his donation letter that "They look a lot like little watermelons with fine 'peach-fuzz' that washes off easily." First offered by SSE to its members in the 2017 Yearbook.

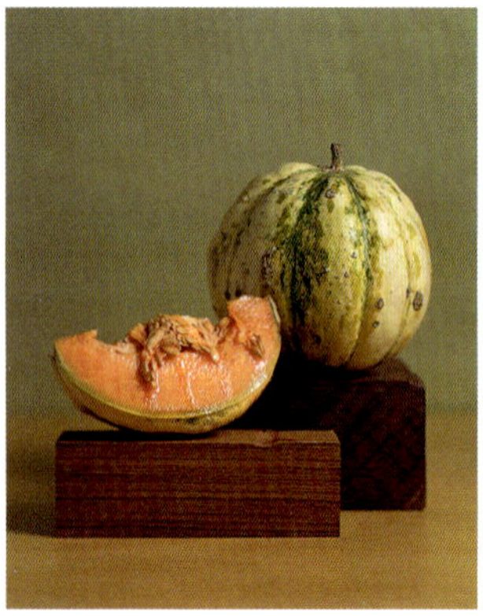

EARLY FRAME PRESCOTT

CUCUMIS MELO **CANTALUPENSIS GROUP**
SIZE: 7" LONG BY 6½" WIDE; WEIGHT: 6 POUNDS
SUGAR CONTENT: 12° BRIX
MATURITY: MIDSEASON
SYNONYMS: EARLY LARGE PRESCOTT, PRESCOTT À CHÂSSIS, PRESCOTT PETIT HÂTIF À CHÂSSIS
SEED SOURCE: SSE

Delectable is the word for Early Frame Prescott. Its aroma doesn't hit you in the face as the aromas of some muskmelons do; rather, it is more subtle and fruity. Within the creamy pumpkinlike shell, marked by streaks of green and occasional warts, lies what tastes like melon confit preserved in its own sweetness.

The Prescott family of cantaloupes is named for the English gardener who evidently first brought them to Paris around 1800. Early Frame Prescott is well suited for forcing a crop under the frost protection of cold frames. To get fruit to grow within such confines of space and time, Early Frame Prescott must first be cut short. If you prune it into a double cruciform, the plant will produce an early crop of four melons. Remove the growing tip after at least four true leaves have formed, and allow a pair of laterals, or side branches, to grow, trained in opposite directions. Nip again after the vines run the length of the frame, and permit one sublateral to grow from each side. Since flavor is in direct proportion to foliage—remember, photosynthesis happens in the leaves—don't sacrifice too much. Cut judiciously.

Frame culture is also conducive to getting a main-season crop when heat and light are in the ascendant in northern latitudes. Set the stage in April or May by digging a hotbed: a mass of fermenting manure and unfinished compost mounded over with topsoil. The warmth and sustenance generated will speed the melon on its way when planted out in late spring. Cover young transplants with cloches (glass bell jars), paper hot caps, Reemay tents, or frames, to guard against cold. Protection can be removed as the weather warms, vines spread, and fruits form.

Many techniques involving hotbeds and/or frame culture were used in the nineteenth century to force this variety to produce fruit unnaturally early. But it doesn't need to be that difficult.

Early Frame Prescott (by various names) was well known to Sageret (1826 and 1827), Jacquin (1832), and Vilmorin (from 1856 on). They knew that its fruit quality is almost always excellent. With its thick and sweet flesh—the color of ripe apricot—

it stands the test of time.

If you haven't seen "Still Life with Melon," painted by Claude Monet, in 1872, then check it out online. It depicts an Early Frame Prescott, exquisitely rendered. Thanks to Derek Fell, the award-winning garden writer and photographer, for letting me in on this.

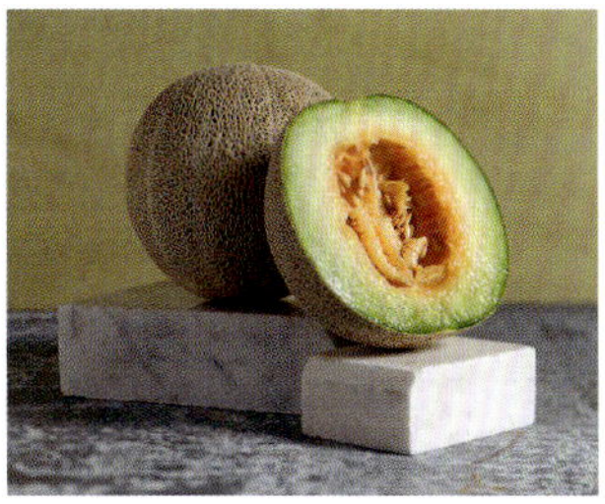

EDEN GEM

CUCUMIS MELO **RETICULATUS GROUP**
SIZE: 5¼" LONG BY 5½" WIDE; WEIGHT: 2½ POUNDS
SUGAR CONTENT: 11° BRIX
MATURITY: MIDSEASON
SYNONYMS: EDEN'S GEM, GEM OF EDEN, GOLD LINED ROCKY FORD, NETTED ROCKS, ROCKY FORD EDEN GEM, ROCKY FORD, ROCKY FORD GREEN FLESH(ED)
SEED SOURCES: BAKE, BERL, BURR, CLIF, EDEN, FED, GRETA, HEIR, HIGH, IRISH, JOHN, LAND, MARY, MOR, PEACE, PINE, REST, SAND, SESE, SOLA, SOTE, SSE, SSEC, STC, SUS, UPRI, URB, VER, VICT, WIL

One of my all-time favorites. May cause drooling. Best to harvest at half-slip when salmon-lined flesh is still green and firm (not yellowish green and soft). Spicy, complex flavor and melting flesh make it utterly addictive. Reminds me of Fordhook Gem (see page 204) in quality and appearance. Vigorous plants are highly prolific—so share the bounty.

Eden Gem was introduced in 1905 by Philo Kneeland Blinn (1869–1950) of the Colorado Agricultural Experiment Station as a rust-resisting crate melon (for shipping) with solid netting and green flesh tinted salmon at the seed cavity. Blinn had mass selected for disease resistance and fruit quality from the Pollock strain of Rocky Ford or Netted Gem. He made an argument in favor of the salmon-tinted type because it seemed to have a spicier flavor and was more attractive at the table than the solid green-fleshed type, especially when cut a little unripe.

Blinn couldn't have developed Eden Gem without donated seed from practical breeder J. P. Pollock. Pollock had carefully selected from the Rocky Ford or Netted Gem variety over a period of about ten years for rounder, more heavily netted, and less lobed melons that, in his words, "averaged well to crate nicely." In the process, he unwittingly selected for rust resistance.

Eden Gem is part of the Netted Gem group that includes Rocky Ford, Pollock 25, Pollock 10-25, Edwards Perfecto, Abbott Pearl, and Honey Ball. Similar to Green Machine (introduced in 1998).

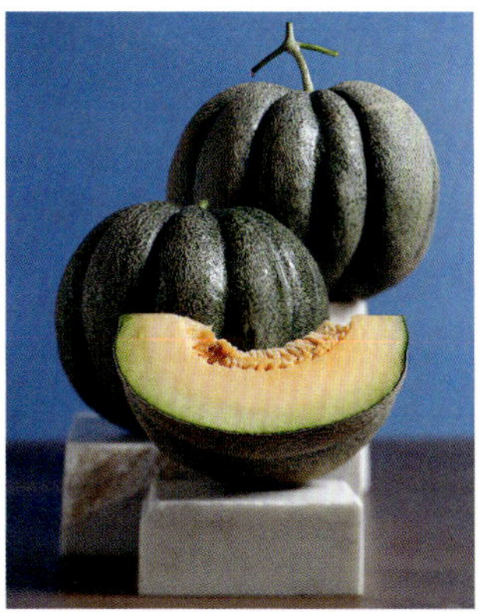

EMERALD GEM

CUCUMIS MELO **RETICULATUS GROUP**
SIZE: 5" LONG BY 6" WIDE; WEIGHT: 2½ POUNDS
SUGAR CONTENT: 11° BRIX
MATURITY: LATE
SYNONYMS: BURPEE'S EMERALD GEM, EARLY EMERALD GEM, EMERALD GREEN, GIANT WONDER GEM, MARY DAISY, AND MORE
SEED SOURCES: GRETA, HEIR, HIGH, NATURE, ROHR, SOLA, SSE, SSEC, SUS, TSK, URB

When harvested at forced slip, cut in cross section with a sliver of emerald remaining inside, Emerald Gem is the perfect vehicle for a scoop of vanilla ice cream. (See more on page 70.)

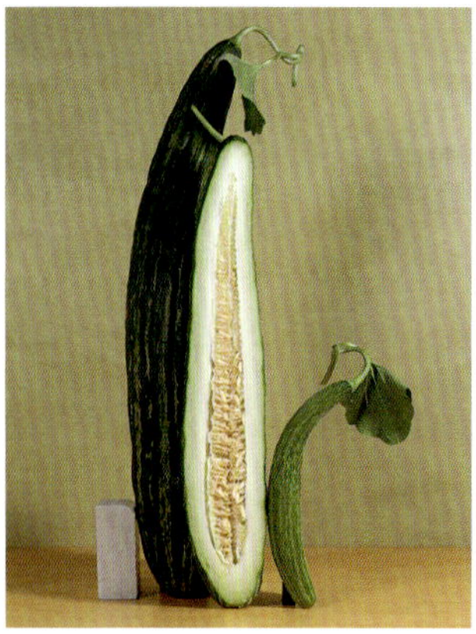

FAKOUSIA AKAPNITIKI

CUCUMIS MELO **FLEXUOSUS GROUP**
SIZE: 25" LONG BY 4" WIDE; WEIGHT: 6½ POUNDS
SUGAR CONTENT: 5° BRIX
MATURITY: EARLY
SYNONYMS: AKAPNOU, CYPRIOT CUCUMBER MELON
SEED SOURCE: SSE

A fascinating dark green snake melon, furrowed and bumpy, that is variously striped and spotted in greenish white. I've never seen another snake quite like it. Best eaten when young and crisp (ten inches long by one inch wide, weighing ½ pound); the skin is still fuzzy at that stage, so you'll need to rub off the soft hairs. Sweeter than your average cucumber.

Akapnitiki is a snake melon landrace from Akapnou, a small village in the Limassol District of Cyprus. Brought to America by the Nicolaou family of Famagusta after the 1974 Turkish invasion of Cyprus. Seeds circulated among Cyprian expatriots for years. First made available to members of Seed Savers Exchange by William Woys Weaver in the 2018 SSE Yearbook.

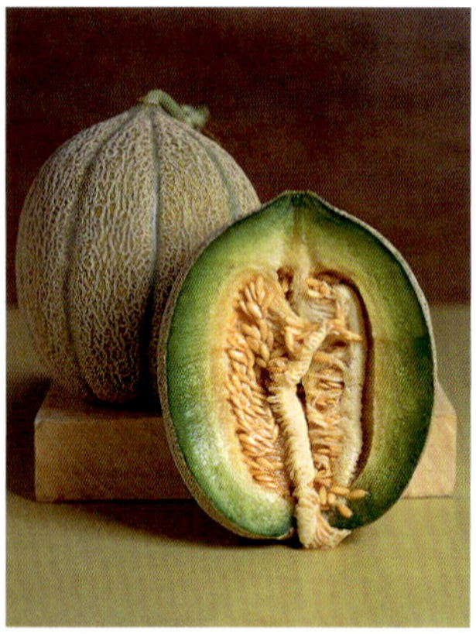

FORDHOOK GEM

CUCUMIS MELO **RETICULATUS GROUP**
SIZE: 6" LONG BY 5" WIDE; WEIGHT: 2 POUNDS
SUGAR CONTENT: 12° BRIX
MATURITY: EARLY
SYNONYM: BURPEE'S FORDHOOK GEM
SEED SOURCE: SSE

An extraordinary dessert melon with sweet and tender green flesh, salmon tint around the seed cavity, and fine silvery netting. In a word: *Yum*! Eating is believing—don't take just my word for it. Heron Breen of Fedco Seeds in Maine, who knows his melons, wrote about Fordhook Gem in the 2011 Seed Savers Exchange Yearbook: "The best green-fleshed melons I've ever eaten. Syrupy, musky, sensual, wow. I am not exaggerating. And so small, it was kinda hard to believe they could hold such yummy stuff."

Introduced by Burpee in 1967 as a distinctive melon, usually prolific, that fares well even in the north. Dropped by Burpee in 1983—why, I do not know—and extremely rare ever since. I have offered seed to Listed Members of SSE since 2006 via the annual Yearbook but cannot supply the whole world. (See more on page 60.)

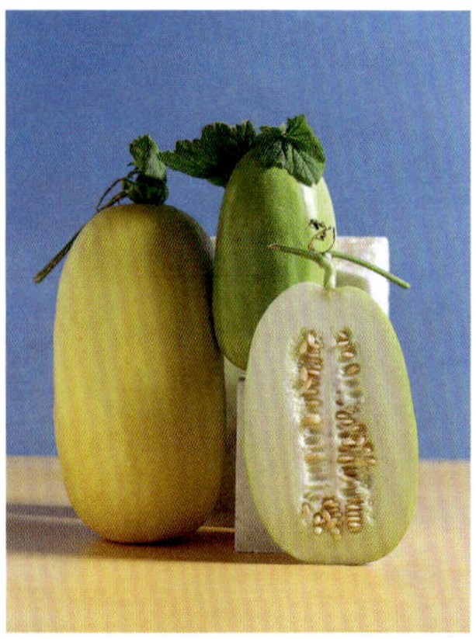

FREEMAN'S CUCUMBER

CUCUMIS MELO **CONOMON GROUP**
SIZE: 9¼" LONG BY 4½" WIDE; WEIGHT: 3 POUNDS
SUGAR CONTENT: 4° BRIX
MATURITY: EARLY
SEED SOURCE: NONE IN THE UNITED STATES OR CANADA

Years ago, Dr. Noriyuki Fujishita, retired Professor at the University of Osaka, sent me a photo illustrating Conomon's diverse shapes, sizes, and colors. But until I grew Freeman's, I'd never seen one with skin so intensely yellow. At first I thought it might be an overgrown Ogon type of Makuwa. The flesh is firm and mild with a cucumbery scent. Not as agreeable as snake or chate melons when eaten raw.

Freeman's Cucumber was used by Zvi Karchi of Israel's Agricultural Research Organization to introgress resistance to Cucumber mosaic virus (CMV) into Galia-type melons.

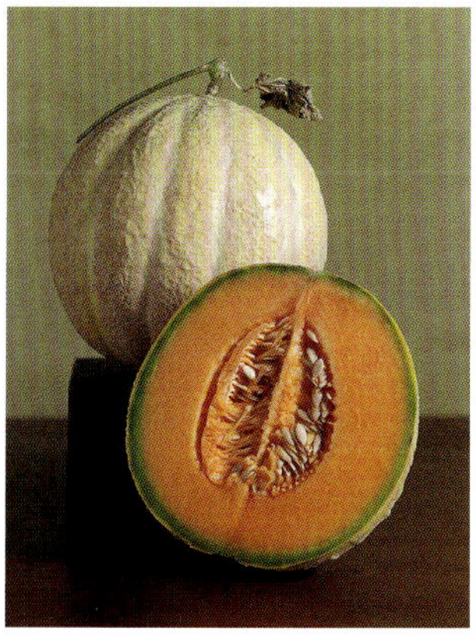

GOLDEN GOPHER

CUCUMIS MELO **RETICULATUS GROUP**
SIZE: 7" LONG BY 7" WIDE; WEIGHT: 5½ POUNDS
SUGAR CONTENT: 10½° BRIX
MATURITY: EARLY
SYNONYMS: GOPHER, POP-OPEN MELON
SEED SOURCES: FED, HER, SAND, SSE, TUR

Doesn't do especially well in my garden but has some fine attributes. Early and hardy. Adapted to the far north, where it performs much better; Brix readings up to 14° have been reported in Maine. CR Lawn, founder of Fedco Seeds—a source for cold-hardy selections—says, "This is the one gopher I won't be without. Each year it reminds me just how good a melon can be." Gopher was bred at the University of Minnesota, St. Paul, in the 1930s, and introduced by Farmer Seed and Nursery Company in 1939. Parentage: possibly Benders Surprise x Honeydew. Fusarium wilt resistant. Catalogued by Seeds Blüm in 1987; apparently saved from oblivion by Glenn Drowns of Sand Hill Preservation Center.

Farmer selections from a recent Cornell cross (Charentais PM x Golden Gopher), known as OSP (Organic Seed Project) Golden Charentais, look promising, according to Ken Ettlinger of the Long Island Seed Project. Gopher is also in the pedigree of Trifecta (see Trifecta, page 223).

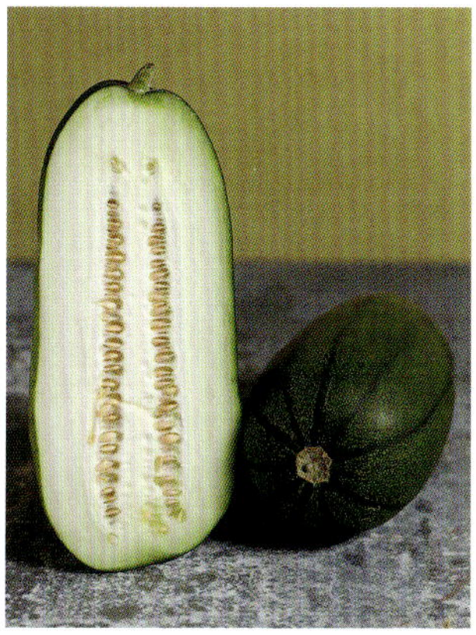

GREEN STRIPED

CUCUMIS MELO **CONOMON GROUP**
SIZE: 8½" LONG BY 3¾" WIDE; WEIGHT: 2 POUNDS, 5 OUNCES
SUGAR CONTENT: 4° BRIX
MATURITY: EARLY
SYNONYMS: AO URI, GREEN STRIPE
SEED SOURCE: KIT

A traditional Japanese pickling melon characterized by dark green penciled stripes running top to bottom on grayish-green skin. Fruit quality is excellent even when the melon is eaten young and raw—the greenish-white flesh is crisp and mild, more like a zuke than a cuke. Neither bitter nor astringent.

Not the same as Shimauri Stripe, which has white penciled stripes. The folks at Kitazawa Seed Company, in Oakland, California, told me they've catalogued Green Striped (Ao uri) for decades.

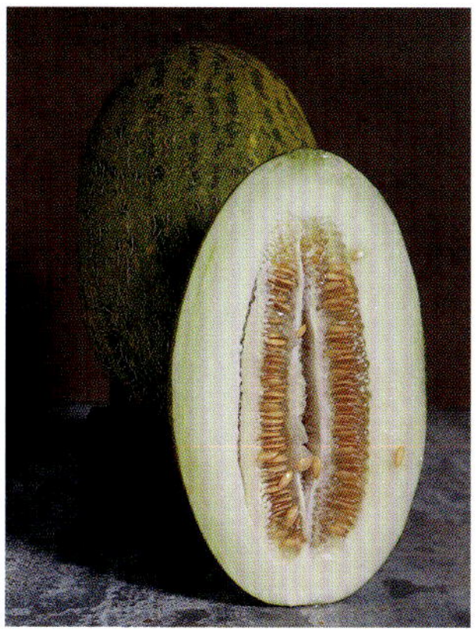

HAMI

CUCUMIS MELO **AMERI GROUP (SUB-GROUP MACULATI)**
SIZE: 12" LONG BY 8" WIDE; WEIGHT: 10½ POUNDS
SUGAR CONTENT: 11° BRIX
MATURITY: MIDSEASON
SYNONYM: HAMI SWEET
SEED SOURCE: NONE IN THE UNITED STATES OR CANADA

This melon derives its name from a town called Hami situated on the northern branch of the Silk Road, in Xinjiang Province, China, where the climate is arid. Hami melons come in a range of shapes, sizes, skin patterns, and flesh colors. Modern hybrid derivatives are popular in China and Taiwan.

This particular open-pollinated Hami is of excellent quality—with plenty of firm, sweet, pale-green flesh to go around. I'm fond of the dark green spots and broken stripes adorning the skin.

Similar sorts can be found in desert environments in Turkmenistan and Uzbekistan. Jim McCreight of the USDA, who participated in a collecting expedition to Turkmenistan in 2008, gave high praise to one particular melon generically referred to as Waharman (Vaharman): "It's like a large Hami melon that can reach about 35 pounds, but it is tastier, in my opinion, than the Hami. It translates roughly as 'Wow! What a melon!' This melon can be stored for six months or more. Somewhat immature Waharman fruit consistently had soluble solids of 18 degrees Brix in Turkmenistan, but we did not see the same levels in the Imperial Valley in 2012, due to powdery mildew, most likely. The Turkmen farmers stated that Waharman becomes sweeter in storage."

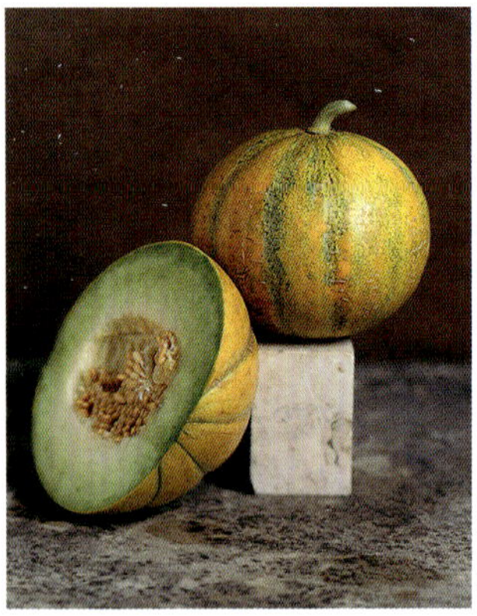

HA'OGEN

CUCUMIS MELO **CANTALUPENSIS GROUP**

SIZE: 6" LONG BY 6" WIDE; WEIGHT: 3 POUNDS

SUGAR CONTENT: 11° BRIX

MATURITY: MIDSEASON

SYNONYMS: DEVASH HA'OGEN NO. 2, HA'OGEN HONEY NO. 2, OGEN, (OLD) ISRAEL

SEED SOURCES: BAKE, BOT, BURR, EDEN, HIGH, IRISH, PEACE, PINE, REDS, REST, SALT, SAND, SHUM, SISK, SOC, SOLA, SSE, SSEC, TSK, WIL

Ha'Ogen is Hebrew for "the anchor." And while we're naming names: Hemed is the name of a Ha'Ogen-type hybrid with resistance to fusarium wilt races 0 and 2. Noy Yizre'el is a fifth backcross generation, powdery mildew race 1–resistant derivative of Ha'Ogen bred by Zvi Karchi and Anneke Govers of the Agricultural Research Organization at Newe Ya'ar, Israel. Noy Yizre'el is one of the parents (the other parent was a Crimean muskmelon) of the famed Galia F1 melon (also bred by Karchi and Govers), released in 1974. (See more on page 42.)

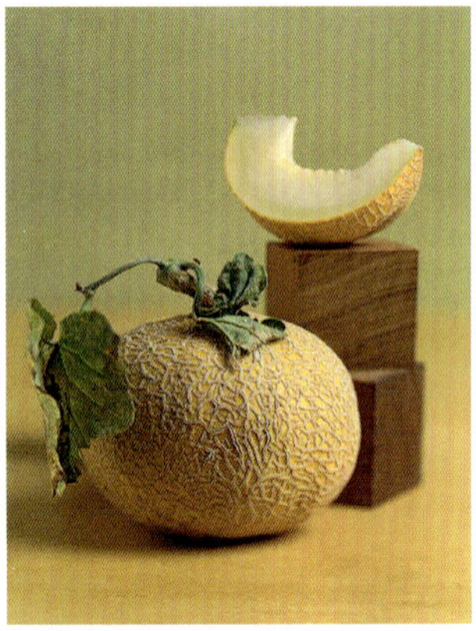

HERO OF LOCKINGE

CUCUMIS MELO **RETICULATUS GROUP**

SIZE: 5½" LONG BY 6½" WIDE; WEIGHT: 3½ POUNDS

SUGAR CONTENT: 11° BRIX

MATURITY: MIDSEASON

SYNONYMS: LOCKINGE PARK, SUTTON'S HERO OF LOCKINGE

SEED SOURCE: SSE

I can't wait to feast—and feast my eyes—on Hero of Lockinge every year. What keeps me coming back for more (apart from Lockinge's beauty) is its very finely perfumed, sweet and juicy yet firm white flesh. I imagine it inherited those flesh qualities from one of its parents, Colston Bassett. The other known parents are Read's Hybrid and Victory of Bath.

Hero of Lockinge was introduced by Sutton's Seeds of England in 1881. It met with immediate acclaim, becoming the most popular white-fleshed melon in the land, whether grown in cold pits, frames, or glasshouses, and infrequently in the open ground. Vilmorin-Andrieux proclaimed it one of the best of melons in 1885.

Hero of Lockinge has always been rare in the United States. In 1903, there was only one vendor: Weeber & Don, New York, NY. Today there are none. (See more on page 58.)

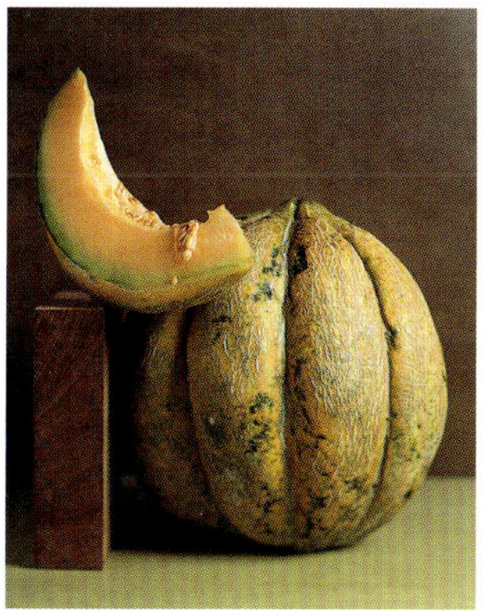

HOLLYBROOK LUSCIOUS

CUCUMIS MELO **RETICULATUS GROUP**
SIZE: 9½" LONG BY 8" WIDE; WEIGHT: 8 POUNDS
SUGAR CONTENT: 9° BRIX
MATURITY: LATE
SYNONYM: LEWIS PERFECTION
SEED SOURCES: SAND, SSE

Impressive muskmelons with prominent lobes and furrows. Can become enormous—but never luscious for me. Flesh is mildly sweet, juicy, and soft when fully mature; best to harvest at half-slip before melons split a gut.

Hollybrook Luscious was introduced in 1905 by T. W. Wood and Sons of Richmond, Virginia, and named for the firm's seed farm. Closely resembles Lewis Perfection, introduced in 1906 by James J. H. Gregory & Son.

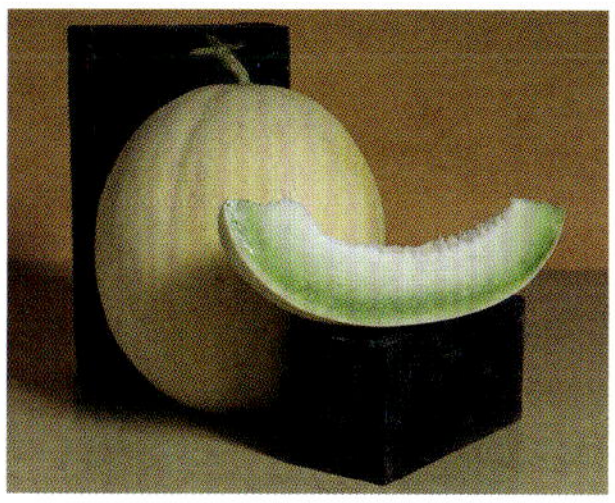

HONEYDEW

CUCUMIS MELO **INODORUS GROUP**
SIZE: 9" LONG BY 7" WIDE; WEIGHT: 7½ POUNDS
SUGAR CONTENT: 11½ DEGREES BRIX
MATURITY: LATE
SYNONYMS: GAUGER'S HONEY DEW, GREEN-FLESHED HONEYDEW, GREEN HONEYDEW, GREEN MEATED HONEY DEW, HONEY DEW, MELON D'ANTIBES BLANC D'HIVER, WHITE AFRICAN, WHITE ANTIBES
SEED SOURCES: ANN, BERL, BUR, BURR, CLIF, DEB, EDEN, EVER, GOU, HEIR, JOR, JSKIT, LEJ, MAIN, MOR, NAT, PEACE, REIM, RICH, ROHR, SHUM, SOWT, STC, SUS, TRADE, TSK, TTS, URB, VICT, WIL

While paging through early twentieth-century Johnson & Musser (later Aggeler & Musser, of Los Angeles) seed offerings in search of an introduction date for the Santa Claus melon (see page 218), I did a double-take when I saw the White African casaba in their 1909 catalogue. The accompanying text and photograph, and entries in later years, leave no doubt that this was the French heirloom White Antibes melon later considered synonymous with Honey Dew. The African point of origin would probably have been Algeria, where White Antibes melon was known to be raised for export. One acre of White African had been grown in 1908 by Charles Richardson near San Gabriel, California (not far from Los Angeles), and such was the reception that he didn't have enough seed to meet demand and orders for the following year. Johnson & Musser wrote, "The melon is entirely white, symmetrical in shape, with smooth rind, hard and firm, a good keeper and good shipper. . . . The flavor cannot be excelled by any melon of the Casaba family." The company knew their casabas and made them a specialty. The 1917 catalogue validated the French provenance of White African (it was not listed as Honey Dew or White Antibes then): This melon was "a great favorite in Southern France, where it is grown in every home garden and stored in cellars until eaten. When wanted it is placed at a sunny window for a few days. It then becomes ripe and of delicious flavor. Why not do so here?"

In another footnote, Roland Morrill, who was a well-known fruit expert from Benton Harbor, Michigan, told the *American Florist* (September 23, 1922) that "the Honeydew melon was first imported for [me], with other novelties, by Vaughan's Seed Store, Chicago, from Vilmorin-Andrieux and Co., of Paris, France, about 25 years ago. There is considerable variation in the quality and color of the flesh of this melon." (Perhaps that's why Vaughan didn't introduce it.)

Whether the original seed that got the Honeydew ball rolling came off a plate in a fancy New York City hotel in 1911; from the Cornell Agricultural Experiment Station, Vilmorin-Andrieux, or Roland Morrill in the 1890s; or from Charles Richardson and the Johnson and Musser Seed Company, it took the professional breeder John E. Gauger of Swink, Colorado, to clean up the variety and bring in a couple of new bulls—in other words, add two outstanding green-fleshed varieties to its pedigree. And it took H. D. Garwood and E. S. Woodside of Rocky Ford, Colorado, as exclusive agents, to introduce Honey Dew in 1915.

Honey Dew has many open-pollinated and F1 hybrid descendants. It is now an established international market-type. (See more on page 86.)

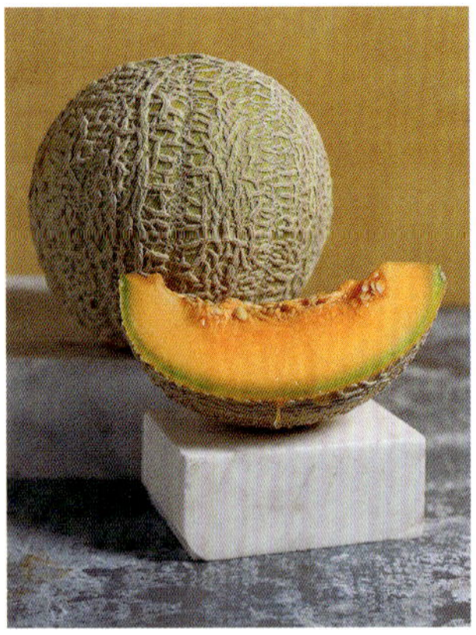

HONEY ROCK

CUCUMIS MELO **RETICULATUS GROUP**
SIZE: 3½" LONG BY 3½" WIDE; WEIGHT: 3 POUNDS
SUGAR CONTENT: 10° BRIX
MATURITY: MIDSEASON
SYNONYMS: HONEYROCK, NEW SUGAR ROCK, SUGAR ROCK, SUPERBA, WHITE SEEDED DELICIOUS
SEED SOURCE: REIM

These are adorable (some might say "scar face") and diminutive honey balls characterized by thick skin covered with coarsely interlaced netting, enclosing sweet, musky, juicy flesh. Reminiscent of the French heirloom melon Sucrin de Tours (see page 221) but smaller, and the modern-day Sugar Cube F1 melon.

Honey Rock was developed by F. Watt Richardson, owner of Wartenbee Farm near Hicksville, Ohio, in collaboration with his son, G. A. Richardson, and named in 1925. It was apparently produced by serendipitous crossing of Irondequoit, Golden Champlain, and Honey Dew and selection for a rock-hard, ropy-netted melon of fine quality. A Gold Medal Winner, All-America Selections, 1933.

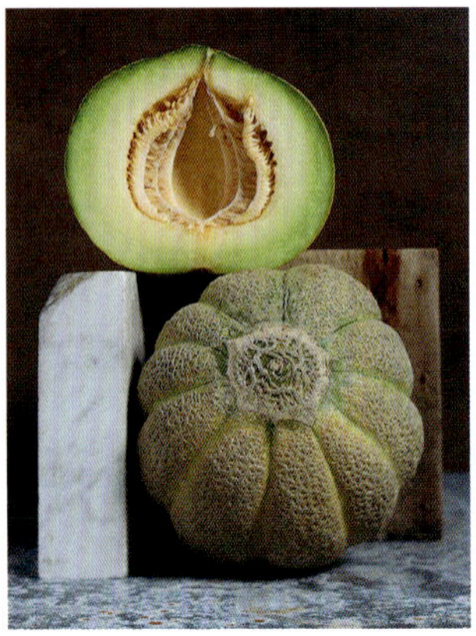

JENNY LIND

CUCUMIS MELO **RETICULATUS GROUP**
SIZE: 7" LONG BY 8" WIDE; WEIGHT: 5½ POUNDS
SUGAR CONTENT: 11° BRIX
MATURITY: EARLY
SYNONYMS: EARLY JENNY, EXTRA EARLY JENNY LIND, JENNY LYNN, JERSEY BUTTON, NORFOLK BUTTON, SHIPPER'S DELIGHT—AND MANY MORE NAMED SELECTIONS
SEED SOURCES: BAKE, BERL, ERS, FED, HEIR, HER, HVSC, LAND, PINE, REIM, ROHR, SHUM, SSE, SSEC, STC, TSK, VER

Jenny Lind melon owes her continued existence to the preservation efforts of garden writer Allen Lacy (1935–2015). In a *Wall Street Journal* column on the Leisure and Arts page in 1982, and in "The Narrow Escape of Jenny Lind," a lost-and-found story contained in *Homeground: A Gardener's Miscellany* (1984), Lacy lamented his favorite melon's seeming demise. In his opinion, "the spicy sweetness of its luscious pale-green flesh" couldn't be beat. Local New Jersey farmers had stopped growing it, there were no commercial seed sources, and neither the Seed Savers Exchange (SSE) nor the National Seed Storage Laboratory had any Jenny Lind in stock.

Then one day in May 1982, Lacy walked into a nursery and garden store (enticed by some unusual perennials on display) a couple of counties away from his home in Linwood, New Jersey, and found a small, plain envelope marked 'Jenny Lind Melon–30 cents' on a seed rack at the back of the store. The pack was so old that the store's owner gave it to Lacy for free. "He used to get a few packs a year from an old farmer nearby who saved the seed. But the farmer had died, and there really wasn't much demand anymore for the melon. There were sixteen seeds in the pack."

Lacy sent the Jenny Lind seeds to plant breeder Claude Hope in Costa Rica who grew them out in isolation and returned two thousand seeds to Lacy four months later. Lacy "pressed [one hundred seeds] into the hands of a local farmer in the hope that henceforth he will be able to supply me and a few other souls who don't mind green melons with strange protuberances with our annual share of Jenny Linds." The remaining seeds were sent to Kent Whealy at Seed Savers Exchange (SSE) for preservation and distribution to members. After reading Lacy's *Wall Street Journal* article on Jenny Lind, Larry Hollar contributed an additional sample of seeds to SSE from Hollar Seeds' collection.

Glenn Drowns was the first SSE member to offer Jenny Lind seed in limited quantities in 1983 through the Winter Yearbook. Dick Meiners of Pinetree Garden Seeds was probably the first to reintroduce it commercially in 1985 (it was initially introduced circa 1858). Now more than a dozen vendors carry the variety. In addition, Jenny Lind has been bred into at least three cultivars developed by Merlyn Niedens that were introduced by Southern Exposure Seed Exchange: Golden Jenny (1997) (not the same as the Golden Jenny of yore), Sleeping Beauty (1997), and Green Machine (Ice Cream Melon) (1998).

In its heyday toward the end of the nineteenth century, the Jenny Lind was grown in Maryland, Delaware, and New Jersey (along with Anne Arundel [see page 192] and Hackensack) to supply high-quality green-fleshed muskmelons for the New York and other eastern markets during midsummer; New Yorkers consumed two to three carloads daily at the height of the season in the nineties, according to Liberty Hyde Bailey (1927).

Today Jenny Lind is grown primarily in home gardens. At my homeplace in upstate New York—especially when the growing season is hot and dry—plants produce an early and prolific crop of sweet and savory, juicy, thick-meated, well-netted melons with yellowish-green flesh. And, oh, those belly buttons! In rare cases, especially with smaller melons, the belly button can be as long as Jenny Lind itself, making for a heart shape when cut in cross section. (See more on page 66.)

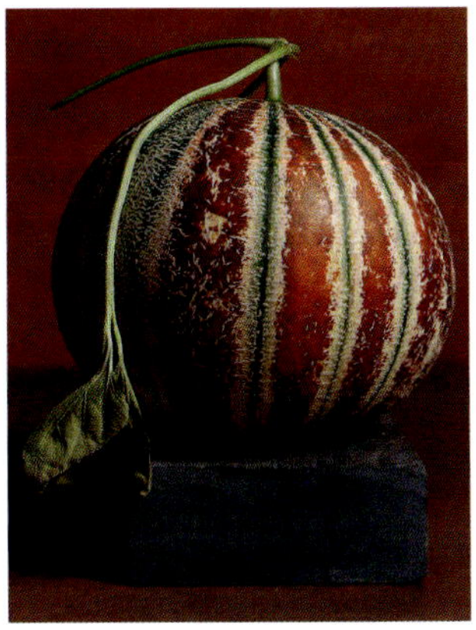

KAJARI

CUCUMIS MELO **KHANDALAK GROUP**

SIZE: 6¼" LONG 7½" WIDE; WEIGHT: 6½ POUNDS

SUGAR CONTENT: 11° BRIX

MATURITY: MIDSEASON

SEED SOURCES: BAKE, TRADE

Unique among melons I've grown. Except for the orange flesh color, it resembles the Zargaldog Khandalak, which is featured in *Melons of Uzbekistan* and is a local variety widespread in the Samarqand region. Appears similar to the Sary kandelek and Sary zamcha illustrated in the *Turkmen Melon Atlas.*

Flesh is smooth, sweet, and honeyed. My champ (featured on page 120) was quite hefty (6½ pounds), but others average about two pounds. Prolific fruit set. When grown in a greenhouse by Yosef Burger of the Agricultural Research Organization in Israel, Kajari proved highly resistant to powdery mildew.

Introduced commercially in the United States in 2015 by Baker Creek Heirloom Seeds. Their original seed was sourced by Joseph Simcox.

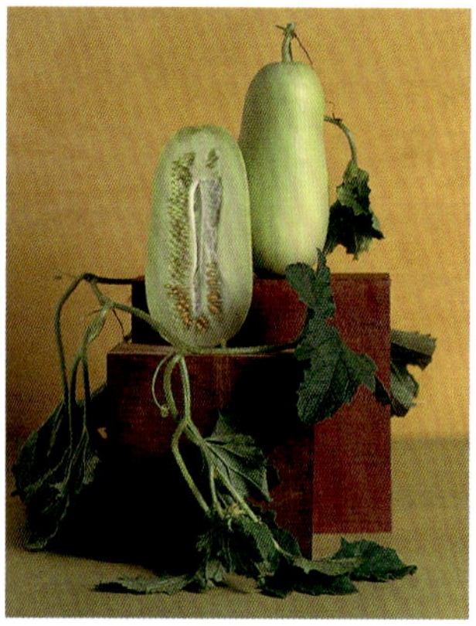

KATSURA GIANT

CUCUMIS MELO **CONOMON GROUP**
SIZE: 9" LONG BY 3½" WIDE; WEIGHT: 2 POUNDS, 2 OUNCES
SUGAR CONTENT: 3° BRIX
MATURITY: EARLY
SYNONYMS: GIANT KATSURA, KATSURA, KATSURA GIANT PICKLING, KATSURA-URI, MELON BLANC DU JAPON, PICKLING MELON
SEED SOURCES: KIT, SSE

An old favorite in Japan for pickles and *kasu-zuke* (involves soaking melon in the lees of sake). Can be bitter or astringent when young. When fully ripe it turns whitish and sweetly fragrant. Has gone out of favor but may find new life as a low-calorie juice made from the milder fully ripened fruit. Preservation efforts are underway in Kyoto Prefecture, where the Katsura Imperial Villa is located. May have been catalogued by Kitazawa Seed Company in the United State since 1917. Resembles Numane (see page 212). (See more on page 134.)

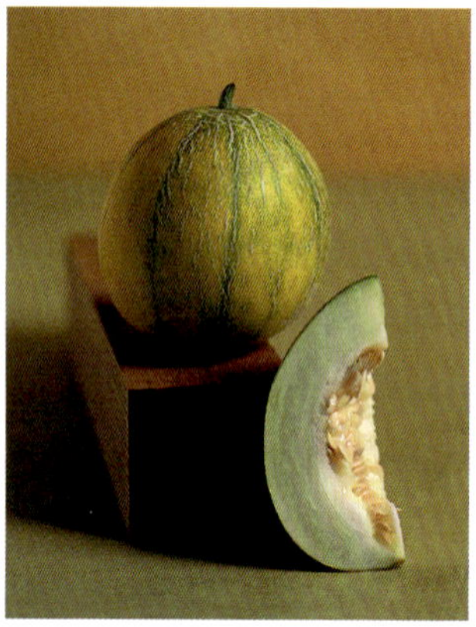

MANGO

CUCUMIS MELO **CHITO GROUP**
SIZE: 4" LONG BY 3" WIDE; WEIGHT: 11 OUNCES
SUGAR CONTENT: 5° BRIX
MATURITY: MIDSEASON
SYNONYMS: GARDEN LEMON, MELON APPLE, VINE PEACH, GLASS MELON, LEMON CUCUMBER, MELON PEACH, MELON DE QUITO, MELON DE GRENADE, CHITO DE LA HAVANE, ORANGE MELON, VINE ORANGE, VEGETABLE ORANGE, QUITO, MOTHER MARY'S PIE
SEED SOURCES: BAKE, JUNG, REIM, REST, TERR, TRADE, WHA

Mature fruit aroma is delicious, but the flesh is often sour. That doesn't seem to deter dogs, who can make fast work of them in the garden according to M. DeFays Du Monceau, an acquaintance of Charles Morren, the man who named the variety *Cucumis Chito*. Use Mango to perfume a room, for ornament, and for pickling when young. (See more on page 103.)

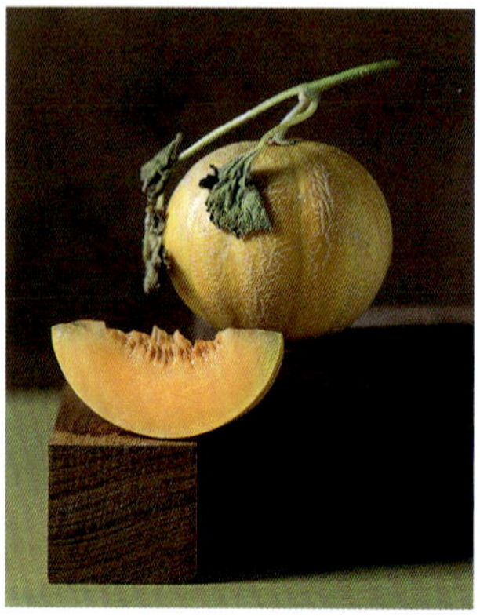

MINNESOTA MIDGET

CUCUMIS MELO **RETICULATUS GROUP**
SIZE: 3¼" LONG BY 3½" WIDE; WEIGHT: 12 OUNCES
SUGAR CONTENT: 11° BRIX
MATURITY: EARLY
SYNONYMS: MINN MIDGET, MINN 356
SEED SOURCES: ANN, BOT, EDEN, ERS, EVER, GOOD, GOU, GRETA, HEIR, HER, HOPE, HUD, HUME, IRISH, JOR, PINE, REST, RICH, ROHR, SALT, SAMP, SAND, SHUM, SOLA, SOWT, SSE, SSEC, STC, SUS, TERR, TERRA, TERRIT, TRADE, TSK, TTS, TUR, URB, VICT, WHA

High quality, finely netted, small, early melon cultivar adapted to short-season areas. Will reliably produce two crops on short vines. Can be grown in containers. Bred at the Minnesota Agricultural Experiment Station, St. Paul, Minnesota. Released in 1948; commercially introduced by Farmer Seed and Nursery Company. Resistant to fusarium wilt. Minnesota Honey is a larger-fruited selection of Minnesota Midget, with possible Honeydew parentage, that was released in 1954.

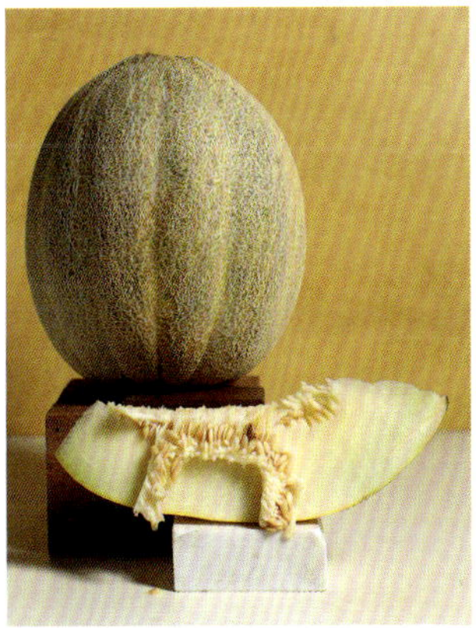

MONTREAL MARKET

CUCUMIS MELO **RETICULATUS GROUP**
SIZE: 10½" LONG BY 8" WIDE; WEIGHT: 12½ POUNDS
SUGAR CONTENT: 11½° BRIX
MATURITY: MIDSEASON
SYNONYMS: CANADIAN MARKET, LARGE MONTREAL MARKET, MAMMOTH MELON, MELON DE MONTREAL, MONTREAL NUTMEG, OUTREMONT BEAUTY, AND MANY MORE.
SEED SOURCES: FED, GRETA, HER, SOLA, TERRA

Another lost and found soul from the Far North. I am partial to green-fleshed muskmelons, and this one is special. Montreal Market has a beautiful symmetry and an abundance of sweet and melting flesh. It was known to grow giants as large as thirty-nine pounds (by J. B. Dabney, Esq., of Savannah, Iowa) in its heyday. As the story goes, W. Atlee Burpee saw the melons at St. Ann's market in Montreal in August 1880, and took home some seeds. (He liked to spend summers sleuthing for novelties.) W. introduced it in his 1881 catalogue as Montreal Green Nutmeg, and illustrated it with an engraved image made from a photograph taken on that August visit to Montreal. The melon's popularity soared. Tracy (1903) listed more than 100 vendors. By 1926, as noted by Beattie, it was no longer grown extensively in the United States, but was still cultured in Montreal: "The plants of this variety are started and practically grown in hot-beds, the sash and frames being removed after the weather becomes warm." Montreals were packed in fine hay and shipped in large wicker baskets without covers, each containing a dozen melons.

Times and tastes inevitably changed. Burpee dropped the variety from its catalogue in 1942 and, by 1991, it was only a dim memory. And a glint in the eyes of a small group of Canadian preservationists: journalist Barry Lazar publicized it, Mark Abley retrieved seeds from the USDA's melon collection in Ames, Iowa, and farmer Ken Taylor grew it out and selected for trueness to type. As the authors of *The Cucurbits of New York* put it in 1937, "The story of the success of this melon is as much a record of continued selection for type as it is of extreme care in its cultivation."

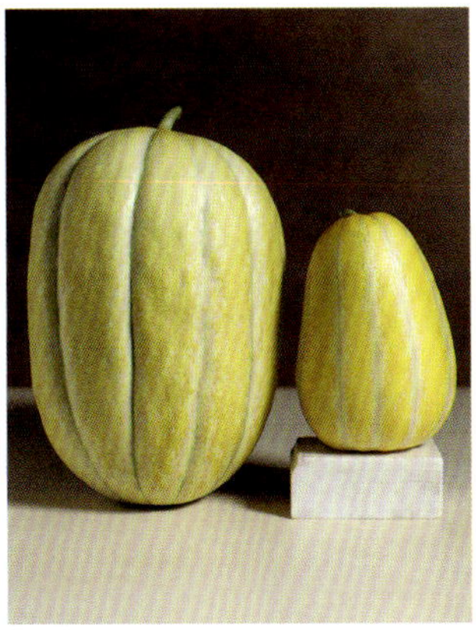

NAVAJO YELLOW

CUCUMIS MELO **MAKUWA GROUP**
SIZE: 10¼" LONG BY 7" WIDE; WEIGHT: 7½ POUNDS
SUGAR CONTENT: 13° BRIX
MATURITY: MIDSEASON
SEED SOURCES: HUD, MARY, NATIVE, PINE, SSE, SUS

With its yellow skin and silvery vein tracts, Navajo Yellow looks like an overblown Chamoe or Early Silver Line, with which it is pictured. On the inside it's sugary and delicious with the orange color and melting texture of a crenshaw—neither crisp nor white-fleshed like Chamoe, but just as sweet. It's a welcome addition to the melon patch.

Native Seeds/SEARCH (NS/S), of Tucson, Arizona, introduced Navajo Mix into their Seedlisting in 1991. The mix originated from seeds taken from a single melon entered in the Navajo Fair at Shiprock, New Mexico. Three fruit types were produced from grow-outs: ribbed ovals, smooth ovals, and elongated melons, with orange or green mild-flavored flesh. Navajo Yellow was selected from the mix and introduced by NS/S in their 1992 Seedlisting: "Round fruits have slightly ribbed yellow skin and orange, yellow or white flesh. Originally purchased at the Navajo Nation Fair. Good keeper if unbruised."

Native Seeds/SEARCH is a nonprofit seed conservation organization working to preserve traditional crops and crop wild relatives from the southwestern United

States and northwestern Mexico. SEARCH stands for Southwestern Endangered Aridland Resource Clearing House. NS/S was founded in 1983 as an offshoot of Gary Nabhan and Mahina Drees' Meals for Millions project, which assisted the Tohono O'odham Nation in establishing gardens for their sustainable food needs.

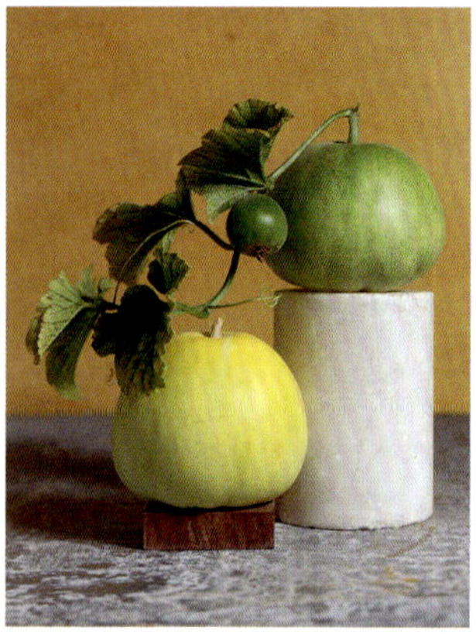

NEW MELON

CUCUMIS MELO **MAKUWA GROUP (SUB-GROUP OGON)**
SIZE: 3¾" LONG BY 4½" WIDE; WEIGHT: 1 POUND, 7 OUNCES
SUGAR CONTENT: 14° BRIX
MATURITY: EARLY
SEED SOURCE: KIT

Like an Asian Pear, which it resembles, New Melon can be tied up in ribbons and bows and presented as a gift Japanese-style. Its very firm and sweet yellowish-green flesh is best eaten when skin is pale green and fruit mature. The yellow fruits are riper, slightly sweeter, and more fragrant—but less firm. It's a trade-off. Why not give, or eat, one of each?

Developed in the 1950s for its sweetness and fragrance. Added to Kitazawa Seed Company's lineup in 2003.

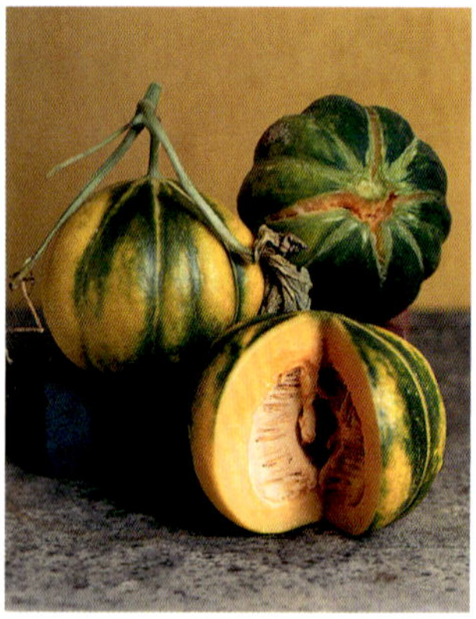

NOIR DES CARMES

CUCUMIS MELO **CANTALUPENSIS GROUP**
SIZE: 4" LONG BY 6" WIDE; WEIGHT: 2 POUNDS
SUGAR CONTENT: 11° BRIX
MATURITY: EARLY TO MIDSEASON
SYNONYMS: CANTALOUP SUCRIN DE MONTREUIL, CARMELITER, DES CARMES, EARLY BLACK ROCK, NOIR DE CARMES
SEED SOURCES: BAKE, GRETA, HER, HUD, NAT, PEACE, PINE, PRAI, REST, SAMP, SESE, SOLA, SSE, SSEC, THOM, TRADE, WHA

There is nothing quite like Noir des Carmes in America today, where the market is dominated by netted melons. It looks different, tastes different, and smells different. Indescribably delicious, it should easily find a niche. (See more on page 48.)

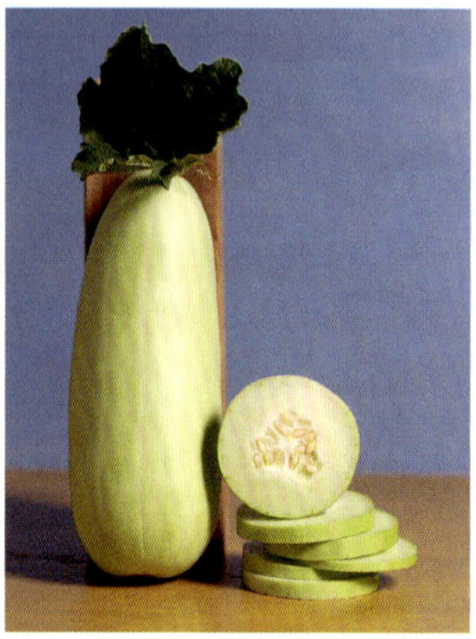

NUMANE

CUCUMIS MELO **CONOMON GROUP**
SIZE: 9½" LONG BY 4" WIDE; WEIGHT: 2½ POUNDS
SUGAR CONTENT: 3½° BRIX
MATURITY: EARLY
SYNONYMS: NUMAME, NUMANE EARLY, NUMAME PICKLING
SEED SOURCE: KIT

A pickling melon known as *oshiro uri*; native to Japan. It resembles the venerable Katsura Giant in most respects (see page 210). Numane is of better quality: crisp and cucumbery when immature, and never bitter.

Offered by Johnny's Selected Seeds in the United States as early as 1982 and catalogued in these words: "The earliest-maturing Japanese selection. . . . These small oblong melons have been used traditionally in the Orient for many types of tasty, healthful pickles. Well adapted to recipes for melon and rind preserves and cucumber pickles."

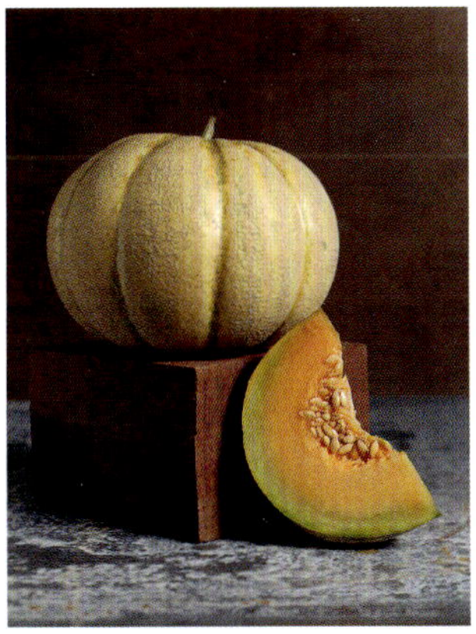

OKA

CUCUMIS MELO **RETICULATUS GROUP**

SIZE: 5¾" LONG BY 7¼" WIDE; WEIGHT: 6 POUNDS

SUGAR CONTENT: 9° BRIX

MATURITY: LATE

SYNONYMS: OKA BIZARD, OKA BIZARD ISLAND STRAIN

SEED SOURCES: GRETA, HER, HOPE, SAND, SOLA, SSEC, TERRA

A Canadian heirloom muskmelon, recently reclaimed on Ile-Bizard, a St. Lawrence River island between Montreal and LeValle. Beautifully lobed and flattened with melting, mildly sweet, light orange flesh. Because of their oblate shape, Fedco Seeds calls them the "cheese pumpkins of muskmelons." Bred around 1912 by the Reverend Father Athanase of the Trappist monastery at Oka, Montreal, Quebec. Oka is the result of a deliberate cross between the green-fleshed Montreal Market and the orange-fleshed Banana (see pages 211 and 194, respectively). Originally distributed in the United States by Joseph Breck & Sons in 1924.

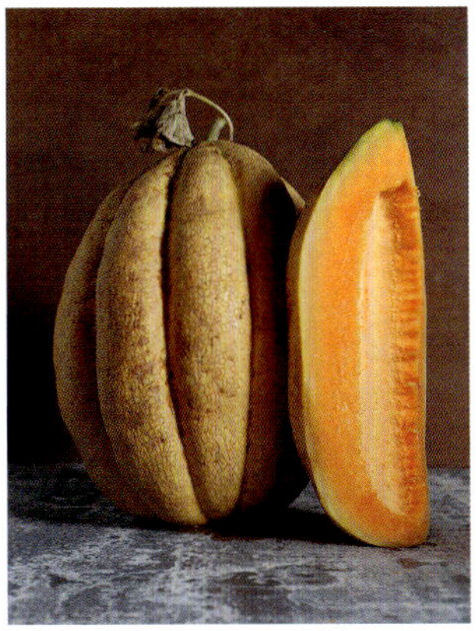

OLD-TIME TENNESSEE

CUCUMIS MELO **RETICULATUS GROUP**

SIZE: 13" LONG BY 8" WIDE; WEIGHT: 14½ POUNDS

SUGAR CONTENT: 12½° BRIX

MATURITY: MIDSEASON

SYNONYMS: GIANT TENNESSEE, JUMBO TENNESSEE

SEED SOURCES: BAKE, SAND, SESE, VICT

This melon is among the most wonderful I've ever grown. Old-Time Tennessee leaves you dumbfounded. I'll never forget watching my daughter, Sara—who was weaned on heirloom melons and watermelons—take her first bite, open her eyes wide in amazement, and utter, "Oh my God." I know exactly how she felt.

But you really need to time it right. Old-Time Tennessee is strictly garden-to-table. Keep an eye on it as it ripens (from smooth and white with green furrows) to netted and yellow, and pounce. It's so highly perfumed that you could probably find it by just sniffing around in the melon patch. When harvested at the peak of perfection, it tastes like lichee and roses and pepper.

Stephen Kolpan, coauthor of *Exploring Wine* (2010), says that Old-Time Tennessee would make a nice, light sweet wine, at about 6 to 7 percent alcohol, in the style of a muscat from Italy.

Southern Exposure Seed Exchange offered seed of a green-fleshed Old-Time Tennessee in 1984. SESE wrote that most melons in trials a year earlier were eaten by rabbits, but this variety was largely ignored: "Any variety that stands up to rabbits deserves some attention."

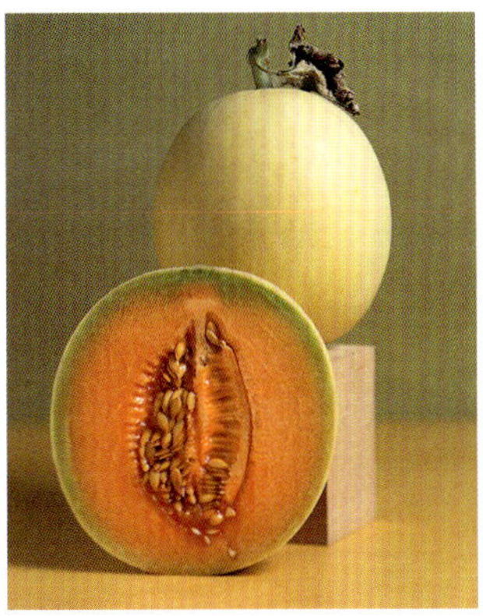

ORANGE FLESH HONEYDEW

CUCUMIS MELO **INODORUS GROUP**

SIZE: 6" LONG BY 5½" WIDE; WEIGHT: 3 POUNDS

SUGAR CONTENT: 13° BRIX

MATURITY: LATE

SYNONYMS: HONEYDEW PINK FLESH, SALMON TINTED HONEYDEW, YELLOW-FLESHED HONEYDEW

SEED SOURCES: BAKE, BUR, BURR, EDEN, GOU, HEIR, MAIN, MOR, PEACE, REDW, REIM, STC, SUS, TRADE, VICT

Honeydew (see page 207) remains the most popular winter melon in America. The orange-fleshed version is marvelous, too, but seems to defy most people's conception of Honeydew. I remember how surprised I was to cut my first one open and find it was flaming orange inside. If asked to choose between green and orange, I'd be stumped. Can't a person love more than one?

In early days there were multiple orange-fleshed Honeydews introduced—I'm not sure which one came first. In 1923, the Aggeler and Musser Seed Company introduced into their catalogue the Yellow Fleshed Honey Dew, pronouncing it positively more delicious than the Honeydew. Its color, character, and carotenoids came from a liaison with Tip Top (see page 222), a leading variety of muskmelon at the time. The introducer of Honey Dew, Garwood and Woodside, introduced "SOMETHING NEW" in their 1927 catalogue—SALMON TINTED HONEYDEW, noting their supply of seed was very short. "This is a Honey Dew melon in every respect, except in color of the flesh and in flavor. The flesh is SALMON TINT to the rind. It has a rich, spicy flavor, which is entirely different from the insipid flavor of the Honey Dew. In size it will average smaller than the green-meated Honey Dew, and it matures several days EARLIER."

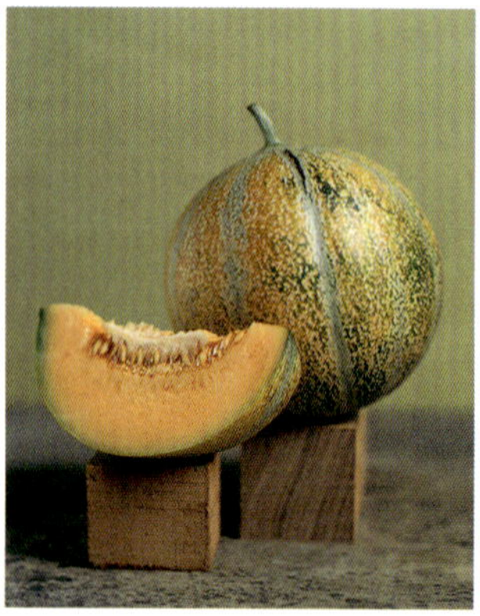

PETIT GRIS DE RENNES

CUCUMIS MELO **CANTALUPENSIS GROUP**
SIZE: 4½" LONG BY 4½" WIDE; WEIGHT: 2 POUNDS
SUGAR CONTENT: 14° BRIX
MATURITY: MIDSEASON
SYNONYMS: PETIT GRIS, PETIT RENNAIS
SEED SOURCES: BAKE, GRETA, HAR, JSKIT, PRAI, REST, ROHR, SAND, SOLA, SSE, SSEC, TERRA, TSK, WHA

I am grateful to Bruno Defay of Laprade, France, and to John Wynkoll of Bolton, England, for introducing me to this melon, and many other French and English heirloom melons, in the 1990s, via the Seed Savers Exchange. (See more about this melon on page 42.)

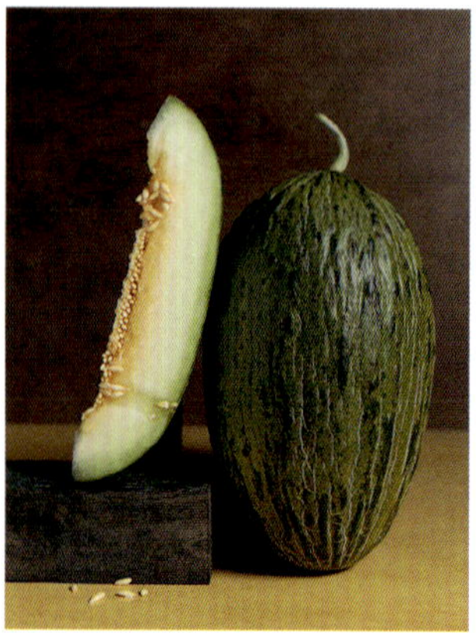

PIEL DE SAPO

CUCUMIS MELO **INODORUS GROUP**
SIZE: 11" LONG BY 7" WIDE; WEIGHT: 9 POUNDS
SUGAR CONTENT: 12° BRIX
MATURITY: MIDSEASON
SYNONYM: FROG SKIN
SEED SOURCES: GOU, PEACE, PINE, REIM, REST, SAND, SESE, SOWT, SSE, SUS, URB, VICT

From a sensational group of Spanish casaba melons, Tipo Piel de Sapo, with rough spotted skins that call to mind some toad species. These melons vary in shape from oval to elliptical to elongate; many are deeply wrinkled, others are smooth, netted, or rarely exhibit bands over vein tracts. Their deep green background color yellows during ripening, but prominent spots remain dark green. The yellowish-green flesh is crisp and in a good year in my garden tastes as sugary as dates.

Piel de Sapo is the favored casaba in much of Spain, where consumers know it by name, says Harry Paris. Piel de Sapo is his all-time favorite too. Americans sometimes call it Santa Claus, but that's a different kettle of fish (or frog). (See Santa Claus, page 218.)

POLLOCK ROCKY FORD

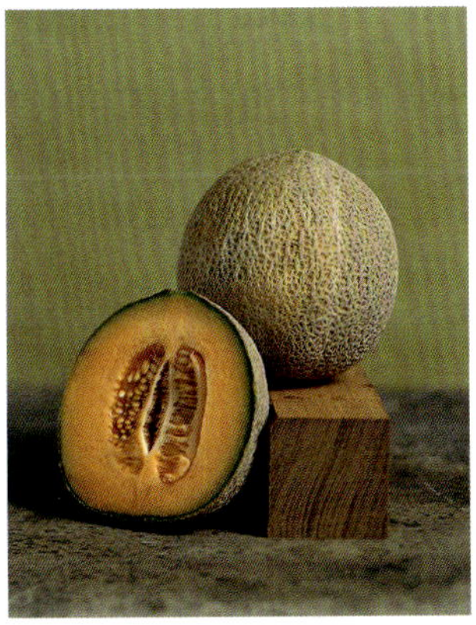

CUCUMIS MELO **RETICULATUS GROUP**

SIZE: 5" LONG BY 5" WIDE; WEIGHT: 2 POUNDS

SUGAR CONTENT: 11½° BRIX

MATURITY: LATE

SYNONYMS: GOLDEN POLLOCK, PERFECTO, IMPROVED PERFECTO, PERFECTED PERFECTO, POLLOCK 10-25, SUPERFECTO

SEED SOURCES: REIM, SAND

In 1895, J. P. Pollock began developing a new type of crate melon for shipping from Netted Gem (which was renamed Rocky Ford in 1897 after the place in Colorado where melons were grown). The original Netted Gem was oblong, decidedly lobed, with smooth furrows and green flesh. According to the authors of *The Cucurbits of New York* (1937), the original Pollock strain had color variations: "In some specimens the salmon coloring showed as a clean cut zone around the seed cavity with a green layer near the skin, in others the green and salmon colors were mottled or blended together and in still others the colors were in irregular patches, possibly the green on one side and the salmon color on the other. In some melons the whole flesh was a solid salmon color and in others the flesh was entirely green." Philo K. Blinn and James B. Ryan subsequently separated the variable stock: "In 1909 the Rocky Ford Cantaloupe Seed Breeders Association introduced these strains as Rust-resistant Pollock No. 25, Green Meat and Salmon Tint." Eden Gem is a green-fleshed strain with salmon tint at the cavity introduced by Blinn in 1905 (see page 203).

Orange selections came into vogue and were widely grown in Colorado and the Imperial Valley of California with irrigation. These included Salmon Tint Pollock 10-25, aka Pollock Rocky Ford, which was introduced in 1916 by the Rocky Ford Cantaloupe Seed Breeders Association. Selections included the various Perfectos. It goes to show, you *can* improve on perfection. Glenn Drowns of Sand Hill Preservation Center aptly describes Pollock Rocky Ford as "superb—not overly large but delightfully sweet."

PRESCOTT FOND BLANC

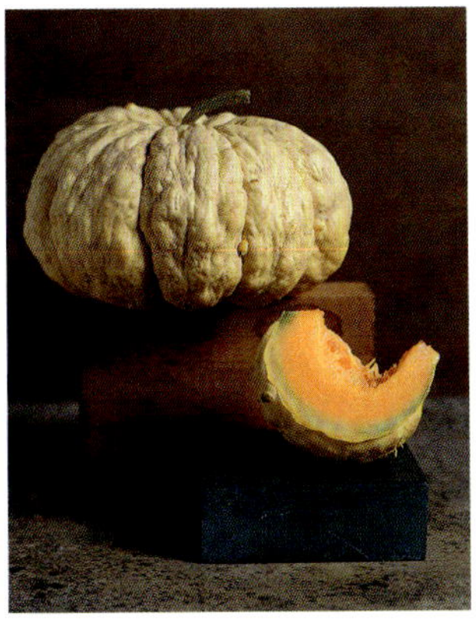

CUCUMIS MELO **CANTALUPENSIS GROUP**

SIZE: 5" LONG BY 9" WIDE; WEIGHT: 6½ POUNDS

SUGAR CONTENT: 12½° BRIX

MATURITY: LATE

SYNONYMS: LARGE PRESCOTT WHITE-SKINNED ROCK MELON, MELON EXCELSIOR. MY FAVORITE SYNONYM FOR PRESCOTT FOND BLANC IS AN ENGLISH ONE, HARDY RIDGE, LISTED AS SUCH BY FEARING BURR IN 1863, AND CATALOGUED UNDER THAT NAME BY JAMES J. H. GREGORY IN 1875. WE NOW KNOW THAT THE HARDY-RIDGED CANTALOUPES HAVE THE *RI* (RIDGED SURFACE) GENE.

SEED SOURCES: FED, REST, SOLA, SSE, SSEC, TERR, UPRI

It's always so good to see this old friend in the garden. Two hundred years ago there were many Prescott types in cultivation in France. Jacquin (1832) described and illustrated the following Prescotts, which had prominent lobes, and warts or galls: petit Prescott fond noir, Cantaloup rosé (a sub-variety of the former), gros Prescott fond noir, Prescott fond blanc, Prescott à ombilic saillant (Prescott cul de singe), gros Prescott fond blanc, and Prescott fond gris (the result of a cross between Prescott fond blanc and Prescott fond noir). In 1856, Vilmorin-Andrieux said Cantaloup Prescott was one of the most exquisite melons they knew—and just about the only one grown by Parisian gardeners. There were so many sub-varieties that it seemed every gardener had her own namesake Prescott. And then there were the cousins Hybride de Louviau (1850) with orange or green flesh and D'Ispahan à chair blanche (1835) with yellow flesh. I can't help but ask: *Où sont les neiges d'antan?* (See more on page 40.)

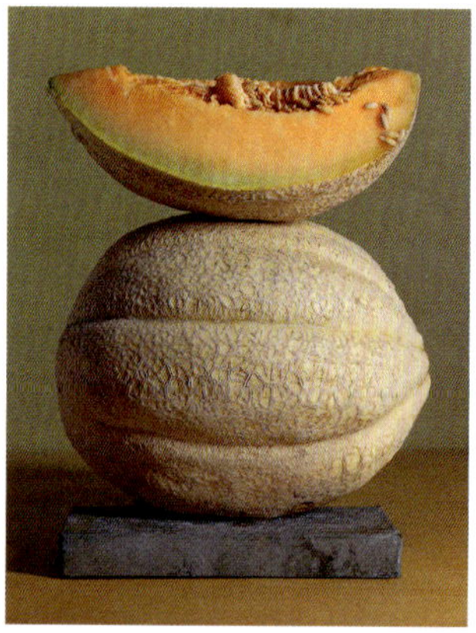

PRIDE OF WISCONSIN

CUCUMIS MELO **RETICULATUS GROUP**

SIZE: 9½" LONG BY 8" WIDE; WEIGHT: 9½ POUNDS

SUGAR CONTENT: 12½° BRIX

MATURITY: EARLY

SYNONYMS: HARVEST QUEEN, MARKET KING, PRIDE OF FORDHOOK, QUEEN OF COLORADO

SEED SOURCES: BAKE, FED, GRETA, HAR, HEIR, JUNG, NATURE, REST, ROHR, SAND, SSE, SSEC, WHA, WHITE

An excellent early musky melon of respectable size and heft. Adapted to short-season areas. Its ropy hard shell and flesh quality remind me of Schoon's (see page 219). CR Lawn, founder of Fedco Seeds, based in Maine, isn't shy about this cultivar: "Welcome to melon nirvana. . . . Move over, Minnesota [Midget]! Pride is destined to take its place alongside Golden Gopher as the best . . . extremely high quality." (See pages 210 [Minnesota] and 205 [Gopher], respectively.)

I don't know the identity of Pride of Wisconsin's breeder or its parentage, but the nerd in me keeps searching. The Robert Buist Company apparently introduced the melon in 1937. Pride (we're on a first-name basis) is described in J. C. Robinson's *The Cucurbits Illustrated* (1937) as "a new introduction which appears to have considerable merit." (Curiously it was not listed in *The Cucurbits of New York* published that same year.) Burpee catalogued it in 1938; F. H. Woodruff and Sons listed it a year later as a recent introduction selling in many sections. It won an All-America Selections Honorable Mention in 1939 under the name of Queen of Colorado (which was an apparent selection out of Pride of Wisconsin by J. C. Robinson).

Pride of Wisconsin is similar to Pride of Fordhook (a selection from the Murray strain of Pride of Wisconsin; introduced by Burpee in 1943) as well as to Harvest Queen (a fusarium wilt–resistant melon bred by Wilbur Scott, from a cross between Market King and Iroquois, at Harris Seeds; introduced in 1954). Market King was an exclusive Harris strain, similar to Pride of Wisconsin and Queen of Colorado. They are all more or less the same thing.

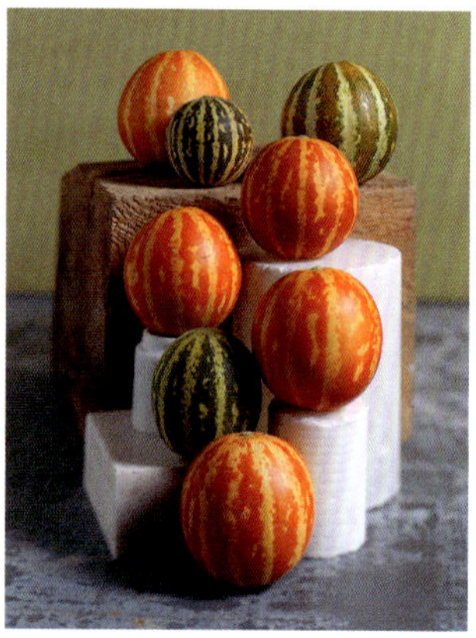

QUEEN ANNE'S POCKET MELON

CUCUMIS MELO **DUDAIM GROUP**

SIZE: 2½" LONG BY 2½" WIDE; WEIGHT: 4½ OUNCES

SUGAR CONTENT: 4½° BRIX

MATURITY: EARLY

SYNONYMS: ABU SHAMMAM, DASTANBOUYA, DUDAIM FRAGRANT; DWARF, ORNAMENTAL, OR VINE POMEGRANATE OR JUST PLAIN POMEGRANATE; KHORSAN, KHORASAN, PLUM GRANNIE, PLUM GRANNY, QUEEN'S POCKET, QUEEN ANNE'S SWEET SCENTED POCKET

SEED SOURCES: BAKE, HER, SAND, SESE, SSE, SSEC, THOM, TRADE, VICT

Odoratissimus! Fills your head with sweet thoughts.

A gift from Persia, where it was first selected by ancient farmers. Makes total sense that "QAPM" is also known as Pomegranate: both, similar in size, shape, and color (reddish) originated in the area corresponding to modern-day Iran. The Dudaim melon is still grown as a crop in Iran and neighboring Turkmenistan. Has been used throughout history as a child's toy (better when green and unripe I'd think), and also prescribed as medicine to relax the stomach and aid digestion.

Did you know? "The Persian-language version of the *Tora* (*Pentateuch*), dating to 1319, translates the Hebrew *duda'im* (mandrakes of Genesis 30: 14–16) as *dastambouha*" (Paris, Amar, and Lev 2012b).

Not always fragrant when grown in the North, said Liberty Hyde Bailey. (See more on page 118.)

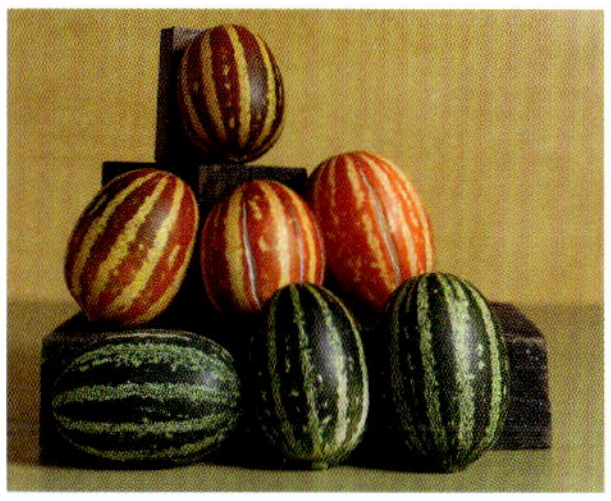

RICH SWEETNESS 132

CUCUMIS MELO **DUDAIM GROUP**

SIZE: 3½" LONG BY 2½" WIDE; WEIGHT: 8 OUNCES

SUGAR CONTENT: 5° BRIX

MATURITY: MIDSEASON

SYNONYMS: RICH SWEETNESS, SWEET RUSSIAN STRIPED

SEED SOURCES: BAKE, WHA

Attractive, perfumed, and richly colored—but sweet? No. Seedy, juicy, soft, and insipid like Queen Anne's Pocket? Yes. The skin colors and striping sequence of QAPM are maintained, though Rich Sweetness is more oval and prone to splitting. (See QAPM, page 216.)

Introduced in 2011 in the United States by Baker Creek Heirloom Seeds. Sourced in the former Soviet Union. Not the same as Rich Sweetness 305.

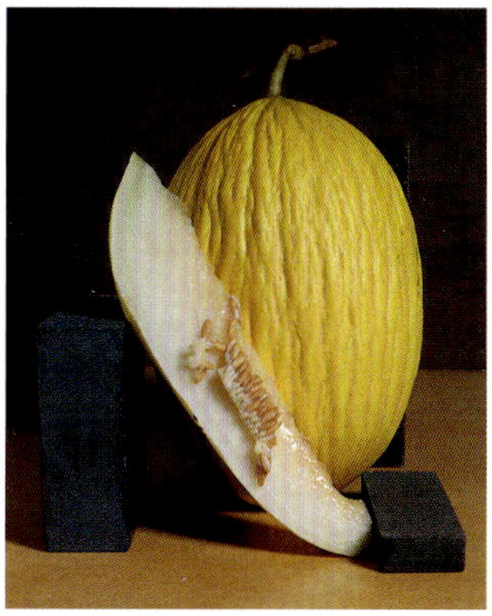

RUGOSO DI COSENZA GIALLO

CUCUMIS MELO **INODORUS GROUP**

SIZE: 13¼" LONG BY 7½" WIDE; WEIGHT: 11 POUNDS

SUGAR CONTENT: 13½° BRIX

MATURITY: MIDSEASON

SYNONYM: RUGOSO DI CONSENZA

SEED SOURCES: HALI, NEWE, REIM

Rugoso di Cosenza Giallo is a big, beautiful, heavily wrinkled yellow casaba of the highest quality. A favorite of mine. Its crunchy cream-colored flesh (sometimes with pale green and orange shades) is intensely sweet with nuances of flavor that will knock you for a loop. From Cosenza in southern Italy.

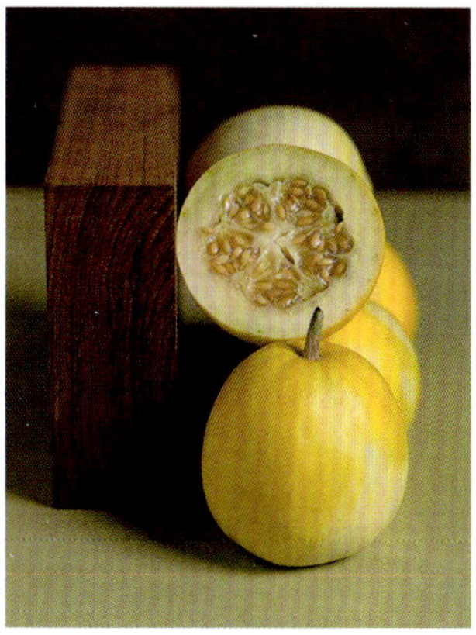

SAKATA'S SWEET

CUCUMIS MELO **MAKUWA GROUP (SUB-GROUP OGON)**

SIZE: 3" LONG BY 3½" WIDE; WEIGHT: 8 OUNCES

SUGAR CONTENT: 14½° BRIX

MATURITY: EARLY

SEED SOURCES: BAKE, EDEN, GRETA, PINE, REDS, SAND, SSEC, SUS, TRADE, TSK, UPRI, URB

These yellow apple look-alikes make a fine presentation on a dessert plate. Good and sweet—like candy. I prescribe (at least) one a day in season.

Introduced by Sakata Seeds of Yokohama. Carried by Johnny's Selected Seeds in the United States as early as 1982. Rob Johnston of Johnny's recommended emulating the Japanese—especially when space is at a premium—by trellising Sakata's Sweet as one would cucumber.

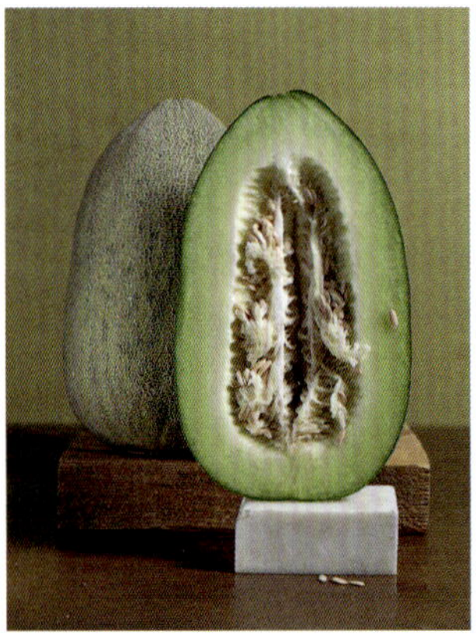

SAMARRA

CUCUMIS MELO **AMERI GROUP (SUB-GROUP MACULATI)**
SIZE: 10½" LONG BY 5¾" WIDE; WEIGHT: 7 POUNDS
SUGAR CONTENT: 5° BRIX
MATURITY: MIDSEASON
SYNONYMS: BATEEKH SAMARA, BATIK SAMARRA, SAMARRAN LEMON
SEED SOURCES: BAKE, SSE, WHA

This is a green-fleshed melon landrace, perhaps centuries old, grown under dryland conditions by Arab farmers in the vicinity of the ancient city of Samarra, Iraq. UNESCO named the city, with its archeological treasures, a World Heritage site in 2007.

Samarra is not alone in the world—many similar Ameri melons of the Maculati type are grown in today's Turkmenistan and Uzbekistan. I imagine it's of good quality in Iraq. When grown in my garden, it becomes unpleasantly juicy and insipid.

Introduced commercially in the United States by Baker Creek Heirloom Seeds in 2014. Their original seed source was Iraqi seedsaver Nael Aziz.

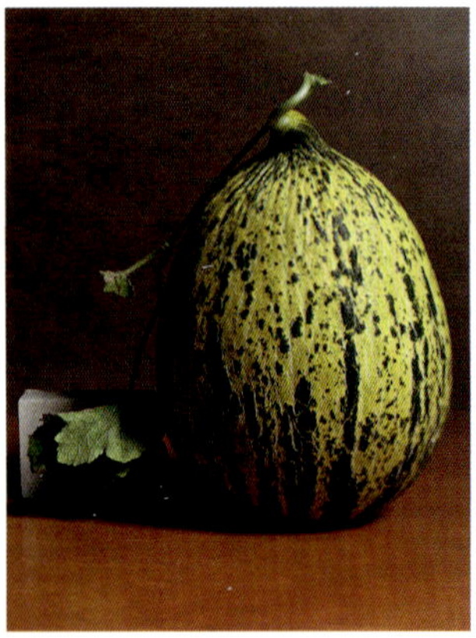

SANTA CLAUS

CUCUMIS MELO **INODORUS GROUP**
SIZE: 11" LONG BY 7¾" WIDE; WEIGHT: 9¼ POUNDS
SUGAR CONTENT: 9° BRIX
MATURITY: LATE
SYNONYMS: BURSA TURKEY, BURSA TURQUIE, CHRISTMAS, JOHNSON AND MUSSER'S SANTA CLAUS, J. & M. SANTA CLAUS, KIRKAGAC, XMAS
SEED SOURCES: JSKIT, TRADE

The Santa Claus is a keeper. This casaba can hold its quality for weeks or even months, long after the melons of summer are gone, and with a bit of luck it will still be around in December to grace your holiday table. The flavor and texture seem to improve in storage. Fruits are characteristically wrinkled, either elongate (aka long strain), or necky and pear-shaped (aka short strain), with thick and sweet green or white flesh. The skin color pattern—bright yellow embossed with dark green spots and streaks—distinguishes this melon from other types. Santa Claus is often confused with the Piel de Sapo by Americans because the patterned skins are similar; the spots and streaks are the same kind and color, but the predominant skin color differs (see Piel de Sapo, page 214).

Santa Claus was commercially introduced in the United States in 1908 by the Johnson & Musser Seed Company (later Aggeler & Musser) of Los Angeles. It appears in living color on that catalogue cover, and on the back cover the following year. The company had a naming contest in 1907 for a new melon, illustrated with a black-and-white photograph, with this copy: "Hearing of this marvelous muskmelon we procured at great cost ten pounds of seed, all there were in the world, as well as all the melons. . . . The rind has the appearance of the mottled Pomegranite Muskmelon. The shell is hard. . . . There is no seed cavity. . . . The light green flesh is absolutely stringless. Its flavor is as nearly like the Ambrosia of the Gods as anything we know. It is beyond description. To taste it is to realize the truth of this statement." Five dollars was offered to the lucky person who proffered the most appropriate name for the new melon (after having bought a melon for a dollar or seed packets at fifty cents each) suggesting its excellence and long-keeping qualities.

Santa Claus has always been of secondary importance in the market in the United States, but it is front and center in its native Turkey, where it is known as Kirkagac. Turkey is a country "of immense genetic wealth in many crops" including sugar beets, lentils, chickpeas, beans, flax, wild progenitors of several cereals, and tree fruits like peach, pear, sweet cherry, and apricot (Fowler and Mooney, 1990). Let me add melon to that list.

Mehmet Oztan, a Turkish member of the Seed Savers Exchange, and owner of Two Seeds in a Pod Heirloom Seed Company (Reedsville, West Virginia), confirmed that Santa Claus and Kirgagac are synonymous. He steered me to the Aggeler & Musser catalogues. The Kirkagac melon, named for the town of the same name, is the most popular casaba melon in Turkey, the long form being preferred.

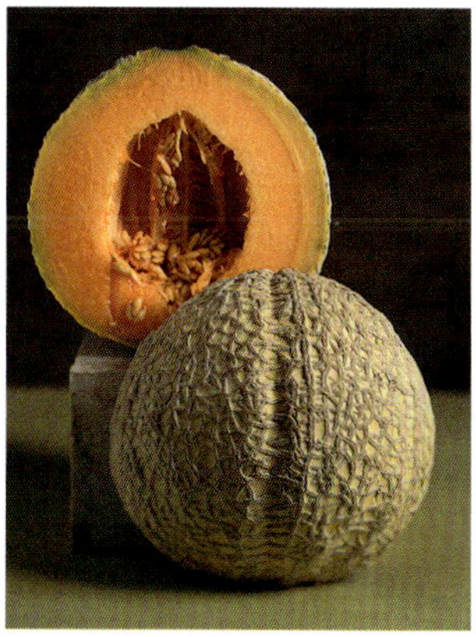

SCHOON'S HARD SHELL

CUCUMIS MELO **RETICULATUS GROUP**
SIZE: 8½" LONG BY 7½" WIDE; WEIGHT: 8½ POUNDS
SUGAR CONTENT: 12° BRIX
MATURITY: MIDSEASON
SYNONYMS: NEW YORKER, SCHOON'S HARD SHELL BENDER, SCHOON'S MUSK
SEED SOURCES: BAKE, BURR, ERS, SAND, SOLA, SSE, SSEC, SUS, VICT, WHITE

Reliable year after year and good as gold: Schoon's juicy, firm, orange flesh is abundant and intensely sweet. Does indeed have a hard shell. Not all melons develop heavy, coarse roping, but netting is always prominent at maturity (skin is not smooth), which is a sign of quality. Vigorous and prolific plants. Adapted to New York state but not Maine or points farther north. Shares features with Spear and Pike melons (see Spear, page 220).

Introduced as Schoon's Hard-Shell Bender by F. H. Woodruff and Sons in 1939. Catalogue copy reads: "A new and excellent flavored melon, similar to Bender's Surprise in shape and general appearance. It is valuable especially for its extra hard shell, which makes it fine for shipping. It will stand in the field or market better than most varieties. It is deeply lobed, well netted over its yellow skin. Melons weigh 5 to 8 lbs. and we've seen one that weighed 14 lbs." (See more on page 54.)

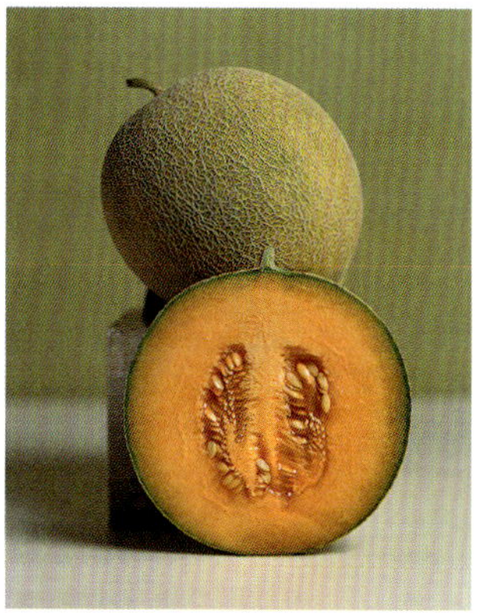

SMALL PERSIAN

CUCUMIS MELO **RETICULATUS GROUP**
SIZE: 5½" LONG BY 5½" WIDE; WEIGHT: 3 POUNDS
SUGAR CONTENT: 10° BRIX
MATURITY: MIDSEASON
SYNONYMS: BABY PERSIAN, PERSIAN SMALL
SEED SOURCES: BURR, REDW, VICT

This type apparently hails from Western Asia and the Middle East. Yet Persian may be a misleading name, as in the case of the Congo watermelon or the Tahitian squash, which originated elsewhere. Small Persian is characteristically round, finely and uniformly netted, lacking lobes and furrows, with thick orange flesh that is sugary and delicious. Does not slip from the vine when ripe. Persians come in small, medium, and large. Larger cultivars offer even more of a good thing.

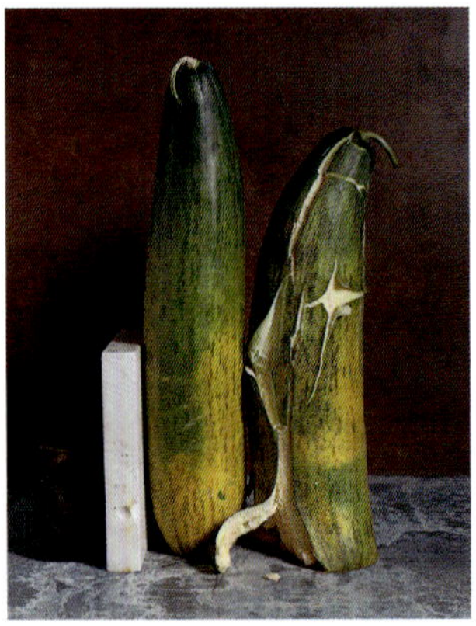

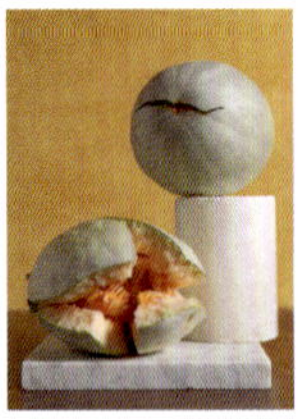

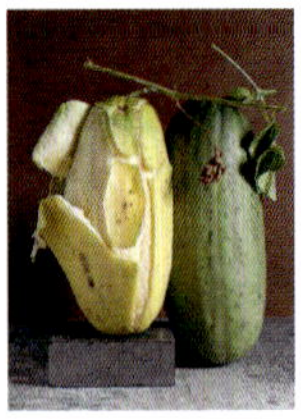

SNAP MELON (SNAP 1, SNAP 2, AND SNAP 3)

CUCUMIS MELO **MOMORDICA GROUP**

SNAP MELON 1: 8¼" LONG BY 7¾" WIDE; 6½ POUNDS; 3½° BRIX; MIDSEASON MATURITY

SNAP MELON 2: 11½" LONG BY 5" WIDE; 4 POUNDS; 8° BRIX; MIDSEASON MATURITY

SNAP MELON 3: 22" LONG BY 6" WIDE; 8 POUNDS; 3° BRIX; MIDSEASON MATURITY

SYNONYMS: PHUT, PHOOT

SEED SOURCE: NONE IN THE UNITED STATES OR CANADA

I present to you three different snap melon landraces grown in three separate years from seed sourced commercially in the United States, all labeled just plain old Snap. So I assigned them numbers. Apparently there aren't many named varieties available. My small sample size illustrates the point that, not only is there diversity in size, shape, and mature fruit color, but "distinct genetic polymorphism for fruit cracking" among snap melon landraces. These include blossom-end cracking (round types only), longitudinal or random cracking starting in the middle, and longitudinal peeling (Dhillon et al., 2007). Harvesting and photographing all three landraces at fruit-splitting stage had its challenges.

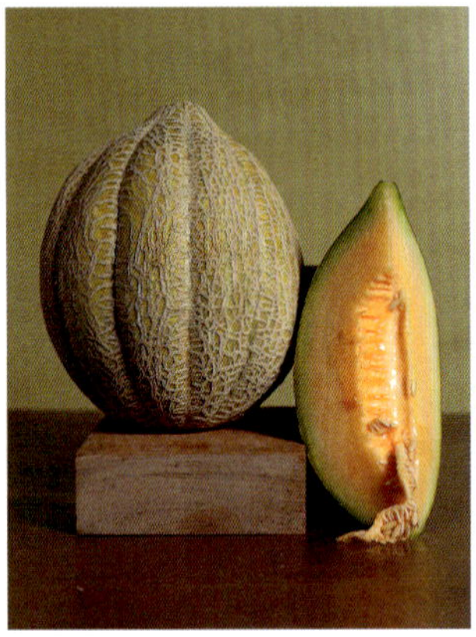

SPEAR

CUCUMIS MELO **RETICULATUS GROUP**

SIZE: 9¼" LONG BY 6¾" WIDE; WEIGHT: 6½ POUNDS

SUGAR CONTENT: 9° BRIX

MATURITY: MIDSEASON

SYNONYMS: GOLDEN HEART, OREGON DELICIOUS, OREGON SPEAR, SPEARMELON, SPEER

SEED SOURCES: SAND, SSE

A local and longtime favorite in western Oregon. Adapted to the Pacific northwest. Catalogued by Abundant Life Seeds in 1976 (then in Seattle) and for many subsequent years. Attractively lobed and netted fruit with a pronounced taper at the stem end. Spear-shaped, you might say—as are its seeds. Melon-yellow flesh is pleasing and soft and easy to scoop. Shares some features with Schoon's Hard Shell (see page 219) and Pike (developed around 1935 from a cross between Hearts of Gold and Spear by Aaron Pike of Pike and Young, Monmouth, Oregon).

According to the authors of *The Cucurbits of New York* (1937), Speer [Spear] melon was named for a gardener by the name of Speer in Albany, Oregon, who had grown it for many years. Gill Brothers Seed Company of Portland, Oregon, speculated that Banana melon might have been one of its parents. Oregon Delicious, introduced by Gill Brothers in 1929, is a selection from an improved version of Speer, with a more rounded stem end.

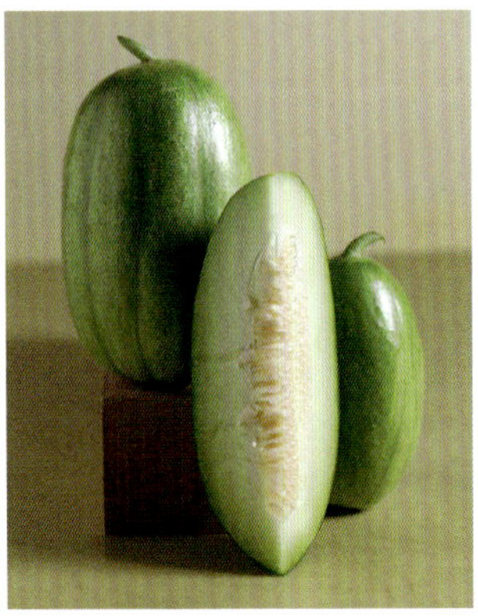

SPUREDDA LECCESE BIANCO

CUCUMIS MELO **ADZHUR GROUP**

SIZE: 4¾" LONG BY 2½" WIDE; WEIGHT: 9 OUNCES

SUGAR CONTENT: 3½° BRIX

MATURITY: EARLY

SYNONYM: BIANCO LECCESE

SEED SOURCES: CUC, RICH, SEEDSF, SOLA, SSE

The *spuredde leccesi* are a group of traditional carosello melons widespread in the Italian province of Leccese. They're also known as *meloncella, cummarazzi,* and *cucumbarazzi.* Spuredda Leccese Scuro has a darker green skin than Bianco Leccese, which is actually light green, not white. Mild and crisp; more barrel-shaped than Barese (see page 194).

STRIPED ARMENIAN

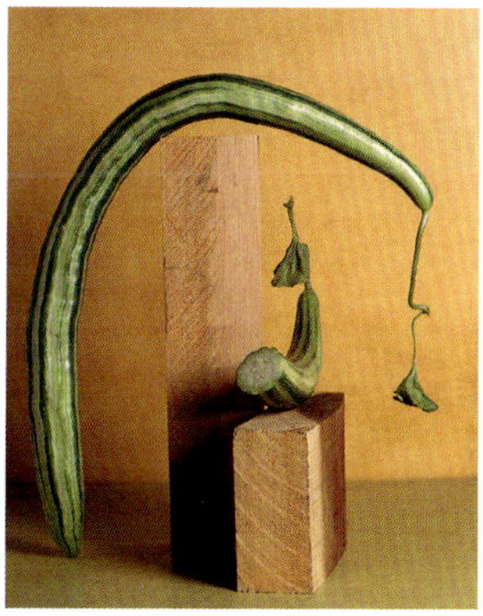

CUCUMIS MELO **FLEXUOSUS GROUP**
SIZE: 25" LONG BY 2½" WIDE; WEIGHT: 2¾ POUNDS
SUGAR CONTENT: 3° BRIX
MATURITY: MIDSEASON
SYNONYMS: ARMENIAN, ARMENIAN CUCUMBER, METKI PAINTED SERPENT, LEBANESE CUCUMBER, PAINTED SERPENT, SERPENT OR SNAKE CUCUMBER, STRIPED SNAKE, WHITE STRIPED SNAKE
SEED SOURCES: CUC, EDEN, FED, HEIR, HUD, HVSC, JOHN, KIT, NAT, PEACE, PINE, REST, SOC, SUS, TTS

Another excellent snake melon. Harry Paris (2011) says that Striped Armenian is a dead ringer for snake melons illustrated in a partially preserved fresco at the site of Ercolano (Herculanum), a city that was destroyed together with Pompeii during the eruption of Mount Vesuvius in 79 C. E. "In this painting, serpent-like striped melon fruits . . . are depicted in and adjacent to a large glass jar. The representation of a glass jar does suggest usage of snake melons for pickling."

SUCRIN DE TOURS

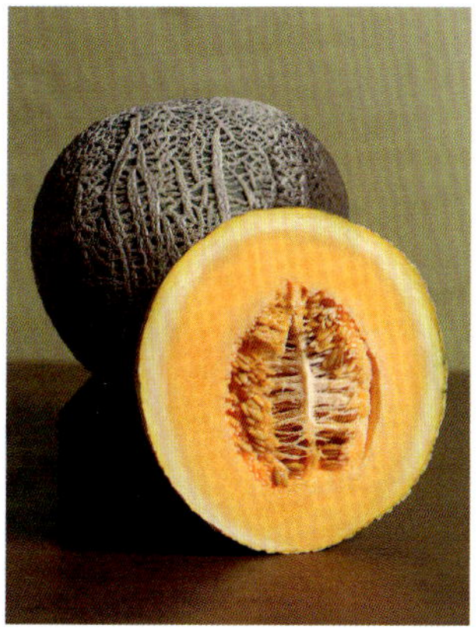

CUCUMIS MELO **RETICULATUS GROUP**
SIZE: 8" LONG BY 8" WIDE; WEIGHT: 6½ POUNDS
SUGAR CONTENT: 10° BRIX
MATURITY: LATE
SYNONYMS: MELON CABOULE DE TOURS, TOURS NETTED SUGAR, TOURS SUGAR
SEED SOURCE: LEJ

Not sugary enough in my garden . . . but it hardly matters. Sucrin de Tours is a magnificent specimen of melonhood: its round form reticulated to the max with overlapping layers of ropy netting. Cut open longitudinally, it's reminiscent of other heirloom melons painted by George Brookshaw (1751–1823). Sucrin de Tours has been around a long time in France in one form or another. Jacquin (1832) described two variants (Sucrin de Tours and Petit Sucrin de Tours); Vilmorin (1856) said the melon was practically covered with coarse netting but was variable in form. The best of the lot had emerged by 1883: Vilmorin described and illustrated the round form that has come down to us today. Burpee carried it in their catalogue as early as 1885. Honey or Sugar Rock is quite similar (see page 208).

SWEET FRECKLES

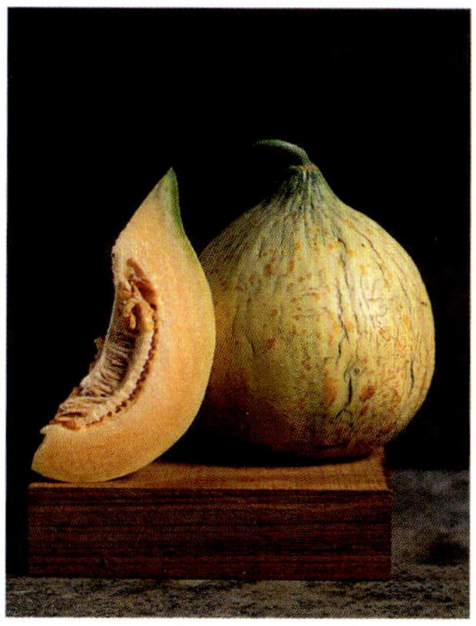

CUCUMIS MELO **INODORUS GROUP**
SIZE: 9½" LONG BY 7" WIDE; WEIGHT: 7½ POUNDS
SUGAR CONTENT: 11° BRIX
MATURITY: MIDSEASON
SEED SOURCES: ADAP, SSE

A Crenshaw (Cranshaw) melon bred by Tim Peters of Peters Seed and Research (PSR Breeding), Myrtle Creek, Oregon, from Farthest North and Casaba crosses. Has a fragrance reminiscent of a rose, says Peters. When ripe its true colors come shining through: lots of charming green and orange freckles (more than Crane) and the wrinkled, necky stem-end of a true casaba. Skin can be glossy or dull with short streaks of color at the blossom end. Exquisite flavor.

Plants are prolific and vigorous. I hesitate to say this because it's not my usual practice: Even the smaller melons that ripen in a second wave on dying vines at the end of the season are a gift to be gleaned (not as good as the main crop, but good enough).

Similar to Crane (see page 200), Eel River (from northern California), possibly Tularosa Market, and Northern Arizona. According to 1998 catalogue copy from

Bountiful Gardens, John Jeavons, gardening guru and co-founder of Ecology Action, "remembers this spectacular [Northern Arizona] fondly from his childhood in Arizona . . . [it was] said to have been brought out West with the pioneers," according to the seed donor.

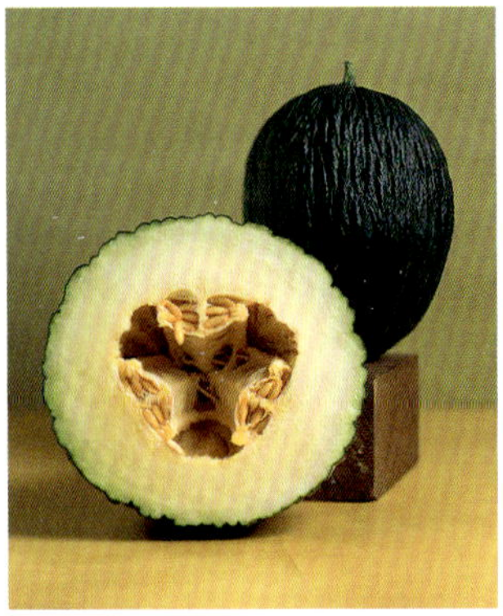

TENDRAL VERDE

CUCUMIS MELO **INODORUS GROUP**
SIZE: 9" LONG BY 7" WIDE; WEIGHT: 8 POUNDS
SUGAR CONTENT: 14° BRIX
MATURITY: LATE
SYNONYMS: TENDRAL VALENCIANO, TENDRAL VERDE TARDIF
SEED SOURCES: REIM, SAND, SEEDSF, SSE, VICT

Way up there on the sweetness scale; every increase over 11° Brix seems exponential. There are five major kinds of Spanish melons: Amarillo, Blanco, Piel de Sapo, Rochet, and Tendral. This is a typical Tendral, with deeply wrinkled hard skin that is black-green (in contrast, Rochet types have medium dark green skin) and firm flesh that is light green with yellowish white accents. Tendral Verde is a melon of mystery, displaying no visible signs of ripeness, except a crack or two.

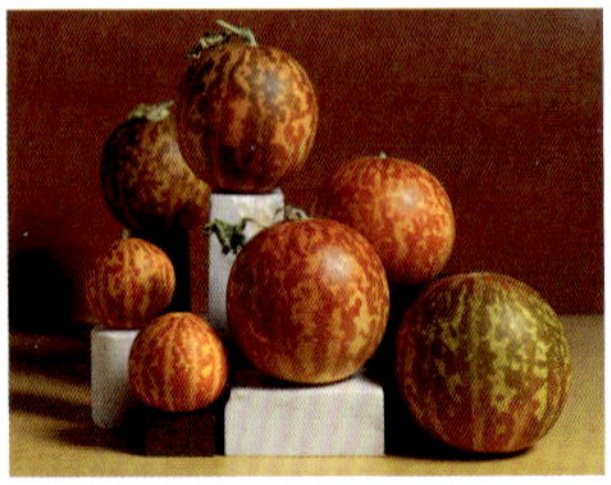

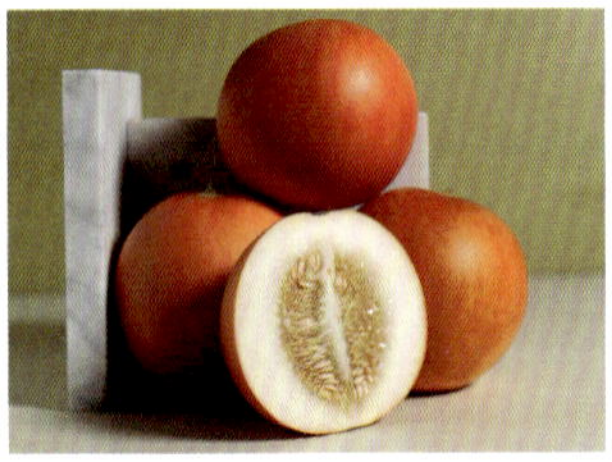

TIGGER AND TIGGER SPORT

CUCUMIS MELO **DUDAIM GROUP**
SIZE: 4½" LONG BY 4½" WIDE; WEIGHT: 1½ POUNDS
SUGAR CONTENT: 6½° BRIX
MATURITY: MIDSEASON
SYNONYM: ARMENIAN TIGGER
SEED SOURCES: BAKE, HUD, JSKIT, REIM, SAMP, SOLA, SSE, SSEC, TERRIT, TRADE

An overblown Queen Anne's Pocket Melon type. Has harder skin and thicker, mealier flesh than Queen Anne's, but is just as insipid. Emits that same transcendent perfume. Harvest a bit green so they last longer in the house. Sometimes the plants grow paprika-red anomalies without stripes: Tigger Sports.

Introduced in 2003 by Baker Creek Heirloom Seeds. Collected at a market in an Armenian mountain valley. Not to be confused with other melons that go by the name of Tiger.

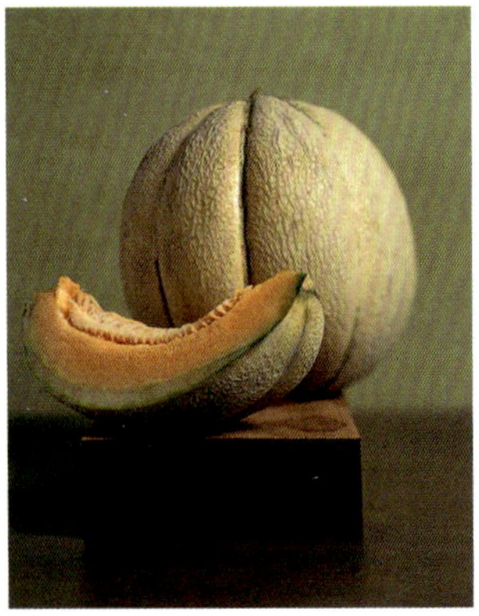

TIP TOP

CUCUMIS MELO **RETICULATUS GROUP**
SIZE: 8" LONG BY 7" WIDE; WEIGHT: 4½ POUNDS
SUGAR CONTENT: 10¾° BRIX
MATURITY: LATE
SYNONYMS: LIVINGSTON'S TIP TOP, MOSLEY TIP TOP, NUTMEG TIP TOP
SEED SOURCES: HER, SSE, VICT

A sweet, juicy melon with musky flavor, characterized by prominent lobing and furrowing. Still tip-top after all these years. A keen-eyed gentleman from Lancaster County, Pennsylvania, with a nose for peculiar melons discovered one in the 1880s by chance in a neighbor's field (where Surprise, Hackensack, Jenny Lind, and other melons were growing for market). (See Jenny Lind, page 208.) All present were astonished by its excellent flavor. The gentleman in question saved seed and grew them out when he moved to Ohio. A. W. Livingston's Sons found the melon at a fancy vegetable stand in Columbus. They named and introduced the melon in 1892 and it quickly caught on. The "Tiptop group" of orange-fleshed melons for home and market gardens includes Surprise, Bender, Baker, Indiana, Ohio Sugar, and Milwaukee Market.

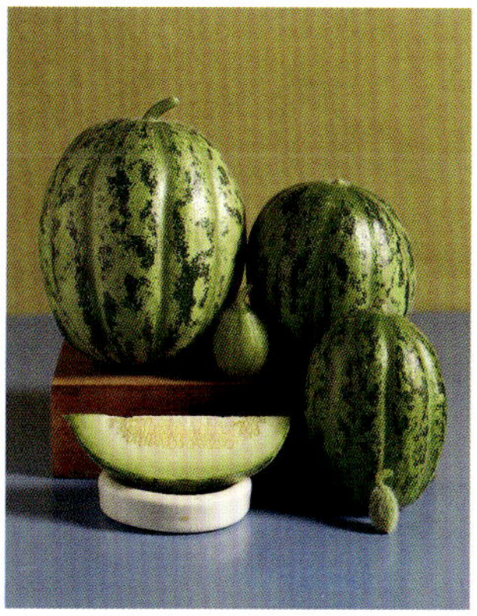

TONDO DI MANDURIA

CUCUMIS MELO **ADZHUR GROUP**
SIZE: 3¾" LONG BY 3" WIDE; WEIGHT: 11 OUNCES
SUGAR CONTENT: 3° BRIX
MATURITY: EARLY
SYNONYM: TONDO LISCIA MANDURIA
SEED SOURCES: CUC, SEEDSF, SOLA

A round-fruited landrace from Manduria in Puglia, Italy. Blotches on young fruits disappear with time. Tondo should be harvested and enjoyed when immature and still crisp. Can be pleasantly fragrant when fully ripe but is of poor quality at that age and stage.

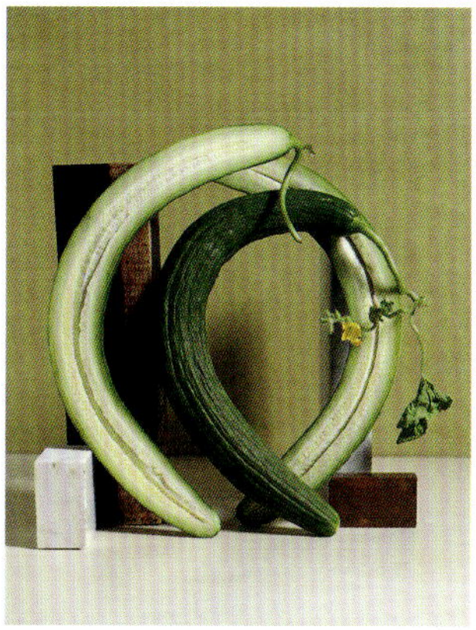

TORTARELLO

CUCUMIS MELO **FLEXUOSUS GROUP**
SIZE: 23" LONG BY 2½" WIDE; WEIGHT: 2 POUNDS
SUGAR CONTENT: 4½° BRIX
MATURITY: EARLY
SYNONYMS: METKI DARK GREEN SERPENT, TORTARELLO ABRUZZESE SCURO, TORTARELLO BARESE, TORTARELLO SCURO, SCURO
SEED SOURCES: BAKE, CUC, HUD, REIM, SOLA, TRADE, URB

A great Italian cucumber-melon (*cetriolo melone*) that is dark green, narrow, furrowed, and curved. Very crisp and at its best when less than twenty inches long. Pubescent enough to require rubbing off the short hairs—*piqqus* of the *faqqous*—before eating.

The *Carrara Herbal,* made between 1390 and 1404, was written in a Paduan dialect of Italian, and illustrated with magnificent, realistic images. Folio 162r shows three snake melons on a vine labeled *Del cogombaro.* They resemble Tortarello Scuro, aka Tortarello Barese, to a T (Paris, Janick, and Daunay 2011).

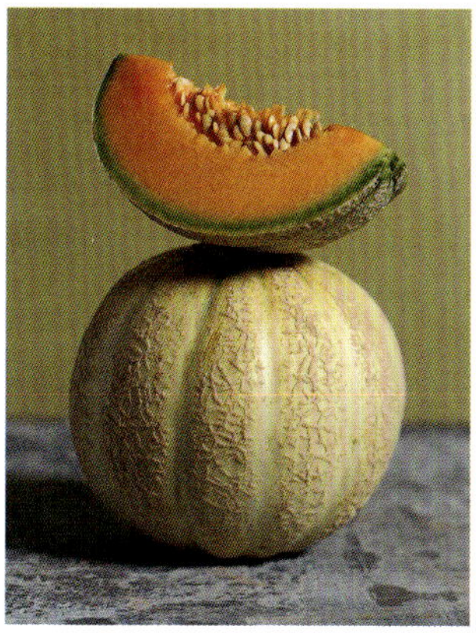

TRIFECTA

CUCUMIS MELO **RETICULATUS GROUP**
SIZE: 6" LONG BY 6½" WIDE; WEIGHT: 4¾ POUNDS
SUGAR CONTENT: 11½° BRIX
MATURITY: MIDSEASON
SEED SOURCE: SESE

Sugary and richly flavored orange-fleshed muskmelon. Bred by Dr. Michael Mazourek, of Cornell University, and selected for use in organic farming systems. The melon's pedigree is Golden Gopher x ZBC5TUF14. (See Golden Gopher, page 205.)

Trifecta got its start as part of a project funded by USDA-SARE (Sustainable Agriculture Research and Education) to produce a more disease-resistant, high-yielding melon that maintains the best traits of the heirloom parents. Dr. Molly Jahn led the research program at Cornell and trained Mazourek (and she indirectly trained me by graciously permitting me to take a tutorial on hand pollination of melon from Cornell research plant breeder George Moriarty in July 2000). She is now a professor at the University of Wisconsin–Madison. When Jahn moved to Wisconsin, Mazourek continued the breeding projects she had initiated—like this one—and started his own. Mazourek is the breeder of August Ambrosia watermelon, Honeynut squash, and Habanada pepper.

In a collaborative project with Mazourek, Edmund Frost of Twin Oaks Seed Farm named Trifecta and distributed it in 2015 via Common Wealth Seed Growers. Mazourek explained to me that Frost is the first to offer his downy mildew–resistant cultivars and deliver those genetics to farmers who need them right away.

Southern Exposure Seed Exchange introduced the variety commercially in

2016. Trifecta has at least three virtues: sweetness, keeping quality, and resistance to downy mildew.

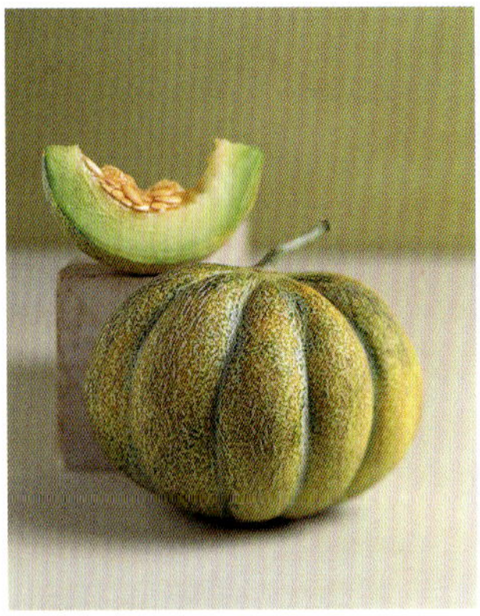

VERT GRIMPANT

CUCUMIS MELO **RETICULATUS GROUP**
SIZE: 4½" LONG BY 5¼" WIDE; WEIGHT: 2 POUNDS
SUGAR CONTENT: 13½° BRIX
MATURITY: MIDSEASON
SYNONYMS: GREEN CLIMBING, MELON VERT À RAMES
SEED SOURCE: SSE

Vert Grimpant, or Green Climbing, is one of my all-time favorite melons. Its sweet, melting, green flesh is hard to beat. Yet, perversely, it fares least well in hot, dry summers. The plant is rampant and prolific when left to sprawl on the ground. Green Climbing's small fruit size, relative earliness, and growth habit also make it suitable for trellis culture when space is at a premium and fewer fruits are wanted. (I'm guilty of pulling plants out of the ground prematurely because I couldn't eat or share the bounty fast enough.) This cultivar was first listed by Vilmorin-Andrieux in 1883 as Melon Vert à Rames; the company catalogued it for decades. Their beautiful engraved illustration of Green Climbing, on a trellis, is engraved in my brain.

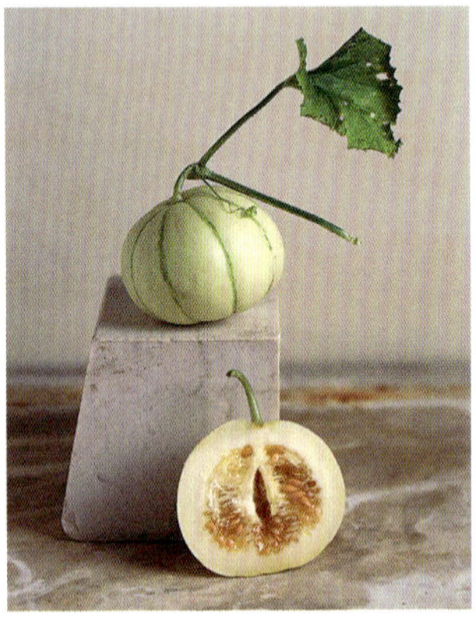

YUKE

CUCUMIS MELO **MAKUWA GROUP (SUB-GROUP YUKI)**
SIZE: 3" LONG BY 3½" WIDE; WEIGHT: 9 OUNCES
SUGAR CONTENT: 8° BRIX
MATURITY: EARLY
SYNONYMS: GIANT CHRYSANTHEMUM, KIKU CHRYSANTHEMUM, LARGE SNOW, OGATAKIKU, YUKI LARGE
SEED SOURCE: BAKE

Not as sweet as other Makuwas but rather citrus-like. Would be nice in a fruit salad with oranges. Sweeter and more delicious when eaten with its seeds.

Catalogued as Yuke Muskmelon by Glecklers Seedmen in 1972, and appeared as Yuki Large (Large Snow) in 1982. They described it as a pure-white melon, imported from Japan, distinctly lobed, midget-sized (a half-pound), sweet, fragrant, and prolific (over twenty fruits per plant). Its snowy skin turns creamy white at maturity.

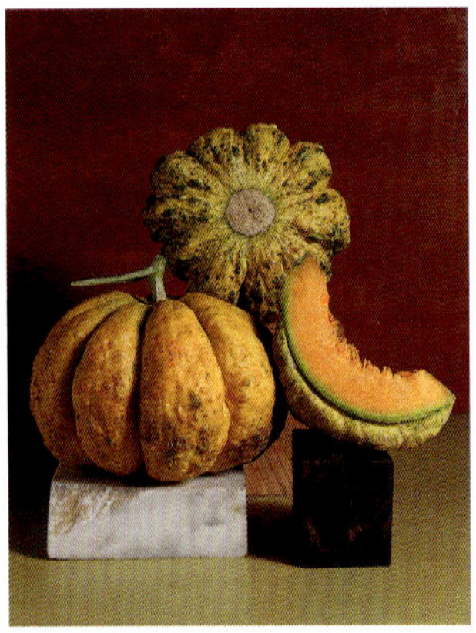

ZATTA

CUCUMIS MELO **CANTALUPENSIS GROUP**
SIZE: 6" LONG BY 6" WIDE; WEIGHT: 6¾ POUNDS
SUGAR CONTENT: 12½° BRIX
MATURITY: MIDSEASON TO LATE
SYNONYMS: BRUTTO MA BUONO, MELONE COTOGNO, MR. UGLY, ZATTA DI MASSA, ZATTA ARANCINA. BROTMABON IS AN F1 HYBRID OF ZATTA.
SEED SOURCES: EDEN, GOU, SEEDSF, SOLA, SSE, THOM, TSK

There were three types of zatte described by Moretti and Chiolini in 1828: Zatta dal collo grosso, Zatta commune, and Zatta arancina. In their opinion, zatte fruit flesh was the best of all species of melon.

Bernard M'Mahon of Philadelphia catalogued Early Small Zatte in 1806 and for decades thereafter through 1857. It's a wonder this cantaloupe never caught on in the United States until now. (See more on page 46.)

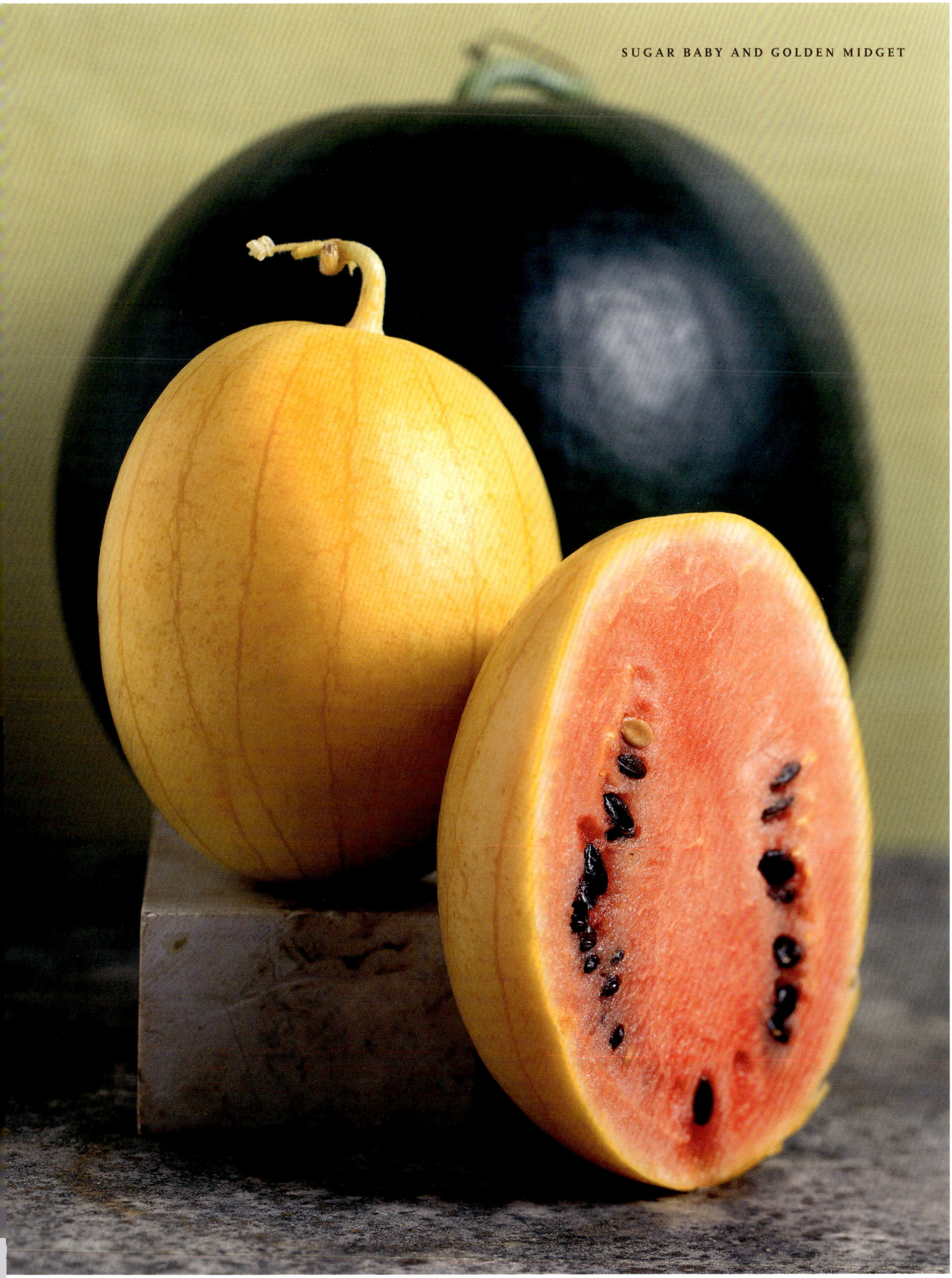

Watermelon Varietal Descriptions

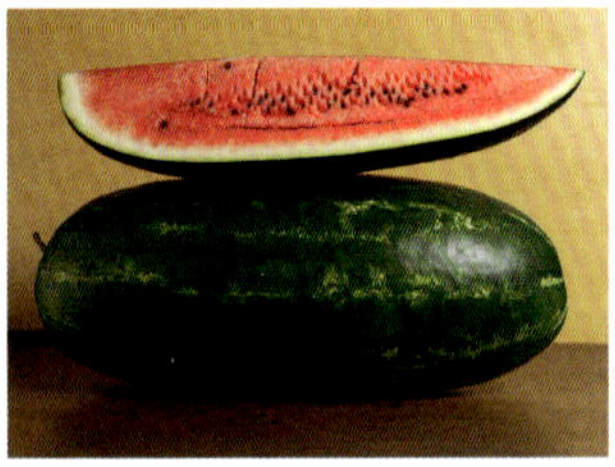

ALLSWEET

CITRULLUS LANATUS

SIZE: 21" LONG BY 10" WIDE; WEIGHT: 36 POUNDS

SUGAR CONTENT: 12½° BRIX

MATURITY: LATE

SYNONYMS: KANSAS 70-1, 2, 3

SEED SOURCES: BERL, BUR, BURR, CLIF, EDEN, ERS, EVER, JOR, PEACE, REDW, REIM, ROHR, STC, SUS, TRADE, TSK, URB, VICT, WIL

Absolutely scrumptious red flesh. One of the best watermelons out there, whether for the home gardener, consumer, or shipper. Its trademark: broad and jagged dark green stripes running lengthwise. But then again, watermelon stripes are always jaggedly edged, a feature that helps distinguish them from striped melons, squash, and pumpkins.

Allsweet was bred by Dr. Charles V. Hall at Kansas State University, Kansas AES, in Manhattan, Kansas, and released in the early 1970s. Derived from the same breeding line that produced those other stars of the watermelon patch: the round, medium-striped Crimson Sweet, Petite Sweet, and Super Sweet. Parentage: [(Miles x Peacock) x Charleston Gray]. (See Crimson Sweet, page 230.) Resistant to fusarium wilt and anthracnose (except race 2). Serves as a parent for many commercial cultivars.

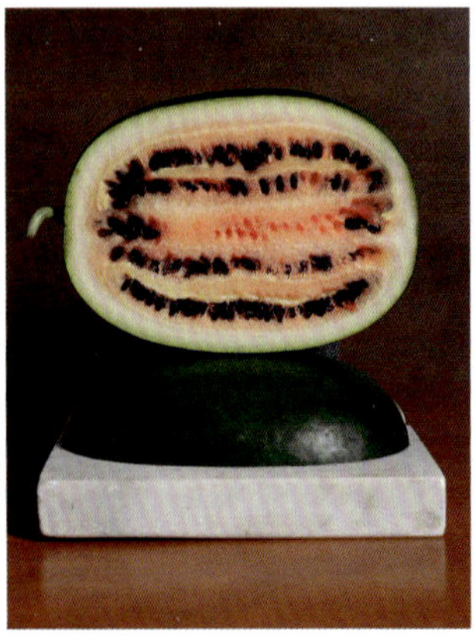

ARIKARA

CITRULLUS LANATUS

SIZE: 9¼" LONG BY 6¾" WIDE; WEIGHT: 8 POUNDS

SUGAR CONTENT: 7° BRIX

MATURITY: EARLY

SYNONYM: ARIKARA WILD

SEED SOURCES: HER, SSE, VICT

Oscar H. Will & Co., of Bismarck, North Dakota, introduced this watermelon in 1925, and claimed it was the one that the Lewis and Clark expedition found growing among the Arikara, Mandan, and Hidatsa tribes when they made their famous journey. On October 10, 1804, expedition member Sergeant John Ordway visited an Arikara village at the mouth of the present-day Grand River "and viewed about sixty earth lodges compactly arranged. He also found a considerable crop of corn, beans, pumpkins, squashes, watermelons, and tobacco. Warmly accepted in one lodge, he enjoyed a bowlful of beans and corn and three more servings of different foods before returning to camp" (Moulton 2018). (See more on page 166.)

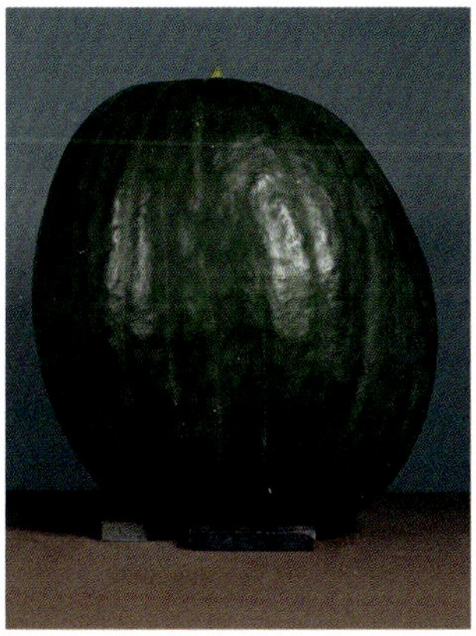

BLACK DIAMOND

CITRULLUS LANATUS

SIZE: 22" LONG BY 19" WIDE; WEIGHT: 98 POUNDS
SUGAR CONTENT: 9½° BRIX
MATURITY: LATE
SYNONYMS: BLACK BOULDER, BLACK CANNONBALL, CANNONBALL, CANNON BALL, CLARA LEE, DIAMOND, FLORIDA GIANT, IMPROVED BLACK DIAMOND, JUMBO BLACK DIAMOND, SELECT BLACK DIAMOND
SEED SOURCES: ANN, BAKE, BERL, BURR, CLIF, EDEN, ERS, EVER, GIANT, HEIR, HUD, JOR, MAIN, MOR, NEWE, REIM, ROHR, SHUM, SSE, STC, SUS, TRADE, TSK, URB, VICT, WHITE, WIL

2016 was the year I set out to grow giants just for fun: pumpkins, marrows, long gourds, tomatoes, melons, and watermelons. I raised the 98-pound Black Diamond watermelon illustrated in this book from Charlie Bob Kemp's "107 Kemp 2012" (a 107-pound melon grown in 2012). Kemp's Jumbo Black Diamond or "Blue Rind" was the result of a cross between a "178 Bright 1977" and a "156 Bright 1981." Charlie says it's a great eating melon and one you can take to the fair: he won Best of Show for it at the DuQuoin (Illinois) State Fair. (See more on page 172.)

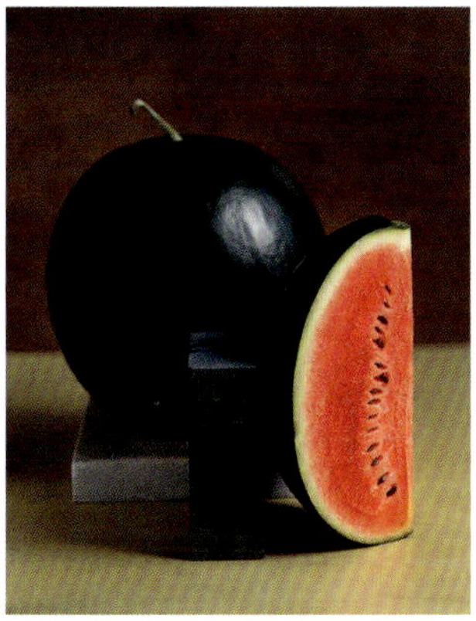

BLACKTAIL MOUNTAIN

CITRULLUS LANATUS

SIZE: 9" LONG BY 9" WIDE; WEIGHT: 14 POUNDS
SUGAR CONTENT: 10½° BRIX
MATURITY: EARLY
SYNONYM: BLACK TAIL MOUNTAIN
SEED SOURCES: AMI, BAKE, BUR, FED, GOOD, GRETA, HAR, HEIR, HIGH, MARY, REST, ROHR, SAND, SESE, SHUM, SSE, SSEC, SUS, TERRIT, TRADE, TUR, UPRI, WHA, WHITE

A Glenn Drowns special. Developed by him in 1977; initially offered to members of Seed Savers Exchange via the 1983 Winter Yearbook; made its commercial debut in Glenn's 1994 Sand Hill Preservation Center catalogue.

"Nonsense," says Glenn to the notion that he named the watermelon for its seeds, which were once rumored to have black tails. The seeds do, however, have black hilums or tips, which could be misconstrued as tails. The whimsical notion of a seed spouting (or sprouting?) a tail seems to have taken root in Southern Exposure Seed Exchange's 1996 catalogue: "The small brown seeds have a 'black tail.'"

BTM was named for Glenn's homeplace: He once lived on Blacktail Mountain Road at the foot of Blacktail Mountain in Idaho.

I'm not sure which version of the truth I like better. (See more on page 182.)

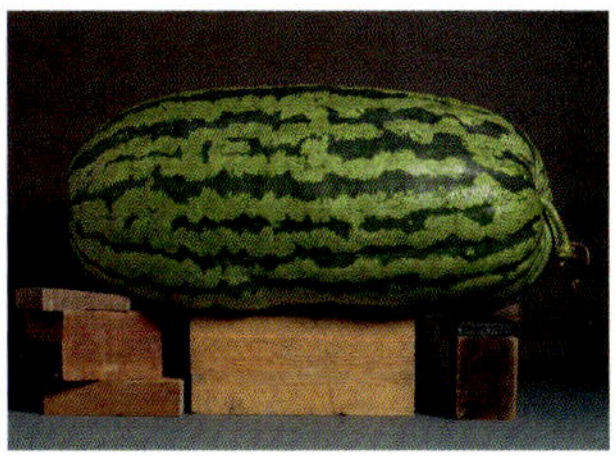

CAROLINA CROSS

CITRULLUS LANATUS

SIZE: 31" LONG BY 16½" WIDE; WEIGHT: 129 POUNDS
SUGAR CONTENT: 11° BRIX
MATURITY: LATE
SYNONYMS: GIANT CAROLINA CROSS, CAROLINA CROSS NO. 180, CAROLINA CROSS NO. 183, NC GIANT, WEEKS, WEEKS NORTH CAROLINA GIANT, WEEK'S NORTH CAROLINA GIANT
SEED SOURCES: BAKE, CLIF, ERS, GIANT, MOR, PPS, REIM, ROHR, SHUM, SSE, SUS, TSK, TTS

Growing a world-record watermelon takes extraordinary skill and vigilance—in addition to the right climate and genetics. In 1979, the Advertisement and Tourism Commission of Hope, Arkansas, issued a challenge and a ten-thousand-dollar reward for the first local grower to deliver a 200-pounder during the Annual Hope

Watermelon Festival. They were determined to snatch back the title for the world-beater from Edward Weeks, of Tarboro, North Carolina. The Bright family delivered, but not without some anxious moments: During the festival the specimen was growing at the rate of six pounds a day, reaching 198 pounds by Saturday (the second-to-last day). Certified scales were brought in and placed under the melon on the vine. As told by Lloyd Bright: "The watermelon's weight was checked throughout the night and the weight remained unchanged until daylight. The scale bar began to rise quickly after daylight on Sunday morning. The watermelon reached 200 pounds at approximately 9:00 am on August 26, 1979 and festival authorities were notified. After witnesses arrived and verified the weight, the 200-pound watermelon was [cut from the vine and] taken to the festival." A controversy ensued regarding the exact date the melon attained 200 pounds. It was finally certified as a world record and awarded a listing in the Guinness Book of World Records. But Ivan and Lloyd Bright received only half of the purse (five thousand dollars in total). (See more on page 182.)

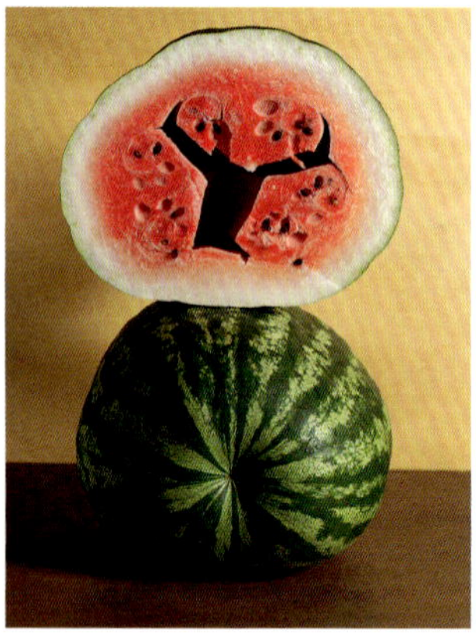

COBB GEM

CITRULLUS LANATUS

SIZE: 20" LONG BY 14¾" WIDE; WEIGHT: 62 POUNDS
SUGAR CONTENT: 11½° BRIX
MATURITY: LATE
SYNONYMS: AMERICAN CHAMPION, COB GEM, GIANT COB GEM, KOLB, KOLB GEM, KOLB'S GEM, MISSISSIPPI COBB GEM, STATE FAIR
SEED SOURCES: REIM, SSE

A big, handsome devil, broadly and boldly striped, with a thick, tough rind, and good eats. Subject to hollow heart—but who's perfect?

Cobb Gem is an exhibition watermelon predisposed to win contests for weight. And it's one of only five cultivars known to produce fruits over a hundred pounds. The others are Black Diamond, Tom Watson, Weeks North Carolina Giant, and Carolina Cross (the old Triumph is apparently extinct). (See Black Diamond, Tom Watson, and Carolina Cross, pages 227, 242, and 227, respectively.)

Cobb Gem has come down through the years without much change in appearance and fruit quality. D. M. Ferry and Company, of Detroit, Michigan, marketed the watermelon in its 1885 seed catalogue as Kolb's Gem or American Champion, and is generally credited with its introduction. They wrote: "A new and very distinct variety originating [in] the south, and becoming very popular with southern growers because of its size, beauty, and remarkable shipping and keeping qualities. Specimens have been shipped from the extreme south to New York and from there back to Philadelphia, yet they arrived in the best possible condition. . . . Fruit of the largest size, round or slightly oval with a cavity, at the blossom end marked with irregular mottled stripes of dark and very light green in sharp contrast, which gives it a bright and attractive appearance. Outer rind or shell exceedingly hard and firm resembling that of a winter squash. Flesh bright red extending to within half an inch of the rind; always solid, very firm, a little coarse but sweet and tender. Seeds dark grey." It was exhibited as Cobb Gem at the 1904 St. Louis World's Fair.

Kolb's Gem was developed by Reuben Francis Kolb (1839–1918) of Montgomery, Alabama. According to a Kolb descendant by the name of Annie *Crenshaw* (can you believe?), R. F. Kolb was a planter by trade. After creating his eponymous watermelon, he was elected Commissioner of Agriculture—twice—and became an influential spokesman for farmers' rights.

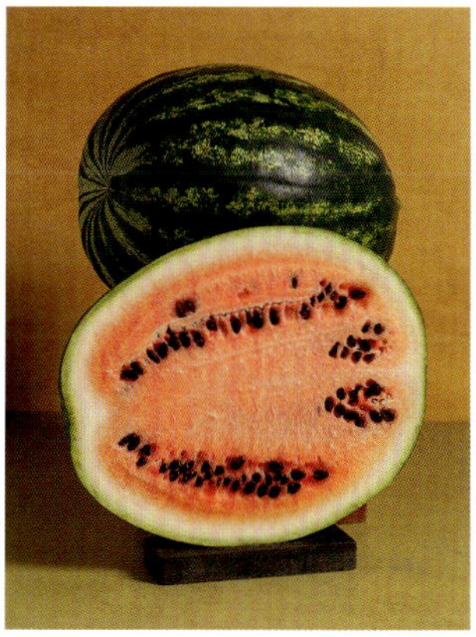

COLE'S EARLY

CITRULLUS LANATUS

SIZE: 11" LONG BY 9½" WIDE; WEIGHT: 16 POUNDS
SUGAR CONTENT: 8½° BRIX
MATURITY: MIDSEASON
SYNONYMS: HARRIS' EARLIEST, HARRIS' EARLY
SEED SOURCES: SAND, SSE

Once a favorite in the Northern tier states. Introduced by Cole's Seed Store, of Pella, Iowa, in 1892, and offered free to paying customers that year. The catalogue copy reads: "This fine new melon, we have the pleasure to introduce to our customers for trial only this season, is worthy of an extended distribution. A few years ago one of our customers sent us a small lot, seed of this variety, to plant in our trial grounds. . . . He had been growing this variety for nearly twenty years, and was always the first to have ripe melons for the market or family use. A good many would buy his melons with the object of obtaining seed, but never did succeed in getting pure seed, as those brought on the market were grown along side of other varieties, and so kept this variety in his own family. The melons are of medium size, nearly round in shape, rind green, striped with lighter shades. It is very early and hardy, vines grow vigorous and is a sure cropper. Flesh deep red color, of fine delicious flavor, perfectly stringless, rind thin and rather brittle. It is not recommended as a shipper."

Oscar H. Will and Company, of Bismarck, North Dakota, catered to the northern grower. Cole's Early was a staple in their stable of watermelons from 1918 through 1959. Subsequently the watermelon was catalogued by Abundant Life Seed Foundation, Port Townsend, Washington, in 1980, then kind of fizzled out. That's when Glenn Drowns of Sand Hill Preservation Center came to the rescue: He has a fondness for old Oscar Will sorts and North Dakota melons.

Cole's Early is lovely in shape and size, boldly and beautifully striped, prolific if not early in my garden. Yet its pastel red (pinkish) flesh lacks sugary flavor, leaving much to be desired—don't grow it unless you love living history or have a hankering for watermelon rind pickles or preserves.

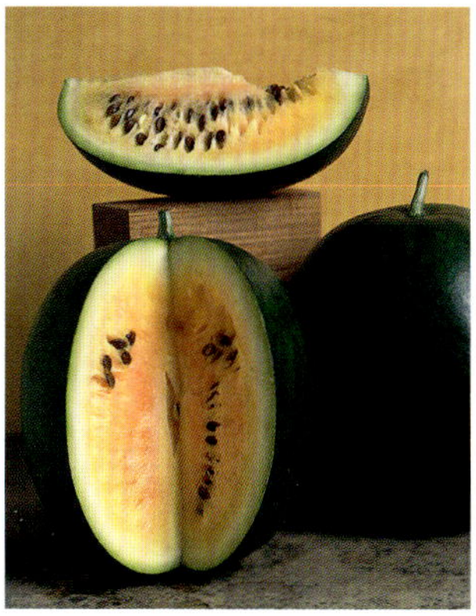

COLORADO STRIPED TARAHUMARA

CITRULLUS LANATUS

SIZE: 8" LONG BY 7" WIDE; WEIGHT: 7 POUNDS
SUGAR CONTENT: 10° BRIX
MATURITY: MIDSEASON
SEED SOURCE: NONE IN THE UNITED STATES OR CANADA

Surprisingly sweet and tasty watermelon if seedy. Has extraordinary tricolor flesh—yellow, orange, and pink—with a rainbow effect. Adapted to dryland farming. Associated with the Tarahumara Indians, native to Mexico, renowned for their prowess as farmers and runners or foot throwers. As explained by Native Seeds/SEARCH in 1986, the Tarahumara traditionally occupied "the remote slopes and deep canyons of the Sierra Madre of Chihuahua and Durango. . . . This isolation has allowed them to maintain one of the most intact traditional agricultural systems in the world. Because of the variety of elevations utilized, their crops include amazing diversity."

Sourced by Glenn Drowns from folks at Native Seeds/SEARCH (it evidently did not appear in any NS/S catalogues as Colorado Striped Tarahumara) and in turn offered by him to members of Seed Savers Exchange via the 1989 Winter Yearbook. Carried almost singlehandedly by Glenn's Sand Hill Preservation Center until 2015. We need another runner to step up to the plate.

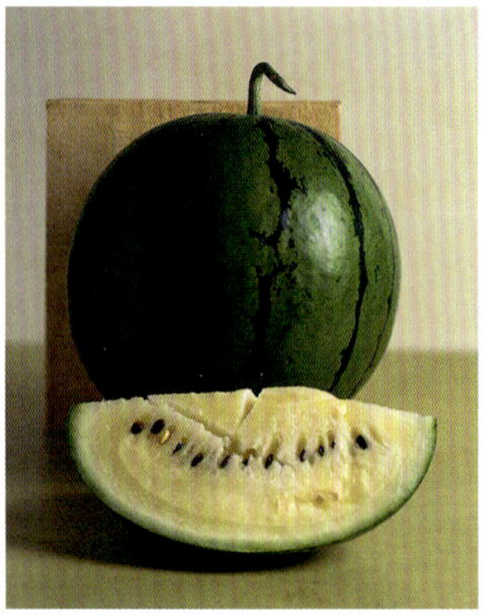

CREAM OF SASKATCHEWAN

CITRULLUS LANATUS

SIZE: 8" LONG BY 8" WIDE; WEIGHT: 10 POUNDS
SUGAR CONTENT:10° BRIX
MATURITY: EARLY
SYNONYM: HONEY CREAM
SEED SOURCES: BAKE, ERS, FED, GOU, GRETA, HEIR, HER, HUD, HVSC, IRISH, MARY, PRAI, REIM, SAMP, SAND, SOTE, SSEC, SUS, TERRA, TRADE, TSK

Cream of Saskatchewan was introduced commercially by Jan Blum of Seeds Blüm—pronounced "bloom" because, really, all seeds do bloom!—in Boise, Idaho, in 1987. Touted as truly white fleshed. With her early involvement in Seed Savers Exchange and its Growers Network, Jan had access to the cream of the crop including, well, Cream of Saskatchewan. Seeds Blüm reintroduced the Moon and Stars watermelon in its 1985/6 catalogue. (See Moon and Stars, page 237.)

I learned a lot from Jan's catalogues in the 1990s. She organized her seed listings by botanical family. The Cucurbitaceae family tree—home to melon, watermelon, and squash—was a bit of whimsy drawn by Jan: melons and pepos dangled from vine-like branches delineating the genera and species. I cut and tacked those to my bulletin board. And her two charts of "How and When to Plant How Much of What," for cool- and warm-weather crops, were my polestar for years. (See more on page 168.)

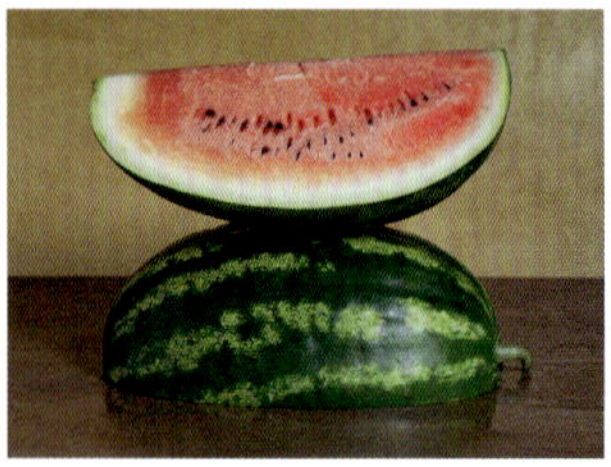

CRIMSON SWEET

CITRULLUS LANATUS

SIZE: 11½" LONG BY 13" WIDE; WEIGHT: 46 POUNDS
SUGAR CONTENT: 14° BRIX
MATURITY: MIDSEASON
SYNONYMS: BIG CRIMSON, CRIMSON SWEET VIRGINIA SELECT, GREEN AND WHITE RUBY, KAN. 60-6
SEED SOURCES: ANN, BAKE, BERL, BOT, BUR, BURR, CLIF, DEB, EDEN, ERS, EVER, FED, GOU, HAR, HEIR, HIGH, HOPE, HUME, IRISH, JOHN, JOR, JUNG, MAIN, MARY, MOR, NAT, NATIVE, NEWE, NIC, OSC, PEACE, REDS, REDW, REIM, REST, RICH, ROHR, SEEDSF, SESE, SHUM, SISK, SOC, SOLA, SOWT, STC, SUS, SWAL, TRADE, TSK, TTS, TUR, URB, VER, VICT, WHA, WIL

Incredibly good.

Without further ado the 1966 catalogue copy from Harris Seeds (Rochester, New York) announcing the debut of Crimson Sweet: "A valuable introduction from Kansas State University. Crimson Sweet has superior quality plus resistance to anthracnose and fusarium wilt. Its rounded blocky fruit weigh 15-25 lbs. and are striped with dark green. The texture and crispness of its sweet red flesh are delightful. Not adapted where frost comes early, but for full-season growing its disease resistance can give it extra importance in many sections." (See more on page 160.)

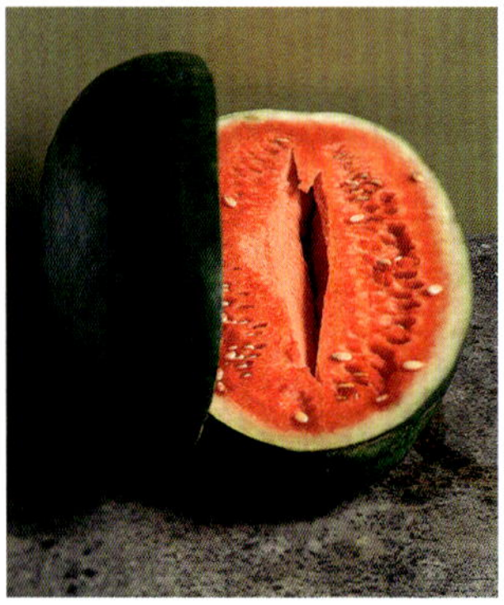

EARLY JUMBO

CITRULLUS LANATUS

SIZE: 18½" LONG BY 9" WIDE; WEIGHT: 23½ POUNDS
SUGAR CONTENT: 11° BRIX
MATURITY: MIDSEASON
SEED SOURCES: SAND, SSE

I didn't do this cultivar justice in my first melon book, and am so glad I trialed it again. Early Jumbo turns out to be a champ—especially when skies are clear and temps high. I'm wowed by its sweet, crisp-textured, deep pink (madder red) flesh. The white seeds add a nice accent. EJ was a real hit at a Rhinebeck Farmers Market taste test I conducted about five years ago. Deserves to be more widely grown.

Offered by Glenn Drowns in the 1992 Seed Savers Exchange Yearbook. Introduced commercially in 2006 by Glenn's Sand Hill Preservation Center. Glenn's original seed source was Tom Knoche (1938–2017), an early and dedicated member of SSE and an avid collector of heirloom squash and watermelons.

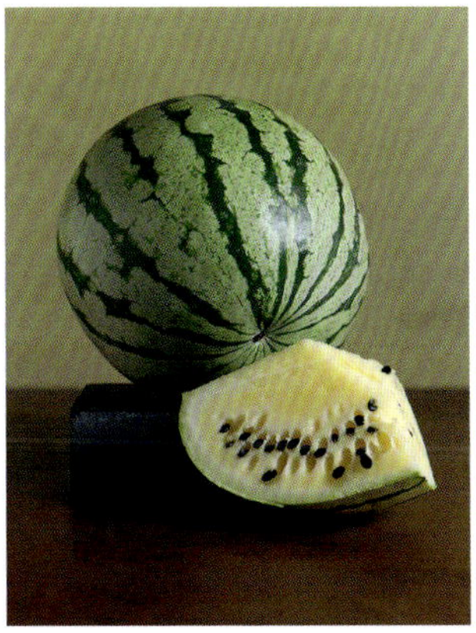

EARLY MOONBEAM

CITRULLUS LANATUS

SIZE: 9" LONG BY 9" WIDE; WEIGHT: 10 POUNDS
SUGAR CONTENT: 10°
MATURITY: EARLY
SYNONYM: YELLOW DOLL OP
SEED SOURCES: ADAP, BAKE, FED, GRETA, IRISH, PEACE, REDS, REST, SESE, SISK, SSE, TRADE, TUR

Don't you just love the name? Says a lot about the man who selected it from the canary-yellow–fleshed Yellow Doll Hybrid or F1, and introduced it as Early Moonbeam in his 1992 Deep Diversity seed list. That would be Dr. Alan M. Kapuler, known as Mushroom. He co-founded Stonebroke Hippie Seeds, which morphed into Peace Seeds, which morphed into Deep Diversity, and Seeds of Change. A planetary peacenik from way back (and Brooklyn like me!) with a degree in molecular biochemistry, who finds daily miracles in the garden.

Early Moonbeam is a standout for crispness and flavor among yellow-fleshed, open-pollinated, icebox watermelons.

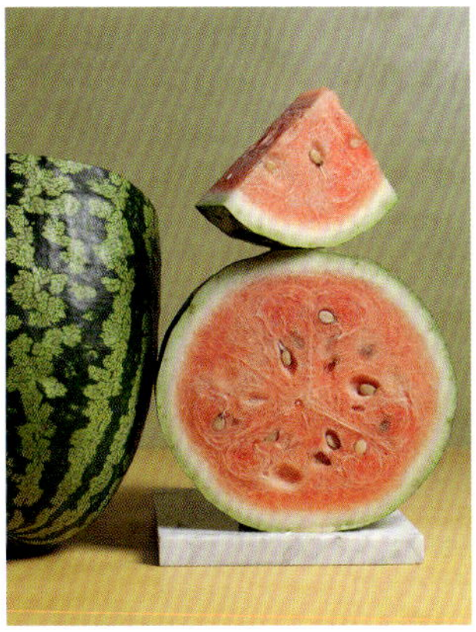

FAIRFAX

CITRULLUS LANATUS

SIZE: 26" LONG BY 10" WIDE; WEIGHT: 42 POUNDS
SUGAR CONTENT: 11° BRIX
MATURITY: LATE
SEED SOURCES: SAND, SSE

A watermelon of the highest caliber and beautiful to boot. I can't resist its bold stripes, symmetry and heft, and superbly sweet, crisp, pink (madder red) flesh. Add to the mix its resistance to fusarium wilt and anthracnose (races 1 and 3) and it's almost unbeatable. Recommended for shipping and home growing.

Fairfax was developed by C. Fred Andrus, a breeder at the USDA, in Charleston, South Carolina, and introduced in 1952. Parentage: Garrison x (African x Iowa Belle) x (Leesburg x Hawkesbury).

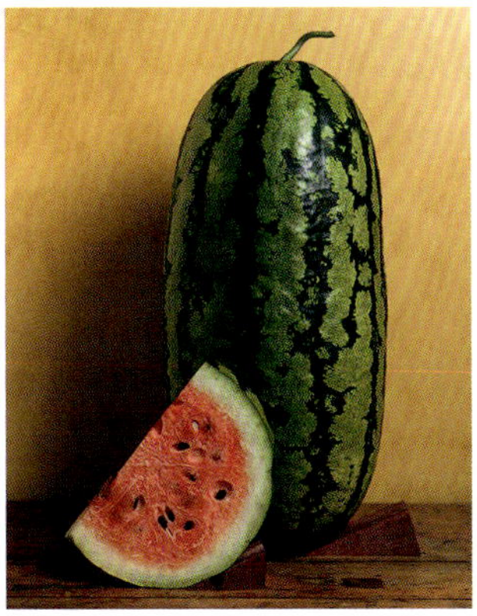

GEORGIA RATTLESNAKE

CITRULLUS LANATUS

SIZE: 22" LONG BY 9" WIDE; WEIGHT: 43 POUNDS
SUGAR CONTENT: 10° BRIX
MATURITY: LATE
SYNONYMS: AUGUSTA RATTLESNAKE, BRANCH'S RATTLESNAKE, GENUINE AUGUSTA RATTLESNAKE, GIPSY, GYPSY, JACKSON, RATTLESNAKE, RATTLE SNAKE, SOUTHERN, STRIPED GYPSEY, STRIPED GYPSY
SEED SOURCES: BAKE, BUR, CLIF, EDEN, ERS, GUR, HEIR, MARY, SHUM, SOWT, SUS, TSK, VICT

A noble watermelon, literally worth its salt (a sprinkle improves the taste of even the sweetest melon), in Allen Lacy's estimation (1984). But "it must be chilled before it's fit to eat, and it's no mean thing to chill a watermelon. . . . A thirty-pound melon can be cooled in a metal washtub with fifty pounds of ice and a bath towel on top for insulation. Four or five melons can be chilled at a time by putting them in a bathtub for a couple of days with three hundred pounds of ice and a little rock salt to hasten things along." Lacy had no taste for modern icebox types: "tiny parvenu rivals, with cute little names, minimal aspirations, and modest claims on space in the refrigerator."

I always wondered why Georgia Rattlesnake wasn't featured in any of Fearing Burr's three books on American garden vegetables (1863, 1865, and 1866). Now I realize it probably just missed the boat: P. J. (Prosper Julius) Berckmans's write-up, "Notes on Melons," in *Southern Cultivator*, which included Rattle Snake, was penned on April 3, 1866.

Here is Berckmans's B-list of watermelons: New Zealand or Tatooed, Ravenscroft, Pomaria, Turner, Odell's Large White (see page 238), Hale or Hales, and Bradford. Rattlesnake belongs to the A-list. (See more on page 148.)

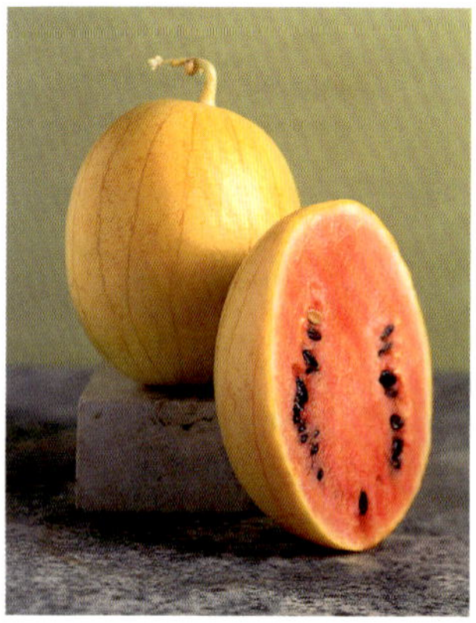

GOLDEN MIDGET

CITRULLUS LANATUS

SIZE: 6" LONG BY 5" WIDE; WEIGHT: 3½ POUNDS

SUGAR CONTENT: 9° BRIX

MATURITY: EARLY

SYNONYMS: EARLY GOLDEN MIDGET, NEW HAMPSHIRE GOLDEN MIDGET

SEED SOURCES: ANN, BAKE, EDEN, ERS, EVER, GOU, GRETA, HEIR, HER, HIGH, HUD, HVSC, MARY, REIM, REST, ROHR, SAND, SESE, SOLA, SSE, SSEC, SUS, TERR, TRADE, TSK, WHA

I get a kick out of this colorful watermelon. It's pleasant and refreshing; makes a beautiful sorbet and salad. Novel in its day (introduced in 1959). Resembles one of its parents in several particulars: New Hampshire Midget (see page 237). Still has a place in northern vegetable gardens. That doesn't mean there isn't room for improvement: I wish it were sweeter, had fewer large seeds, and no chlorosis of leaves. Golden Crown F1—the 1991 AAS Edible/Vegetable Winner—accomplishes all of that. And the fruit is yellow out of the gate (rather than green). (See more on page 178.)

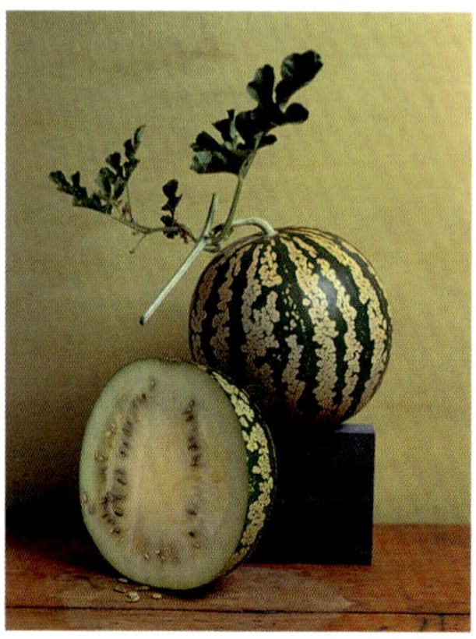

GREEN-SEEDED CITRON

CITRULLUS AMARUS

SIZE: 8" LONG BY 8" WIDE; WEIGHT: 9 POUNDS

SUGAR CONTENT: 2½° BRIX

MATURITY: LATE

SYNONYMS: CALIFORNIA PIE-MELON, COLORADO PRESERVING, PIE MELON, TEXAS PIE MELON

SEED SOURCES: GRETA, LAND, PRAI, SSE, TRADE

Green-seeded citrons, of various shapes and rind-color patterns, were commonly used in nineteenth-century America for making pies and preserves. They were preferred by many over the red-seeded "ordinary" citrons for their larger size and better quality: "The preserving qualities are the very finest; it makes beautiful, clear, nearly transparent preserves, of surpassingly fine flavor," said Cole's Vegetable Seeds in 1892.

Fearing Burr provided a recipe for Pie-melon, also known as California Pie-melon, in 1863: "After removing the rind, cut the flesh into pieces of convenient size, and stew until soft and pulpy. Lemon-juice, sugar, and spices should then be added; after which, proceed in the usual manner of making pies from the apple or any other fruit."

An early member of the Seed Savers Exchange by the name of Susan Cox from Larson, North Dakota, sent an appeal to members in 1979 for pie melon seed, confiding that Old-Timers used to grow them to fool kids who stole watermelons.

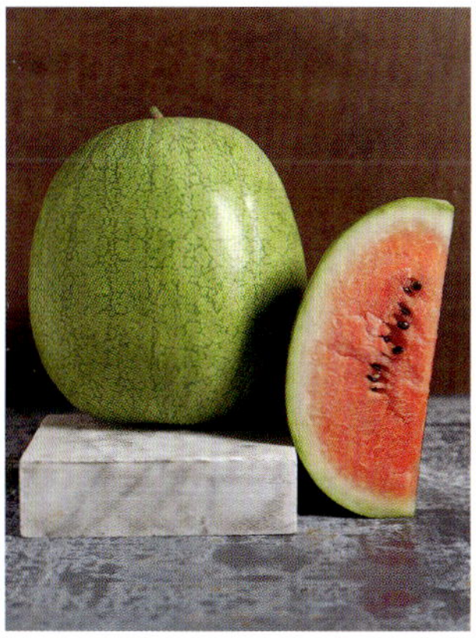

GROVER DELANEY

CITRULLUS LANATUS

SIZE: 9" LONG BY 8" WIDE; WEIGHT: 10 POUNDS

SUGAR CONTENT: 9½° BRIX

MATURITY: EARLY

SEED SOURCE: SSE

This was the last watermelon (or indeed melon) to be photographed for the book. And it almost didn't get included. Having realized the omission (and in a panic), I sowed its seeds on June 22, 2018, harvested the first mature fruits on September 19, and presented one fine fellow to Victor's camera the same day. Do the math: this variety ripens its fruits very early.

Grover has found favor with home growers and seed savers in the Northern tier of states, Canada, and cool foggy areas of the Pacific Northwest. The pink flesh, embedded with but few small brown seeds (the size of Sugar Baby's; see page 241), is sweet and firm enough, if stringy.

Named for Grover Delaney of Baker, Montana, who gave seeds to August F. Schuman of Bassett, Nebraska; August gifted seeds to me in 2006.

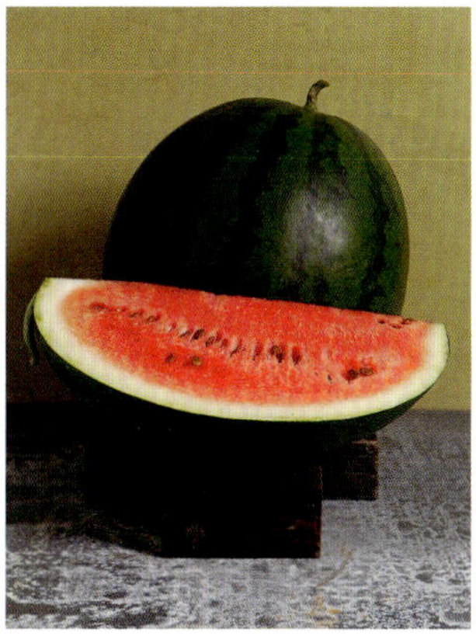

IOPRIDE

CITRULLUS LANATUS

SIZE: 12" LONG BY 11" WIDE; WEIGHT: 24 POUNDS

SUGAR CONTENT: 11° BRIX

MATURITY: LATE

SYNONYM: PRIDE OF IOWA

SEED SOURCES: SAND, SSE, VICT

Iowa *should* be proud. A high-quality watermelon with an attractive rind color pattern (dark green jagged stripes on a medium green background) and sweet, pleasing, firm, orangey red flesh. The kind you almost can't get enough of. When you've tanked up on slices, make juice or ice pops . . . and share the love.

Bred by Lewis Edwin Peterson at the Muscatine Island Research Farm, operated by Iowa State University, located in Fruitland, Iowa. Pedigree: Blackstone x H. R. Garrisonian. Iopride was officially released to the commercial seed trade on May 10, 1976, by the Iowa Agricultural Experiment Station.

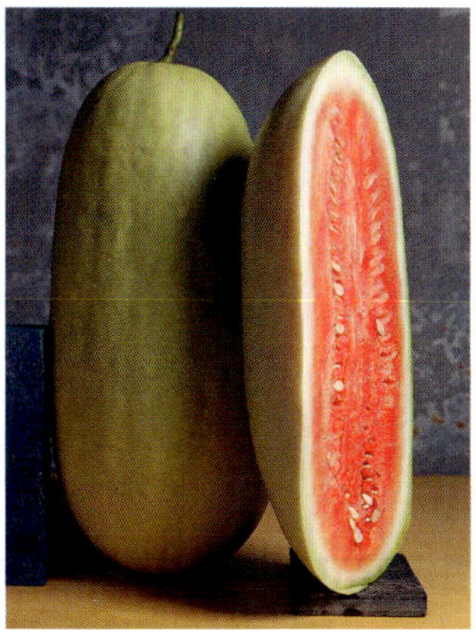

IRISH GREY

CITRULLUS LANATUS

SIZE: 15½" LONG BY 7" WIDE; WEIGHT: 16 POUNDS

SUGAR CONTENT: 12° BRIX

MATURITY: LATE

SYNONYM: IRISH GRAY

SEED SOURCES: BAKE, SSE

A top-notch watermelon with very sweet and crisp pink (madder red) flesh. Plants produce what feels like tons—especially when you're lifting and carrying—of uniformly cylindrical gray (pale green) fruits. Leave them on the vine to catch that last surge of sucrose—they seem impervious to sunscald.

According to Whitaker and Jagger (1937), "The original Irish Gray is reported to have been a volunteer plant found growing in an unoccupied stock-feeding pen by Charles Renew, of Rebecca, Ga., in 1913. Some believe the chance seed from which the variety was increased was brought in with feed imported from South America, but there is no proof of such an origin."

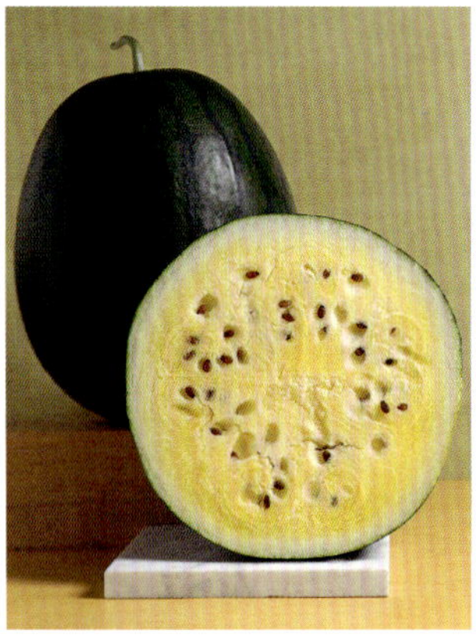	**JANOSIK** *CITRULLUS LANATUS* **SIZE: 13" LONG BY 11" WIDE; WEIGHT: 26 POUNDS** **SUGAR CONTENT: 10½° BRIX** **MATURITY: EARLY** **SEED SOURCES: BAKE, UPRI, SOTE, SSE**

A good-looking watermelon from Poland with medium green rind, narrow dark green stripes, and canary-yellow flesh. Sweet, crisp and juicy, with small brown seeds. Produces its first mature fruit in 60 days from transplant—quite extraordinary considering its size—and keeps on giving. Designed to be grown in the open field or cultured in tunnels. Named for Juraj Janosik, the legendary Polish or Slovak Robin Hood–like figure and hero to the oppressed, born in 1688 in what is now modern Slovakia. Janosik watermelon was introduced to members of Seed Savers Exchange in 2007 by Andrey Baranovski of Minsk, Belarus, who said it was the first sweet yellow-fleshed Polish variety to be developed. The seed source was PNOS, one of the largest seed companies and distributors in Poland, located in Ozarow Mazowiecki.

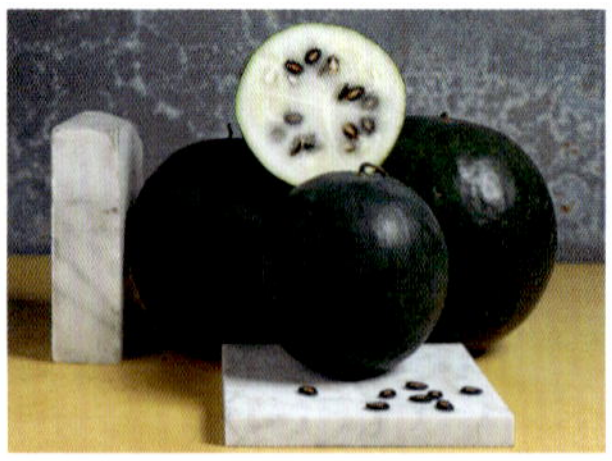

KILLY EDIBLE-SEEDED

CITRULLUS LANATUS

SIZE: 9¾" LONG BY 8¼" WIDE; WEIGHT: 11 POUNDS
SUGAR CONTENT: 6° BRIX
MATURITY: HARVEST LATE FOR MATURE SEEDS
SYNONYMS: KILLY EDIBLE SEED, KILY EDIBLE, WANLI, WANLI EDIBLE SEED F2
SEED SOURCE: SSE

A confectionary type with large seeds for snacking. Try it—you'll like it, if you can find it. (See more on page 160.)

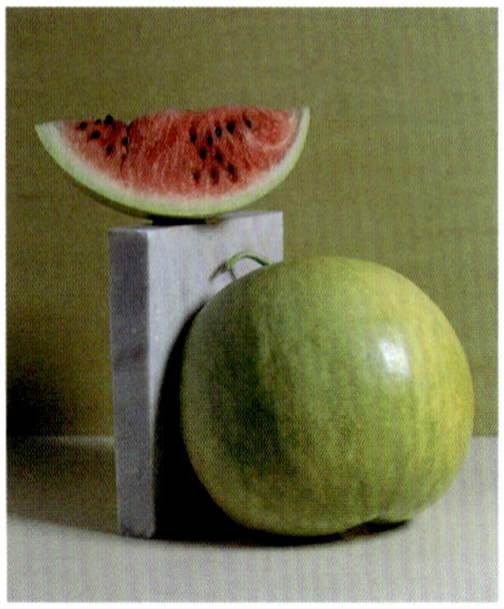

KING AND QUEEN WINTER

CITRULLUS LANATUS

SIZE: 9" LONG BY 9" WIDE; WEIGHT: 12 POUNDS
SUGAR CONTENT: 10° BRIX
MATURITY: MIDSEASON
SYNONYMS: BLACK SEEDED WINTER QUEEN, CHRISTMAS, KING WINTER, QUEEN WINTER, WINTER, WINTER KEEPER, WINTER KING AND QUEEN, WINTER MELON, WINTER QUEEN, WINTER WATERMELON
SEED SOURCES: ADAP, BAKE, GRETA, SAND, SSE, TRADE, VICT

A well-rounded watermelon with gray (yellowish-green) rind (faintly striped) and sweet, crisp, pink flesh embedded with black seeds—perhaps too many. (A variant with pinkish seeds was catalogued in the 1930s.) Cultivars such as this are less susceptible to sunburn than are dark-rinded ones. Noteworthy for its long shelf life: When harvested slightly immature before frost, and stored in cool, dry conditions, King and Queen can remain edible for about three months.

Aggeler & Musser Seed Company of Los Angeles first catalogued it as Winter Melon in 1918: "This new melon is slowly but surely forcing itself into favor because of its many good qualities which are not apparent at first acquaintance. Like the Casaba, which required 12 years to establish its reputation, the Winter Melon might easily be mistaken for a small white citron and the peddler must be convinced he in turn must convert his customers to the knowledge that the Winter Melon is a thing of great luxury. . . . The flavor when cold is sweet, crisp and surprisingly delicious." Catalogued by Oscar H. Will & Co., Bismarck, North Dakota from 1925 to 1959 (first as Winter Watermelon, then Queen Winter).

Not the same as the Winter Sweet watermelon introduced by Oscar H. Will in

1933, which was described as "A fine new sort of Russian origin whose outstanding qualities caught Professor A. F. Yeager's eye and taste in his trial grounds." King and Queen Winter may have been one of the parents of Navajo Sweet, which is synonymous with Osh Kirgizia (see Osh Kirgizia, page 239).

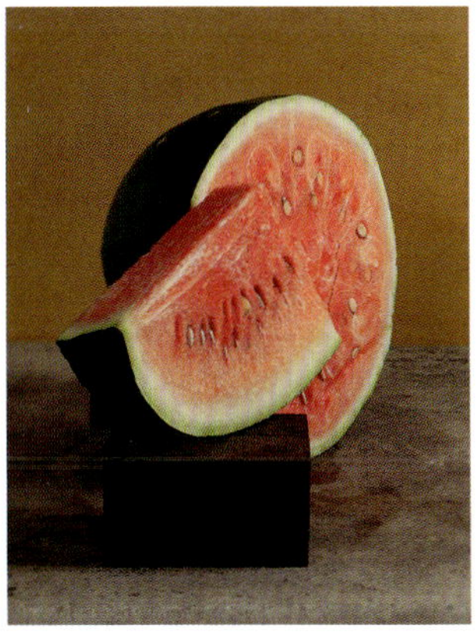

LEDMON

CITRULLUS LANATUS

SIZE: 15" LONG BY 13" WIDE; WEIGHT: 41 POUNDS

SUGAR CONTENT: 13° BRIX

MATURITY: LATE

SYNONYM: LEDNUM

SEED SOURCES: BAKE, SAND, SOWT, SSE

This is one of dozens of rare watermelon varieties whose seeds were sent to me in a shoe box via overnight mail by Seed Savers Exchange in the spring of 2000—giving me some of the raw materials for my first melon book. Yet it wasn't until twelve years later that I revisited SSE Watermelon #11. The verdict, my friends, is *sublime.* Ledmon's pink flesh is sugary, juicy, firm, and addictive. I feel a craving coming on just thinking about it.

Ledmon is a North Carolina heirloom. According to an article written by Jim Schlosser and published in the *Greensboro News and Record* in 2004, "The Lednum is named for farmer J. E. Lednum, who developed it in the early 1900s on his farm five miles south of Greensboro, near Pleasant Garden. An agriculture booklet published in 1938 by the Board of County Commissioners and the old county Board of Agriculture misspelled Lednum's name as J. C. Ledmon. The mistake has perpetuated confusion among watermelon connoisseurs. Some insist on calling the melon a Ledmon. But Roger Smith, 82, who knew J. E. Lednum and grew up in sight of his farm, vouches for the spelling and the fact Lednum created this delicious red fruit. . . . The 1938 county farm booklet hails the Lednum and another crop creation, Greeson wheat, as two of Guilford County's proudest agriculture contributions."

Lednum apparently created his eponymous watermelon by crossing Stone Mountain and Kleckley Sweet. Locals say if it isn't round, white-seeded (tan-tipped), with a brittle rind that makes it suitable only for the home garden, then it isn't a true Lednum or Ledmon. Be advised.

This cultivar was once carried by the Wyatt-Quarles Seed Company, of Raleigh, North Carolina. Glenn Drowns obtained seed from them in 1982 and offered Ledmon to members of SSE a year later. In 1992 it was reintroduced commercially by SSE member Jeff McCormack of Southern Exposure Seed Exchange. He said it was "Rare, and too good to be lost."

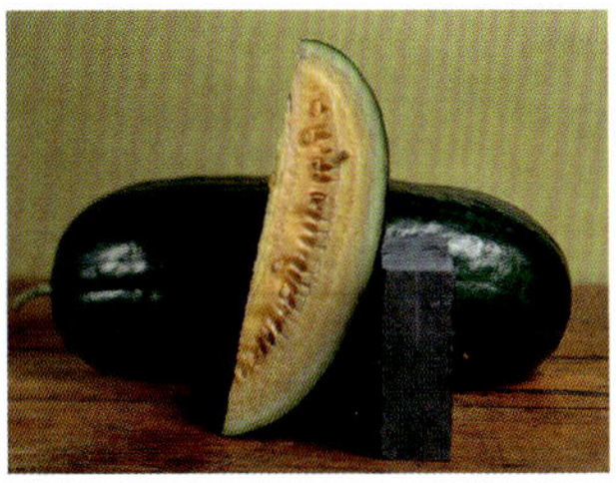

LUSCIOUS GOLDEN

CITRULLUS LANATUS

SIZE: 20" LONG BY 9" WIDE; WEIGHT: 23 POUNDS

SUGAR CONTENT: 11° BRIX

MATURITY: MIDSEASON

SYNONYMS: GOLDEN-SWEET, LUSCIOUS GOLDEN SWEET

SEED SOURCE: SAND

You can't always read a watermelon by its rind: This unassuming green-rinded elongated watermelon holds luscious golden flesh with tan seeds. Makes for good eating and outstanding gazpacho.

Introduced by W. Atlee Burpee & Company in its fortieth anniversary catalogue in 1916 as the Luscious "Golden-Sweet" Watermelon with this copy: "For many years we have tested Watermelons with yellow flesh, but previously have always found the flavor insipid. We were surprised and pleased, therefore, during the season of 1913, to find among the hundreds of trials of Watermelons at our New Jersey SUNNYBROOK FARM one distinct variety with yellow flesh that fully equalled in flavor most of those with red flesh. Golden-Sweet resembles our famous *Kleckley Sweets* in form. . . . The rind is dark green and the bright golden-yellow flesh extends very close to the rind. The meat is of a luscious sugary flavor and is of so firm a texture as to insure the fruits arriving in good condition on distant markets. The melons are of medium size, averaging about twenty pounds in weight. A really delicious yellow-fleshed watermelon is a novelty that will create remark when tasted on the table—just like that "*Chunk of Sweetness*," our famous *Golden Bantam Corn*."

Luscious Golden was eventually dropped from commercial production; in 2001 it was reintroduced by Sand Hill Preservation Center. SHPC proprietor Glenn Drowns obtained seed stock from the USDA seed collection in 1986—proving its value (and his) for the nth time.

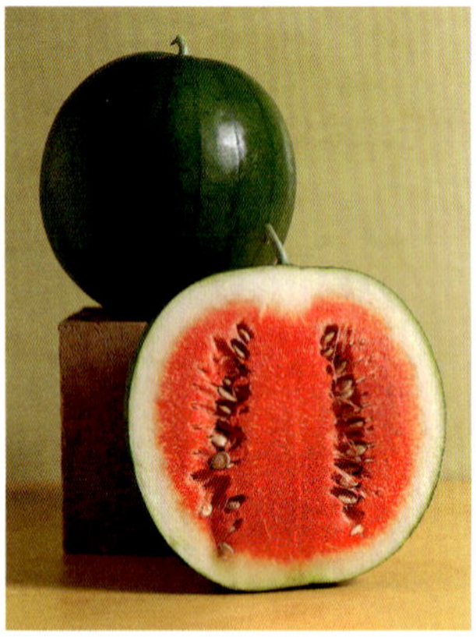

MALALI

CITRULLUS LANATUS

SIZE: 9¾" LONG BY 8¾" WIDE; WEIGHT: 15 POUNDS

SUGAR CONTENT: 10° BRIX

MATURITY: HARVEST MIDSEASON FOR FRESH EATING; LATE FOR MATURE SEEDS

SYNONYMS: GREEN MALALI, MALALI ISRAELI

SEED SOURCE: NONE IN THE UNITED STATES OR CANADA

I shrieked with happiness when I cut open my first Malali for the camera and found a perfect picture: vivid-red coarse flesh with black-tipped tan seeds for all the world to see. When chowing down on it that night, I found the flesh sweet, juicy, and crisp. Later I realized Malali was also grown for "confectionary" purposes: seed consumption. I'm snacking on Malali seeds (unroasted) as I write these lines and can affirm they're good—and easier for the novice to extract with the teeth than Killy's seeds (see page 234).

Malali is an improved variety developed from a rogue in a local landrace in Israel and commercialized around 1940. It is adapted to dryland conditions. According to Paris, Cohen and others (2013), it was "thought to have been selected in a commercial field at Kefar Malal, a semi-private community, *moshav*, in the Sharon Plain of central Israel, by two residents named Silberg and Burnstein." It may "have been derived from a chance cross between a local variety and Tangela . . . or a chance crossing between Mahasni and Chilean Black, which Malali resembles externally."

What's indisputable is that Malali become the most widely grown and popular watermelon variety in Israel in the late 1950s. It was later supplanted by cultivars such as Crimson Sweet (see page 230), which are adapted to full irrigation. Malali is still grown in Israel—primarily for its nutritious seeds.

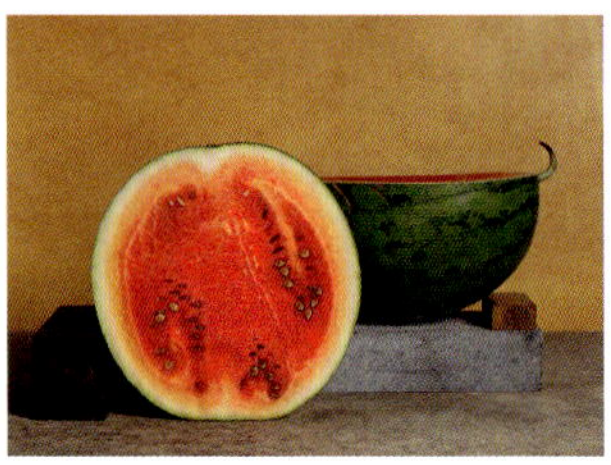

MAR-JOHN

CITRULLUS LANATUS

SIZE: 11" LONG BY 11" WIDE; WEIGHT: 25 POUNDS

SUGAR CONTENT: 12° BRIX

MATURITY: LATE

SYNONYMS: MARJOHN, MAR-JON

SEED SOURCE: NONE IN THE UNITED STATES OR CANADA

A beautiful and delicious dark-pink-fleshed watermelon, from Baku, Azerbaijan, with an intermittently striped rind pattern. Similar in outward appearance to Osh

Kirgizia (see page 239), but Mar-John's dark green stripes are displayed on a medium green background.

Seed was obtained by Kent Whealy of Seed Savers Exchange in 1991 either at or subsequent to a meeting at an environmental conference in Moscow. He met a group of scientists there from the Institute of Genetics and Selection near Baku who told him they had collected endemic seeds from all over Azerbaijan. One of the scientists confessed that he saw those seeds as living books reaching back through time: "Putting his palms together, he opened them slowly and gazed into an imaginary book, then caught himself and dropped his eyes, surprised and slightly embarrassed at having revealed such deep feelings so quickly."

Glenn Drowns offered seeds to members of SSE in the 1990s, and the variety still grows in home gardens like mine. I hope Mar-John will become commercially available in the near future.

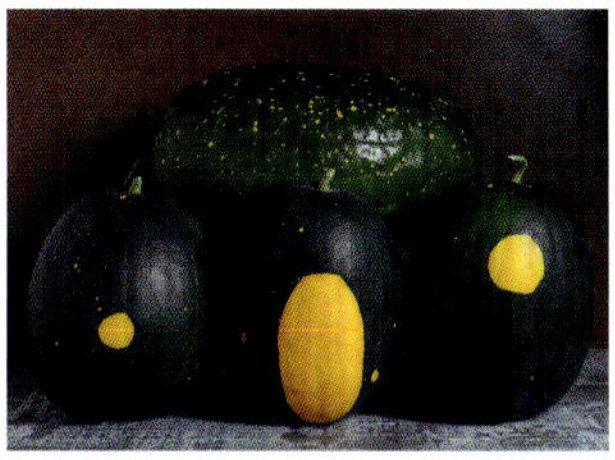

MOON AND STARS

CITRULLUS LANATUS

SIZE: 13" LONG BY 11½" WIDE; WEIGHT: 34 POUNDS

SUGAR CONTENT: 11° BRIX

MATURITY: MIDSEASON

SYNONYMS AND STRAINS: AMISH, CHEROKEE, AND LONG MILKY WAY MOON AND STARS; LUNE-ÉTOILES; MOON AND STARS #1 AND #2; NORTH STAR; PINK-FLESHED MOON AND STARS; PLANETS AND STARS; YELLOW-FLESHED MOON AND STARS; SUN, MOON, AND STARS; VAN DOREN'S MOON AND STARS

SEED SOURCES: ANN, BAKE, BERL, BUR, CLIF, EDEN, ERS, EVER, FED, GOU, HEIR, HER, HIGH, HOPE, IRISH, JUNG, LAND, MOR, NEWE, PEACE, PINE, REDW, REIM, REN, RICH, ROHR, SALT, SAMP, SAND, SESE, SHUM, SISK, SOTE, SOWT, SSE, SSEC, STC, SUS, TERR, TERRIT, THOM, TRADE, TSK, URB, VICT, WHA

My pen has run dry on this one. (See more on page 152.)

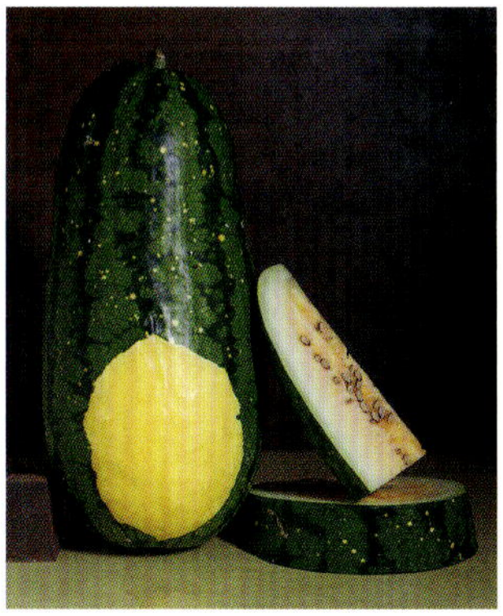

MOON AND STARS AND STRIPES

CITRULLUS LANATUS

SIZE: 15" LONG BY 8" WIDE; WEIGHT: 27 POUNDS

SUGAR CONTENT: 11° BRIX

MATURITY: MIDSEASON

SEED SOURCE: NONE IN THE UNITED STATES OR CANADA

A surprise in my 2012 garden—the progeny of a serendipitous cross between Moon and Stars and Orangeglo (see pages 237 and 239, respectively). Moons and stars overlap both the dark green stripes and the medium green background. Sometimes there's a lollapalooza of a star. Am trying to stabilize the variety so that it breeds true from seed. Divinely sweet and satisfying orange flesh.

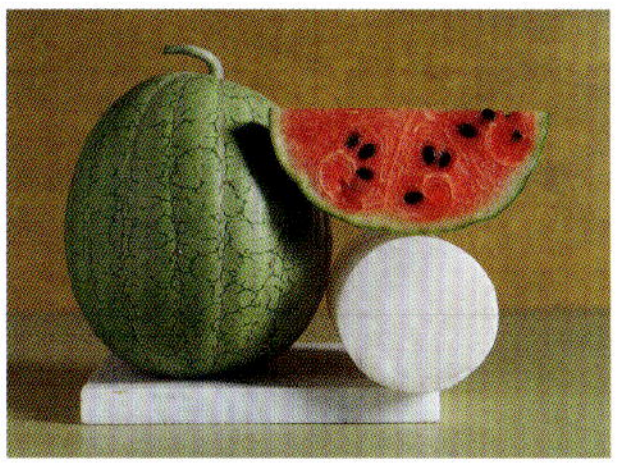

NEW HAMPSHIRE MIDGET

CITRULLUS LANATUS

SIZE: 6½" LONG BY 6" WIDE; WEIGHT: 3½ POUNDS

SUGAR CONTENT: 10° BRIX

MATURITY: EARLY

SYNONYM: N. H. MIDGET

SEED SOURCES: SAND, SSE

One of the earliest and most prolific watermelons I grow. Matures more quickly than you'd imagine (65 to 70 days from seed) and should be harvested promptly before it becomes overripe. The watermelon's breeder, Dr. Albert F. Yeager, at the University of New Hampshire Agricultural Experiment Station, once said it was very easy to determine ripeness: The gray rind of a ripe N. H. Midget is hard and crisp when pressed by the thumb (and not springy like an unripe one). Yeager selected from some of the same breeding stocks used to develop White Mountain watermelon: Favorite Honey x Dakota Sweet. The former is a small, oval, thin-rinded, yellow-fleshed variety of Japanese origin with excellent quality; the latter a red-fleshed variety selected from the Russian introduction Winter Sweet.

New Hampshire Midget made its commercial debut in 1951—winning an AAS Gold Medal in the Edible/Vegetable category that same year. Of historical significance, this cultivar has dwindled in popularity over the years because of its explosive rind (take a knife to it and boom!), numerous large black seeds, so-so flesh quality, and lack of disease resistance. But it was once literally "The Talk of the Town" in *The New Yorker* (February 18, 1950) for its small size "suited to the cramped quarters of the average modern family" and ability "to mature within the length of a Northern summer and thus break the watermelon monopoly of the South." One watermelon was sliced and divided equally among five magazine staff members: they unanimously pronounced it delicious. Yeager suggested that Midgets could be grown "on a New York roof garden if trellised, well watered, and well handled generally." I'm not sure if anyone ever took him up on that.

Offshoots include Golden Midget (see page 232), Early Midget (less furrowed than N. H. Midget), Market Midget (N. H. Midget x Winter Queen), and Petite Sweet (Crimson Sweet x N. H. Midget).

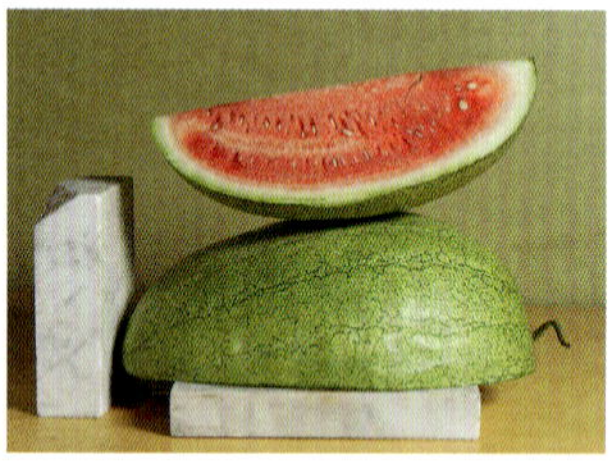

ODELL'S LARGE WHITE

CITRULLUS LANATUS

SIZE: 14½" LONG BY 11¼" WIDE; WEIGHT: 31 POUNDS
SUGAR CONTENT: 12° BRIX
MATURITY: MIDSEASON
SYNONYMS: CALIFORNIA, EARLY CALIFORNIA, IMPROVED ODELLA, ODELL WHITE, ODELL'S WHITE, ODELL'S LARGE ROUND, ODELLA, STONEY MOUNTAIN, WHITE STONEY MOUNTAIN
SEED SOURCES: SESE, SOWT

This variety originated in the mid-nineteenth century on Adam Summer's Ravenscroft plantation, located in central South Carolina. Pomologist William D. Brinckle was Fearing Burr's informant for the varietal description of Odell's that appears in three of his books on American garden vegetables (1863, 1865, and 1866). Contained in a biographical memoir of Brinckle (Gardette 1863), is a list of watermelon varieties either produced, described, or contributed by him to the horticultural world: Mitchell, Lawson, Nelson, Turner, Summer, Pomaria, Orange, Kale, Souter, Bradford, Odell's Large White, Tatooed, Imperial, Stillman, Ravenscroft, Florida, Elerbe's Mountain Sprout, Clarendon, Dark Speckled, and Long Mottled.

Brinckle was also Downing's informant for the description of Odell's Large White that first appears in *The Fruits and Fruit Trees of America* in 1857: "Very large, round, skin grey, with green net-work. Rind three quarters of an inch thick. Flesh pale red, of a 'very good' quality. Keeps a long time after being gathered Originated with Col. A. G. Sumner [*sic*], South Carolina." P. J. Berckmans weighed in on Odell's in 1866: "From S. C. Round, large, white, rind thin; flesh pale red, crisp, sugary, but leaving a sediment in the mouth; seeds large, black, and very few in a fruit; is apt to be hollow at the core."

Odell's is evidently the same as the Odella Water Melon catalogued in 1881 by James J. H. Gregory as early, large, weighing twenty-five to fifty pounds, of light green color and round form. Peter Henderson carried it in 1883 as California or Improved Odella. In 1885, Burpee catalogued it as Early California or Improved

Odella: "Of recent years this has become a very popular market variety. Shaped like Odella; color light green; thin rind, but tough shell, so that there is scarcely any breakage in shipping to market. It is next to *Ice Cream* in fine flavor, and grows readily to 40 lbs. weight; has been grown to weigh 60 lbs. It ripens *very early*."

Similar to the Chelsea watermelon of Iowa; resembles a watermelon that is traditionally dryland-farmed in the Middle East, according to Harry Paris. Fares well when unirrigated and summer rains are scarce, says James E. Kibler, expert on the Summers of Pomaria and Ravenscroft plantations. Odell's White was reintroduced in 2018 by Sow True Seed from Kibler's seed and by Southern Exposure Seed Exchange from seed grown by Rodger Winn and shepherded by the Metze family for generations. The watermelon was boarded onto Slow Food's Ark of Taste in 2018. (See more on page 144.)

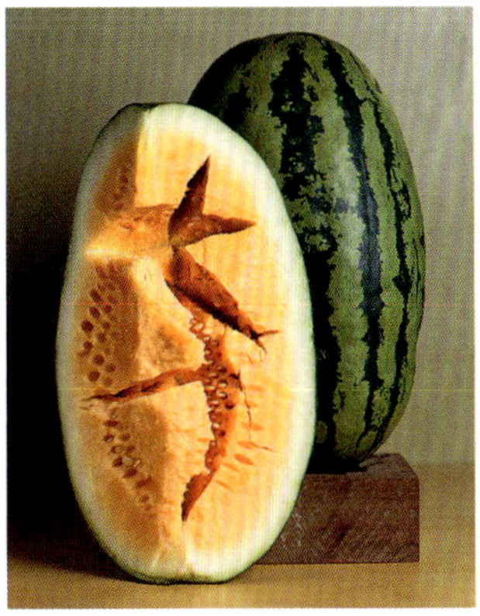

ORANGEGLO

CITRULLUS LANATUS

SIZE: 15" LONG BY 9" WIDE; WEIGHT: 22 POUNDS

SUGAR CONTENT: 10° BRIX

MATURITY: MIDSEASON

SYNONYMS: ORANGEGLO, ORANGE GLO

SEED SOURCES: BAKE, BERL, FED, HIGH, MARY, REST, ROHR, SAND, SESE, SSE, SSEC, TERR, TRADE, TSK, WHA

This watermelon turns heads: Its orange flesh is electrifying—as well as sweet and refreshing. Makes a great mimosa, as you might imagine. Contributed some of its genetics to my Moon and Stars and Stripes (see page 237).

Introduced by the Willhite Seed Company of Poolville, Texas, in the early 1960s, and was long one of their signature offerings.

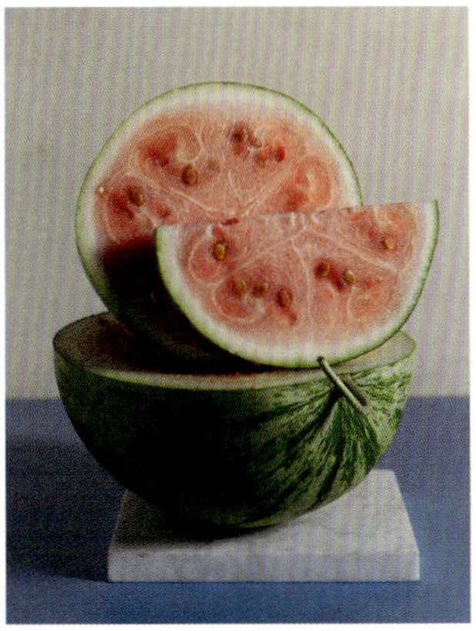

OSH KIRGIZIA

CITRULLUS LANATUS

SIZE: 10¼" LONG BY 10" WIDE; WEIGHT: 18½ POUNDS

SUGAR CONTENT: 11½° BRIX

MATURITY: MIDSEASON

SYNONYMS: NAVAJO, NAVAJO RED, NAVAJO SWEET, OSH KHIRGIZH, OSH KHIRGIZIA, OSH KIRGHIZIA

SEED SOURCES: AMI, SSE

Named for a city in Kyrgyzstan, from whence it presumably came. Seed Savers Exchange imported Osh Kirgizia seeds from Russia and introduced the variety commercially (spelled as Osh Khirgizia) in its spring 1998 catalogue; seeds had circulated as early as 1989 among SSE members as Osh Khirgizh. Osh Kirgizia is synonymous with Navajo Sweet, introduced by D. V. Burrell Seed Growers, of Rocky Ford, Colorado, in 1991.

Both Osh Kirgizia and Navajo Sweet have an unusual striping pattern. Navajo Sweet is the type line for the gene for intermittent stripes (*ins*). This recessive genotype "produces narrow dark stripes at the peduncle [stem] end of the fruit that become irregular in the middle and nearly absent at the blossom end of the fruit" (Gusmini and Wehner 2006). The same striping pattern can be found on Mar-John watermelon (see page 236). Osh Kirgizia's lovely reddish-orange seed coat color (which fades in storage) is an added attraction when spitting seeds from ripe flesh. Fruits are sweet and delicious.

A handwritten note on a plain seed packet saved by Glenn Drowns provides an insight into the development of Navajo Sweet—we don't know the pedigree of Osh

Kirgizia. The packet was sent to Glenn in August of 1989 by Wade Wofford, who was a collector of Native American varieties. In Wade's scrawl: "Navajo Sweet. Bred out of King and Queen [see King and Queen Winter, page 234] crossed to a regular watermelon and then selected since 1973 by a breeder at the Rocky Ford Seed Company. Obtained from a feed and seed store manager in Farmington, New Mexico, who was a friend of the breeder." It seems Wade was on to something. Here's Burrell's 1991 catalogue copy: "A new 10-20 pound melon with unusually sweet flavor exhibiting shiny dark red seeds in firm red flesh. Fruit shape is round, similar to Winter Queen [synonymous with King and Queen], with small dark green striping on a pale green background. Rind thickness ranges from thin to medium. Very good shelf life has been noted in this garden-roadside variety. Disease resistance has not been categorized yet."

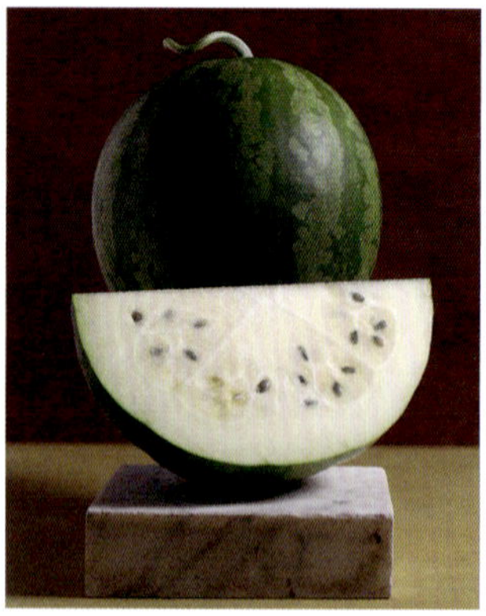

PASTÈQUE À CONFIRE À GRAINE VERTE

CITRULLUS AMARUS

SIZE: 10" LONG BY 10" WIDE; WEIGHT: 18 POUNDS
SUGAR CONTENT: 2° BRIX
MATURITY: MIDSEASON
SYNONYMS: À CONFIRE À GRAINE VERTE, GREEN SEEDED CITRON
SEED SOURCE: SAND

A perfectly round green-seeded French variety with a grayish-green rind and deep green stripes. I use a microwave to soften its flesh to express juice for Brix testing. That's one way to get water from a citron—the other, more time-tested way (for millennia) is to cut it open and pound it.

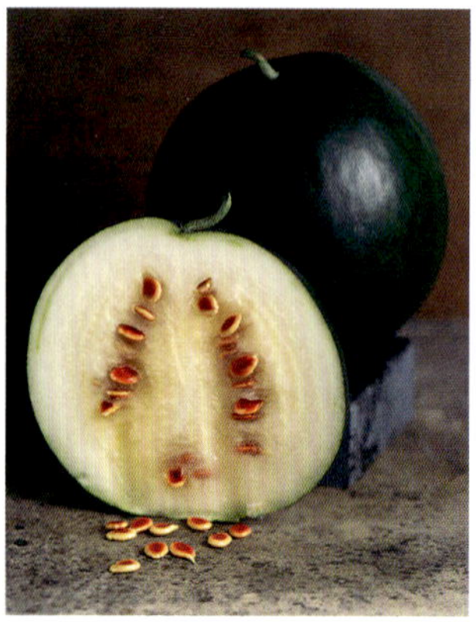

RED-SEEDED CITRON

CITRULLUS AMARUS

SIZE: 7" LONG BY 7" WIDE; WEIGHT: 6 POUNDS
SUGAR CONTENT: 2° BRIX
MATURITY: LATE
SYNONYMS: À GRAINE ROUGE, CALIFORNIA CITRON, CITROUILLE D'ESPAGNE À CONFITURE, COLORADO PRESERVING, PRESERVING MELON, RED-SEEDED ROUND
SEED SOURCES: BAKE, HER, PRAI, REIM, REST, SAND, SSE, SSEC, WHA

Citron watermelons generally don't set fruit until hot weather arrives in mid-August—then they take off like wildfire. Some citron varieties I've grown have seeds that retain their gorgeous red color for many years. Red-seeded citrons, whatever the rind color pattern, have become a relict crop in America, as have their green-seeded counterparts. The number of vendors offering seeds in the United States and Canada is on a downhill slope.

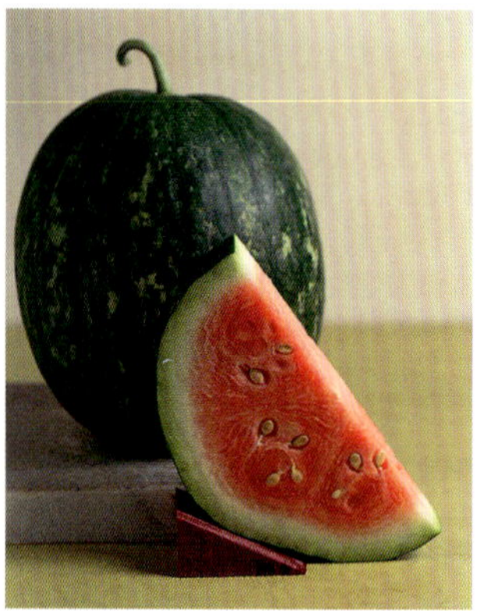

SCALY BARK

CITRULLUS LANATUS

SIZE: 14½" LONG BY 11" WIDE; WEIGHT: 37 POUNDS
SUGAR CONTENT: 9° BRIX
MATURITY: MIDSEASON
SYNONYMS: SCALEY BARK, WHITE SCALEY BARK, WHITE SCALY BARK
SEED SOURCES: BAKE, HER, SAND, SESE, SSE

I grow Scaly Bark for its unusual greenish-white marbled rind pattern, which appears over a darker green pattern, and not for its pink flesh, which is fair at best (although when semi-frozen it makes a delicious slushy). The thick white rind (mesocarp) is used for rind preserves or pickles in the South. The Scaly Bark I grow is white-seeded; it's also known as White Scaly Bark (reintroduced by R. H. Shumway's, of Graniteville,

South Carolina, in 1997). Other variants have black or brown seeds.

Scaly Bark was introduced by Peter Henderson & Company, of New York, in 1883. The catalogue copy reads: "A new variety that reached us too late for insertion under its proper heading of 'Novelties and Specialties.' It was first exhibited at the Atlanta Exhibition in 1881. The skin is dark green and looks as if covered with fish scales although quite smooth. It is said to be an unusually productive sort, the average weight of the melons being 35 pounds. . . . The flesh is light crimson, solid, tender, and of exquisite flavor, but its peculiar value lies in the fact that it remains in choice eating condition from ten to fifteen days after being pulled. This with its very tough yet thin rind will make it a most valuable sort for shipping purposes." It merited special attention in the 1885 catalogue with a large engraved field view of Scaly Bark watermelons, the garden crew in the background, and a notation that these melons could "bear any amount of handling without bursting or cracking open."

Scaly Bark's rind pattern resembles that of Phinney's, introduced by James J. H. Gregory in 1868, and the Nabob, introduced by the Livingston Seed Company in 1895.

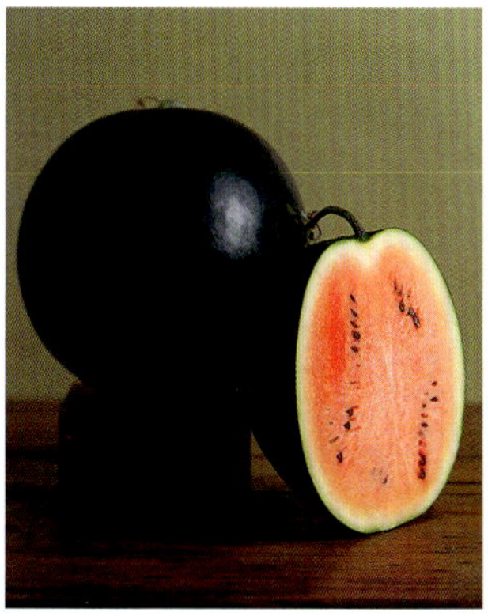

SUGAR BABY

CITRULLUS LANATUS

SIZE: 7" LONG BY 7" WIDE; WEIGHT: 6 POUNDS

SUGAR CONTENT: 9° BRIX

MATURITY: MIDSEASON

SYNONYMS: ICEBOX, ICEBOX MIDGET, SUGAR BABE, TOUGH SWEETS

SEED SOURCES: ANN, BAKE, BERL, BOT, BUR, BURR, CLIF, DEB, EDEN, ERS, EVER, FED, GOU, GRETA, GUR, HALI, HAR, HEIR, HIGH, HOPE, HUD, HUME, HVSC, IRISH, JOHN, JOR, JSKIT, JUNG, LEJ, LIN, MAIN, MARY, MOR, NAT, NEWE, NIC, OSC, PEACE, PINE, PRAI, REST, ROHR, SAMP, SESE, SHUM, SOC, SOLA, SOTE, SOWT, SSE, STC, SUS, SWAL, TERR, TERRA, TERRIT, TRADE, TSK, TTS, TUR, URB, VER, VICT, WHA, WIL

A familiar figure in the icebox: perfectly round, black green, and waxy at maturity (medium green with darker narrow stripes when immature). Fills a need for a small, none-too-heavy watermelon of good enough quality that's easy to manage in the kitchen, and can be eaten without inviting a crowd. A dependable sort that can be grown just about anywhere. Its tough rind makes this baby a natural shipper.

Sugar Baby was introduced by Woodside Seed Company in 1959 and developed in Geary, Oklahoma, by M. Hardin, who also sent Pumpkin Rind watermelon to Dr. Yeager at UNH, enabling the development of Golden Midget (see page 232). Hardin selected from a pre-war Japanese variety once catalogued under the name "Tough Sweets," according to C. Fred Andrus (1955), a breeder at the USDA in Charleston, South Carolina. And Sugar Baby was inbred for thirteen years. It is considered synonymous with Tough Sweets, though I haven't had the pleasure. It's unlikely, as some have claimed, that Sugar Baby was selected from a "picnic" watermelon (oblong-shaped).

In 1987, W. Atlee Burpee Company, of Warminster, Pennsylvania, introduced a novel cultivar for gardeners with limited space that has become quite popular: Bush Sugar Baby, a bush-type of Sugar Baby with dwarf vines about three feet long, yielding an average of two high-quality watermelons per plant, each weighing eight to twelve pounds.

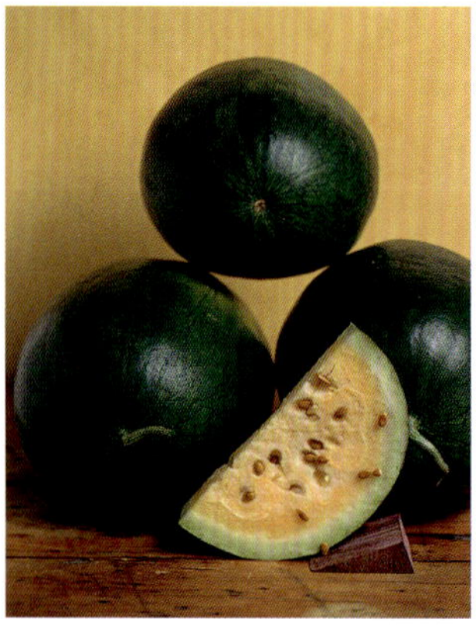

SWEET SIBERIAN

CITRULLUS LANATUS

SIZE: 13" LONG BY 7" WIDE; WEIGHT: 11 POUNDS
SUGAR CONTENT: 10° BRIX
MATURITY: MIDSEASON
SYNONYM: SIBERIAN YELLOW
SEED SOURCES: BAKE, GRETA, HER, HVSC, IRISH, NATURE, PRAI, REIM, SAND, SOLA, SSE, SSEC, TRADE

A fetching watermelon of a different flesh color—Persian orange. It is both sweet and crisp. And adapted to short-season areas.

Peter Henderson & Company of New York held a naming contest in 1898 for their novelty watermelon, known as No. 3. They introduced the watermelon as Sweet Siberian in their 1899 catalogue. The copy reads: "This variety is of Russian origin. We have been many years working it up to its present perfection, and at last have secured a distinct and valuable sort. . . . The fruits are oval in shape, medium in size; the color of the rind is dark green, with faint pencilings of a lighter shade, and though the rind is thin it is very tough. The flesh is very deep, extending to within half an inch of the rind, and is entirely different from anything ever known in a Watermelon. The color can best be described as a *rich orange buff,* and it is as sweet as any variety in cultivation. The Melon grower in our Northern latitudes is often compelled to lose a large proportion of his crop by reason of prematurely cold weather. . . . But this sort has the peculiarity of ripening fruits up quickly."

Reintroduced commercially in 1996 by Glenn Drowns, who sourced seed from the USDA. Sweet Siberian is also known as Siberian Yellow, a variety introduced by Oscar H. Will and Company in 1957 and marketed as an earlier and more flavorful alternative to the old Favorite Honey.

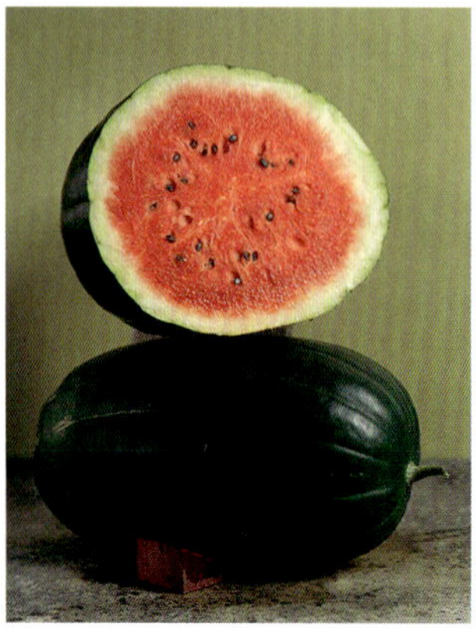

TOM WATSON

CITRULLUS LANATUS

SIZE: 19½" LONG BY 9½" WIDE; WEIGHT: 28½ POUNDS
SUGAR CONTENT: 12° BRIX
MATURITY: LATE
SYNONYMS: ALEXANDER'S SELECTED WATSON, BLUE RIND WATSON, FAMOUS TOM WATSON, RED HEART WATSON, SELECT TOM WATSON, THE WATSON, WATSON
SEED SOURCES: BAKE, ERS, HEIR, PINE, ROHR, SHUM, SSE, VICT

An absolute favorite for its sweet, solid, firm, coarse red flesh—better quality than many of the old-fashioned, furrowed, big lunkers I've cultivated. Known to produce giants. Has small black seeds; there's one with white seeds that I'm itching to try. In 1906, the Alexander Seed Company of Augusta, Georgia, introduced "The Watson" watermelon to a wider audience. "It is our good fortune this year to introduce another new melon from this Augusta section and we think so highly of it that we illustrate it on the front cover page of our catalogue. It is a long, solid, dark green, shining melon of large size with red flesh of the most tender and delicious eating qualities. It matures for main crop and is solid and firm, making it a splendid shipping and market melon as well as a most desirable one for home use." (See more on page 156.)

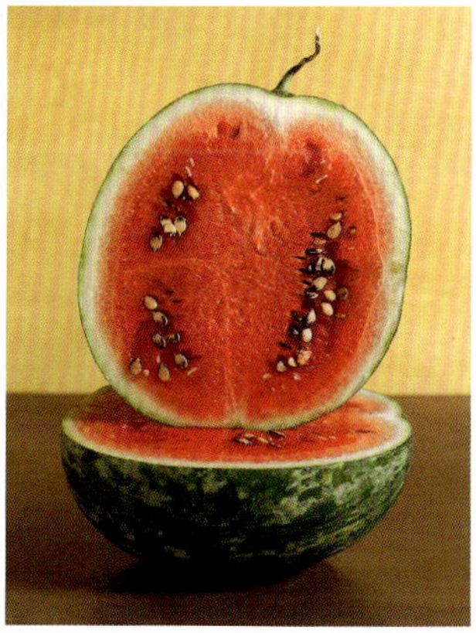

WILSON SWEET

CITRULLUS LANATUS

SIZE: 9¼" LONG BY 8½" WIDE; WEIGHT: 12 POUNDS
SUGAR CONTENT: 9° BRIX
MATURITY: MIDSEASON
SYNONYM: WILSON'S SWEET
SEED SOURCES: BAKE, REST, SAND, SESE, SHUM, SSE, WHA

A truly amazing watermelon. (See more on page 162.)

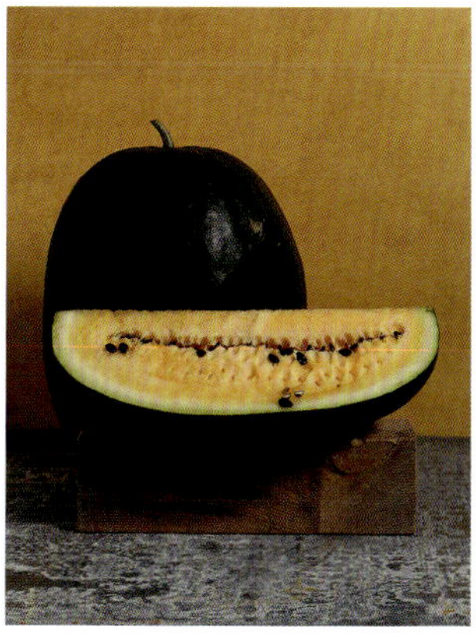

YELLOW FLESH BLACK DIAMOND

CITRULLUS LANATUS

SIZE: 14" LONG BY 10¾" WIDE; WEIGHT: 28 POUNDS
SUGAR CONTENT: 11° BRIX
MATURITY: LATE
SYNONYMS: BLACK DIAMOND, YELLOW FLESH; WILLHITE'S YELLOW FLESH BLACK DIAMOND, YELLOW MEATED BLACK DIAMOND
SEED SOURCE: REIM

This melon is a hit at my house. I've been growing it for more than twenty years. Its overall eating quality is excellent. The flesh is a light orange color I've rarely seen—about the same as the flesh of a newly opened Table Queen acorn squash. Has all the markings and makings of a Black Diamond: intensely dark green rind that is thick and tough; furrowing, potential for large size, and round shape (see Black Diamond, page 227).

Developed and introduced by Willhite Seed Company in their 1958–1959 catalogue and illustrated with a color photograph on the inside back cover. Copy reads: "Strike gold with Willhite's new Yellow Flesh Black Diamond. This new melon is fast becoming a leader. Exactly like red-flesh Black Diamond in shape and size until you cut one and find the tenderest, most delicious yellow flesh you've ever eaten. Mid-season melon reaching 60 to 70 pounds under favorable conditions. The seeds are greyish black, with light markings, and rather small; glossy dark green rind thin and tough for fine shipper."

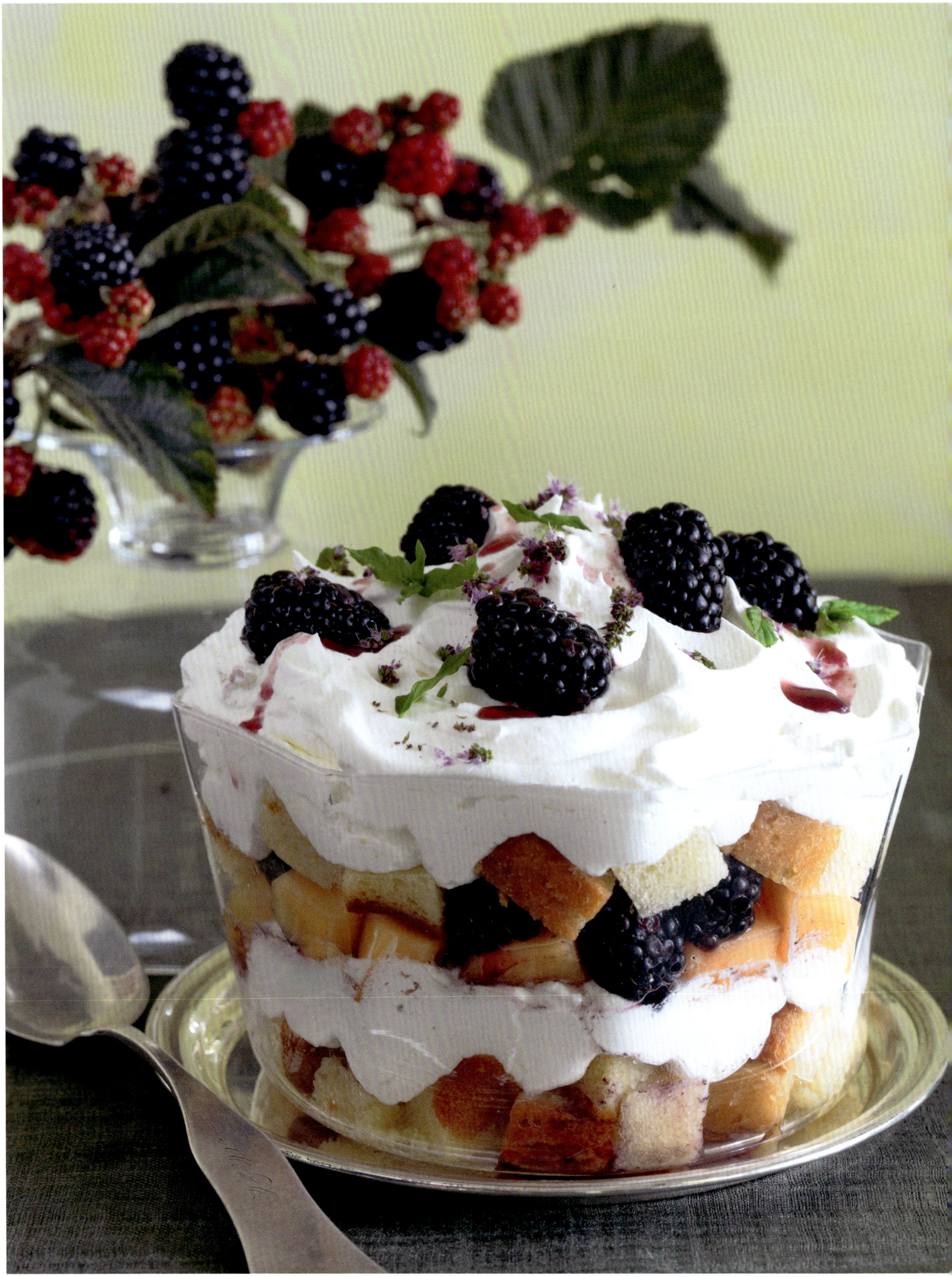

CHAPTER NINE

Recipes

Salads and Starters

Mains

Drinks

Sweets

Fresh Melon Salsa

Serve this vivid salsa as a starter with chips or use it as a topping for grilled chicken or fish. A mix of varieties and colors makes for a striking dish: picture scarlet red, orange, canary yellow, golden, and white-fleshed cubes. For a spicier salsa, leave in the chile seeds.

2 CUPS ¼-INCH CUBES MELON OR WATERMELON

¾ CUP FINELY CHOPPED RED ONION

¾ CUP FINELY CHOPPED SEEDED CUCUMBER OR SNAKE MELON

¾ CUP CHOPPED CILANTRO

2 TABLESPOONS FRESH LIME JUICE

1 MEDIUM SERRANO CHILE, STEMMED, SEEDED, THINLY SLICED

1 TABLESPOON EXTRA-VIRGIN OLIVE OIL

½ TEASPOON FINE SEA SALT

1. In a large bowl, gently stir together all ingredients. Serve immediately or cover and refrigerate for up to 2 days.

MAKES ABOUT 3 CUPS

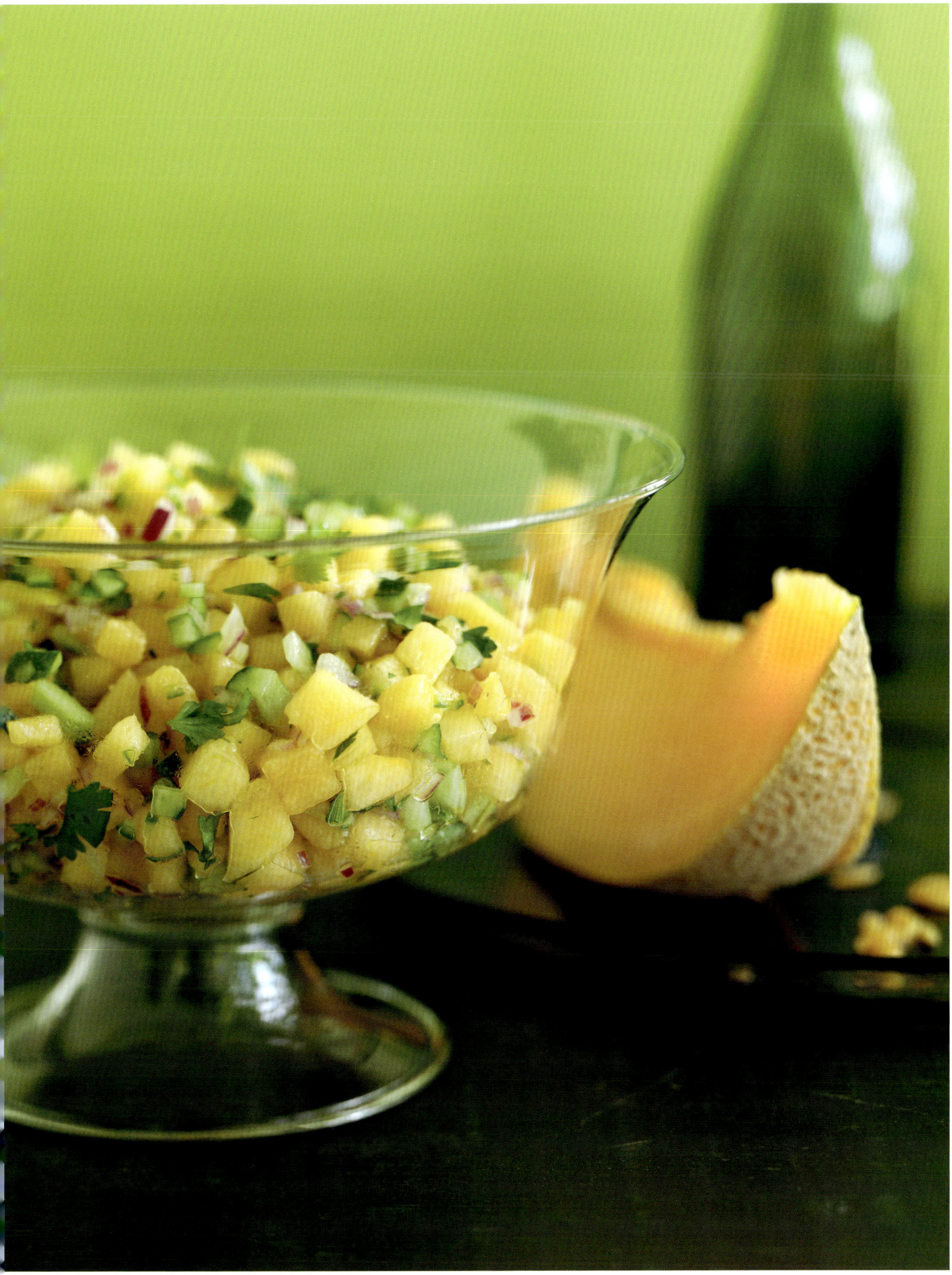

Luscious Golden Watermelon Gazpacho with Toasted Pine Nuts

This classic summer soup is a splendid make-ahead dish. Serve it in shot glasses as a welcoming treat at parties, or as a soup course at lunch or dinner with family and friends. Substitute snake melon or Adzhur melon for cucumber if you have any on hand. Topping the finished gazpacho with a little shaved cheese, such as Dry Jack or Parmesan, makes it even more savory, rich, and creamy.

4 TABLESPOONS PINE NUTS

3 CUPS CUBED ORANGE- OR YELLOW-FLESHED WATERMELON (SUCH AS LUSCIOUS GOLDEN) OR MELON

1 SMALL YELLOW ONION, COARSELY CHOPPED

1 SMALL CUCUMBER, PEELED, SEEDED, COARSELY CHOPPED

1 RED BELL PEPPER, COARSELY CHOPPED

1 SMALL SERRANO PEPPER, SEEDED

3 TABLESPOONS FRESH LIME JUICE

3 TABLESPOONS EXTRA-VIRGIN OLIVE OIL PLUS MORE FOR DRIZZLING

¼ CUP COARSELY CHOPPED CILANTRO PLUS MORE FOR SERVING

FINE SEA SALT

FRESHLY GROUND BLACK PEPPER

½ CUP SHAVED DRY JACK OR AGED PARMESAN CHEESE

1 CUP FINELY DICED WATERMELON (FOR GARNISH; OPTIONAL)

1. Spread pine nuts in a small dry skillet. Heat over medium-low, shaking pan occasionally, until they're fragrant and lightly golden, 3 to 5 minutes. Transfer nuts to a plate and let cool.

2. In the jar of a blender or a food processor, purée watermelon or melon, onion, cucumber, bell pepper, serrano pepper, lime juice, olive oil, and toasted pine nuts until smooth. Add cilantro and blend. Season to taste with salt and pepper. Thin soup with water, if desired. Transfer to a container; refrigerate, covered, until flavors blend and soup is well chilled, about 2 hours.

3. Just before serving, stir gazpacho to combine, and add more salt and pepper if desired. Divide among serving bowls. Top with cheese, cilantro, and a drizzle of oil.

SERVES 4 TO 6

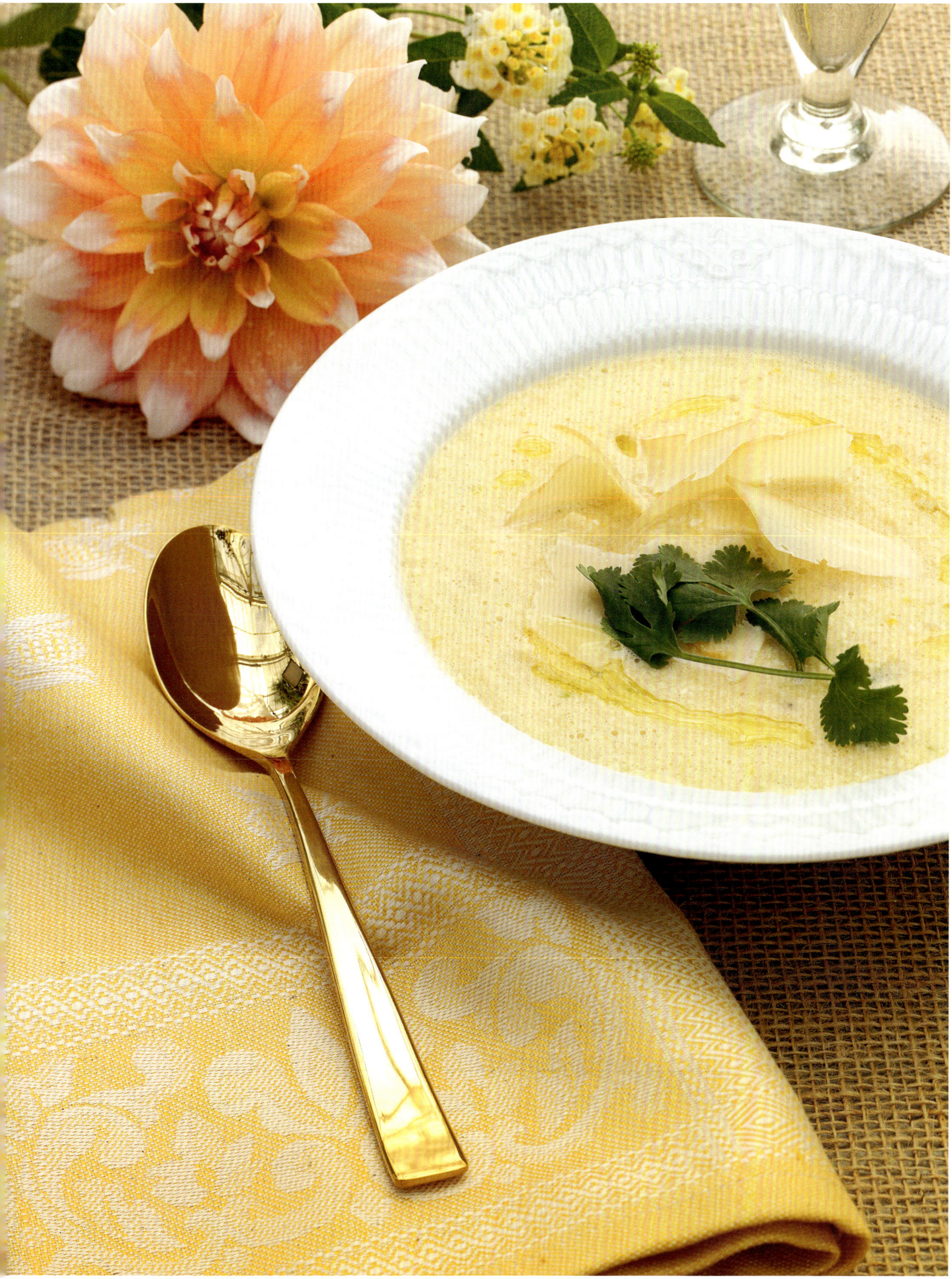

Mixed Melon Balls with Watermelon Aspic

This is not your grandmother's aspic. And no fancy old-style molds are required. Instead, a mix of melon balls, with fresh herbs and toasted hazelnuts, sprinkled on top of a refreshing watermelon base, turns the retro dish into a surprisingly modern salad.

1 LARGE RED-FLESHED WATERMELON

1 CUP WATER, DIVIDED

2 (¼-OUNCE) ENVELOPES UNFLAVORED GELATIN (ABOUT 5 TEASPOONS)

2 LEMONS, JUICED

⅓ CUP HAZELNUTS

1 SMALL MELON OR WATERMELON WITH YELLOW OR ORANGE FLESH

1 CUP MIXED WHOLE HERB LEAVES, SUCH AS TARRAGON, PARSLEY, CHERVIL, AND/OR FINELY CHOPPED CHIVES

2 TABLESPOONS EXTRA-VIRGIN OLIVE OIL

1 TEASPOON FLAKY SEA SALT, SUCH AS MALDON

FRESHLY GROUND BLACK PEPPER

1. Cut off and discard rind from half of the red watermelon, then coarsely chop enough of the peeled fruit to yield 2 quarts; seed the chopped melon. Reserve the remaining watermelon.

2. Pour ½ cup water in a small bowl. Sprinkle gelatin over and let stand 5 minutes to soften. Meanwhile, in the jar of a blender, purée the chopped watermelon and the lemon juice until smooth, then strain through a fine-mesh sieve into a bowl. Discard the solids (you should have about 1 quart juice).

3. In a small saucepan, bring ½ cup water just to a boil, then remove from heat. Add gelatin mixture and stir until dissolved; then stir gelatin mixture into watermelon juice. Divide the mixture into 6 medium serving bowls. Chill until firm, at least 12 hours.

4. Heat oven to 350°F. Spread hazelnuts on a rimmed baking sheet and toast in oven until lightly browned and skins start to blister, 10 to 12 minutes. Rub hot toasted hazelnuts in a clean kitchen towel to remove skins. Let cool completely, then coarsely chop.

5. Using a melon baller, scoop out 30 balls, half from the yellow melon or watermelon, and half from the remaining red watermelon; remove and discard seeds. Place balls in a bowl; cover and refrigerate until ready to use.

6. Just before serving, top aspic with mixed melon balls and herbs. Drizzle with oil, then sprinkle with hazelnuts, salt, and pepper.

SERVES 6

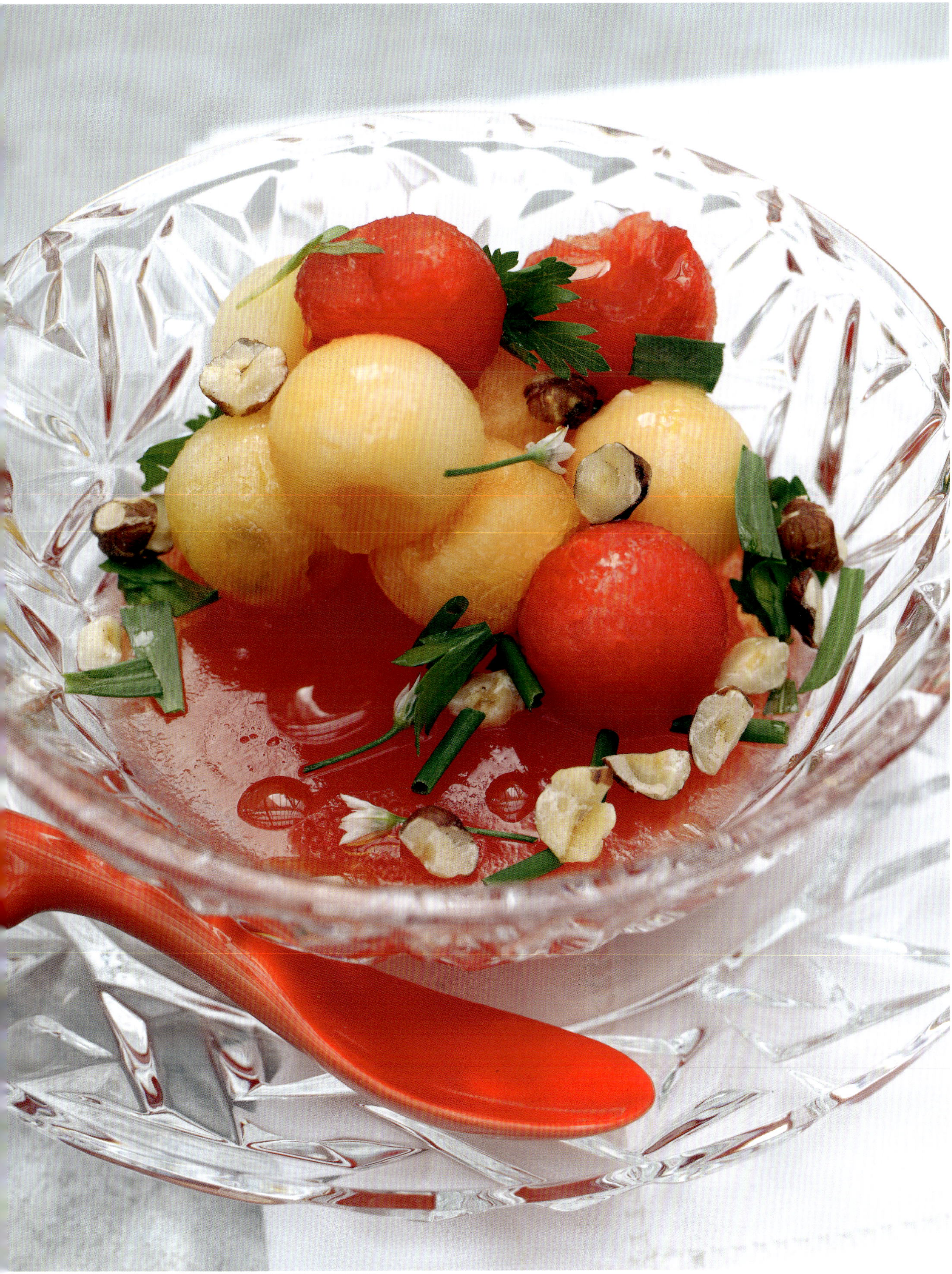

Melon, Farro, and Radish Salad with Fresh Mint and Dill

This Mediterranean salad is more than enough on its own or piled alongside grilled fish, chicken, or steak. Sumac adds gorgeous flecks of deep magenta and a tart, lemony flavor that plays off the sweetness of Santa Claus casaba. Soaking the radishes in ice water crisps them and mellows their peppery bite.

¾ TEASPOON FINE SEA SALT, DIVIDED, PLUS MORE FOR SALTING WATER
¼ CUP PLUS 2 TABLESPOONS FARRO (SEMIPEARLED OR PEARLED)
5 AVERAGE-SIZED RADISHES, TRIMMED, CUT INTO THIN ROUNDS
½ CUP THINLY SLICED RED ONION
4 TEASPOONS RED WINE VINEGAR
2 TABLESPOONS FRESH LEMON JUICE
2 TABLESPOONS EXTRA-VIRGIN OLIVE OIL
4 CUPS LOOSELY PACKED BABY ARUGULA (2 OUNCES)
2 CUPS BALLED OR CUBED WHITE-FLESHED TURKISH OR SPANISH CASABA MELON
½ CUP COARSELY CHOPPED MINT
½ CUP COARSELY CHOPPED DILL
1 TABLESPOON GROUND SUMAC

1. Bring a medium saucepan of salted water to a boil. Add farro and cook until al dente, 15 to 30 minutes. Strain through a fine-mesh sieve (do not rinse) then spread on a large plate; let cool completely.

2. Meanwhile, submerge radishes in a small bowl of ice water; set aside. In a second small bowl, stir together onion, vinegar, and ¼ teaspoon salt. Let stand at least 10 minutes.

3. Drain and pat dry radishes. Whisk lemon juice and olive oil into bowl with onion mixture. In a large salad bowl, gently toss together farro, radishes, arugula, melon, mint, dill, sumac, ½ teaspoon salt, and the onion dressing.

SERVES 4 TO 6

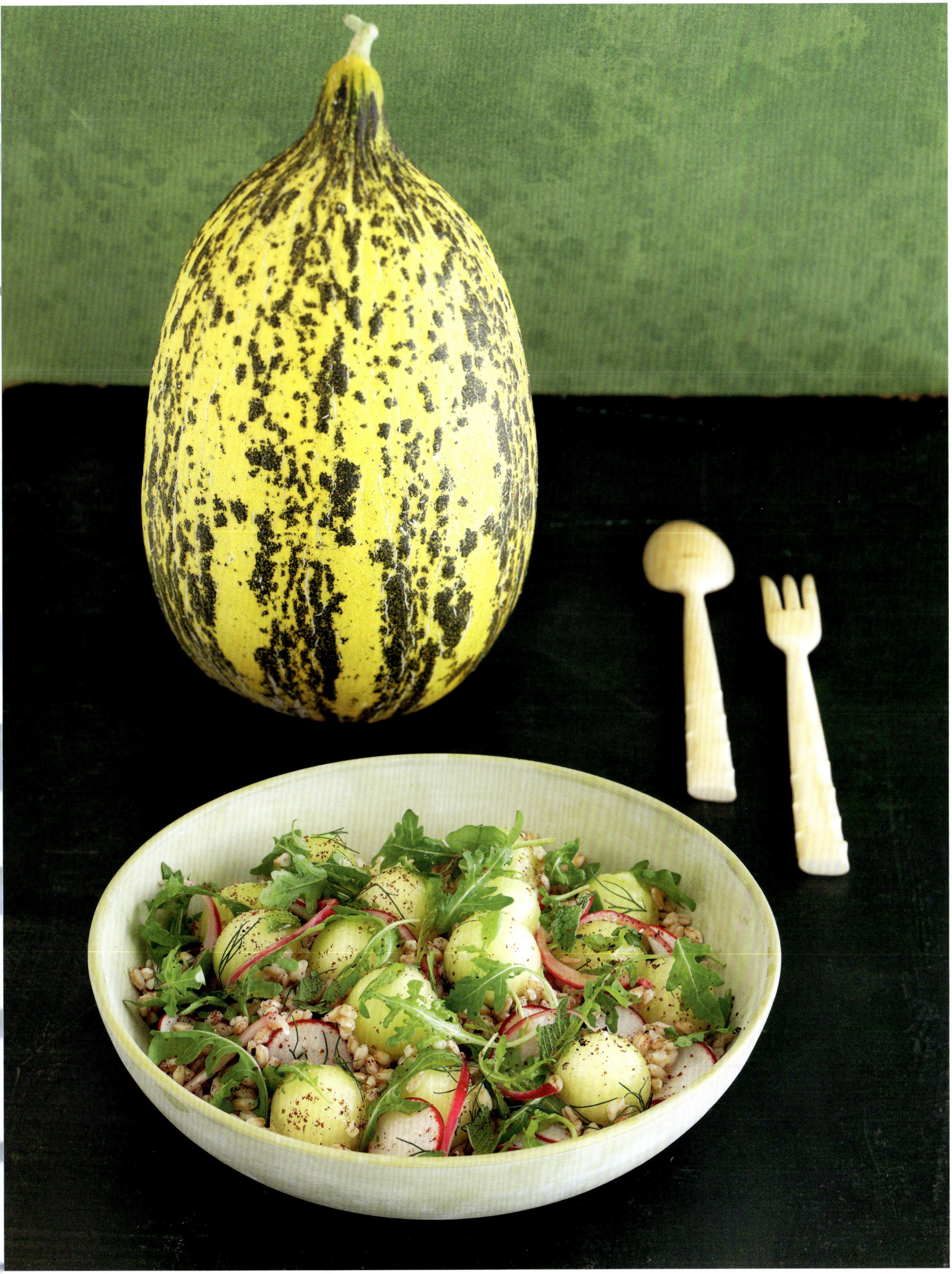

Burrata and Green Melon Salad with Toasted Cumin and Basil

Creamy burrata cheese balances the sweet-tangy flavors of lightly pickled green-fleshed melon, such as Eden Gem, in this simple yet sophisticated summer salad. Snake melon can be used instead of cucumber.

1 POUND GREEN-FLESHED MELON, SEEDED, RIND REMOVED, THINLY SLICED

1½ TABLESPOONS WHITE WINE VINEGAR

¼ TEASPOON FINE SEA SALT

1½ TEASPOONS WHOLE CUMIN SEEDS

8 OUNCES FRESH BURRATA OR FRESH MOZZARELLA CHEESE, TORN

1 YOUNG BEIT ALPHA (BET ALFA) OR PERSIAN CUCUMBER, THINLY SLICED

2 TEASPOONS EXTRA-VIRGIN OLIVE OIL

¼ CUP BASIL LEAVES (FOR SERVING), TORN IF LARGE

FLAKY SEA SALT (FOR SERVING)

FRESHLY CRACKED BLACK PEPPER (FOR SERVING)

1. In a large bowl, toss melon with vinegar and fine sea salt. Let stand 10 minutes.

2. Meanwhile, in a small dry skillet, toast cumin seeds over low heat, tossing occasionally, until fragrant, 3 to 4 minutes. Transfer seeds to a plate and let cool, then lightly crush with a mortar and pestle or on a cutting board, rocking the bottom of a heavy skillet over top.

3. Reserving the vinegar, arrange melon slices on serving plates. Top with cheese and cucumber. Drizzle each plate with ½ teaspoon each oil and reserved vinegar. Sprinkle with cumin seeds, basil, flaky sea salt, and cracked pepper.

SERVES 4

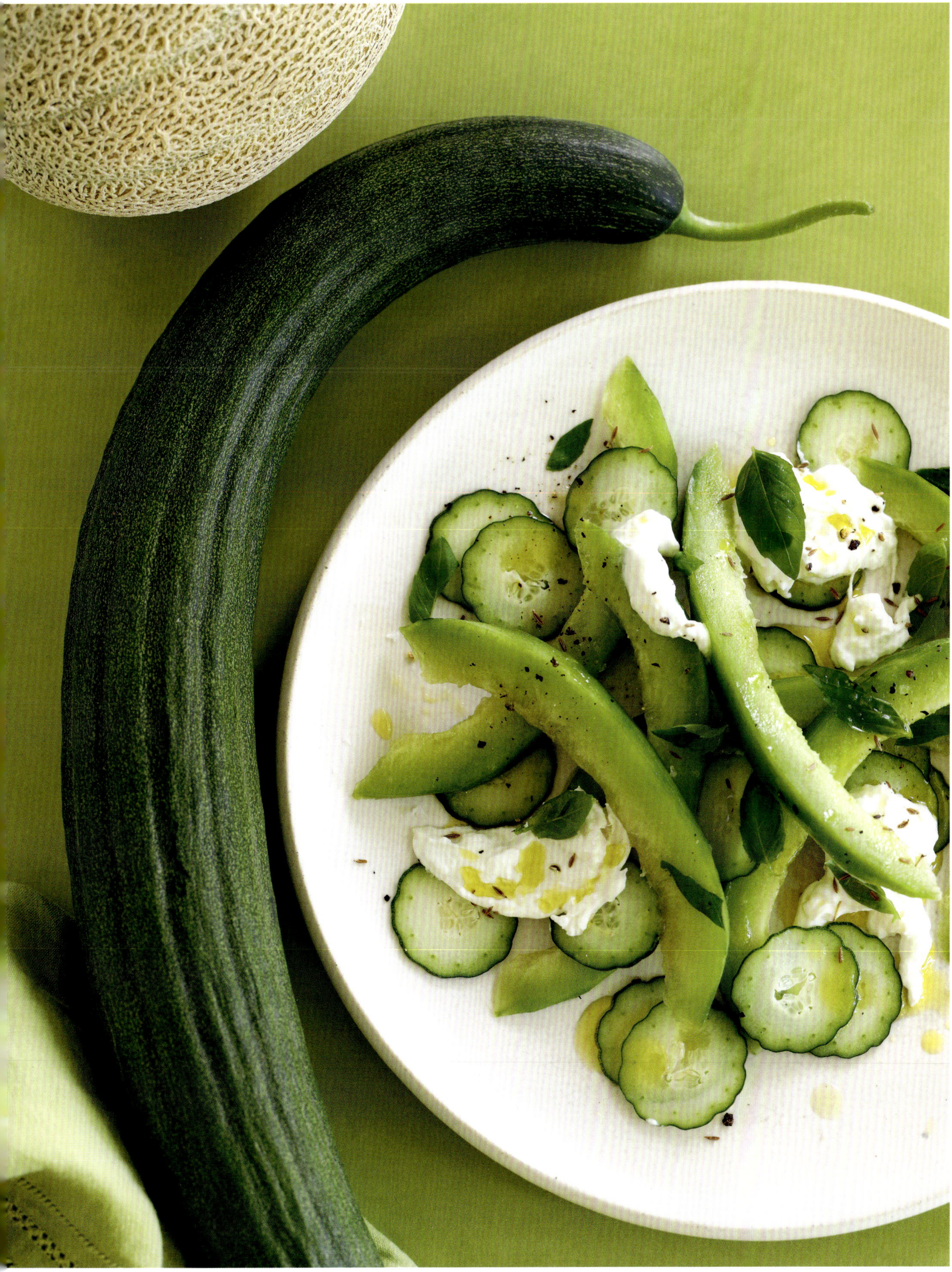

Lobster, Grilled Melon, and Watercress Salad

Melon intensifies lobster's sweeter side while playing off its saltiness. Serve this sumptuous salad for lunch or dinner. If you prefer it as a main course, double the ingredients. You can save the lobster shells to make stock.

1 GARLIC CLOVE, FINELY CHOPPED

1 TEASPOON HOT SAUCE

¼ CUP PLUS 2 TABLESPOONS EXTRA-VIRGIN OLIVE OIL, DIVIDED, PLUS MORE FOR GRILL

2 TABLESPOONS FRESH LEMON JUICE, DIVIDED

1 TEASPOON FINE SEA SALT PLUS MORE FOR SEASONING

½ TEASPOON FRESHLY GROUND BLACK PEPPER PLUS MORE FOR SEASONING

1 SMALL FIRM-RIPE CHARENTAIS OR OTHER ORANGE-FLESHED MELON, PEELED, SEEDED, AND CUT INTO ¾-INCH WEDGES

2 TABLESPOONS DISTILLED WHITE VINEGAR

1 (1½ TO 2 POUND) LIVE LOBSTER OR ½ POUND FRESH OR THAWED FROZEN LOBSTER MEAT, CUT INTO ½-INCH PIECES

2 TABLESPOONS CHOPPED MIXED HERBS, SUCH AS TARRAGON, PARSLEY, CHIVES, AND CHERVIL

2 CUPS WATERCRESS, THIN STEMS AND LEAVES ONLY

¼ CUP THINLY SLICED SHALLOT

2 TABLESPOONS CHAMPAGNE VINEGAR

1. In a large resealable plastic bag, combine garlic, hot sauce, 2 tablespoons oil, 1 tablespoon lemon juice, 1 teaspoon salt, and ½ teaspoon pepper. Add melon wedges and seal bag, pressing out air. Refrigerate 2 hours.

2. Meanwhile, if you are using a fresh live lobster, fill a large bowl with ice and cold water. Add distilled white vinegar to an 8-quart pot of boiling salted water. Plunge in the lobster headfirst and cook exactly 13 minutes, then transfer to ice water and let cool 5 minutes. Drain lobster, then remove meat from claws, joints, and tail. Discard tomalley, roe, and shells (or save for another use). Leave claws whole; cut remaining meat into ½ inch pieces. Refrigerate, covered, until ready to serve.

3. Oil grill rack. Heat grill to high. Remove melon from marinade and pat dry with paper towels; discard marinade. Grill melon until nicely charred, 1 to 2 minutes per side. Transfer to serving plates.

4. In a large bowl, combine lobster pieces and herbs. In a second bowl, combine watercress and shallot.

5. In a small bowl, whisk together champagne vinegar and remaining tablespoon lemon juice. Whisk in remaining ¼ cup oil in a slow and steady stream. Season dressing to taste with salt and pepper. Toss watercress mixture with just enough dressing to lightly coat; repeat with lobster mixture. Top charred melon with watercress and lobster mixtures. Drizzle remaining dressing over top.

SERVES 4 TO 6

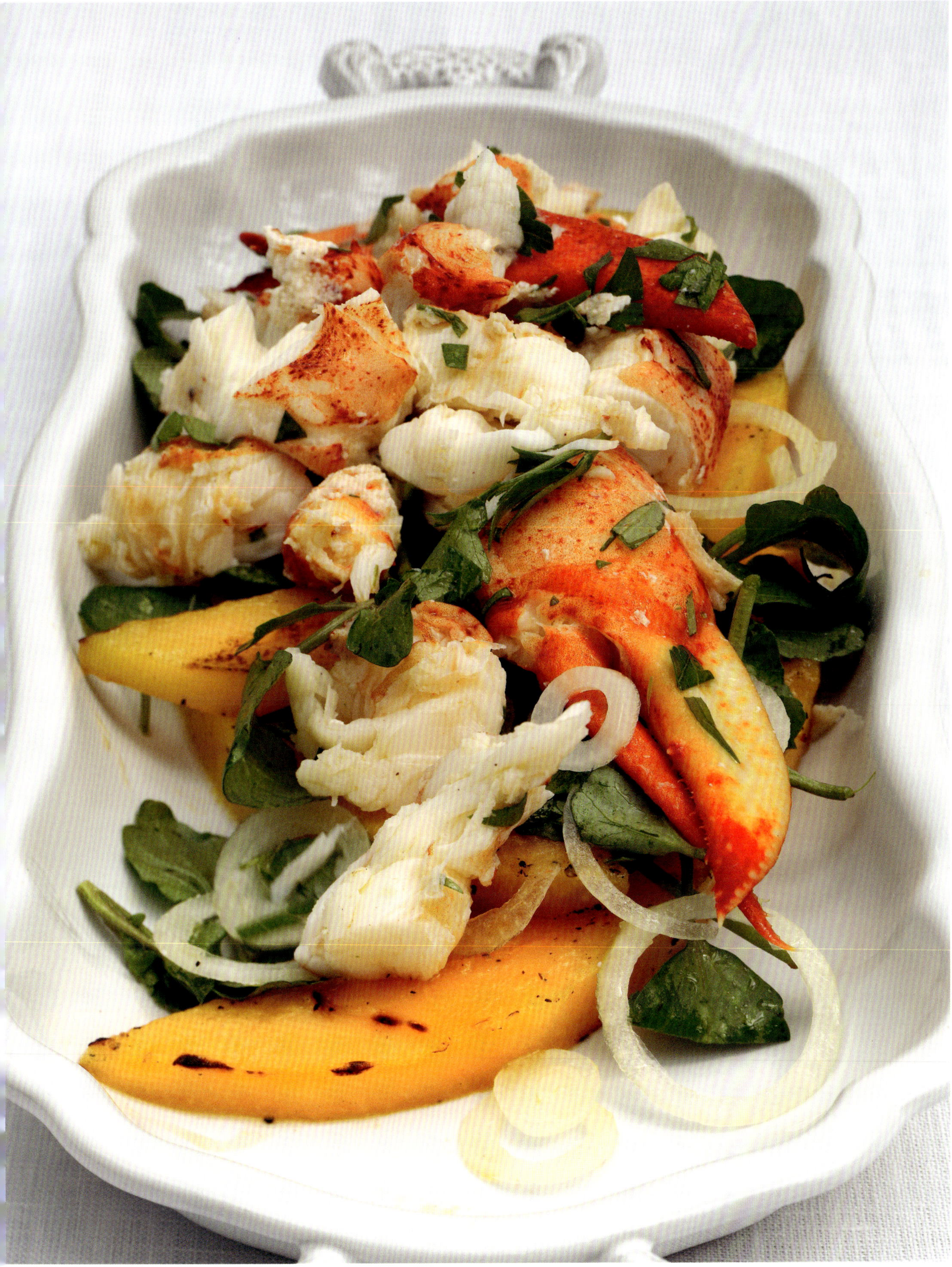

Vietnamese-Style Poached Chicken and Watermelon Salad

Watermelon with intensely red flesh, such as Crimson Sweet, is right at home in this bracing combination of sweet, spicy, and salty Asian flavors. The chicken can be prepared ahead, and the dish assembled in a few moments. Melon balls look especially pretty here but if you don't have a baller, cut the fruit into ½-inch cubes instead.

3 CUPS PLUS TWO TEASPOONS WATER, DIVIDED
2 CUPS LOW-SODIUM CHICKEN BROTH
1 POUND SKINLESS BONELESS CHICKEN BREAST
3 TABLESPOONS LIME JUICE
3 TABLESPOONS FISH SAUCE
2 TEASPOONS SUGAR
1 GARLIC CLOVE, FINELY CHOPPED
1 TEASPOON EXTRA-VIRGIN OLIVE OIL
6 CUPS THINLY SLICED NAPA CABBAGE (ABOUT ½ MEDIUM)
2 BEIT ALPHA (BET ALFA) OR PERSIAN CUCUMBERS, THINLY SLICED INTO ROUNDS
3 CUPS BALLED WATERMELON
1 BUNCH (ABOUT 8 SMALL) RADISHES, THINLY SLICED
4 SCALLIONS, THINLY SLICED
1 MEDIUM SERRANO CHILE, THINLY SLICED
½ CUP COARSELY CHOPPED CILANTRO
½ CUP COARSELY CHOPPED MINT
½ CUP ROASTED, SALTED PEANUTS, ALMONDS, OR CASHEWS, COARSELY CHOPPED
LIME WEDGES (FOR SERVING)

1. Combine 3 cups water and chicken broth in a medium saucepan; bring to a simmer. Add chicken and simmer, uncovered, 6 minutes. Remove pan from heat and cover, then let stand until chicken is cooked through, about 15 minutes. Transfer chicken to a plate and cool 15 to 20 minutes. Shred into small pieces.

2. Whisk together lime juice, fish sauce, sugar, garlic, oil, and 2 teaspoons tepid water in a bowl until sugar dissolves.

3. Arrange cabbage, cucumber, melon, radishes, scallions, chile, and shredded chicken in serving bowls. Whisk the dressing again, then drizzle it over salads. Sprinkle with cilantro, mint, and nuts. Serve with lime wedges.

SERVES 6

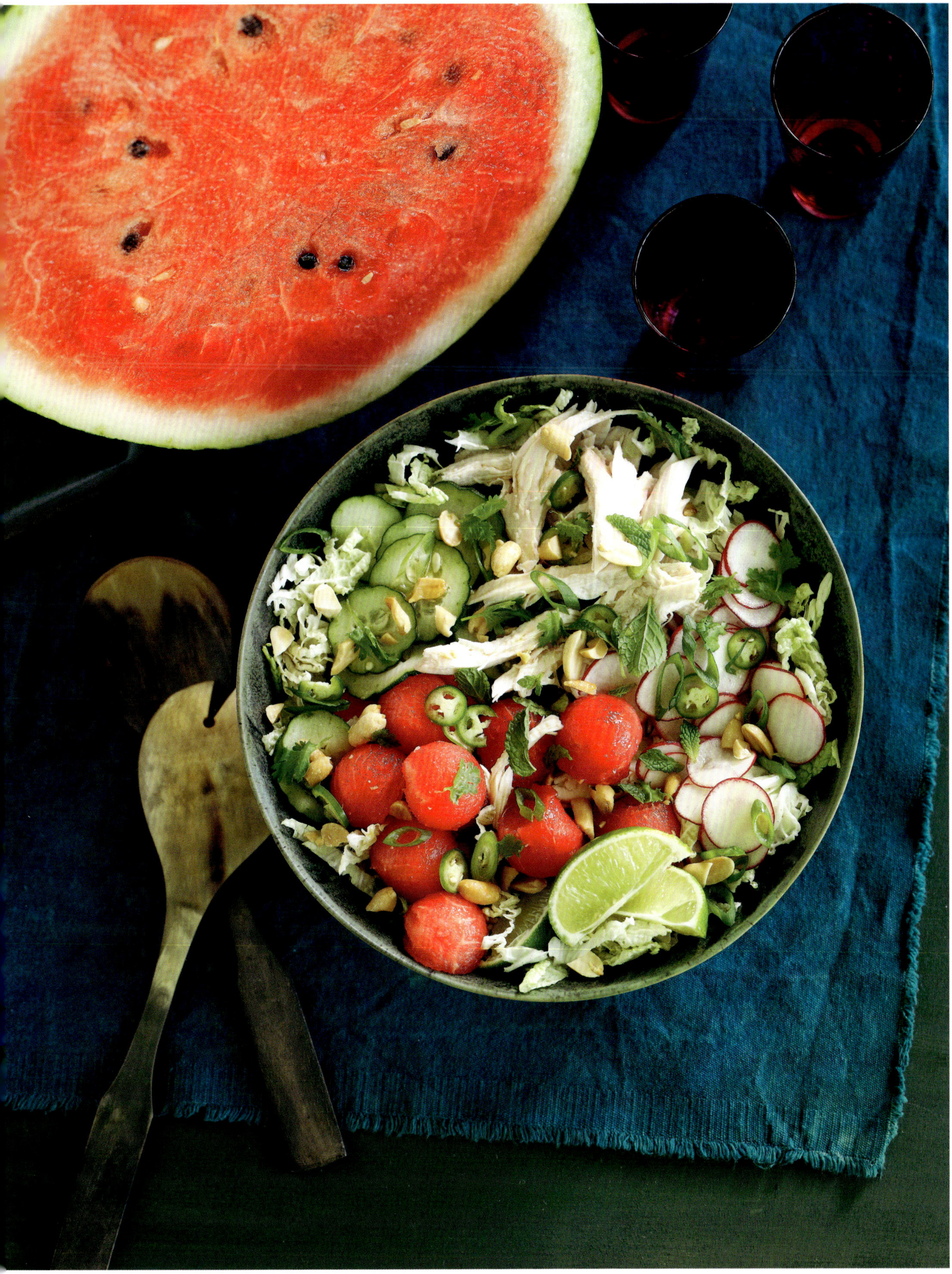

Curried Chicken Salad with Melon and Toasted Almonds

Casabas with firm white or green flesh add unexpected nuance to this salad and balance the sweet and savory flavors of curry. Makuwa melons such as Chamoe would work well, too.

½ CUP RAW UNSALTED ALMONDS, 2 TABLESPOONS RESERVED FOR GARNISH

1 LEMON

¾ POUND COOKED CHICKEN OR TURKEY BREAST OR THIGH MEAT, CUT INTO ½-INCH CUBES (ABOUT 2½ CUPS)

2¾ CUPS ½-INCH CUBES MELON OF YOUR CHOICE

1½ CUPS DICED CELERY (ABOUT 4 RIBS)

4 SCALLIONS, THINLY SLICED, 2 TABLESPOONS RESERVED FOR GARNISH

½ CUP COARSELY CHOPPED CILANTRO, 2 TABLESPOONS RESERVED FOR GARNISH

½ CUP COARSELY CHOPPED DRIED CHERRIES

⅓ CUP PLUS 2 TABLESPOONS MAYONNAISE

⅓ CUP PLAIN GREEK YOGURT

2 TEASPOONS CURRY POWDER

½ TEASPOON FINE SEA SALT

1 GARLIC CLOVE, CUT IN HALF

4 SLICES MICHE OR OTHER RUSTIC COUNTRY BREAD, HALVED, TOASTED

EXTRA-VIRGIN OLIVE OIL (FOR DRIZZLING)

4 LEAVES BIBB OR OTHER SOFT LETTUCE, TORN

FLAKY SEA SALT

FRESHLY CRACKED BLACK PEPPER

1. Heat oven to 350°F. Spread nuts on a rimmed baking sheet and toast until fragrant and lightly golden, 10 to 12 minutes. Transfer nuts to a plate and let cool completely, then coarsely chop.

2. Finely grate zest from lemon, then squeeze 1 tablespoon juice. Place zest and juice in a large mixing bowl. Combine all of the ingredients through and including the fine sea salt, except those reserved for garnish. Mix well.

3. Rub cut sides of garlic on toasts, then arrange toasts on serving plates and drizzle with oil. Top with lettuce leaves, melon salad, and reserved cilantro, scallion, and nuts; sprinkle with flaky sea salt and black pepper.

SERVES 4

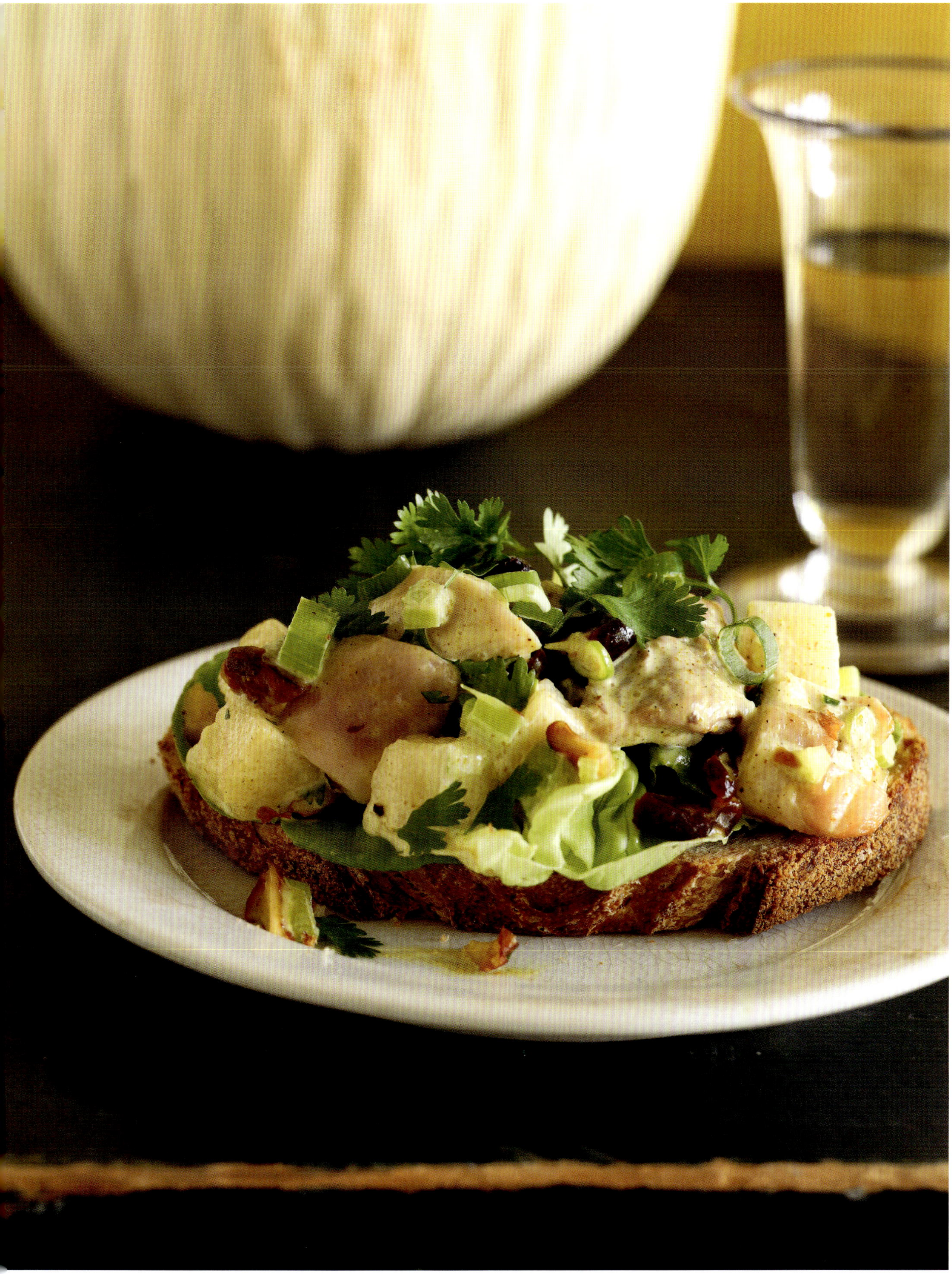

Grilled Tuna Skewers with Celery, Melon, and Olive Salad

Olives, crisp celery heart, and sweet green-fleshed melon make an invigorating green salad to pair with spiced grilled tuna.

2 TABLESPOONS WHOLE CORIANDER SEEDS

2 TABLESPOONS WHOLE FENNEL SEEDS

½ TEASPOON FINE SEA SALT

½ TEASPOON FRESHLY GROUND BLACK PEPPER

½ TEASPOON SMOKED PAPRIKA

⅛ TEASPOON CAYENNE PEPPER

2 TABLESPOONS PLUS 1 TEASPOON EXTRA-VIRGIN OLIVE OIL PLUS MORE FOR GRILL AND DRIZZLING

1½ POUNDS FRESH TUNA STEAK, CUT INTO 1-INCH CUBES

2 CUPS VERY THINLY SLICED CELERY HEART AND INNER RIBS, LEAVES RESERVED

1½ CUPS ½-INCH CUBES GREEN-FLESHED MELON

¾ CUP COARSELY CHOPPED OLIVES, PREFERABLY CASTELVETRANO AND GAETA

FINELY GRATED ZEST FROM 1 LARGE LEMON

1½ TEASPOONS FRESH LEMON JUICE

FLAKY SEA SALT (FOR SERVING)

1. In a spice grinder, pulse coriander and fennel until coarsely ground. Transfer to a bowl; stir in fine sea salt, black pepper, smoked paprika, and cayenne.

2. Oil grill rack. Heat grill to medium-high. Thread tuna onto skewers, leaving ¼ inch between each piece. Season tuna all over with spice mixture.

3. While the grill heats up, gently toss together in a large bowl sliced celery, ⅓ of reserved celery leaves, and the melon, olives, oil, lemon zest and lemon juice.

4. Grill tuna skewers, turning occasionally, 4 minutes total for medium-rare.

5. Transfer skewers to serving plates. Scatter plates with salad, then sprinkle with remaining celery leaves. Drizzle with oil and sprinkle with flaky sea salt.

SERVES 4

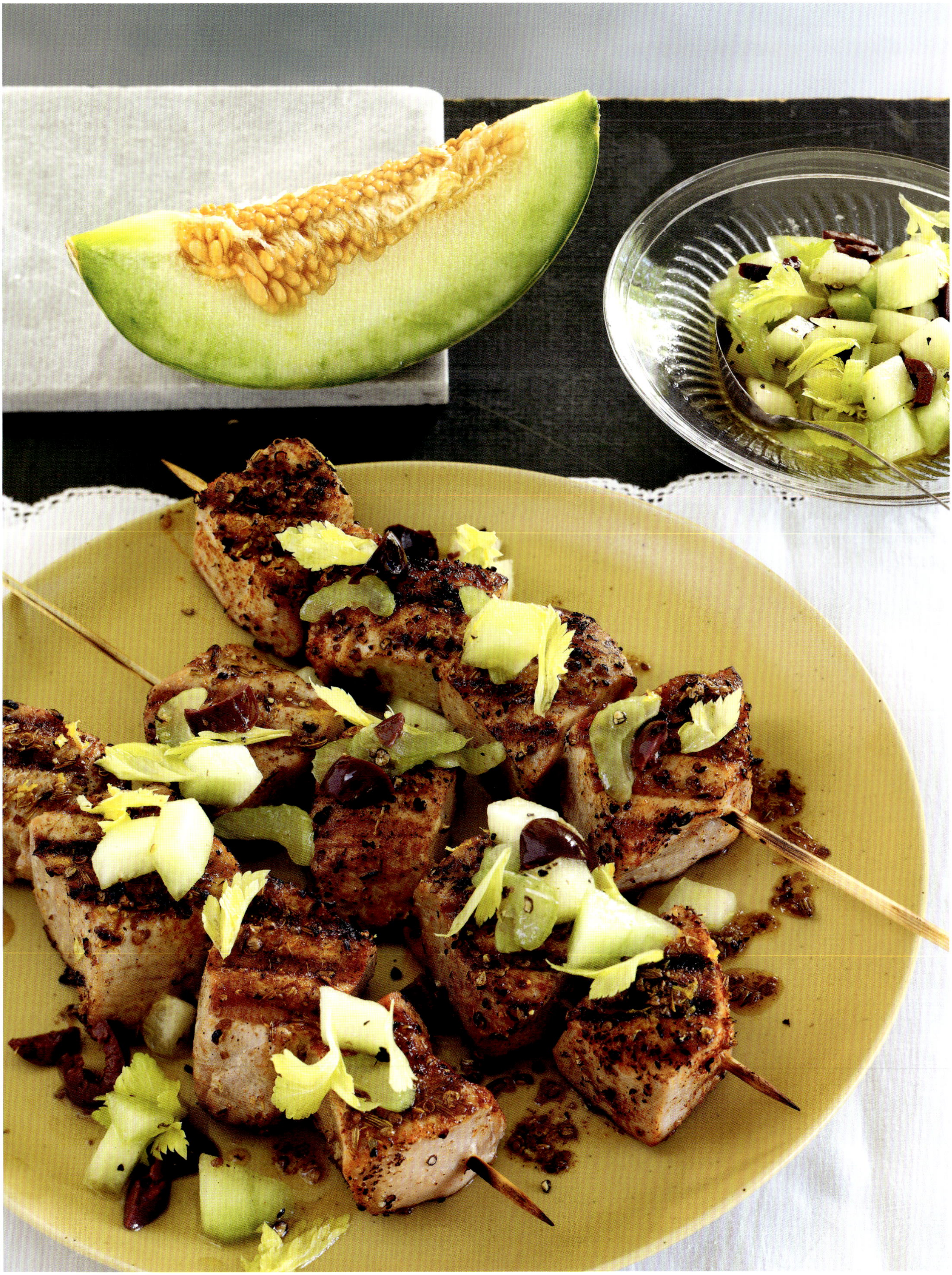

Spatchcocked Cornish Hens with Watermelon Barbeque Sauce

Melon is a favorite breakfast, salad, and dessert ingredient, but it's a great player in barbeque sauce, too. The fruit teams up with molasses to provide sweet balance to the savory, spicy sauce. Cornish hens are delicious grilled and make an impressive presentation. You can also use the sauce with grilled chicken or shrimp.

BARBEQUE SAUCE

1 3-POUND WATERMELON, PEELED, SEEDED, CUBED (7 TO 8 CUPS)
2 TABLESPOONS EXTRA-VIRGIN OLIVE OIL
1 SMALL YELLOW ONION, FINELY CHOPPED
½ RED BELL PEPPER, SEEDED, FINELY CHOPPED
½ MEDIUM JALAPEÑO PEPPER, FINELY CHOPPED
2 GARLIC CLOVES, FINELY CHOPPED
1 TABLESPOON TOMATO PASTE
2 TABLESPOONS MOLASSES
1 TABLESPOON DIJON MUSTARD
1 TEASPOON CHILI POWDER
1 TEASPOON SMOKED PAPRIKA
½ TEASPOON CUMIN
¼ CUP DRY SHERRY
¼ CUP APPLE CIDER VINEGAR
1½ TEASPOONS FRESH LIME JUICE
FINE SEA SALT AND FRESHLY GROUND BLACK PEPPER

HENS

2 TABLESPOONS EXTRA-VIRGIN OLIVE OIL
2 TABLESPOONS FRESH LIME JUICE
1 TEASPOON CHILI POWDER
1 TEASPOON FRESHLY GROUND BLACK PEPPER
4 CORNISH HENS (EACH ABOUT 1 POUND), SPATCHCOCKED (SEE NOTE)
FINE SEA SALT

1. For barbeque sauce: In the jar of a blender, purée watermelon until smooth. Pass purée through a fine-mesh sieve into a bowl; discard solids. In a medium saucepan, simmer watermelon juice over medium heat, stirring occasionally, until reduced to 1 cup, 15 to 20 minutes. Set aside.

2. Meanwhile, in a large skillet, heat oil over medium-high. Add onion, bell pepper, jalapeño, and garlic; cook, stirring occasionally, until tender and lightly golden. Add tomato paste; continue cooking, stirring occasionally, until browned, about 3 minutes. Add molasses and mustard; cook 2 minutes more, then stir in chili powder, paprika, and cumin. Cook, stirring, 1 minute more.

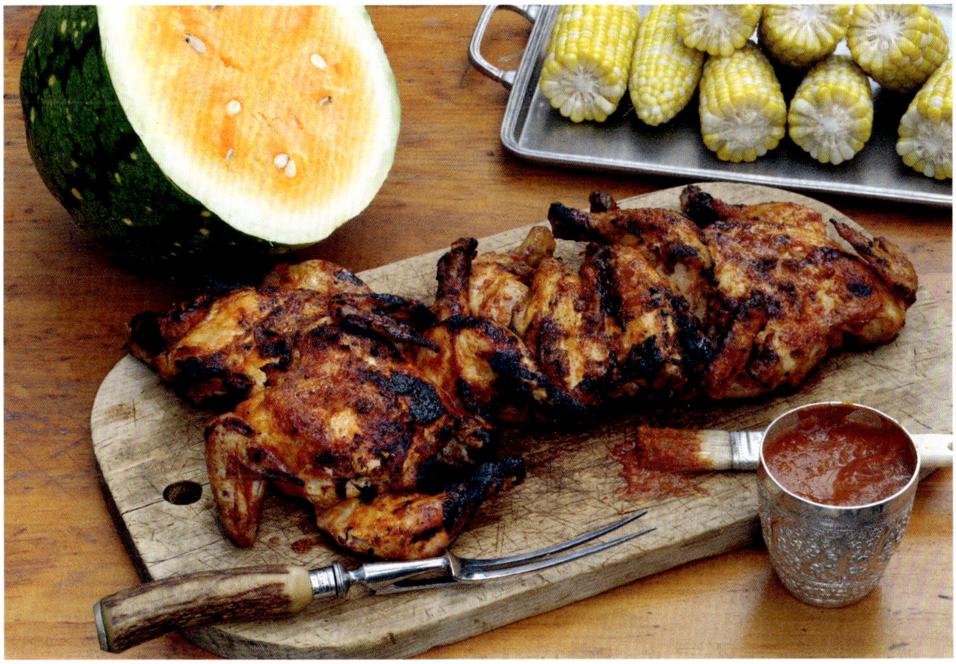

3. Stir in sherry and vinegar; cook until reduced by half, then add the reduced watermelon juice. Gently simmer over low heat, stirring and scraping down sides of pan, as needed, until sauce is reduced by half, about 20 minutes.

4. Remove sauce from heat. Stir in lime juice and season with salt and pepper to taste. Let cool to room temperature, then purée until smooth. Cover and refrigerate until ready to use.

5. For hens: In a shallow baking dish, whisk together oil, lime juice, chili powder, pepper, and 1 cup watermelon barbeque sauce. Add hens and toss to coat. Cover and refrigerate at least 1 hour or overnight.

6. Divide remaining barbecue sauce into 2 bowls (one for basting, the other for serving). Let hens stand at room temperature 15 minutes. Meanwhile, prepare grill for medium heat.

7. Oil grill rack. Remove hens from marinade, season generously with salt, and arrange skin-side down on the rack. Cook until skin starts to brown, about 6 minutes, then rotate 90 degrees. Continue cooking, basting occasionally with barbecue sauce, 6 minutes more, then flip. Continue cooking and occasionally basting, until juices run clear when meat is pierced with the tip of a knife, rotating 90 degrees halfway through, 12 to 15 minutes more.

8. Serve hens with reserved barbeque sauce.

Note: Ask your butcher to spatchcock hens, or do it yourself, following these steps: Place a hen on a cutting board, breast side down. Using kitchen scissors or poultry shears, cut down each side of the backbone, from neck to tailbone, to remove backbone. At base of breastbone, make small slit in cartilage to reveal keel bone. Using both hands, open up hen like a book. Pull out and remove keel bone. Cut off and discard wing tips. Repeat with remaining hens.

SERVES 4 TO 6

Melon-Coconut Lassi with Pistachios and Pink Peppercorns

This classic Indian drink, designed to cool the system on a hot summer day, does its job extremely well. My version is lightly sweetened with a hint of salt. You can adjust both sugar and salt to suit your taste and the sweetness of your melon. Pink peppercorns add edge. Lightly crush the spice with your fingertips as you sprinkle it over the drink.

¼ CUP UNSWEETENED COCONUT FLAKES

2 CUPS CHILLED PLAIN WHOLE-MILK YOGURT (SEE NOTE)

2 CUPS CHILLED ½-INCH CUBES HA'OGEN OR OTHER GREEN-FLESHED MELON

2 TEASPOONS SUGAR

FINE SEA SALT

2 TABLESPOONS CHOPPED UNSALTED SHELLED PISTACHIOS AND

1 TEASPOON PINK PEPPERCORNS (FOR SERVING)

1. Heat oven to 325°F. Spread coconut on a rimmed baking sheet and toast until fragrant and lightly golden, about 5 minutes. Transfer toasted flakes to a plate and let cool.

2. In a blender, purée yogurt, melon, sugar, and a generous ¼ teaspoon salt until well blended and frothy. Adjust salt and sugar to taste. Divide lassi among 4 glasses. Top with cooled toasted coconut flakes, pistachios, and pink peppercorns. Serve immediately.

Note: If you use Greek yogurt, add a splash of chilled water or milk to dilute.

SERVES 4

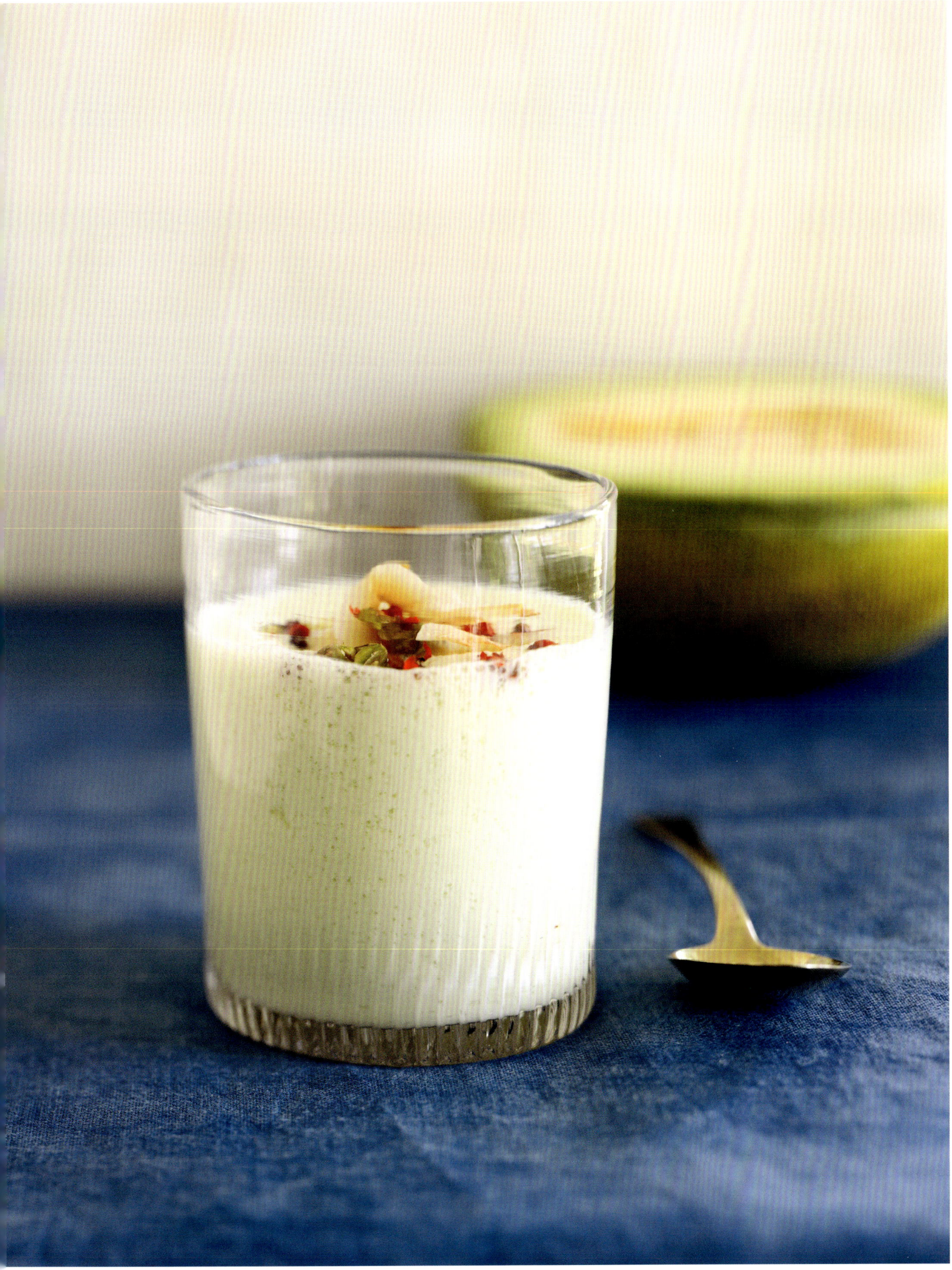

Watermelon Gin Rickey

Watermelon brings welcome fruity complexity to this effervescent gin drink. Choose watermelon that is saturated with color, like Blacktail Mountain, Orangeglo, or Early Moonbeam.

1 CUP WATERMELON CUBES, SEEDED

2 LIMES, CUT INTO QUARTERS

1 LIME, CUT INTO WEDGES (FOR SERVING; OPTIONAL)

12 OUNCES OF YOUR FAVORITE GIN

CHILLED CLUB SODA

1. In the jar of a blender, purée watermelon until smooth. Strain through a fine-mesh sieve into a bowl; discard solids.

2. Fill 4 10-ounce Collins glasses with ice. Squeeze 2 lime quarters over the ice in each glass, then add 3 ounces gin and 1 ounce watermelon purée to each. Top with club soda. Stir drink and garnish with a lime wedge, if desired, before serving.

MAKES 4 DRINKS

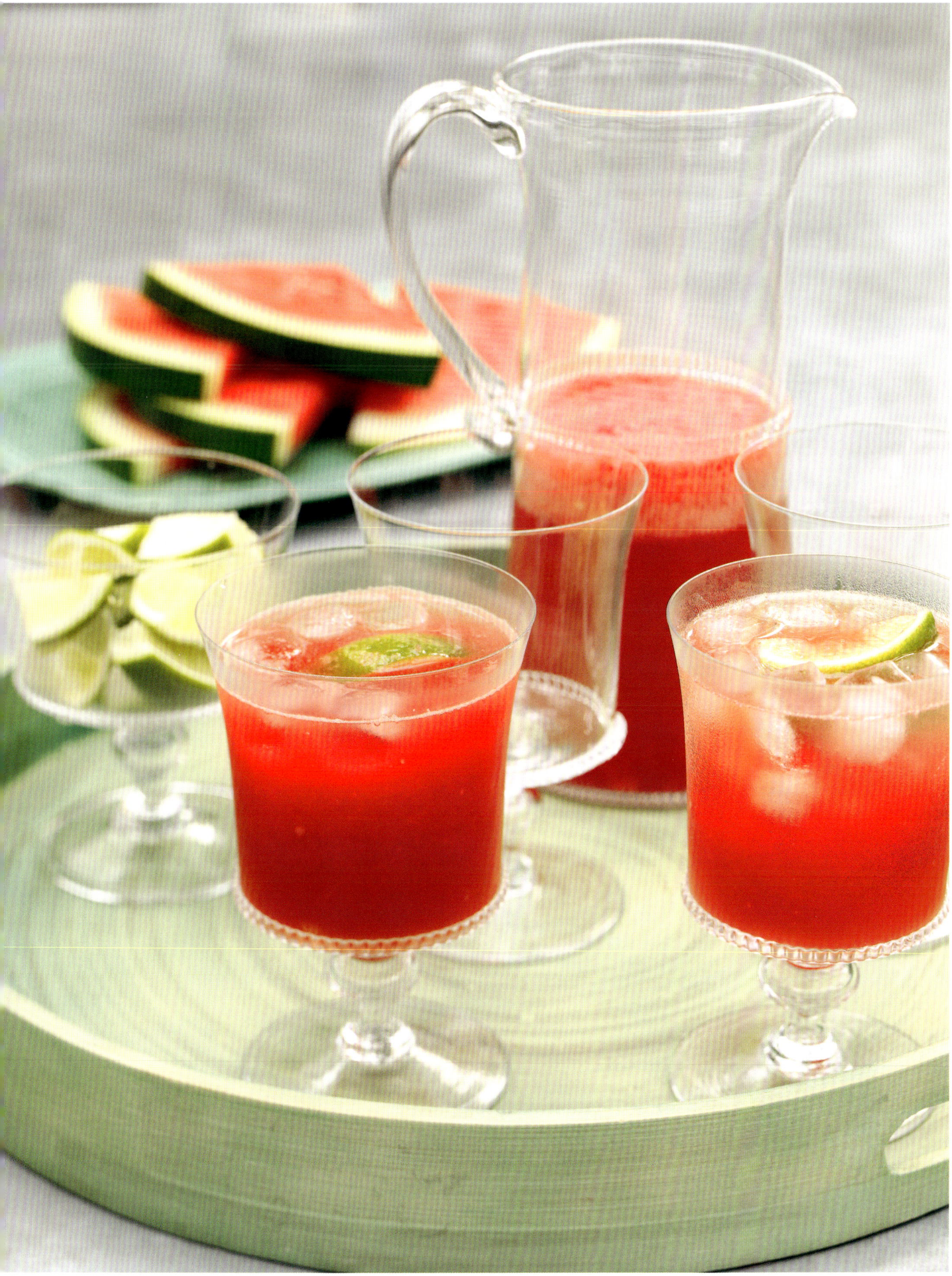

Mint Julep with Watermelon Ice Cubes

Watermelon infuses the base of this classic Derby drink with sweet fruity flavor. It also provides a welcome icy chill, in the form of frozen watermelon cubes. It's easy to make extra ones. Pop them into a glass of still or sparkling water, or lemonade. They make addictive treats on their own.

2 CUPS ¾-INCH CUBES RED-FLESHED WATERMELON

¼ CUP WATER

1 TABLESPOON SUGAR

2 TABLESPOONS PACKED TORN MINT LEAVES, DIVIDED

1 CUP YELLOW-FLESHED WATERMELON CUBES, SEEDED

2 CUPS CRUSHED ICE

10 OUNCES OF YOUR FAVORITE BOURBON

4 MINT SPRIGS

1. Arrange red watermelon cubes on a small baking pan lined with parchment paper. Freeze 2 to 3 hours.

2. In a small bowl, microwave ¼ cup water on high for 30 seconds. Add sugar and 1 tablespoon mint leaves, stirring until sugar is dissolved. Let stand until cooled to room temperature, then strain the mixture through a mesh sieve into a bowl. Discard mint.

3. In the jar of a blender, purée yellow-fleshed watermelon until smooth. Transfer to a large bowl; add mint-infused syrup, then muddle with remaining tablespoon mint.

4. Divide frozen watermelon cubes among 4 mint julep cups or highball glasses; top with ice. Divide bourbon, then watermelon-mint mixture among the glasses. Top with more ice or a splash of chilled water, if desired. Garnish with mint sprigs.

SERVES 4

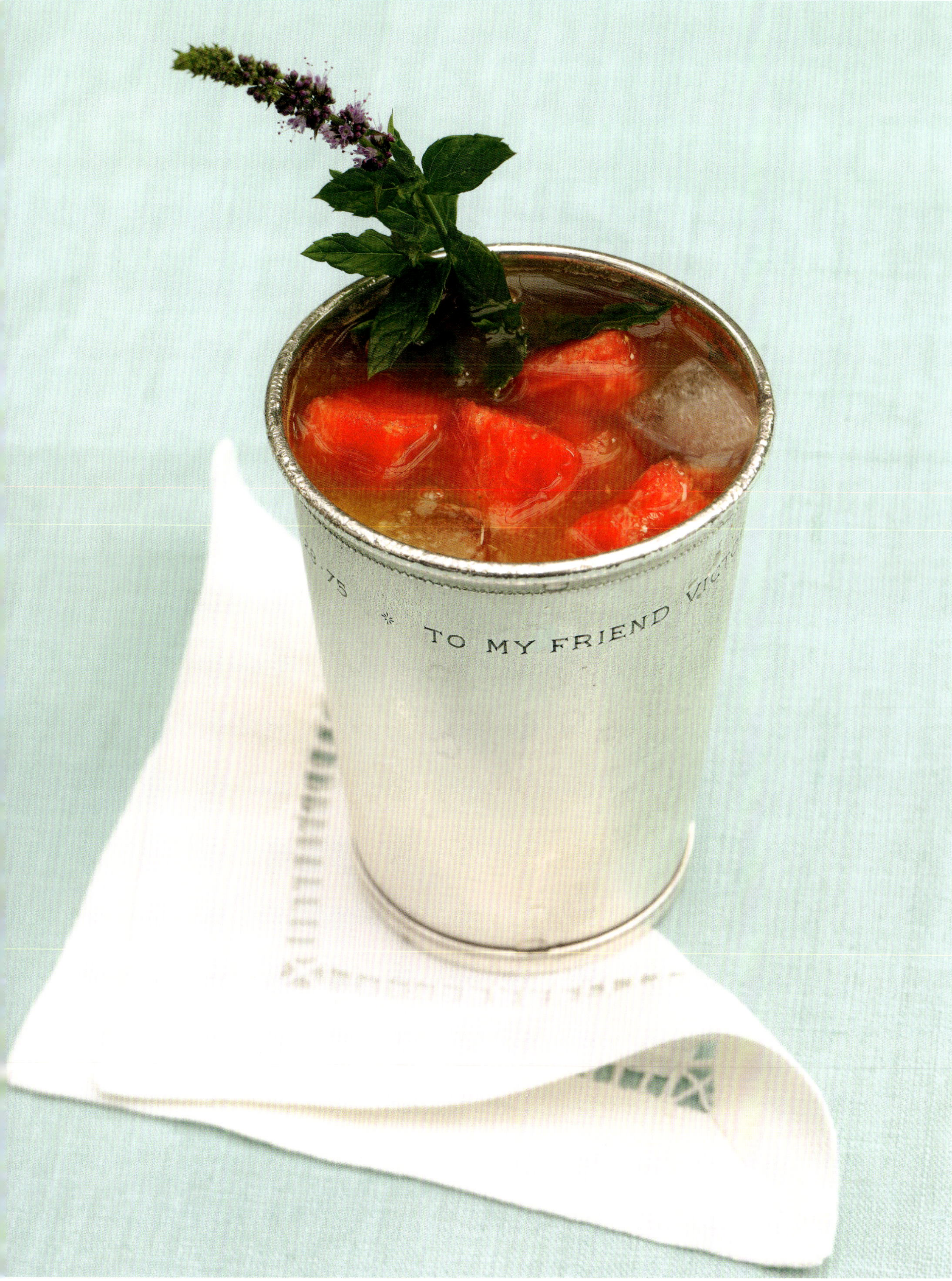
TO MY FRIEND

Frozen Watermelon Margarita

This classic summer favorite takes a novel turn with its dazzling layers of color. Rimming the glass with pink sea salt adds to the beauty, but regular sea salt works just as well. Your freezer will do most of the work—just remember to leave ample time for chilling.

2 LIMES

2 CUPS ½-INCH CUBES RED-FLESHED WATERMELON, SEEDED

1 CUP TEQUILA, DIVIDED

½ CUP TRIPLE SEC, DIVIDED

2 CUPS ½-INCH CUBES YELLOW-FLESHED WATERMELON, SEEDED

3 CUPS ICE, DIVIDED

FINE SEA SALT, PREFERABLY PINK (FOR RIMMING GLASSES)

4 WATERMELON WEDGES (FOR SERVING)

1. Cut 1 small wedge from 1 lime, cover it with plastic wrap, and refrigerate until ready to use. Juice the remaining limes.

2. In the jar of a blender, purée red-fleshed watermelon, ½ cup tequila, ¼ cup triple sec, and half of the lime juice until smooth. Transfer to a container with lid; cover and freeze until well chilled, at least 4 hours or overnight. Repeat with yellow-fleshed watermelon and remaining lime juice, tequila, and triple sec.

3. Separately purée each frozen watermelon mixture in a blender with 1½ cups ice. In a medium bowl, stir together a third of each of the two watermelon mixtures to form an orange purée.

4. Just before serving, pour some sea salt into a small dish. Rub reserved lime wedge over the rims of 4 serving glasses. Gently dip rims into salt to coat lightly.

5. Spoon a quarter of the yellow-fleshed watermelon mixture into each glass, or enough to fill it about a third full (keep any leftover mixture in the freezer until ready to use). Repeat with the orange mixture and then the red. Garnish with thinly sliced watermelon wedges.

SERVES 4

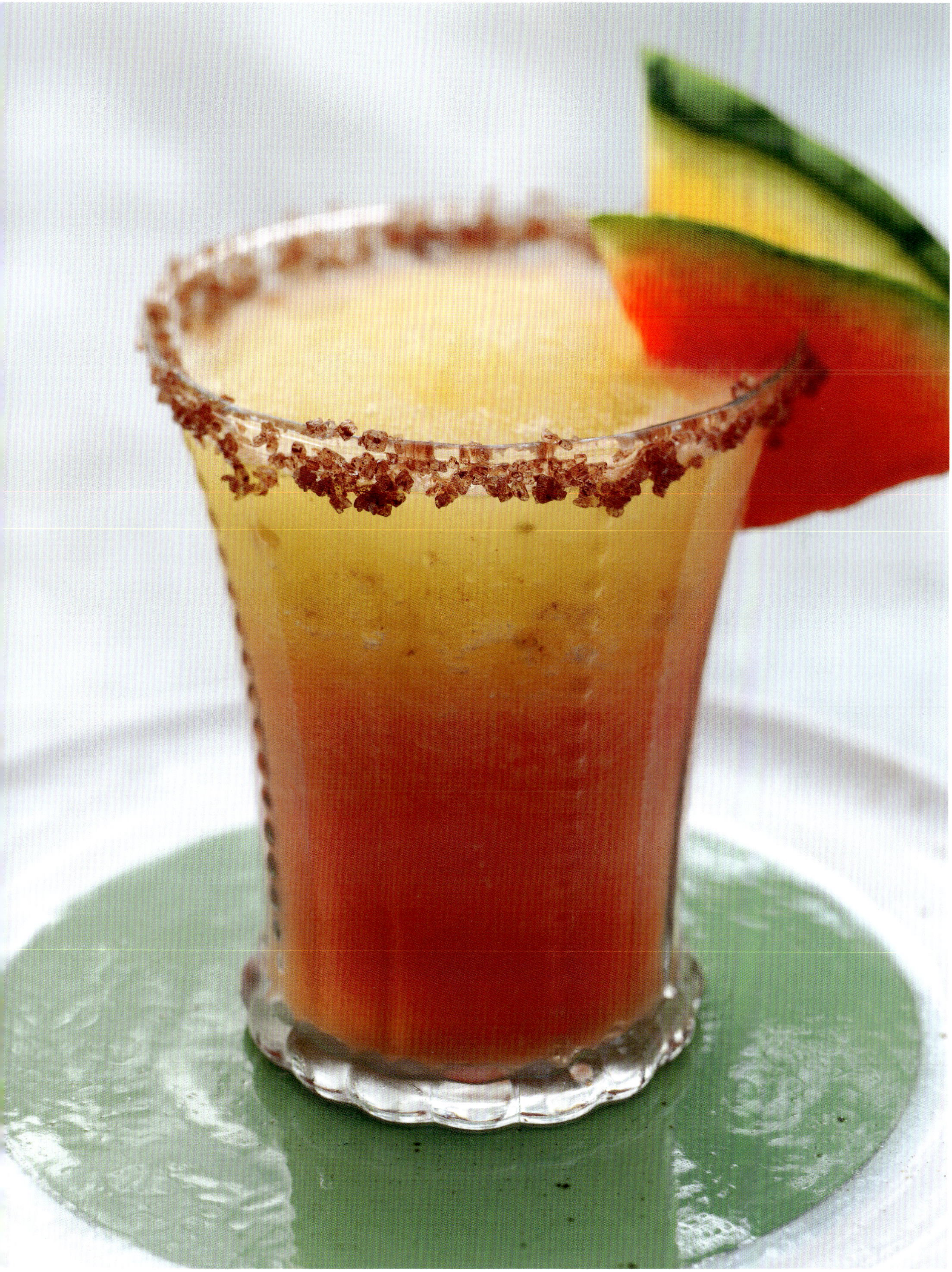

Yogurt with Melon, Fresh Cherries, and Toasted Buckwheat

Any orange-fleshed melon would work well in this dish, which makes a lovely breakfast or summer dessert. Buckwheat groats—the hulled seeds of the buckwheat plant—lend a toasted nuttiness.

2 TABLESPOONS BUCKWHEAT GROATS
2 TABLESPOONS HONEY
½ TEASPOON GRATED LEMON ZEST
⅛ TEASPOON GROUND CARDAMOM
PINCH FINE SEA SALT
1½ CUPS PLAIN GREEK YOGURT (2% MILKFAT)
1 SMALL ORANGE-FLESHED MELON, SKIN AND SEEDS REMOVED, CUT INTO WEDGES
½ CUP HALVED PITTED FRESH SWEET CHERRIES
¼ CUP LOOSELY PACKED MINT LEAVES, TORN IF LARGE

1. Toast buckwheat in a large, dry skillet over medium heat, tossing often, until fragrant, 3 to 4 minutes. Transfer to a plate; let cool.

2. Meanwhile, stir together honey, lemon zest, cardamom, and salt.

3. Arrange yogurt, melon, and cherries on 4 serving plates. Drizzle with citrus-cardamom honey and sprinkle with toasted buckwheat and mint.

SERVES 4

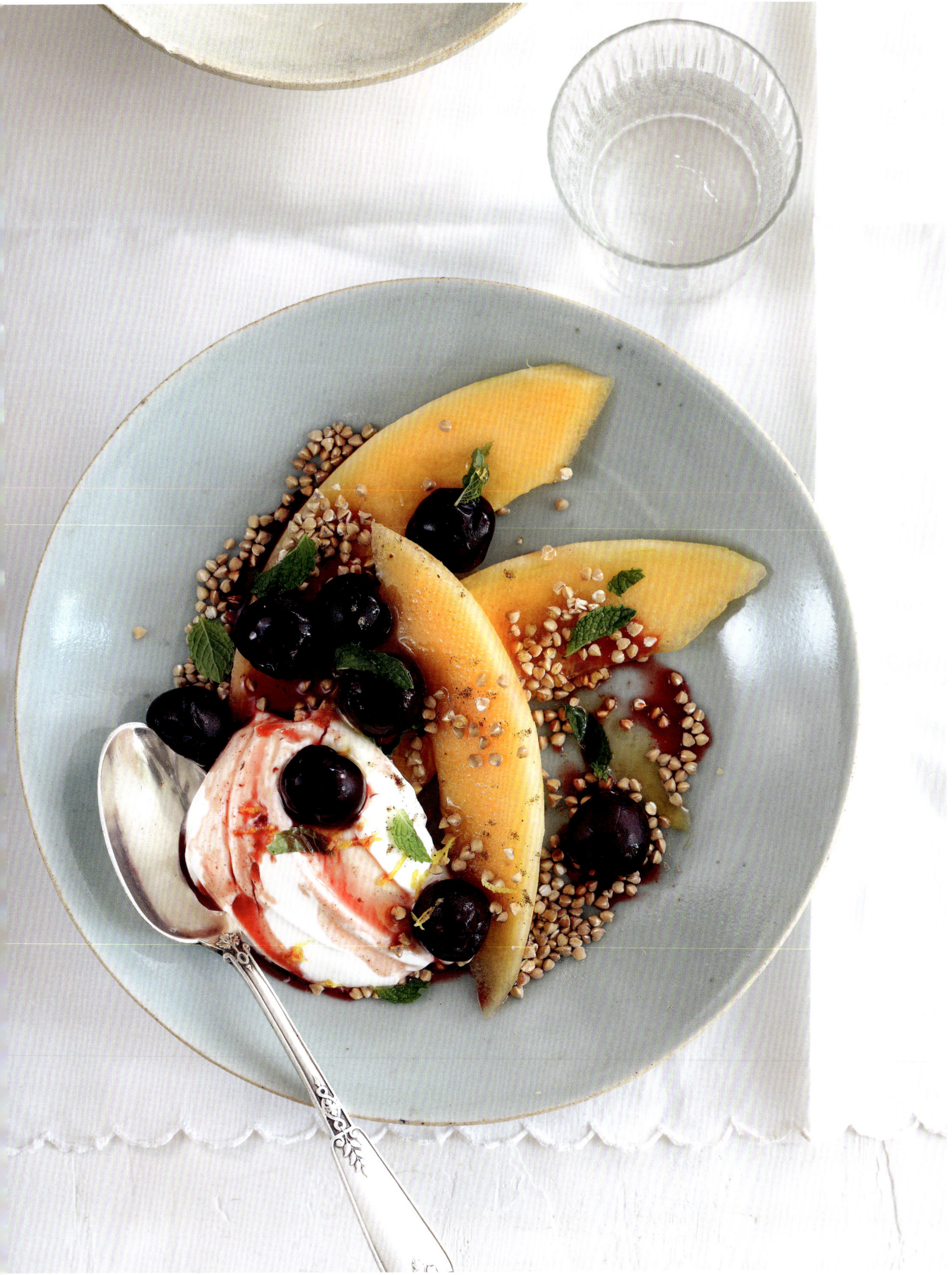

Watermelon Mojito Granita

This wonderfully icy treat, spiked with a splash of rum and lime juice, recalls one of my favorite Cuban cocktails. Use any variety of watermelon that you like. Chilling serving bowls or glasses in the refrigerator or freezer before serving helps keep the finished granita frosty.

3 CUPS ½-INCH CUBES WATERMELON, SEEDED

2 TABLESPOONS FRESH LIME JUICE

2 TABLESPOONS WHITE RUM

2 TABLESPOONS SIMPLE SYRUP (SEE NOTE)

¼ CUP FINELY CHOPPED FRESH MINT PLUS MORE FOR GARNISH

1. In the jar of a blender or a food processor, purée watermelon, lime juice, rum, and simple syrup until smooth. Pour mixture into an 8- or 9-inch square metal baking pan; stir in mint. Freeze for 1 hour. Stir mixture together well with a fork, mashing up any frozen lumps. Return pan to freezer and freeze until mixture is solid, 2 to 3 hours.

2. Using a fork, scrape granita into icy flakes. Granita can be made up to 3 days ahead; cover tightly with foil and keep frozen. Scrape with fork again, just before serving.

3. Scoop granita into chilled serving bowls; sprinkle with additional mint, if desired, and serve.

Note: To make ¾ cup simple syrup, combine ½ cup granulated sugar and ½ cup water in small saucepan. Bring just to a boil, whisking to dissolve sugar. Remove pan from heat, let cool, then transfer to a jar. Simple syrup keeps, covered and refrigerated, for up to 1 month. Leftover syrup can be used to make more granita, or to sweeten iced teas, lemonades, or cocktails.

MAKES ABOUT 2 QUARTS / SERVES 8 TO 10

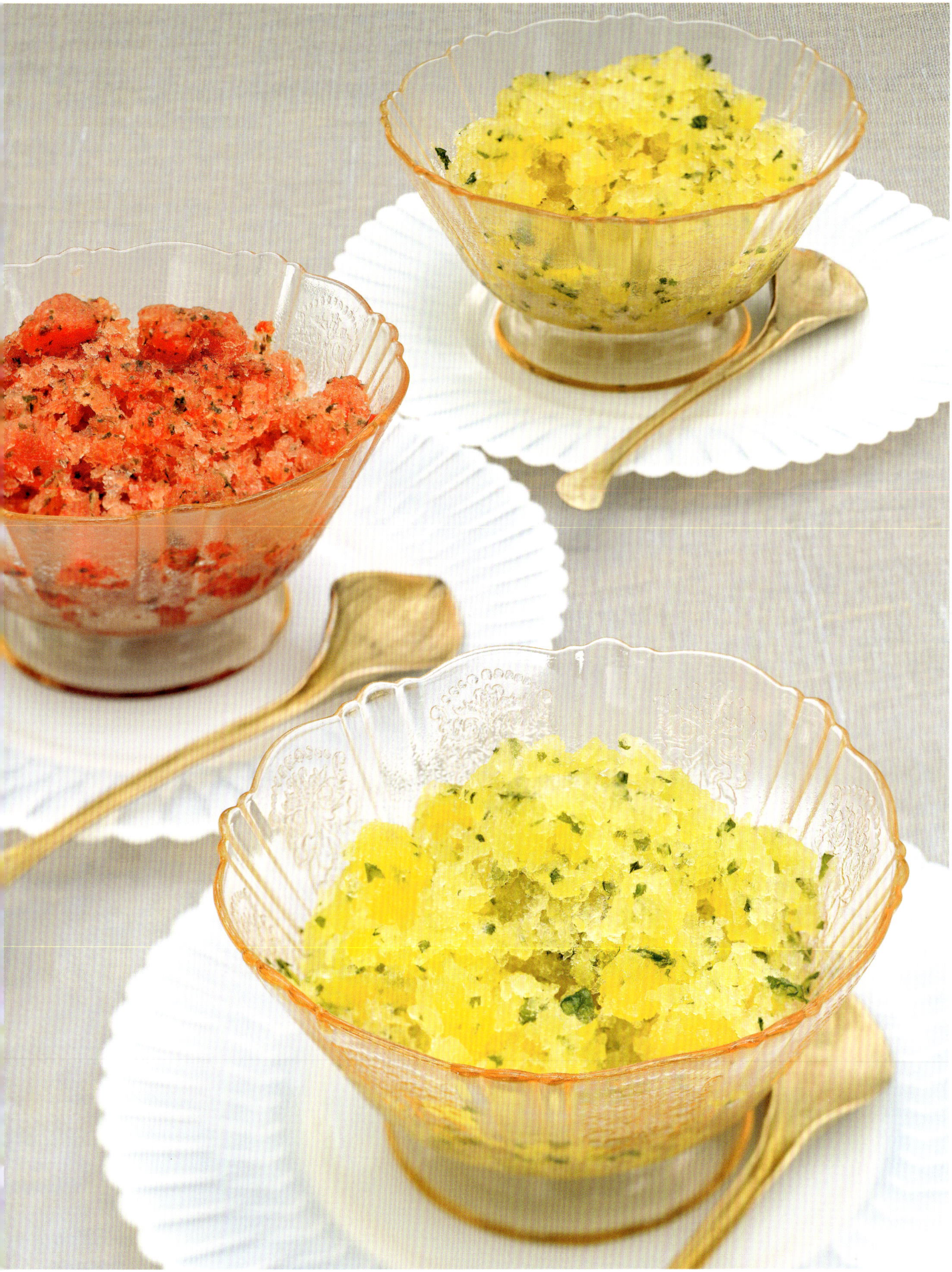

Melon, Ginger, and Lime Ice Pops

Cilantro is the secret ingredient that brings pizazz to these lightly sweetened melon pops—but if you're not a fan of this herb, it's okay to leave it out. Use a fine grater for the ginger, grating the root over a bowl to collect any juices before adding it to the mix.

6 CUPS ½-INCH CUBES ANNE ARUNDEL OR OTHER GREEN-FLESHED MELON

½ CUP LOOSELY PACKED CILANTRO LEAVES (OPTIONAL)

⅓ CUP SUGAR

⅓ CUP FRESH LIME JUICE

2½ TEASPOONS FINELY GRATED PEELED FRESH GINGER (FROM A 2-INCH PIECE)

⅛ TEASPOON FINE SEA SALT

1. Purée all ingredients in a blender until smooth. Pour mixture into 10 3-ounce or 8 4-ounce molds, dividing evenly. Insert ice-pop sticks and freeze, at least 6 hours.

MAKES 8 TO 10 POPS

Cantaloupe–Coconut Milk Ice Pops

These two-tone ice pops mix sweet melon with creamy cardamom-flecked coconut—a taste treat that is anything but commonplace. Remaining mixture, if any, can be used for smoothies.

5 CUPS ½-INCH CUBES PERSIAN OR OTHER ORANGE-FLESHED DESSERT MELON

½ CUP SUGAR, DIVIDED

1 (13- TO 14-OUNCE) CAN FULL-FAT UNSWEETENED COCONUT MILK, WELL SHAKEN

¾ TEASPOON GROUND CARDAMOM

½ TEASPOON PURE VANILLA EXTRACT

⅛ TEASPOON FINE SEA SALT, DIVIDED

1. In a blender, purée melon and ¼ cup sugar until smooth; set aside.

2. In a bowl, vigorously whisk together coconut milk, cardamom, vanilla, salt, and remaining ¼ cup sugar.

3. Add melon purée to 10 3-ounce or 8 4-ounce ice pop molds, filling a third of the way, about 2 tablespoons. Whisk coconut mixture well, then slowly add to molds, filling another third of the way. Fill the remaining space with melon purée.

4. Insert ice-pop sticks and freeze at least 6 hours.

MAKES 8 TO 10 POPS

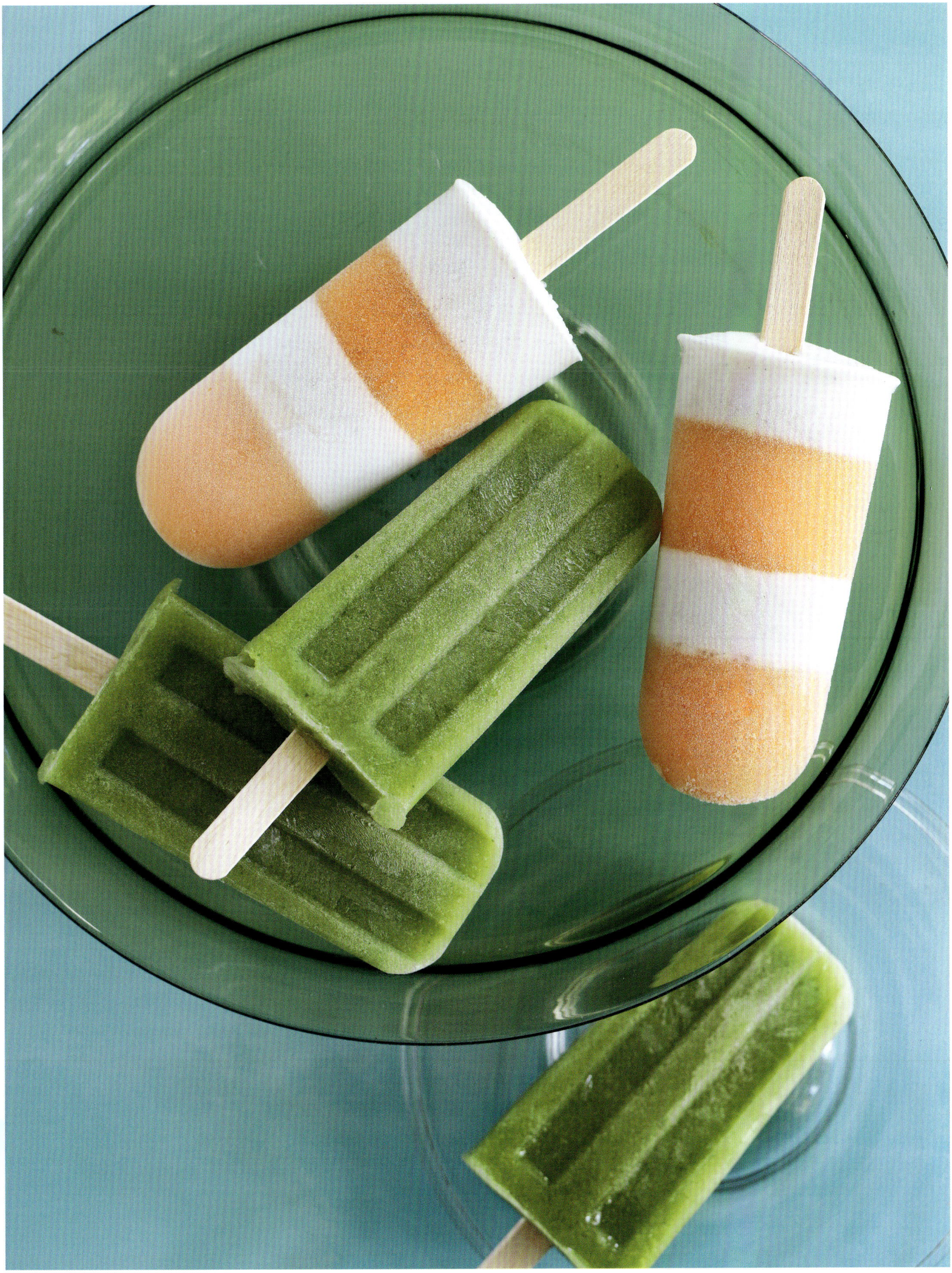

Pavlova with Rosé-Poached Melon and Berries

This marshmallowy meringue, piled high with freshly whipped cream and rosé-poached fruit, is an elegant dessert. The trio of white-, green-, and orange-fleshed melon, set off by fresh blueberries, is at once colorful and nuanced. Prepare this dessert on a clear, dry day; humidity and moisture make it challenging to ace, as does any bit of fat on whisk or bowl.

1¾ CUPS PLUS 1½ TEASPOONS SUPERFINE GRANULATED SUGAR, DIVIDED

4 LARGE EGG WHITES, AT ROOM TEMPERATURE

½ TEASPOON PURE VANILLA EXTRACT

1 TABLESPOON CORNSTARCH

1 TEASPOON DISTILLED WHITE VINEGAR

1 VANILLA BEAN, SPLIT IN HALF LENGTHWISE

2 CUPS DRY ROSÉ WINE

⅛ TEASPOON FINE SEA SALT

1½ CUPS MIXED GREEN-, WHITE-, AND ORANGE-FLESHED MELON

¾ CUP FRESH BLUEBERRIES

1 CUP HEAVY CREAM, CHILLED

1. Heat oven to 300°F with rack in middle. Line a baking sheet with parchment paper; draw an 8-inch circle on the paper, then turn it over, circle-side down.

2. Measure out 1 cup sugar; set aside. In the bowl of an electric mixer fitted with the whisk attachment, beat egg whites on medium speed until soft opaque peaks form, about 3 minutes. Increase speed to high, then slowly add the 1 cup sugar, 1 tablespoon at a time, beating to fully incorporate each before adding the next spoonful and scraping bowl, as needed, until meringue holds very stiff shiny peaks and mixture feels smooth when a little bit is rubbed between two fingers, about 3 minutes more. Beat in vanilla extract, then sprinkle cornstarch and vinegar over top and gently fold to combine.

3. Using the back of a large spoon, spread meringue inside the circle. Form a well in the center, leaving the edges slightly higher.

4. Place meringue in the oven, then reduce oven temperature to 250°F. Bake 50 to 60 minutes, or until outside is dry and a very pale cream color. Turn off oven. Prop open door with handle of a wooden spoon. Let meringue cool completely, at least 2 to 3 hours (walls of meringue should be dry; center will be a bit soft).

Meanwhile, scrape seeds from half of the vanilla bean. Combine seeds and scraped bean, wine, salt, and ¾ cup sugar in a medium saucepan. Bring to a boil, whisking occasionally to dissolve sugar. Remove from heat, then add melon and let stand 2 minutes. Add blueberries and gently stir just to coat with liquid. Using slotted spoon, immediately transfer the fruit to a plate and let cool.

5. Return syrup to a boil and cook until thickened and reduced to about ½ cup, 18 to 20 minutes. Remove from heat and let cool completely.

6. When meringue is cooled, carefully peel off parchment paper and transfer to a serving plate.

7. In the bowl of an electric mixer fitted with the whisk, combine cream and remaining 1½ teaspoons sugar. Beat on medium-high speed until soft peaks form, about 1½ minutes. Dollop the cream mixture into the well of the meringue. Top with fruit, then drizzle with 2 to 3 tablespoons of the syrup. Serve immediately, with remaining syrup on the side.

SERVES 8

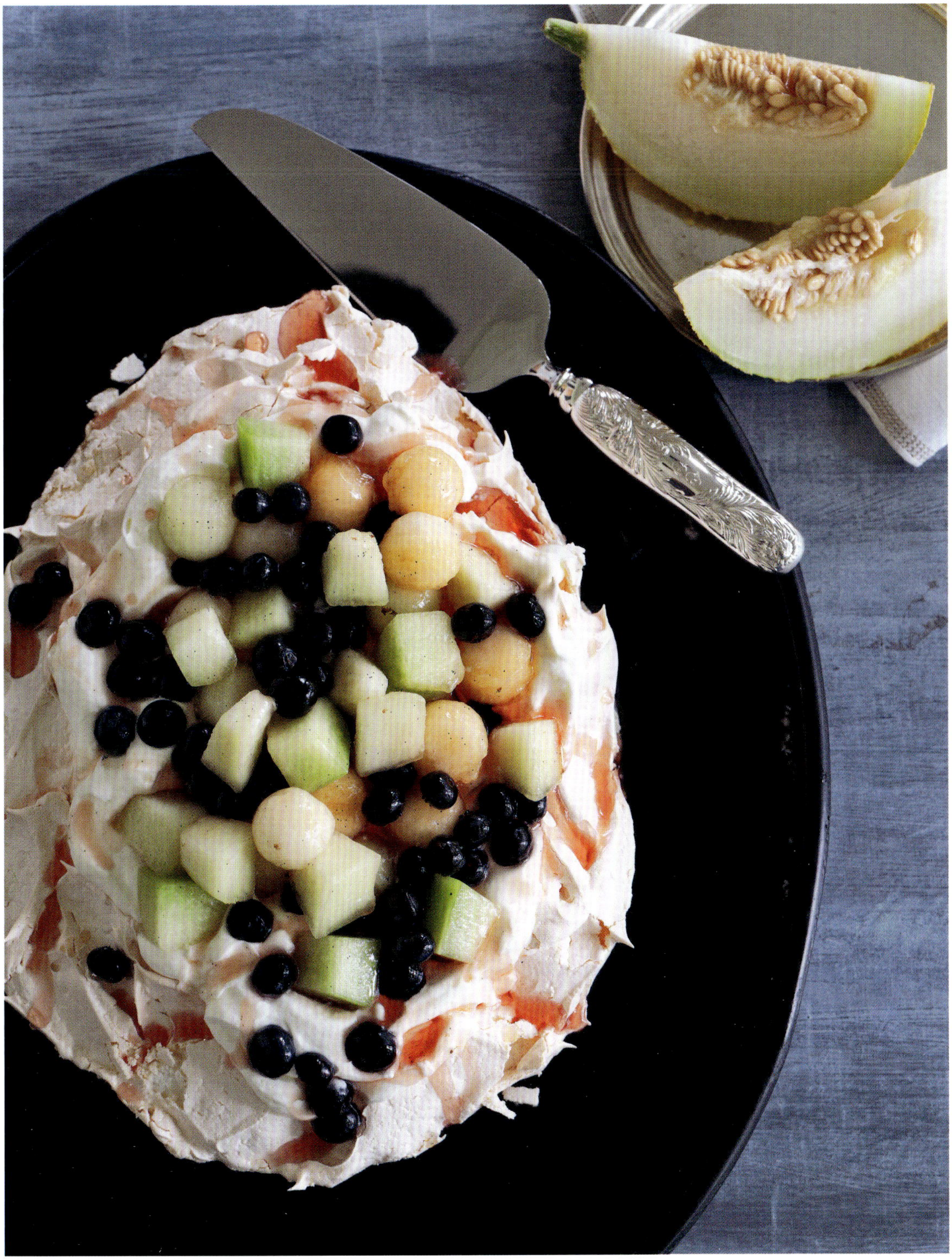

Charentais, Blackberry, and Limoncello Trifle

This beautiful dessert is easier to make than you'd imagine—and messier, in a good way. A glass trifle bowl is a great vessel in which to showcase the fruits of the earth. The Limoncello lends a tart note, a touch of booze, and a little Italian flair.

1¾ CUPS CHILLED HEAVY CREAM

2 TABLESPOONS PLUS ¼ TEASPOON SUGAR

5 TABLESPOONS LIMONCELLO, DIVIDED, PLUS MORE FOR SERVING

3½ CUPS FRESH BLACKBERRIES (ABOUT 14 OUNCES)

3 CUPS ½-INCH CUBES CHARENTAIS OR OTHER FIRM ORANGE-FLESHED MELON

1 STORE-BOUGHT POUND CAKE (10 ¾ OUNCES), CUT INTO ½-INCH CUBES

FRESH MINT AND MINT FLOWERS IF AVAILABLE (FOR SERVING)

1. Whip cream and 2 tablespoons sugar to soft peaks. Add 2 tablespoons Limoncello and continue to whip until it just forms medium peaks.

2. Set aside 10 blackberries for garnish.

3. Place about half of the melon and half of the remaining blackberries in the bottom of a 3-quart trifle dish or glass bowl. Arrange about half of the cake cubes on top. Drizzle with 1½ tablespoons Limoncello. Spoon about half of the whipped cream on top. Repeat to make a second layer.

4. Refrigerate, covered, at least 4 hours or overnight. Before serving, let stand at room temperature 30 to 45 minutes. Meanwhile, in a small bowl, mash together 4 of the remaining blackberries and ¼ teaspoon sugar. Let stand until you're ready to serve the trifle.

5. Just before serving, press the mashed berries through a mesh sieve into a small bowl to extract the juices; discard any solids left behind. Top the trifle with the remaining whole berries, then drizzle the berry juices. Garnish with fresh mint. Serve with chilled Limoncello, if desired.

SERVES 8

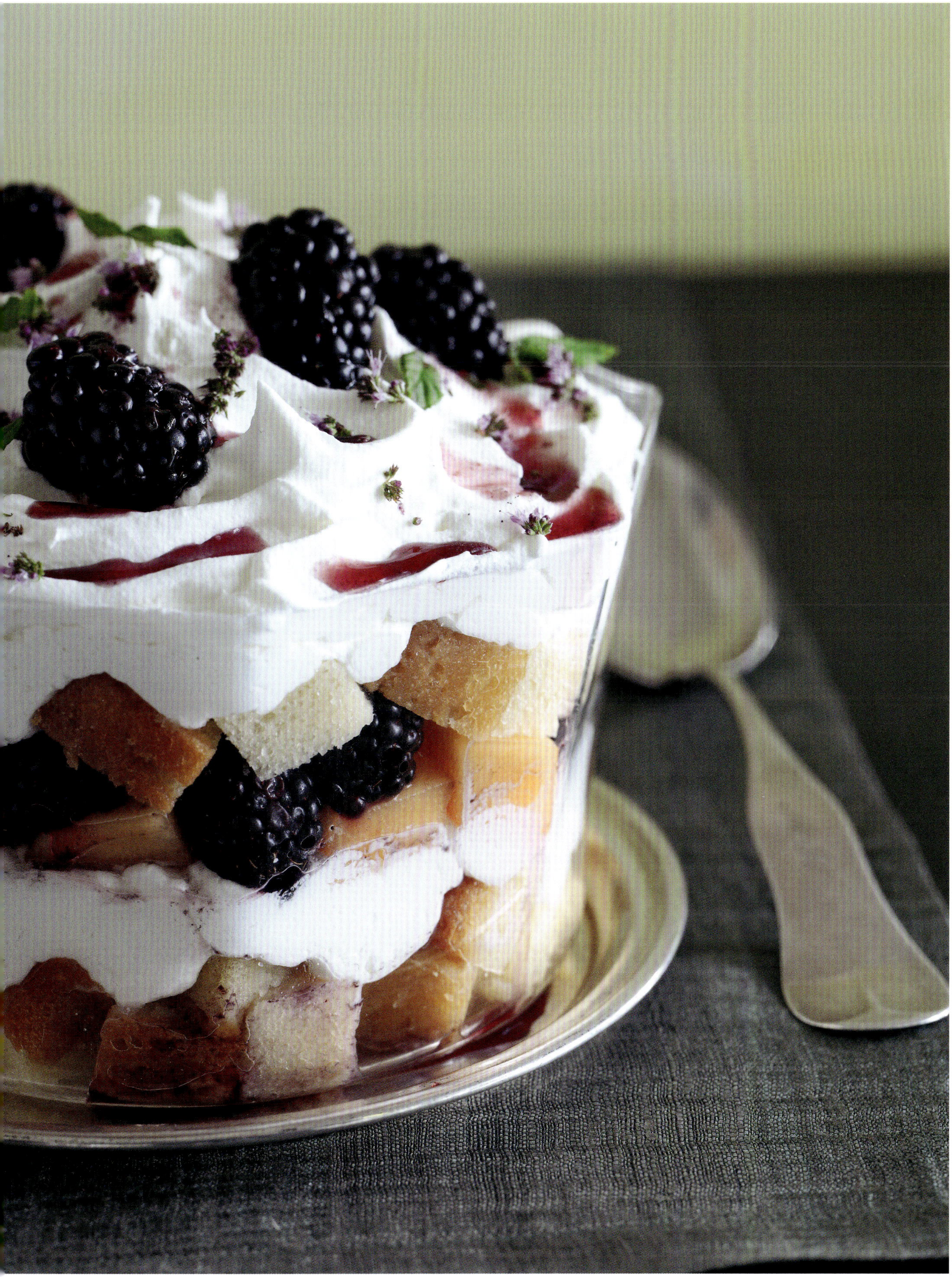

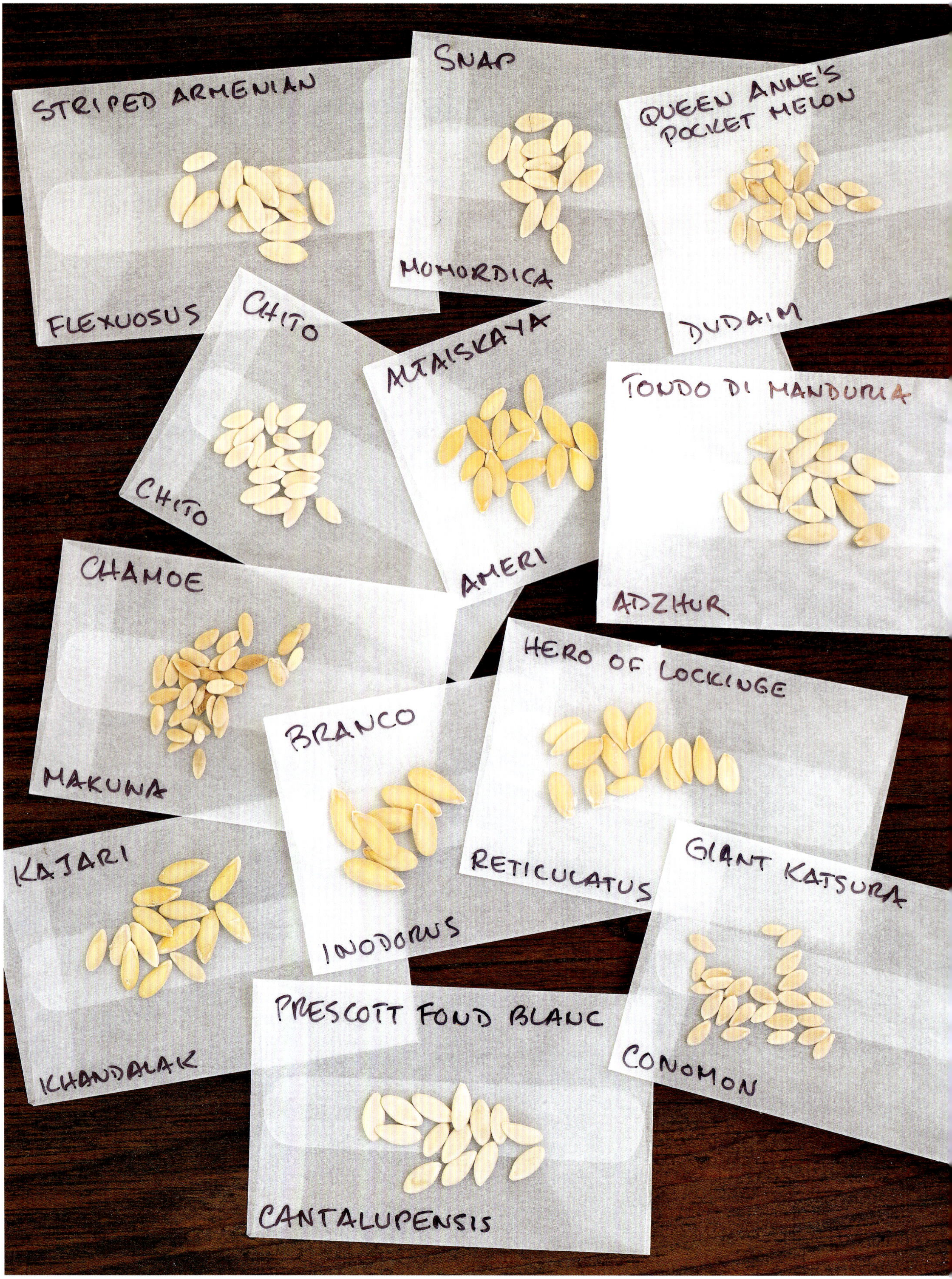
STRIPED ARMENIAN
FLEXUOSUS
SNAP
MOMORDICA
QUEEN ANNE'S POCKET MELON
DUDAIM
CHITO
CHITO
ALTAISKAYA
AMERI
TONDO DI MANDURIA
ADZHUR
CHAMOE
MAKUWA
HERO OF LOCKINGE
RETICULATUS
BRANCO
INODORUS
KAJARI
KHANDALAK
GIANT KATSURA
CONOMON
PRESCOTT FOND BLANC
CANTALUPENSIS

Seed Sources

ADAP—ADAPTIVE SEEDS
WWW.ADAPTIVESEEDS.COM
25079 BRUSH CREEK ROAD
SWEET HOME, OR 97386
Free catalogue to U.S. and Canadian addresses. Farm-based seed company established in 2009 by Andrew Still and Sarah Kleeger. Offers Pacific Northwest grown, open pollinated, and organic seeds, including rare and heritage varieties, public domain and modern varieties, and diverse gene pool mixes and landraces.

AMI—AMISHLAND HEIRLOOM SEEDS
WWW.AMISHLANDSEEDS.COM
BOX 365
REAMSTOWN, PA 17567-0365
Online catalogue. Grows and offers organically grown seeds, including local central Pennsylvania heirlooms, and landrace, unusual, and exotic seeds from around the world.

ANN—ANNIE'S HEIRLOOM SEEDS
WWW.ANNIESHEIRLOOMSEEDS.COM
4646 RANSOM STREET
HUDSONVILLE, MI 49426
PHONE: 800-313-9140
Free catalogue. Grows and offers organically grown heirloom vegetable seeds. Retail and bulk.

Opposite: Dried melon seeds representing twelve horticultural groups. Melon seeds aren't as flashy as watermelon seeds; they're typically tan or yellow and have a pointed radicle end.

BAKE—BAKER CREEK HEIRLOOM SEED CO.
WWW.RARESEEDS.COM
2278 BAKER CREEK ROAD
MANSFIELD, MO 65704
PHONE: 417-924-8917
Free catalogue to U.S. addresses; can be sent to international customers for shipping costs. Sells only nonhybrid seeds. Specializes in heirloom varieties, including many rare types and Asian and European varieties. Outstanding melon collection. Retail, bulk, and wholesale offerings.

BERL—BERLIN SEEDS
5335 COUNTY ROAD 77
MILLERSBURG, OH 44654
PHONE: 877-464-0892
Free catalogue. Ohio's largest mail order seed company. Retail store and gardening and supply catalogue includes vegetable seeds, fruit trees, and garden tools and supplies. Retail and bulk.

BOT—BOTANICAL INTERESTS, INC.
WWW.BOTANICALINTERESTS.COM
660 COMPTON STREET
BROOMFIELD, CO 80020
PHONE: 877-821-4340
Free catalogue. United States sales only. High-quality seeds, including many heirlooms and organically grown varieties. Beautifully illustrated seed packets include variety information, history and botanical facts, growing and harvesting instructions. Retail and wholesale.

BUR—W. ATLEE BURPEE & COMPANY
WWW.BURPEE.COM
300 PARK AVENUE
WARMINSTER, PA 18974
PHONE: 800-888-1447
Free catalogue. Offering home gardeners fine-quality varieties for over 140 years.

BURR—BURRELL SEED GROWERS, LLC
WWW.BURRELLSEEDS.US
PO BOX 150
ROCKY FORD, CO 81067
PHONE: 844-254-7333 / 719-254-3318
Free catalogue. Family-owned business founded in 1900. Fine collection of cantaloupe, watermelon, vegetable, flower, and herb seeds.

CLIF—CLIFTON SEED COMPANY
WWW.CLIFTONSEED.COM
PO BOX 206
FAISON, NC 28341
PHONE: 800-231-9359
Free catalogue. Family business founded in 1928. Offering a variety of high-quality vegetable seeds for home gardeners and commercial growers. Retail and wholesale.

CUC—CUCUMBER SHOP
WWW.CUCUMBERSHOP.COM
Online catalogue. Hobby seed business offering high-quality, burpless, bitter-free, rare Carosello seeds.

DEB—DE BRUYN SEED CO.
WWW.DEBRUYNSEED.COM
101 E. WASHINGTON
ZEELAND, MI 49464
PHONE: 616-772-2316
Online catalogue. Family owned and operated business offering a full line of quality seeds since 1915. Retail and wholesale.

EDEN—EDEN BROTHERS
WWW.EDENBROTHERS.COM
2099 BREVARD ROAD
ARDEN, NC 28704
PHONE: 828-633-6338
Offers hundreds of heirloom, organic, open-pollinated, and rare and hard-to-find vegetable, herb, and flower seeds. Retail and bulk.

ERS—E & R SEED, LLC
1356 E. 200 STREET
MONROE, IN 46772
PHONE: 260-692-6827
Free catalogue. Separate retail and wholesale catalogues, each offering more than 1,000 varieties of vegetable and flower seeds, including many open-pollinated and heirloom varieties.

EVER—EVERWILDE FARMS, INC.
WWW.EVERWILDE.COM
PO BOX 40
SAND CREEK, WI 54765
PHONE: 888-848-3837
Online catalogue. Family-owned company specializing in wildflower and grass seeds. Also offers heirloom and organic vegetable and herb seeds.

FED—FEDCO SEEDS
WWW.FEDCOSEEDS.COM
PO BOX 520
CLINTON, ME 04927
PHONE: 207-426-9900
Free catalogue. Cooperative seed company offering a large selection of certified organic cultivars and regional heirloom varieties, including vegetable, herb, flower, cover crop, and green manure seeds. Specializes in selections for short growing seasons. Retail and wholesale.

GIANT—GIANT WATERMELONS
WWW.GIANTWATERMELONS.COM
PO BOX 141
HOPE, AR 71802
Online listing. Small hobby seed business offering quality seeds of four giant varieties grown on their farm that are potential contest winners.

GOOD—THE GOOD SEED CO.
WWW.GOODSEEDCO.NET
PO BOX 4706
WHITEFISH, MT 59937
Online catalogue. Offers regionally adapted organic and ecologically farmed heirloom herb, vegetable, and flower seeds.

GOU—GOURMET SEED INTERNATIONAL
WWW.GOURMETSEED.COM
743 SHORE ROAD
HOLLISTER, CA 95023-9427
PHONE: 831-637-2411
Online catalogue. Small, family-owned seed company providing high-quality heirloom seeds, including many certified organic varieties. Seeds are imported from many sources, including Bavicchi, one of Italy's finest seed houses.

GRETA—GRETA'S ORGANIC GARDENS
WWW.SEEDS-ORGANIC.COM
399 RIVER ROAD
GLOUCESTER, ON K1V 1C9 CANADA
PHONE: 613-521-8648
Offering organically certified, open-pollinated vegetable and garden seeds.

HALI—HALIFAX SEED COMPANY LTD.
WWW.HALIFAXSEED.CA
5860 KANE STREET
PO BOX 8026
HALIFAX, NS B3K 5L8 CANADA
PHONE: 902-454-7456

Free catalogue, Canada only. Established in 1866; Canada's oldest family-owned seed company. Retail and wholesale.

HAR—HARRIS SEEDS
WWW.HARRISSEEDS.COM
355 PAUL ROAD
PO BOX 24966
ROCHESTER, NY 14624
PHONE: 800-544-7938

Free catalogue. Providing the finest-quality vegetable and flower seeds, plants, and supplies to large and small growers and gardeners throughout the United States. Retail and wholesale.

HEIR—HEIRLOOM SEEDS
WWW.HEIRLOOMSEEDS.COM
287 E. FINLEY DRIVE
WEST FINLEY, PA 15377

Online orders to U.S. addresses only. Small, family-run seed house offering high-quality heirloom varieties at fair prices since 1988.

HER—HERITAGE HARVEST SEED
WWW.HERITAGEHARVESTSEED.COM
PO BOX 279
CARMAN, MB R0G 0J0 CANADA
PHONE: 204-745-6489

Free catalogue; sales to Canada only. Small farm-based business offering rare and endangered open-pollinated, untreated, heirloom vegetable, flower, and herb seed.

HIGH—HIGH MOWING ORGANIC SEEDS
WWW.HIGHMOWINGSEEDS.COM
76 QUARRY ROAD
WOLCOTT, VT 05680
PHONE: 802-472-6174 / 866-735-4454

Free catalogue. Providing home gardeners and commercial growers with certified organic, open-pollinated and hybrid vegetable, flower, and herb seeds since 1996. Ships to the United States and Canada. Retail and wholesale.

HOPE—HOPE SEEDS
WWW.HOPESEED.COM
324 ST. GEORGE STREET, BOX 460
ANNAPOLIS ROYAL, NS B0S 1A0 CANADA
PHONE: 902-286-4673

Free catalogue. Offering a wide range of organic and open-pollinated garden seed, including rare and heritage varieties.

HUD—J. L. HUDSON, SEEDSMAN
WWW.JLHUDSONSEEDS.NET
BOX 337
LA HONDA, CA 94020-0337

Free catalogue. No telephone. Public access seed bank specializing in seeds of rare and unusual plants from around the world. Established in 1911, successors to Harry Saier. Stresses open-pollinated, nonhybrid seeds and preservation of biological and cultural diversity.

HUME—ED HUME SEEDS
WWW.HUMESEEDS.COM
PO BOX 73160
PUYALLUP, WA 98373

Online catalogue. Family owned and operated since 1977. Specializing in seeds for cool-climate and short-season areas.

HVSC—HUDSON VALLEY SEED COMPANY
WWW.HUDSONVALLEYSEED.COM
484 METTACAHONTS ROAD
ACCORD, NY 12404
PHONE: 845-204-8769

Free catalogue. The Hudson Valley Seed Library has evolved into a seed company that offers certified organically grown heirloom and open-pollinated vegetable, flower, and herb seeds.

IRISH—IRISH EYES GARDEN SEEDS
WWW.IRISHEYESGARDENSEEDS.COM
5045 ROBINSON CANYON ROAD
ELLENSBURG, WA 98926
PHONE: 509-933-7150

Online catalogue. Specializes in vegetables that are trialed and tasted to ensure quick maturity, good flavor, and vigor. Sales to U.S. addresses only. Retail and wholesale.

JOHN—JOHNNY'S SELECTED SEEDS
WWW.JOHNNYSEEDS.COM
955 BENTON AVENUE
WINSLOW, ME 04901
PHONE: 877-564-6697

Free catalogue. An employee-owned company serving growers since 1973.Vegetable, flower, herb, and farm seed for northern climates. Many heirlooms and new introductions. Extensive trial grounds throughout North America. Retail and commercial.

JOR—JORDAN SEEDS, INC.
WWW.JORDANSEEDS.COM
6400 UPPER AFTON ROAD
WOODBURY, MN 55125-1146
PHONE: 651-738-3422

Free catalogue. Third-generation vegetable growers offering full range of vegetable and herb seeds. Retail and bulk sales.

JSKIT—JOHN SCHEEPERS KITCHEN GARDEN SEEDS
WWW.KITCHENGARDENSEEDS.COM
23 TULIP DRIVE
PO BOX 638
BANTAM, CT 06750
PHONE: 860-567-6086
Free catalogue. Family-owned business offering high-quality, unique varieties from around the world. Illustrated catalogue includes recipes. In business since 1908.

JUNG—J. W. JUNG SEED COMPANY
WWW.JUNGSEED.COM
335 SOUTH HIGH STREET
RANDOLPH, WI 53956
PHONE: 800-247-5864
Free catalogue. Quality seeds since 1907. Specializes in seed and nursery stock suitable for northern climates. Retail and wholesale.

KIT—KITAZAWA SEED COMPANY
WWW.KITAZAWASEED.COM
201 4TH STREET, #206
OAKLAND, CA 94607
PHONE: 510-595-1188
Free catalogue. Source for oriental vegetable seeds for home gardeners, retailers, and commercial growers since 1917. Retail and bulk.

LAND—LANDIS VALLEY HEIRLOOM SEED PROJECT
WWW.LANDISVALLEYMUSEUM.ORG
2451 KISSEL HILL ROAD
LANCASTER, PA 17601
PHONE: 717-569-0401
Free brochure. Specializes in heirloom Pennsylvania German varieties.

LEJ—LE JARDIN DU GOURMET
WWW.ARTISTICGARDENS.COM
PO BOX 75
ST. JOHNSBURY CENTER, VT 05863-0075
PHONE: 800-659-1446
Free catalogue. Specializes in shallots and gourmet seeds. Offers forty-cent sample packets and larger packets of vegetable, herb, and flower seeds. Retail.

LIN—LINDENBERG SEEDS LTD.
WWW.LINDENBERGSEEDS.CA
803 PRINCESS AVENUE
BRANDON, MB R7A 0P5 CANADA
PHONE: 204-727-0575
Free catalogue to Canadian addresses. Sales within Canada only. Specializes in seeds selected for success during the short, sunny prairie growing season.

MAIN—MAIN STREET SEED & SUPPLY
WWW.MAINSTREETSEEDANDSUPPLY.COM
401 MAIN STREET
BAY CITY, MI 48706
PHONE: 866-229-3276
Online catalogue. Offering a variety of high-quality vegetable seeds for the farmer and home gardener. Retail and bulk.

MARY—MARY'S HEIRLOOM SEEDS
WWW.MARYSHEIRLOOMSEEDS.COM
PO BOX 3763
RAMONA, CA 92065
PHONE: 760-870-4555
Family business offering over 500 varieties of heirloom, open-pollinated garden seeds.

MOR—MORGAN COUNTY SEEDS
WWW.MORGANCOUNTYSEEDS.COM
18761 KELSAY ROAD
BARNETT, MO 65011-3009
PHONE: 573-378-2655
Free catalogue. Large selection of open-pollinated seeds, many untreated. Retail and wholesale.

NAT—THE NATURAL GARDENING COMPANY
WWW.NATURALGARDENING.COM
PO BOX 258
DIAMOND SPRINGS, CA 95619-0258
PHONE: 707-766-9303
Free catalogue. The oldest U.S. certified organic nursery offering organically certified seeds and plants. Retail and bulk.

NATIVE—NATIVE SEEDS/SEARCH
WWW.NATIVESEEDS.ORG
3061 N. CAMPBELL AVENUE
TUCSON, AZ 85719
PHONE: 520-622-5561
Nonprofit organization dedicated to conserving the rich agrobiodiversity of the arid Southwest. Providing farmers and gardeners with a large selection of rare and unusual indigenous heirloom vegetable seeds. Retail and bulk.

NATURE—NATURE AND NURTURE SEEDS
WWW.NATUREANDNURTURESEEDS.COM
7100 MARSHALL ROAD
DEXTER, MI 48130
PHONE: 734-929-0802
Free catalogue. Specializing in open-pollinated, heirloom seeds that are grown locally and ecologically. Retail sales.

NEWE—NEW ENGLAND SEED COMPANY (NESEED)
WWW.NESEED.COM
122 PARK AVENUE, BUILDING H
EAST HARTFORD, CT 06108
PHONE: 800-825-5477
Free catalogue. Providing home gardeners and commercial growers with high-quality seeds. Retail and wholesale.

NIC—NICHOLS GARDEN NURSERY
WWW.NICHOLSGARDENNURSERY.COM
1190 OLD SALEM ROAD NE
ALBANY, OR 97321
PHONE: 800-422-3985
Free catalogue. Providing vegetable, herb, and flower seeds and herb plants to gardeners and market growers for over 65 years.

OSC—OSC SEEDS (ONTARIO SEED CO. LTD.)
WWW.OSCSEEDS.COM
PO BOX 7
WATERLOO, ON N2J 3Z6 CANADA
PHONE: 519-886-0557
Catalogue and sales to Canada only. Specializing in untreated, heirloom, open-pollinated seeds for over 125 years.

PEACE—PEACEFUL VALLEY FARM & GARDEN SUPPLY
WWW.GROWORGANIC.COM
PO BOX 2209
GRASS VALLEY, CA 95945
PHONE: 888-784-1722
Free catalogue with 2,000 items, including fertilizers, vegetable, herb, and cover crop seeds. Providing tools and supplies for organic gardeners and farmers since 1976.

PINE—PINETREE GARDEN SEEDS
WWW.SUPERSEEDS.COM
PO BOX 300
NEW GLOUCESTER, ME 04260
PHONE: 207-926-3400
Free catalogue. Specializes in flavorful varieties for home gardeners. Offers gardening books, tools, and bulbs.

PPS—P & P SEED CO.
WWW.PANDPSEED.COM
56 E. UNION STREET
HAMBURG, NY 14075-5007
Online catalogue. Offering seeds that produce world-record giant varieties. Retail and bulk sales.

PRAI—PRAIRIE GARDEN SEEDS
WWW.PRSEEDS.CA
BOX 2758
HUMBOLDT, SK S0K 2A0 CANADA
PHONE: 306-682-1475
Specializes in open-pollinated vegetable and flower varieties grown without agricultural chemicals and in short-season dryland growing conditions. Offers many heirlooms.

REDS—REDWOOD SEEDS
WWW.REDWOODSEEDS.NET
PO BOX 431
MANTON, CA 96059
PHONE: 530-524-5537
Online catalogue. Family seed farm offering organic, open-pollinated seeds, including many heirlooms.

REDW—REDWOOD CITY SEED COMPANY
WWW.ECOSEEDS.COM
PO BOX 361
REDWOOD CITY, CA 94064
PHONE: 650-325-7333
Free catalogue to United States, Canada, and Mexico; $2.00(US) to other countries. Two annual catalogue supplements online, listing seasonal or rare items when only a dozen packets are available. Specializes in seeds of endangered traditional varieties, all open-pollinated, heirloom vegetables and herbs, including an extensive selection of hot peppers. Order by mail, phone, or fax; no online ordering.

REIM—REIMER SEEDS
WWW.REIMERSEEDS.COM
PO BOX 206
SAINT LEONARD, MD 20685-0206
Online catalogue. Providing vegetable, flower, and herb seeds to home gardeners and market growers.

REN—RENEE'S GARDEN
WWW.RENEESGARDEN.COM
6060 GRAHAM HILL ROAD
FELTON, CA 95018
PHONE: 888-880-7228
Online catalogue. Family-owned company offering vegetable, herb, and flower seeds, including open-pollinated, heirloom, and gourmet varieties. Ships to the United States and Canada. Retail and wholesale.

REST—RESTORATION SEEDS
WWW.RESTORATIONSEEDS.COM
9969 WAGNER CREEK ROAD
TALENT, OR 97540
Online catalogue. Farmer owned. Promoting food sovereignty by offering all open-pollinated varieties to gardeners and farmers. Retail and bulk.

RICH—RICHTERS HERBS
WWW.RICHTERS.COM
357 HIGHWAY 47
GOODWOOD, ON LOC 1AO CANADA
PHONE: 905-640-6677
Free catalogue to Canada and United States. Offering herb and vegetable seeds and plants, dried herbs, books, and more. Retail and wholesale.

ROHR—P. L. ROHRER & BRO., INC.
WWW.ROHRERSEEDS.COM
PO BOX 250
SMOKETOWN, PA 17576
PHONE: 717-299-2571
Free catalogue. Quality farm and garden seeds since 1919. Seeds and bulbs for over 900 varieties of vegetables, flowers, herbs, cover crops, and lawn grasses. Many organic and heirloom seeds. Retail and wholesale.

SALT—SALT SPRING SEEDS
WWW.SALTSPRINGSEEDS.COM
BOX 444, GANGES P.O.
SALT SPRING ISLAND, BC V8K 2W1 CANADA
PHONE: 250-537-5269 (INQUIRIES ONLY; NO PHONE ORDERS)
Catalogue $2.00. No sales to the United States. Grows and sells organically grown, open-pollinated vegetable, herb, and grain seeds.

SAMP—THE SAMPLE SEED SHOP
WWW.SAMPLESEEDS.COM
978 RIVERVIEW BLVD.
TONAWANDA, NY 14150
Online catalogue. Offers sample packets of quality open-pollinated vegetable seeds in simple packaging at fair prices.

SAND—SAND HILL PRESERVATION CENTER HEIRLOOM SEEDS & POULTRY
WWW.SANDHILLPRESERVATION.COM
1878 230TH STREET
CALAMUS, IA 52729
PHONE: 563-246-2299
Free catalogue. All varieties are open-pollinated and grown on-site. Operated by Glenn and Linda Drowns. Excellent melon selection. Orders accepted by regular mail only; no credit card orders. Retail only.

SEEDSF—SEEDS FROM ITALY
WWW.GROWITALIAN.COM
PO BOX 3908
LAWRENCE, KS 66046
PHONE: 785-748-0959
Free catalogue. Sales within United States only. U.S. distributor of Italy's Franchi seeds, including heirloom Italian vegetables, flowers, and herbs.

SESE—SOUTHERN EXPOSURE SEED EXCHANGE
WWW.SOUTHERNEXPOSURE.COM
PO BOX 460
MINERAL, VA 23117
PHONE: 540-894-9480
Free catalogue. Offering more than 900 varieties of heirloom and traditional vegetables, flowers, and herbs. Many are suited to hot, humid, and disease-prone areas. Seed is untreated, and much of it is grown organically. Retail and bulk.

SHUM—R. H. SHUMWAY'S COMPANY
WWW.RHSHUMWAY.COM
334 W. STROUD STREET
RANDOLPH, WI 53956
PHONE: 800-342-9461
Free catalogue. Specializes in heirloom and open-pollinated seeds.

SISK—SISKIYOU SEEDS
WWW.SISKIYOUSEEDS.COM
3220 EAST FORK ROAD
WILLIAMS, OR 97544
PHONE: 541-415-0877 (NO PHONE ORDERS; SEED AVAILABILITY OR ORDER QUESTIONS ONLY)
Free catalogue. Providing growers with organically certified, open-pollinated, vigorous seeds.

SOC—SEEDS OF CHANGE
WWW.SEEDSOFCHANGE.COM
PO BOX 4908
RANCHO DOMINGUEZ, CA 90220
PHONE: 888-762-7333
Free catalogue. Specializes in vegetables, flowers, and herbs that are 100 percent certified organic and open pollinated. Wide selection of traditional and heirloom varieties. Retail and wholesale.

SOLA—SOLANA SEEDS
HTTPS://SOLANASEEDS.NETFIRMS.COM
17 PLACE LEGER
REPENTIGNY, QC J6A 5N7 CANADA
Online catalogue. Small Canadian seed company offering vegetable and flower seeds, including many heirlooms, and rare and unusual exotic plant seeds.

SOTE—SALT OF THE EARTH SEED COMPANY
WWW.SALTOFTHEEARTHSEEDS.COM
PO BOX 226
MATTITUCK, NY 11952
Specializes in Long Island regional open-pollinated and heirloom varieties. Collaborates with chefs, local breeders, and nonprofit organizations to promote a sustainable, biodiverse food culture for all.

SOWT—SOW TRUE SEED
WWW.SOWTRUESEED.COM
243 HAYWOOD STREET
ASHEVILLE, NC 28801
PHONE: 828-254-0708
Free catalogue. Offers over 500 varieties of heirloom, organic, and open-pollinated vegetable, flower, and herb seeds.

SSE—SEED SAVERS EXCHANGE
ONLINE SEED EXCHANGE
HTTPS://EXCHANGE.SEEDSAVERS.ORG
3094 NORTH WINN ROAD
DECORAH, IA 52101
PHONE: 563-382-5990
The Seed Exchange is a gardener-to-gardener seed swap that takes place online and through the annual print Seed Savers Exchange Yearbook, where hundreds of gardeners offer thousands of open-pollinated and heirloom varieties of flowers, herbs, and fruit.

SSEC—SEED SAVERS EXCHANGE CATALOGUE
WWW.SEEDSAVERS.ORG
3094 NORTH WINN ROAD
DECORAH, IA 52101
PHONE: 563-382-5990
SSE is the nation's most prominent nonprofit seedsaving organization. Founded in 1975 by Kent and Diane Whealy. Free annual color catalogue includes over 600 varieties of vegetables, flowers, and herbs, including many heirlooms from the Seed Savers Exchange Heritage Farm Collection. Project-related revenue from seed sales is used to permanently maintain Heritage Farm's vast seed collections of rare varieties. Retail catalogue contains bulk prices. Wholesale price list available on request. Available to everyone.

STC—ST. CLARE HEIRLOOM SEEDS
WWW.STCLARESEEDS.COM
PO BOX 556
GILLETT, WI 54124
Online catalogue. Small family business offering over 500 organically certified, open-pollinated, and heirloom seeds.

SUS—SUSTAINABLE SEED COMPANY
WWW.SUSTAINABLESEEDCO.COM
355 E. 20TH STREET
CHICO, CA 95928
Online catalogue. Certified organic seed farm offering gardeners and farmers heirloom vegetable, grain, and herb seeds. Retail and bulk.

SWAL—SWALLOWTAIL GARDEN SEEDS
WWW.SWALLOWTAILGARDENSEEDS.COM
122 CALISTOGA ROAD, #178
SANTA ROSA, CA 95409
PHONE: 877-489-7333 (UNITED STATES ONLY) OR 707-538-3585
Online catalogue. Offering over 1,700 varieties of annual and perennial flower seeds, and herb and vegetable seeds. Retail and bulk.

TERR—TERROIR SEEDS
WWW.UNDERWOODGARDENS.COM
PO BOX 4995
CHINO VALLEY, AZ 86323
PHONE: 888-878-5247
Online catalogue. Printed Master Seed List $5.00. "Home of Underwood Gardens" since 2008. Continues in the preservation of genetic diversity by saving and distributing open-pollinated and heirloom seeds.

TERRA—TERRA EDIBLES
WWW.TERRAEDIBLES.CA
BOX 164
535 ASHLEY STREET
FOXBORO, ON K0K 2B0 CANADA
PHONE: 613-961-0654
Online catalogue. Specializes in heirlooms, edible landscaping plants, and varieties that require little space or are extra nutritious. Retail only.

TERRIT—TERRITORIAL SEED COMPANY
WWW.TERRITORIALSEED.COM
PO BOX 158
COTTAGE GROVE, OR 97424
PHONE: 800-626-0866
Free catalogue. Specializes in varieties for maritime climate (west of the Cascades), all grown without chemical pesticides and using only organic fertilizers.

THOM—THOMAS JEFFERSON CENTER FOR HISTORIC PLANTS
WWW.MONTICELLOSHOP.ORG
PO BOX 318
CHARLOTTESVILLE, VA 22902
PHONE: 800-243-1743
Free catalogue. Specializes in vegetables and flowers that were grown by Thomas Jefferson. Retail only.

TRADE—TRADE WINDS FRUIT
WWW.TRADEWINDSFRUIT.COM
PO BOX 9396
SANTA ROSA, CA 95405
Online catalogue. Offering seeds of exotic and rare edibles from around the world, including heirlooms and ornamentals.

TSK—THE SEED KINGDOM
WWW.THESEEDKING.COM
PHONE: 806-789-5789
Online catalogue. Providing vegetable seeds to gardeners and market growers since 1999.

TTS—T & T SEEDS LTD.
WWW.TTSEEDS.COM
BOX 1710
WINNIPEG, MB R3P 3C6 CANADA
PHONE: 204-895-9962
Free catalogue; Canada only. Offers seeds of hardy prairie plants, vegetables, herbs, flowers, and nursery stock. In business for over 70 years.

TUR—TURTLE TREE BIODYNAMIC SEED INITIATIVE
WWW.TURTLETREESEED.ORG
10 WHITE BIRCH ROAD
COPAKE, NY 12516
PHONE: 800-930-7009
Free catalogue. Sales within the United States only. Offers open-pollinated, organic, biodynamic vegetable, herb, and flower seeds.

UPRI—UPRISING SEEDS
WWW.UPRISINGORGANICS.COM
PO BOX 5431
BELLINGHAM, WA 98227
PHONE: 360-778-3749
Online catalogue. Certified organic, open-pollinated seeds that are grown in the Pacific Northwest on small family farms.

URB—URBAN FARMER SEED
WWW.UFSEEDS.COM
120 E. 161ST STREET
WESTFIELD, IN 46074
PHONE: 317-600-2807
Free catalogue. Preserving and improving heirloom seeds through natural and organic farming methods. Hundreds of varieties of vegetable, flower, and herb seeds to gardeners, farmers, and commercial growers.

VER—VERMONT BEAN SEED COMPANY
WWW.VERMONTBEAN.COM
334 W. STROUD STREET
RANDOLPH, WI 53956
PHONE: 800-349-1071
Free catalogue. Known for their large selection of beans, plus gourmet specialties, and rare and unusual flowers.

VICT—VICTORY SEEDS
WWW.VICTORYSEEDS.COM
PO BOX 192
MOLALLA, OR 97038
Catalogue $2.00. Farm-based, family owned and operated organization offering open-pollinated, heirloom, and rare vegetable, flower, and herb varieties. Retail and bulk.

WHA—A. P. WHALEY SEED COMPANY, LLC
WWW.AWHALEY.COM
1928 SYVRUD ROAD
MOUNT HOREB, WI 53572
PHONE: 608-437-9081
Online sales. Family owned and operated business committed to sustainable agriculture. Offering the best new selections of heirloom, open-pollinated, and cutting-edge hybrids to home gardeners, the seed trade, and professional growers. Retail and wholesale.

WHITE—WHITE HARVEST SEED COMPANY
WWW.WHITEHARVESTSEED.COM
1744 PYATT ROAD
HARTVILLE, MO 65667
PHONE: 866-424-3185
Free catalogue. Family business offering heirloom vegetable, flower, and herb seeds.

WIL—WILLHITE SEED, INC.
WWW.WILLHITESEED.COM
PO BOX 23
POOLVILLE, TX 76487-0023
PHONE: 800-828-1840
Free catalogue. Extensive collection of melons and watermelons; many are their own introductions. Retail and wholesale.

RED AND GREEN CITRON SEEDS

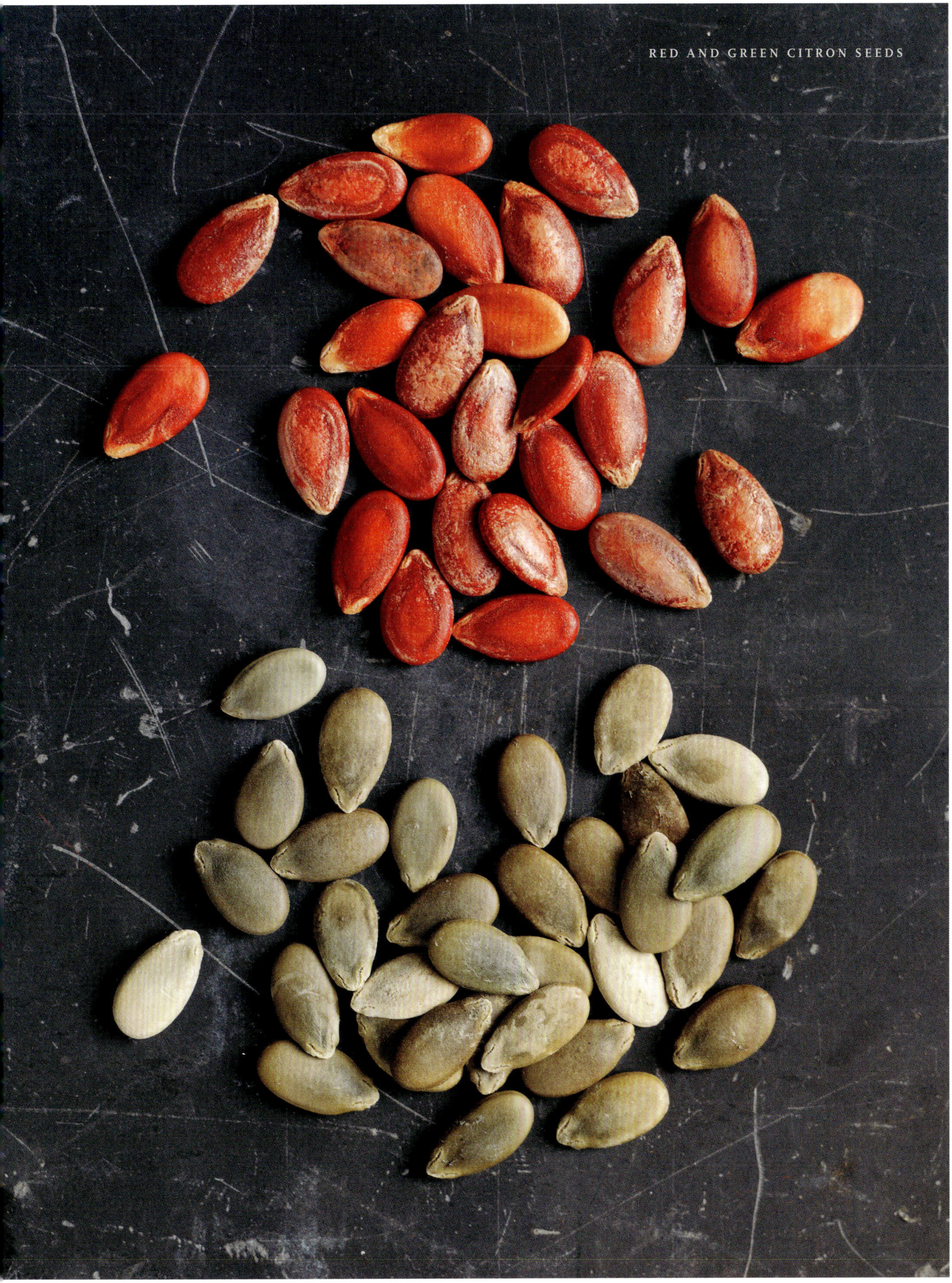

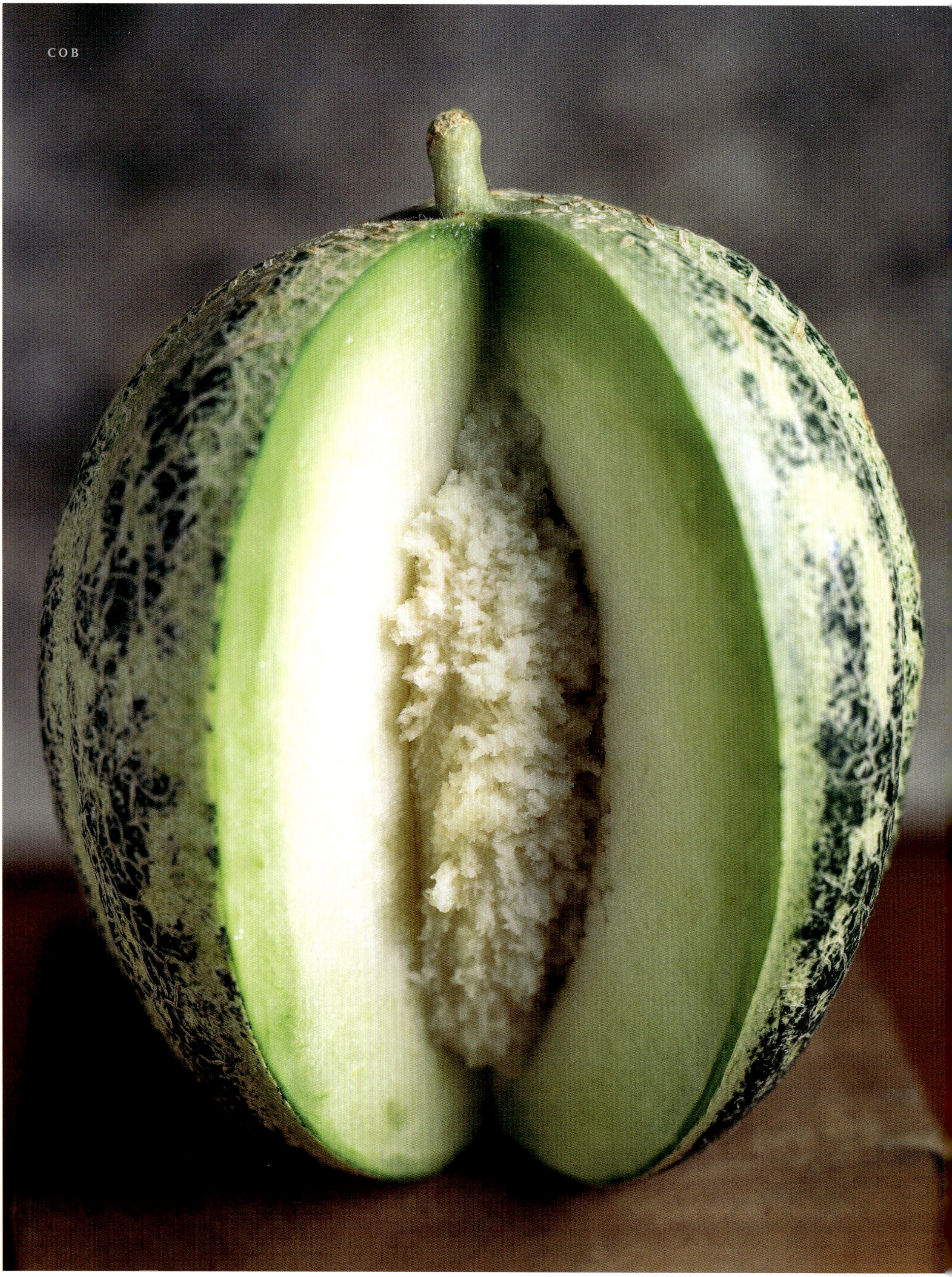

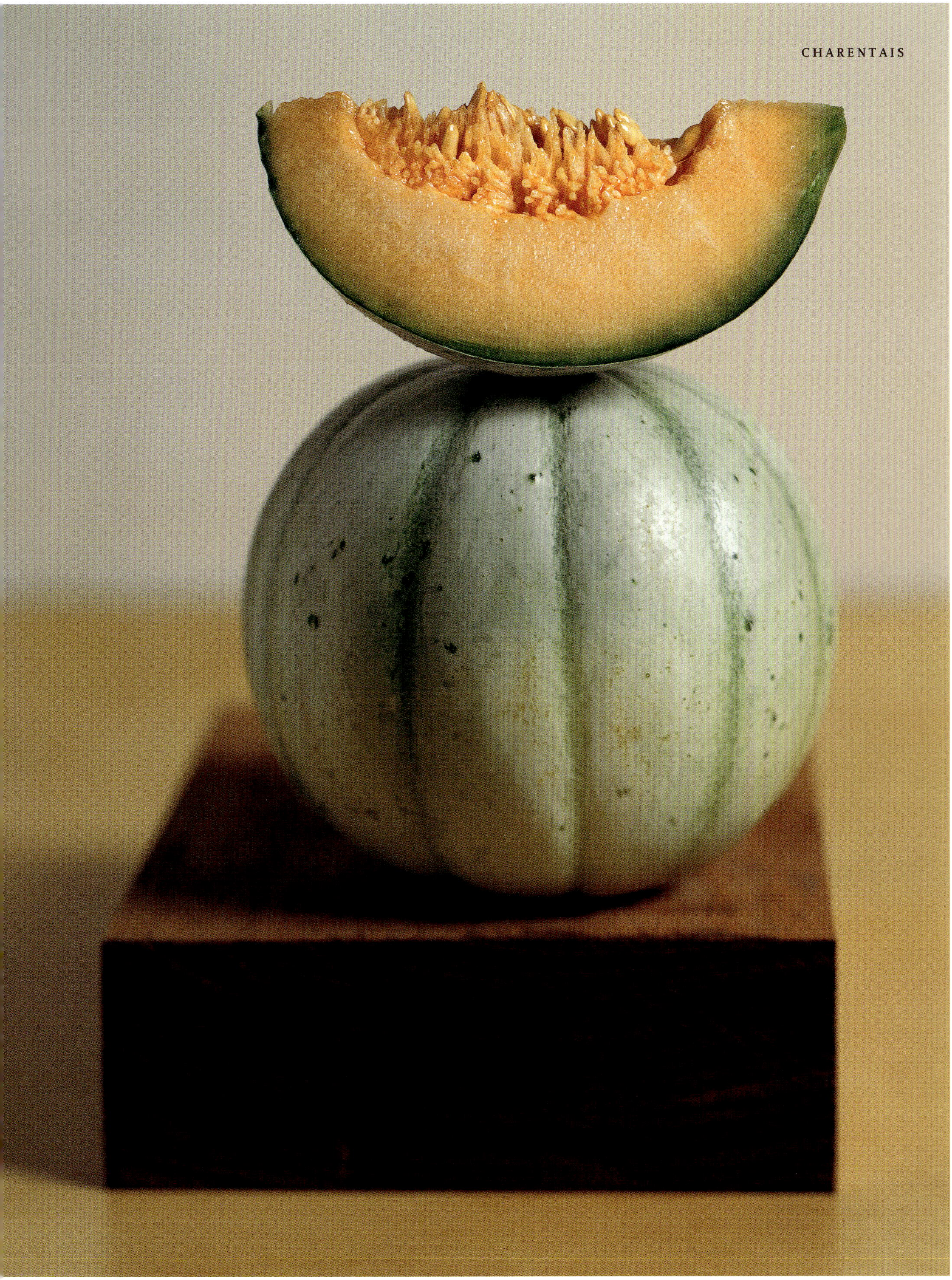

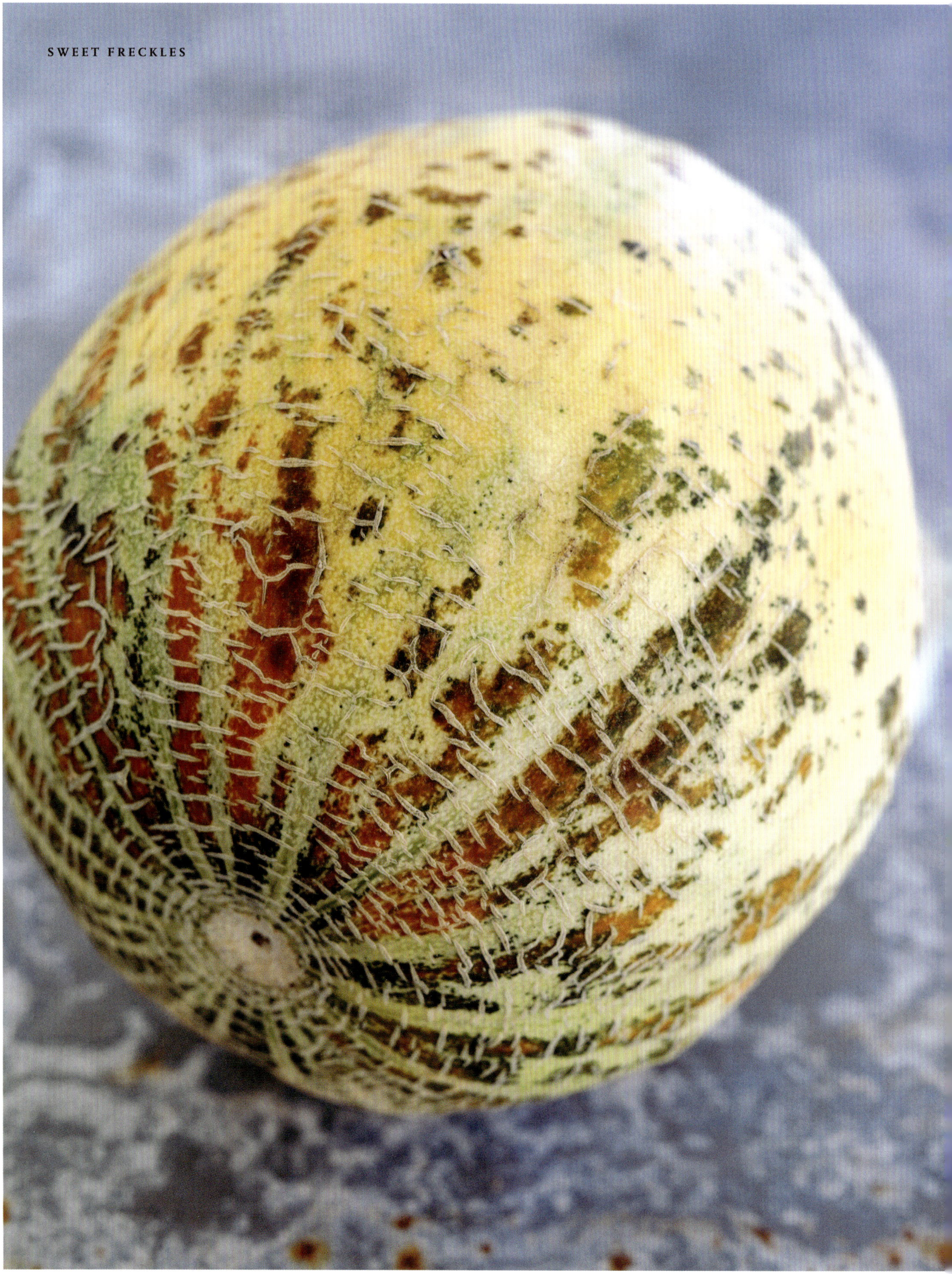

Acknowledgments

Melons and watermelons are my favorite "fruit vegetables" (fruits of herbaceous plants). No one could have done them more justice than Victor Schrager. It's been an honor and a pleasure to work side by side with him on this book for so many years.

M. Mark has also been with me all the way. She is my guiding light. I couldn't imagine a finer editor.

Doyle Partners has outdone itself on book design. Thank you Stephen Doyle, Rosemarie Turk, and Rhoan O'Connell.

I am deeply indebted to David Wilk of City Point Press for managing this project so skillfully.

Accomplished librarians and archivists in special collections provided invaluable assistance. A big thank-you to Stephen Sinon, William B. O'Connor Curator of Special Collections, Research and Archives at the New York Botanical Garden. Gratitude to Seed Historian Sara Straate at the Seed Savers Exchange and to former SSE Librarian Bill Musser, who found treasures in the Robert Becker Memorial Library. Thanks as well to Jennie Oldfield, Cataloger and Archivist at the Cherokee Garden Library of the Atlanta History Center, and Carolyn S. Jackson at Texas A&M University Medical Sciences Library.

Special appreciation to my scientific advisers, Harry S. Paris and James D. McCreight, who taught me so much about melon and watermelon.

My thanks also to the many other "Melonians" who contributed their expertise. Among these were Glenn Drowns, Diane Ott Whealy, Rowen White, Tom Andres, Joseph Burger, Narinder Dhillon, Todd Wehner, David S. Shields, James E. Kibler, Jr., Rodger Winn, Karen Shealy Winn, Mehmet Oztan, Michael Mazourek, Bruce Carle, Christopher Kent, Bruno Defay, James Campbell, Barbara Melera, Ethan Platt, Maya Shiroyama, Mike Dunton, Krystyna Glowiak, Keith Crotz, Erica Kempter, and F. Robert Stein. Kate Petrella assisted with copy editing and Joanne Thuente provided seed sources. Michael Ryan, Brandon Beckingham, Peggy Kontogiannis, and Anthony Ciferri helped in the garden.

Mindy Fox created wonderful recipes; Nate Auchter added a few, and Colin Flynn tested them all. They were illustrated by Victor with help from food stylist team Jamie Kimm and Anna Edqwist, prop stylists Philippa Braithwaite and Ariella Chezar, and photography assistant Megan Senior. Eve Ashcraft created beautiful surfaces for all the melon portraits. My thanks also to Kate Rauner for graciously granting permission to include her lovely poem "Desert Watermelon" in this book.

My husband, Cary Fowler, and my daughter, Sara Goldman Arno, are my favorite melon taste testers. They share my passion for these splendid fruits of the earth.

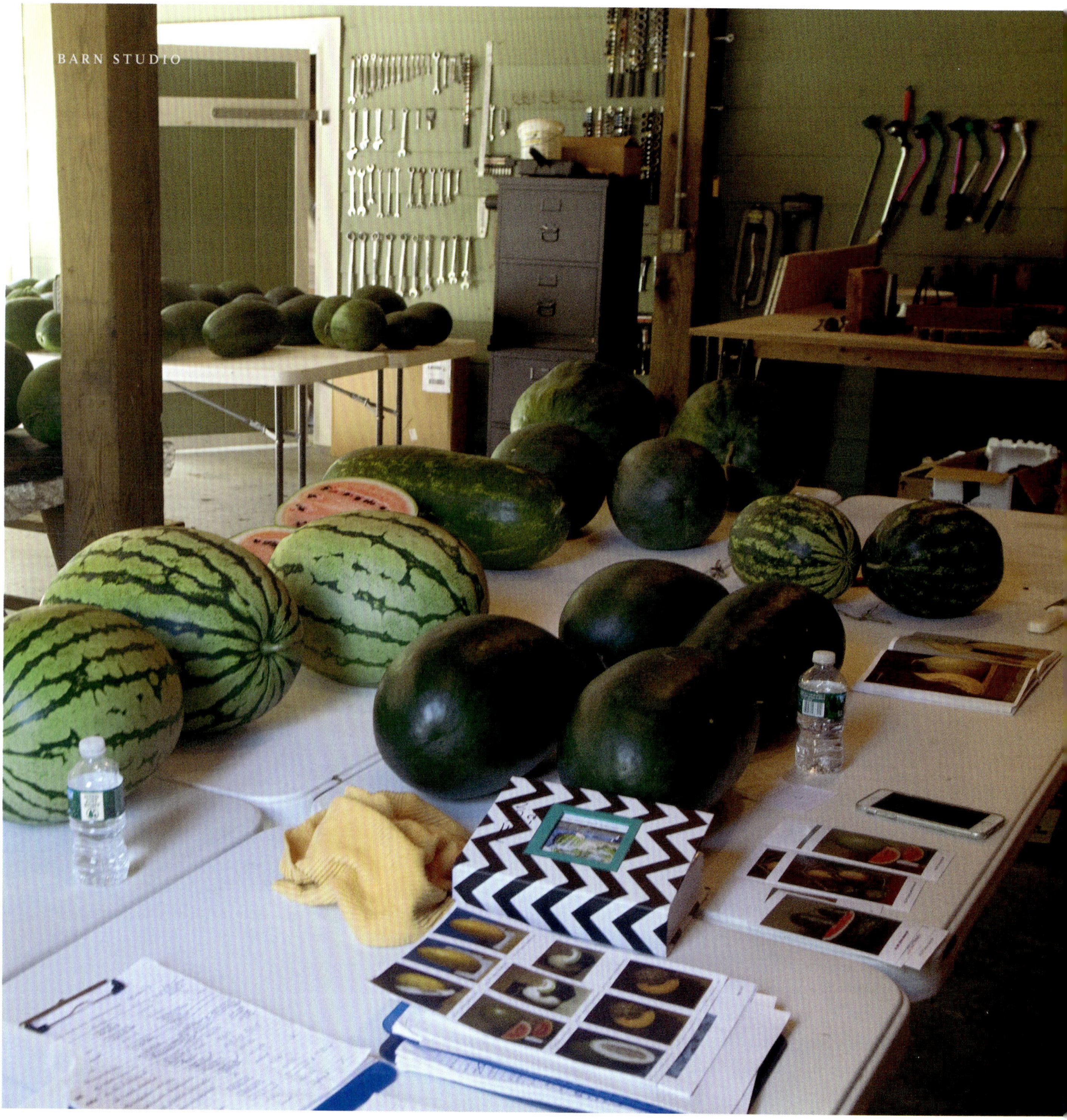

Amy Goldman

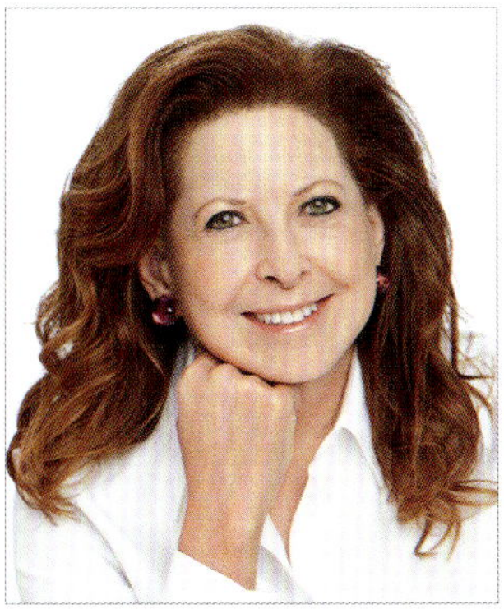

Amy Goldman is a gardener, author, artist, and well-known advocate for seed saving, plant breeding, and heirloom fruits and vegetables. Her mission is to celebrate and catalogue the magnificent diversity of standard, open-pollinated varieties, and to promote their conservation.

Goldman is described as "perhaps the world's premier vegetable gardener" by Gregory Long, President Emeritus of the New York Botanical Garden.

She has authored four books, including three that have earned the American Horticultural Society Book of the Year award: *Melons for the Passionate Grower* (2002), *The Compleat Squash: A Passionate Grower's Guide to Pumpkins, Squashes, and Gourds* (2004), and *The Heirloom Tomato: From Garden to Table* (2008). Most recently she wrote *Heirloom Harvest: Modern Daguerreotypes of Historic Garden Treasures* (2015), which was illustrated by Jerry Spagnoli.

Melons are Goldman's lifelong love and calling. This new book, *The Melon*, illustrated by Victor Schrager and many years in the making, is a comprehensive and definitive work that includes portraits in words and photographs of 125 extraordinary varieties, expert advice on cultivation and seed saving, and a generous selection of delicious melon recipes.

Victor Schrager

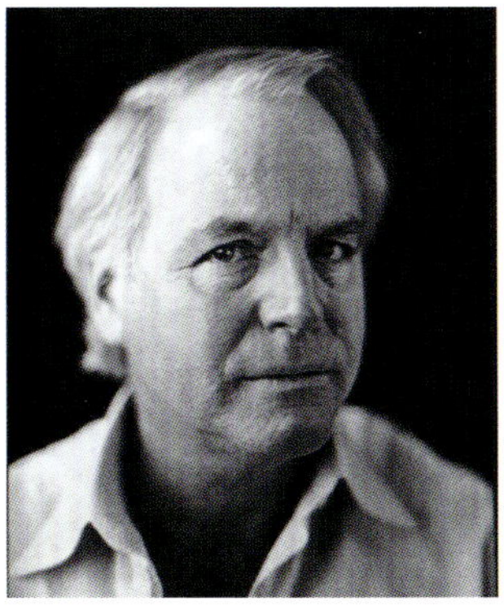

For more than 30 years Victor Schrager has been a master of still-life photography. Born in Maryland, Schrager grew up in New York City. He earned a BA at Harvard in 1972 and an MFA at Florida State University in 1975.

Schrager's photographs explore how different modes of information, especially visual, literary, historical, and scientific, function to produce and communicate knowledge. The works for which he is best known are still lifes composed of layered images and texts—among them reproductions of paintings, maps, magazine pictures, pages from books, and photographic prints—in an investigation of how context structures the meaning of all representations.

His work has been exhibited at galleries and museums around the world and is included in the collections of the Museum of Modern Art, Philadelphia Museum of Art, Maison Européenne de la Photographie, the Whitney Museum of American Art, and the Museum of Fine Arts, Houston, among many others. He has received fellowships from the Guggenheim Foundation, the MacDowell Colony, and the National Endowment for the Arts.

Schrager has been the principal photographer for three of Amy Goldman's previous books on vegetables and fruits. He lives in upstate New York.

Bibliography

Andrus, C. F. 1955. "New Watermelon Varieties." *Seed World*, March 4.

Avital, Anat, and Harry S. Paris. 2014. "Cucurbits Depicted in Byzantine Mosaics from Israel, 350–600 CE." *Annals of Botany* 114: 203–222.

Bailey, L. H. 1895. "Winter Muskmelons." *Cornell University Agricultural Experiment Station Bulletin* 95: 270–292.

———. 1927. *The Standard Cyclopedia of Horticulture*. Vol 2. London: MacMillan.

Baron, Robert C., ed. 1987. *The Garden and Farm Books of Thomas Jefferson*. Golden, CO: Fulcrum.

Beattie, W. R. 1926. "Muskmelons." *Farmers' Bulletin No. 1468*. Washington, DC: USDA.

Beattie, W. R., and S. P. Doolittle. 1951 (1926). "Muskmelons." *Farmers' Bulletin No. 1468*. Revision. Washington, DC: USDA.

Berckmans, P. J. May 1866. "Notes on Melons." *Southern Cultivator* 24 (5): 124– 125.

Blake, Leonard W. 1981. "Early Acceptance of Watermelon by Indians of the United States." *Journal of Ethnobiology* 1 (2): 193–199.

Blinn, Philo K. November 1905. "A Rust-Resisting Cantaloupe." *Colorado Agricultural Experiment Station Bulletin* 104: 3–15.

Bright, Lloyd. 1988. *World Record Giant Watermelons 1917–1987*. Hope, AR: Etter Printing.

———. 2008. *Winner's Guide to Watermelon Festival Giants*. Hope, AR: Southern Printing.

Brookshaw, George. 1817. *Pomona Britannica: or, A Collection of the Most Esteemed Fruits at Present Cultivated in Great Britain. . . .* London: Bensley and Son.

Burger, Yosef, Harry S. Paris, Ron Cohen, Nurit Katzir, Yaakov Tadmor, and Efraim Lewinsohn. 2010. "Genetic Diversity of Cucumis Melo." In *Horticultural Reviews*, edited by Jules Janick. Vol. 36. Hoboken, NJ: John Wiley & Sons.

Burr, Fearing, Jr. 1863. *The Field and Garden Vegetables of America: Full Descriptions of Nearly Eleven Hundred Species and Varieties; with Directions for Propagation, Culture, and Use*. Boston: Crosby and Nichols.

———. 1865. *The Field and Garden Vegetables of America: Full Descriptions of Nearly Eleven Hundred Species and Varieties; with Directions for Propagation, Culture, and Use*. Reprinted with preface by Kent Whealy and introduction by Robert F. Becker. Chillicothe, IL: The American Botanist, Booksellers, 1994.

———. 1866. *Garden Vegetables, and How to Cultivate Them*. Boston: J. E. Tilton.

Chomicki, Guillaume, and Susanne S. Renner. 2015. "Watermelon Origin Solved with Molecular Phylogenetics Including Linnaean Material: Another Example of Museomics." *New Phytologist* 205 (2): 526–532.

Coleman, Hawes H. 1856. "Melons: Cultivation of the Watermelon at the South." *Report of the Commissioner of Patents for the Year 1855: Agriculture*. Washington, DC: Cornelius Wendell.

Colley, Micaela, and Jared Zystro. 2015. *The Seed Garden: The Art and Practice of Seed Saving*. Decorah, IA: Seed Savers Exchange.

Conseil National des Arts Culinaires. 1994. *Bretagne: Produits du terroir et recettes traditionnelles*. Paris: Éditions Albin Michel.

Cotler, Gordon, and Brendan Gill. 1950. "Handy." *New Yorker*, February 18.

Dai Nihon Nokai. 1895. *Useful Plants of Japan: Described and Illustrated*. Tokyo: Agricultural Society of Japan.

Davis, Glen N., Thomas W. Whitaker, G. W. Bohn, and Robert F. Kasmire. 1965. "Muskmelon Production in California." *Circular No. 546*, California Agricultural Extension Service. Davis, CA: University of California.

Decker-Walters, Deena S., S. M. Chung, J. E. Staub, H. D. Quemada, and A. I. López-Sesé. 2002. "The Origin and Genetic Affinities of Wild Populations of Melon (Cucumis melo, Cucurbitaceae) in North America." *Plant Systematics and Evolution* 233: 183–197.

Dhillon, Narinder P. S., R. Ranjana, Kuldeep Singh, I. Eduardo, A. J. Monforte, M. Pitrat, N. K. Dhillon, and P. P. Singh. 2007. "Diversity among Landraces of Indian Snapmelon (Cucumis melo var. momordica)." *Genetic Resources and Crop Evolution* 54: 1267–1283.

Dhillon, Narinder P. S., H. Singh, M. Pitrat, A. J. Monforte, and James D. McCreight. 2015. "Snapmelon (Cucumis melo L Momordica Group): An Indigenous Cucurbit from India with Immense Value for Melon Breeding." *Acta Horticulturae* 1102: 99–108.

Downing, A. J. 1845. *The Fruits and Fruit Trees of America: The Culture, Propagation, and Management, in the Garden and Orchard, of Fruit Trees Generally; with Descriptions of all the Finest Varieties of Fruit, Native and Foreign, Cultivated in this Country*. New York: Wiley and Putnam.

———. 1857. *The Fruits and Fruit Trees of America: The Culture, Propagation, and Management, in the Garden and Orchard, of Fruit Trees Generally; with Descriptions of all the Finest Varieties of Fruit, Native and Foreign, Cultivated in this Country*. Revised and corrected by Charles Downing. New York: Wiley and Halsted.

Dubois, Louis. 1810. *Des melons et de leurs variétés: considérés dans leur histoire, leur physiologie, leur culture naturelle et artificielle, leurs divers usages, etc*. Paris: D. Colas.

Dubuisson-Aubenay, François-Nicolas Baudot. 1898–1902. *Itinéraire de Bretagne en 1636, d'après le manuscrit original*. With notes by Paul de Berthou and Léon Maître. Nantes: Société des bibliophiles Bretons.

Esquinas-Alcazar, Jose Telesforo. 1977. *Alloenzyme Variation and Relationships in the Genus* Cucumis. Ann Arbor: University Microfilms International.

Food Industry Association of Turkmenistan. 2008. *Turkmen gawunlary atlas* [Turkmen Melons Atlas]. 3rd ed. Ashgabat: Ylym.

Fowler, Cary. 2016. *Seeds on Ice*. Westport, CT: Prospecta Press.

Fowler, Cary, and Pat Mooney. 1990. *Shattering: Food, Politics, and the Loss of Genetic Diversity*. Tucson: University of Arizona Press.

Gardette, E. B. 1863. *Biographical Memoir of William Draper Brinckle, M. D.* As read on invitation before the Pennsylvania Horticultural Society, March 24. Philadelphia: C. Sherman, Son & Co.

Gillis, Michael J., and Michael F. Magliari. 2003. *John Bidwell and California: The Life and Writings of a Pioneer, 1841–1900*. Spokane: Arthur H. Clark Company.

Goldman, Amy. 2002. *Melons for the Passionate Grower*. New York: Artisan.

———. 2011. "Luscious Heirloom Watermelons." *Organic Gardening* 58 (5): 34–41.

Goyal, Madhu, and S. K. Sharma. 2009. "Traditional Wisdom and Value Addition Prospects of Arid Foods of Desert Region of North West India." *Indian Journal of Traditional Knowledge* 8 (4): 581–585.

Gusmini, Gabriele, and Todd C. Wehner. 2006. "Qualitative Inheritance of Rind Pattern and Flesh Color in Watermelon." *Journal of Heredity* 97 (2): 177–185.

Hatch, Peter J. 2012. *'A Rich Spot of Earth': Thomas Jefferson's Revolutionary Garden at Monticello*. New Haven: Yale University Press.

Hedrick, U. P. 1950. *A History of Horticulture in America to 1860*. Facsimile of the first edition, with an addendum of books published from 1861–1920 by Elisabeth Woodburn. Portland, OR: Timber Press, 1988.

Henning, Mark J., Henry M. Munger, and Molly M. Jahn. 2005. "PMR Delicious 51: An Improved Open-Pollinated Melon with Resistance to Powdery Mildew." *HortScience* 40 (1): 261–262.

Heurtel, Pascale, and Michelle Lenoir, eds. 2016. *Les vélins du Muséum National D'Histoire Naturelle*. Paris: Muséum National D'Histoire Naturelle.

Hodge, Geoff, and Rosemary Ward. 2011. *Royal Horticultural Society Propagation Techniques*. London: Mitchell Beazley.

Hyams, Edward. 1952. *Melons under Cloches*. London: Faber & Faber.

Jacquin. 1832. *Monographie complète du melon: Contenant la culture, la description et le classement de toutes les variétés de cette espèce. . . .* Paris: Rousselon.

Jagger, I. C., and G. W. Scott. June 1937. "Development of Powdery Mildew Resistant Cantaloup No. 45." *Circular No. 441*. Washington, DC: USDA.

Janick, Jules, and Harry S. Paris. 2006. "The Cucurbit Images (1515–1518) of the Villa Farnesina, Rome." *Annals of Botany* 97: 165–176.

Janick, Jules, Harry S. Paris, and David C. Parrish. 2007. "The Cucurbits of Mediterranean Antiquity: Identification of Taxa from Ancient Images and Descriptions." *Annals of Botany* 100: 1441–1457.

J. C. Robinson Seed Company. 1937. *The Cucurbits Illustrated*. Waterloo, NE: J. C. Robinson Seed Company.

Keinath, Anthony P., William M. Wintermantel, and Thomas A. Zitter, eds. 2017. *Compendium of Cucurbit Diseases and Pests*. 2nd ed. St. Paul, MN: American Phytopathological Society.

Kibler, Jr., James Everett, ed. 2017. *Taking Root: The Nature Writing of William and Adam Summer of Pomaria*. Columbia: University of South Carolina Press.

Kolpan, Steven K., Brian H. Smith, and Michael A. Weiss. 2010. *Exploring Wine*. 3rd ed. Hoboken: John Wiley and Sons.

Kraft, Ken. 1963. *Garden to Order: The Story of Mr. Burpee's Seeds and How They Grow*. Garden City, NY: Doubleday.

Labaree, Leonard W., and Whitfield J. Bell, Jr., eds. 1960. *The Papers of Benjamin Franklin*. Vol. 1. New Haven: Yale University Press.

Lacy, Allen. 1984. *Home Ground: A Gardener's Miscellany*. New York: Farrar Straus Giroux.

Levi, Amnon, Robert Jarret, Shaker Kousik, W. Patrick Wechter, Padma Nimmakayala, and Umesh K. Reddy. 2017. "Genetic Resources of Watermelon." In *Genetics and Genomics of Cucurbitaceae*, edited by Rebecca Grumet, Nurit Katzir, and Jordi Garcia-Mas. New York: Springer.

Lloyd, John William. 1928. *Muskmelon Production*. New York: Orange Judd.

Loisel. n.d., ca. 1853. *Traité complet de la culture des melon: Nouvelle méthode de cultiver ces plantes sous cloches, sur buttes et sur couches*. 3rd ed. Paris: Dusacq.

Malik, Ajaz A., V. K. Vashisht, Kuldeep Singh, Abbishek Sharma, D. K. Singh, Hira Singh, A. J. Monforte, James D. McCreight, and Narinder P. S. Dhillon. 2014. "Diversity among Melon (Cucumis melo L.) Landraces from the Indo-Gangetic Plains of India and Their Genetic Relationship with USA Melon Cultivars." *Genetic Resources and Crop Evolution* 61: 1189–1208

Mavlyanova, R., A. Rustamov, R. Khakimov, A. Khakimov, M. Turdieva, and S. Padulosi. 2005. *O'zbekiston qovunlari* [Melons of Uzbekistan]. Rome: IPGRI.

Maynard, Donald N. 2001. *Watermelons: Characteristics, Production, and Marketing*. Alexandria, VA: ASHS Press.

McCreight, James D., Jack E. Staub, Todd C. Wehner, and Narinder P. S. Dhillon. 2013. "Gone Global: Familiar and Exotic Cucurbits Have Asian Origins." *HortScience* 48 (9): 1078–1089.

Moerman, Daniel E. 1998. *Native American Ethnobotany*. Portland, OR: Timber Press.

Moretti, Giuseppe, and Carlo Chiolini. 1828. *L'Ortolano istruito: Sia precetti teorico-pratici sul coltivamento degli orti*. Milan: Presso Francesco Epimaco Artaria Editore.

Morren, Charles. 1851. "Jardin fruitier: Le chito de la Havane." *Belgique Horticole* 1: 208–211.

Moulton, Gary E. 2018. *Lewis and Clark Expedition Day by Day*. Lincoln, NE: University of Nebraska Press.

Naudin, C. 1859. "Essais d'une monographie des espèces et des variétés du genre Cucumis." *Annuales des Sciences Naturelles: Botanique* 4 (11): 5–87.

Nuez, F., J. Prohens, A. Iglesias, and P. Fernandez de Cordova. 1996. *Catalogo de semillas de melon*. Madrid: Instituto Nacional de Investigación y Tecnologia Agraria y Alimentaria.

Orcutt, Georgia. 1982. "Watermelon Country." *Horticulture* August 1982: 42–49.

Paillieux, A., and D. Bois. 1892. *Le potager d'un curieux*. Paris: Librairie Agricole de la Maison Rustique.

Paris, Harry S. 2011. "Semitic-Language Records of Snake Melons (Cucumis melo, Cucurbitaceae) in the Medieval Period and the 'piqqus' of the 'faqqous.'" *Genetic Resources and Crop Evolution* 59: 31–38.

———. 2015. "Origin and Emergence of the Sweet Dessert Watermelon, Citrullus lanatus." *Annals of Botany* 116: 133–148.

Paris, Harry S., Zohar Amar, and Efraim Lev. 2012a. "Medieval Emergence of Sweet Melons, Cucumis melo (Cucurbitaceae)." *Annals of Botany* 110: 2–33.

———. 2012b. "Medieval History of the Duda'im Melon (Cucumis melo, Cucurbitaceae)." *Economic Botany* 66 (3): 276–284.

Paris, Harry S., Ron Cohen, Menahem Edelstein, Yosef Burger, Yohanan Ma'oz, and Arthur A. Schaffer. 2013. "Origin and History of Old Cucurbit Cultivars in Israel and the Sources of Several Internationally Important Market Types." *Genetic Resources and Crop Evolution* 60: 1629–1640.

Paris, Harry S., Marie-Christine Daunay, and Jules Janick. 2009. "The Cucurbitaceae and Solanaceae Illustrated in Medieval Manuscripts Known as the Tacuinum Sanitatis." *Annals of Botany* 103: 1187–1205.

———. 2012. "Occidental Diffusion of Cucumber (Cucumis sativus) 500–1300 CE: Two Routes to Europe." *Annals of Botany* 109: 11–126.

———. 2013. "Medieval Iconography of Watermelons in Mediterranean Europe." *Annals of Botany* 112: 867–879.

Paris, Harry S., Jules Janick, and Marie-Christine Daunay. 2011. "Medieval Herbal Iconography and Lexicography of Cucumis (cucumber and melon, Cucurbitaceae) in the Occident, 1300–1458." *Annals of Botany* 108: 471–484.

Pitrat, Michel. 2008. "Melon." In *Handbook of Plant Breeding, Vegetables 1: Asteraceae, Brassicaceae, Chenopodicaceae, and Cucurbitaceae*, edited by Jaime Prohens and Fernando Nuez. New York: Springer.

———. 2017. "Melon Genetic Resources: Phenotypic Diversity and Horticultural Taxonomy." In *Genetics and Genomics of Cucurbitaceae*, edited by Rebecca Grumet, Nurit Katzir, and Jordi Garcia-Mas. New York: Springer.

Pitrat, Michel, Peter Hanelt, and Karl Hammer. 2000. "Some Comments on Infraspecific Classification of Cultivars of Melon." *Acta Horticulturae* 510: 29–36.

Range, Willard. 1952. "P. J. Berckmans: Georgia Horticulturist." *The Georgia Review* 6 (2): 219–226.

Renner, Susanne S., and Hanno Schaefer. 2017. "Phylogeny and Evolution of the Cucurbitaceae." In *Genetics and Genomics of Cucurbitaceae*, edited by Rebecca Grumet, Nurit Katzir, and Jordi Garcia-Mas. New York: Springer.

Robinson, R. W., and D. S. Decker-Walters. 1997. *Cucurbits*. New York: CAB International.

Sageret, Augustin. 1826. *Mémoire sur les Cucurbitacées, principalement sur le melon, avec des considérations sur la production des hybrides, des variétés, etc.* Paris: Madame Huzard.

———. 1827. *Deuxième mémoire sur les Cucurbitacées, principalement sur le melon, contenant la culture en pleine terre de celui-ci, etc., et les perfectionnemens dont elle serait susceptible.* Paris: Madame Huzard.

Schaefer, Hanno, Christoph Heibl, and Susanne S. Renner. 2009. "Gourds Afloat: A Dated Phylogeny Reveals an Asian Origin of the Gourd Family (Cucurbitaceae) and Numerous Oversea Dispersal Events." *Proceedings of the Royal Society B* 276: 843–851.

Schlosser, Jim. 2004. "This Watermelon Is Sweet, Fragile, and from Guilford." *Greensboro News and Record* (Greensboro, NC), August 23.

Sebastian, Patrizia, Hanno Schaefer, Ian R. H. Telford, and Susanne S. Renner. 2010. "Cucumber (Cucumis sativus) and Melon (C. melo) Have Numerous Wild Relatives in Asia and Australia, and the Sister Species of Melon is from Australia." *Proceedings of the National Academy of Sciences of the United States of America* 107 (32): 14269–14273.

Sophia-Miralles, Nina. 2018. "How the Pineapple Became So Popular." *The Week*, July 27.

Starobinski, Jean. 1985. *Montaigne in Motion*, translated by Arthur Goldhammer. Chicago: University of Chicago Press.

Tapley, William T., Walter D. Enzie, and Glen P. Van Eseltine. 1937. *The Cucurbits of New York. Volume 1 of The Vegetables of New York*. Albany, NY: J. B. Lyon Company.

Tracy, W. W. 1903. *List of American Varieties of Vegetables for the Years 1901 and 1902*. Washington, DC: Government Printing Office.

Troop, James. 1911. *Melon Culture: A Practical Treatise on the Principles Involved in the Production of Melons*. New York: Orange Judd.

Valder, Peter. 2005. *The Garden Plants of China*. Portland, OR: Timber Press.

Vilmorin-Andrieux. 1856. *Description des plantes potagères*. Paris: Félix Malteste.

———. 1883. *Les plantes potagères: Description et culture des principaux* légumes des *climats tempérés*. Paris: Motteroz.

———. 1885. *The Vegetable Garden: Illustrations, Descriptions, and Culture of the Garden Vegetables of Cold and Temperate Climates*. London: John Murray.

———. 1891. *Les plantes potagères: Description et culture des principaux* légumes des *climats tempérés*. 2nd ed. Paris: Vilmorin-Andrieux.

———. 1904. *Les plantes potagères: Description et culture des principaux* légumes des *climats tempérés*. 3rd ed. Paris: Vilmorin-Andrieux.

———. 1905. *The Vegetable Garden: Illustrations, Descriptions, and Culture of the Garden Vegetables of Cold and Temperate Climates*. 2nd ed. London: John Murray.

———. 1920. *The Vegetable Garden: Illustrations, Descriptions, and Culture of the Garden Vegetables of Cold and Temperate Climates*. 3rd ed. London: John Murray.

———. 1925. *Les plantes potagères: Description et culture des principaux légumes des climats tempérés*. 4th ed. Paris: Vilmorin-Andrieux.

———. 1946. *Dictionnaire Vilmorin des plantes potagères*. Paris: Vilmorin-Andrieux.

Yeager, Albert F. 1950. "Breeding Improved Horticultural Plants." *New Hampshire Agricultural Experiment Station Bulletin* 380: 3–23.

Yeager, Albert F., and Elwyn M. Meader. 1956. "Breeding New Vegetable Varieties." *New Hampshire Agricultural Experiment Station Bulletin* 426: 1–33.

Weaver, William Woys. 2018. *Heirloom Vegetable Gardening: A Master Gardener's Guide to Planting, Seed Saving, and Cultural History*. Minneapolis: Voyageur Press.

Wehner, Todd C. 2008. "Watermelon." In *Handbook of Plant Breeding, Vegetables 1: Asteraceae, Brassicaceae, Chenopodicaceae, and Cucurbitaceae*, edited by Jaime Prohens and Fernando Nuez. New York: Springer.

Whealy, Diane Ott. 2011. *Gathering: Memoir of a Seed Saver*. Decorah, IA: Seed Savers Exchange.

Whealy, Kent. 1991. "Soviet Horticultural and Agricultural Projects." Seed Savers 1991 Harvest Edition. Decorah, IA: Seed Savers Exchange.

Whitaker, Thomas W., and Glen N. Davis. 1962. *Cucurbits: Botany, Cultivation, and Utilization*. New York: Interscience Publishers.

Whitaker, Thomas W., and I. C. Jagger. 1937. "Breeding and Improvement of Cucurbits." *Yearbook of Agriculture*. Washington, DC: USDA.

Wilson, Ernest Henry. 1913. *A Naturalist in Western China*. Facsimile of the first edition, with an introduction by Geoffrey Smith. London: Cadogan Books, 1986.

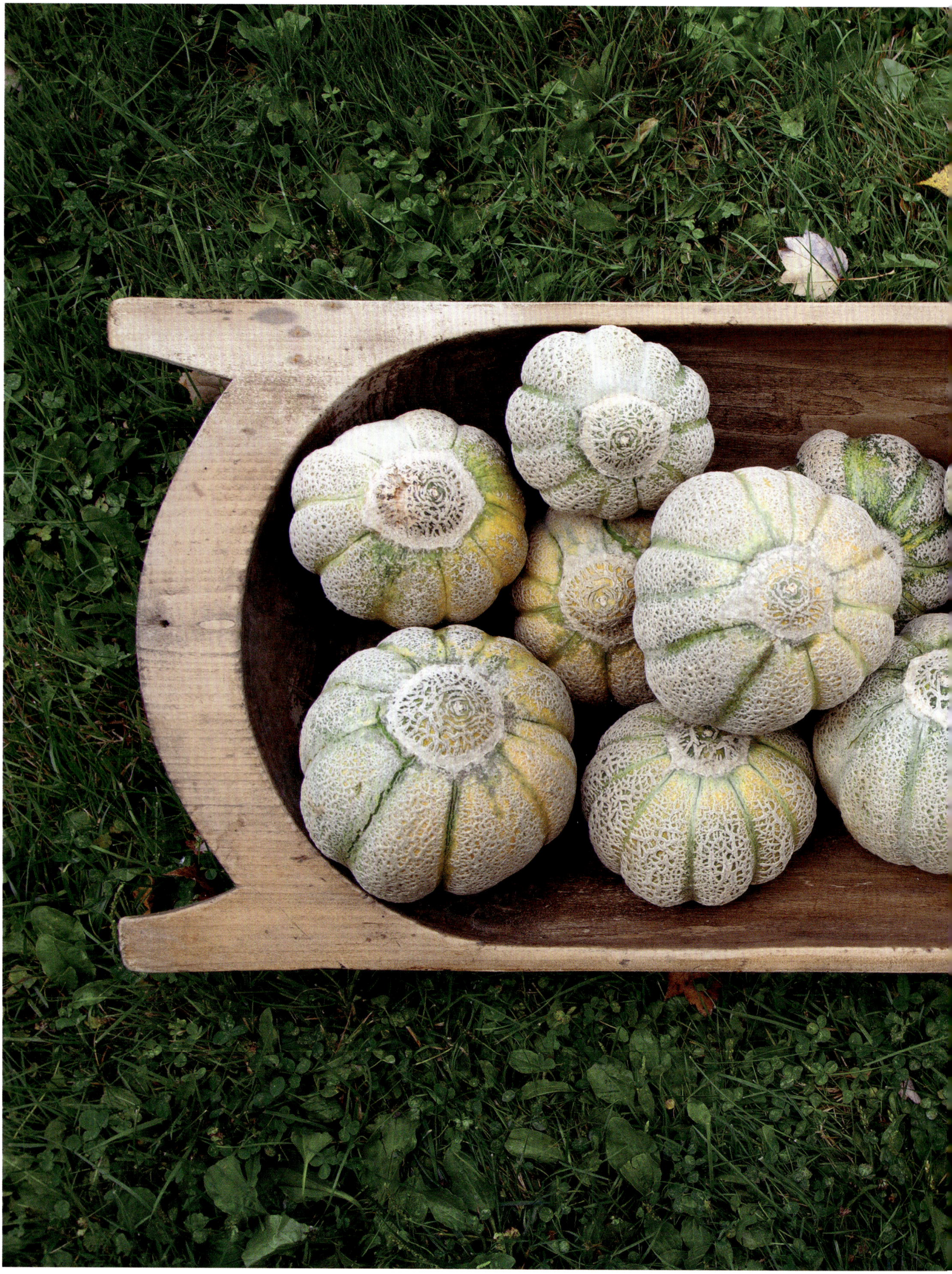

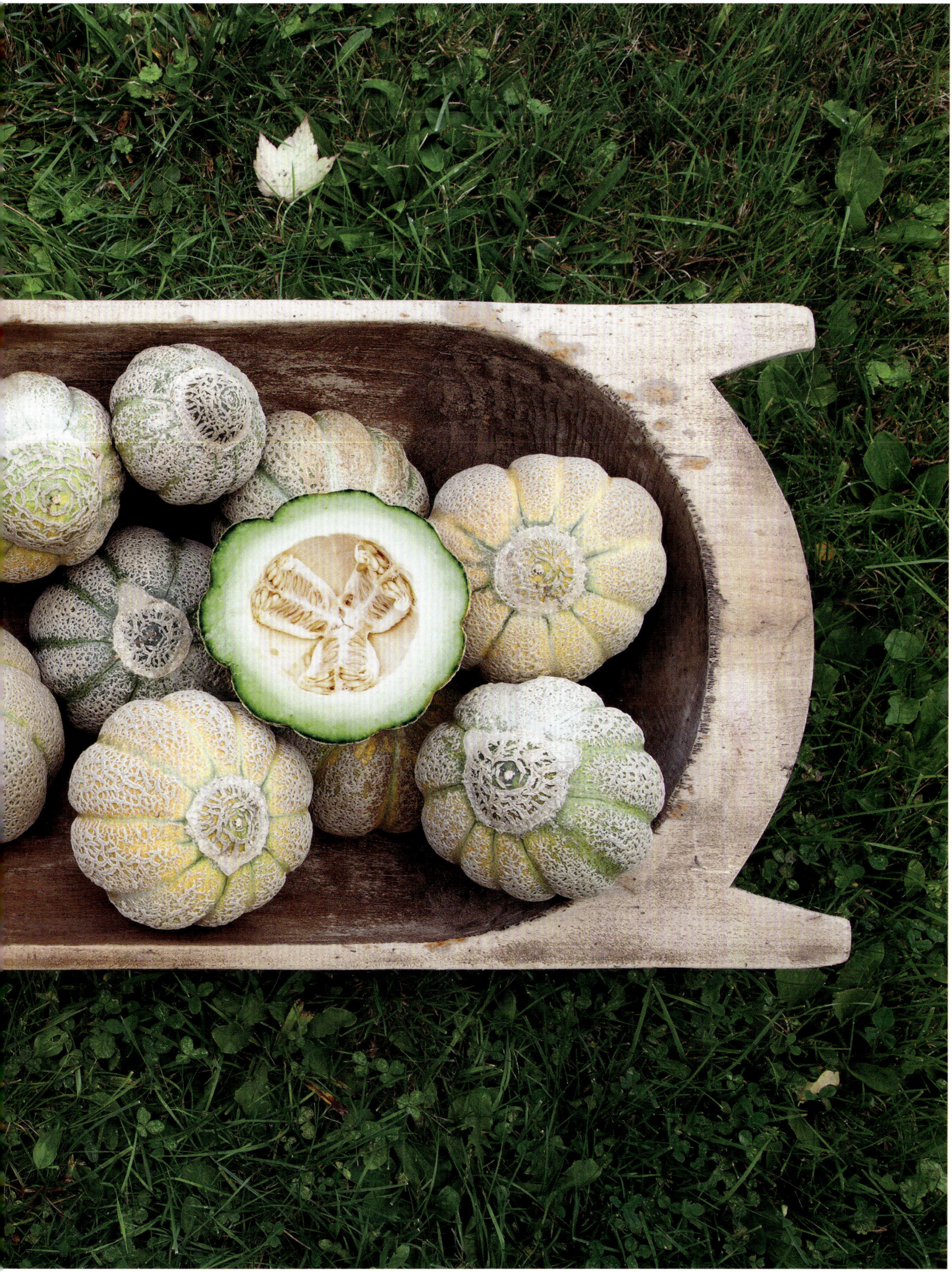

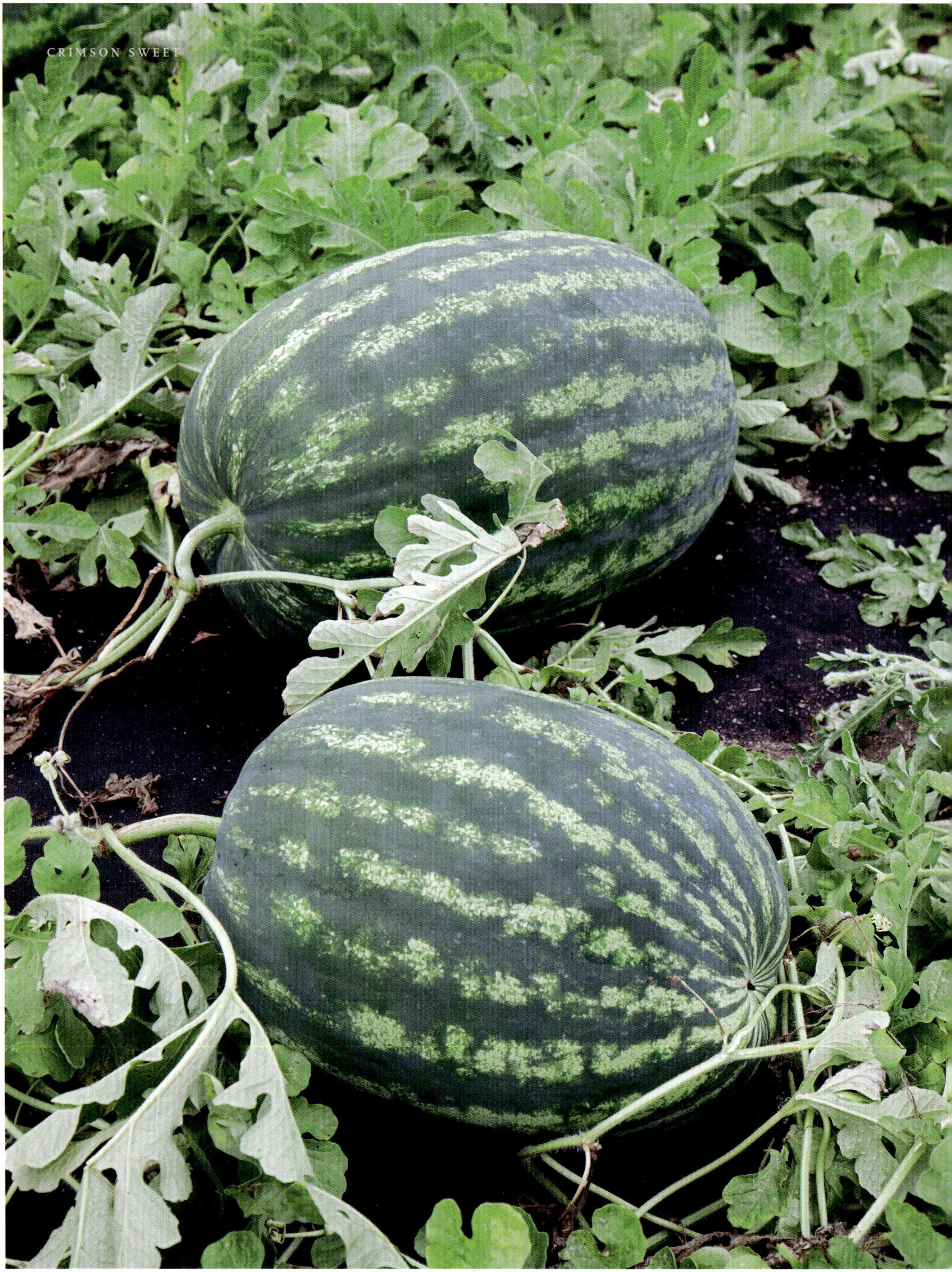
CRIMSON SWEET

Index

Note: Page numbers in **bold** indicate portraits and varietal descriptions of melons/watermelons, and page numbers in *italics* indicate photo captions.